地理信息系统开发与实践丛书

地理信息系统二次开发及案例分析

主　编　柳　林　李万武　毛坤德

副主编　潘宝玉　魏旭晨　杨玉坤　王　恒　董水峰　张　倩

参　编　吴孟泉　满　旺　赵卫东　王振杰　张合兵　王怀洪　吴会胜　赵江洪　李保杰　肖　燕　张连蓬　邵振峰　杨金玲　刘亚静　陆　波　何振芳　曲　畅　陈　黔　张　旗　满苗苗　王小鹏　柳　诚　李　强　吴泉源　陈永刚　窦世卿　张子民　张红日　刘　冰　张　翔　孙琪童　张淑珍　孙　玉　孟德坤　张　鹏　于永辉

WUHAN UNIVERSITY PRESS
武汉大学出版社

图书在版编目(CIP)数据

地理信息系统二次开发及案例分析/柳林,李万武,毛坤德主编.—武汉:武汉大学出版社,2015.9
地理信息系统开发与实践丛书
ISBN 978-7-307-16919-7

Ⅰ.地…　Ⅱ.①柳…　②李…　③毛…　Ⅲ.地理信息系统—系统开发—案例　Ⅳ.P208

中国版本图书馆 CIP 数据核字(2015)第 227538 号

责任编辑:鲍　玲　　责任校对:汪欣怡　　版式设计:马　佳

出版发行:**武汉大学出版社**　(430072　武昌　珞珈山)
(电子邮件:cbs22@ whu. edu. cn 网址:www. wdp. com. cn)
印刷:湖北民政印刷厂
开本:787×1092　1/16　印张:38.75　字数:913 千字　插页:1　插图:1
版次:2015 年 9 月第 1 版　2015 年 9 月第 1 次印刷
ISBN 978-7-307-16919-7　定价:79.00 元

前言

地理信息系统(GIS)技术的应用已经深入到各行各业，其重要性已经得到公认。GIS是一门多学科交叉的边缘学科，属于技术导引型(technology driven)和应用导引型(application driven)学科，所以实践性很强。该学科特点决定了GIS专业的人才培养更要强调实际动手能力，包括GIS软件操作能力和软件研发能力。GIS应用软件的研发和IT界软件研发有所不同，GIS软件的开发一般基于基础平台进行的二次开发，这样更方便、快捷，且周期短、成本低。所以，为满足市场和行业对GIS实践和研发型人才的需求，提高GIS专业学生的研发能力，出版一本“GIS二次开发”相关的教材是非常有必要的。

GIS应用软件研发依赖于专业基础平台。国外GIS软件平台中ArcGIS具有很高的市场占有率，在国内也属于主流平台。同时，在国家支持国产GIS软件发展的大背景下，国内的SuperMap软件平台和MapGIS软件平台也成为GIS领域的重要基础平台。作者在多年GIS教学中发现，市场上现有的GIS开发教材，大多针对某一个平台、某一种开发方式，如ArcEngine开发、基于ArcGIS的Web开发等。缺乏一本集成多个主流GIS基础平台、涵盖多种开发模式的GIS二次开发综合性教材。

GIS二次开发综合性教材，不仅可以使学生掌握多种平台的不同开发模式，而且可以使学生在学习研发的过程中进行对比学习、系统学习，较快掌握不同平台下不同开发模式的开发特点和方法。GIS专业的学生可以从入学到毕业用一本研发教材学习不同的GIS研发课程。对于企事业单位的GIS初级研发人员而言，拥有一本囊括多种平台下不同开发模式的手册式书籍也是有必要的，在开发不同项目时，遇到模糊的地方可以随手查阅。正是出于这样一种想法和目的，编写了本书。

本书涵盖ESRI的ArcGIS平台，国内中地数码集团的MapGIS平台、北京超图软件股份有限公司的SuperMap三种平台的二次开发方法及案例；基于同一软件平台，包括桌面开发、Web开发、移动开发不同开发类别；同一开发类型还包括基于不同的开发框架、接口和技术的不同开发方式。基于此，可以说本书的第一个特点是：内容系统、全面。

全书按照开发平台、开发类型和方式进行组织，共包括6编33章，前26章是基于ArcGIS的二次开发，第27章到第29章是基于MapGIS平台的二次开发，最后4章是基于SuperMap平台的二次开发。第一编ArcGIS Engine开发，包括10章，分别是ArcGIS Engine开发基础、ArcGIS Engine二次开发控件、坐标系与几何对象、地图对象与图层控制、空间数据模型及数据库、栅格数据处理、数据编辑、空间分析、地图制图、“噪音污染分析与决策系统”案例；第二编ArcGIS API for Flex，从第11章到第16章，分别介绍相关技术、环境搭建、应用接口、地图功能开发、查询功能实现、地理处理功能实现；第三编 ArcGIS API for JavaScript，从第 17 章到第 21 章，分别介绍相关技术、开发基础、服务访问、地图操作、任务；第四编 ArcGIS for Android 移动开发，从第 22 章到第 26 章，分别介绍移动开发基础、创建地图工程、地图浏览功能、查询和检索功能、移动开发应用案例；第五编 MapGIS IGserver for Flex 开发，从第 27 章到第 29 章共 3 章，分别介绍 MapGIS for Flex 初级开发、MapGIS for Flex 中级开发、MapGIS for Flex 高级开发；第六编 SuperMap for JavaScript 开发，从第 30 章到第 33 章，分别介绍 SuperMap for JavaScript 初级开发、SuperMap for JavaScript 中级开发、SuperMap for JavaScript 高级开发、"旅印"系统开发案例。

作者多年的教学实践和 GIS 专业人才培养经验表明，通过参加专业竞赛来加深对 GIS 理论的理解，来提高学生的实践动手能力是最有效的教学途径之一，有幸的是 GIS 企业和同行为学生提供了如此多的专业竞赛的机会。GIS 二次开发注重实战，本书给出了大量案例，包括三大竞赛的获奖作品和作者团队的科研案例，随书所赠光盘包括 10 个带有源代码的案例，数据量 2GB。所以，本书的第二个特点是：包含丰富的案例。

本书是作者多年在 GIS 开发与应用教学和专业竞赛指导方面的经验结晶。多年来，作者进行 GIS 教学模式改革和人才实践培养模式创新，主编柳林博士指导学生获得全国 GIS 专业竞赛特等奖 1 项、一等奖 2 项、二等奖 5 项、三等奖 9 项、优胜奖 18 项。在学生实践能力培养方面的丰富经验，全部呈现在书中，这是本书的第三个特点。本书的主要内容是作者在教学实践和项目研发中逐渐形成的，同时也参照了 ArcGIS、SuperMap、MapGIS 的官方帮助文档以及同行博客研发心得等小部分内容，在此一并致谢。

本书的编写得到山东省教改项目(编号：2012251)、山东科技大学群星计划项目(编号：qx2013103)、山东科技大学研究生教育创新计划项目(KDYC14003)、山东省自然科学基金项目(ZR2012FM015)、青岛经济技术开发区重点科技计划项目(2013-1-27)、卫星测绘技术与应用国家测绘地理信息局重点实验室经费资助项目(编号：KLAMTA-201407)、山东省泰山学者建设经费的资助，特此鸣谢！

本书由山东科技大学的柳林老师设计和组织编写，山东科技大学的李万武老师和沈阳市勘察测绘研究院的毛坤德高级工程师组织编写案例部分；山东省地质测绘院的潘宝玉总工、正元地理信息有限责任公司的杨玉坤高工参与框架的设计和指导；山东科技大学的魏旭晨、王恒、董水峰、张倩进行案例相关程序的测试和修正。参与本书编写和指导的人员有：鲁东大学的吴孟泉老师、厦门理工学院的满旺老师、合肥工业大学的赵卫东老师、中国石油大学(华东)的王振杰老师、河南理工大学的张合兵老师、山东省煤田地质规划勘察研究院的王怀洪高级工程师、中国石油大学(华东)的吴会胜老师、北京建筑大学的赵江洪老师、江苏师范大学的李保杰老师、聊城大学的肖燕老师、江苏师范大学的张连蓬老师、武汉大学的邵振峰老师、黑龙江工程学院的杨金玲老师、华北理工大学的刘亚静老师、上海赛华信息技术有限公司的陆波工程师、聊城大学的何振芳老师；山东师范大学的吴泉源老师、浙江农林大学的陈永刚老师、黑龙江科技大学的窦世卿老师、山东建筑大学的张子民老师、山东科技大学的张红日和刘冰老师、Esri 有限公司的张淑珍工程师；山东科技大学的曲畅、陈黔、张旗、满苗苗、王小鹏、柳诚、李强、张翔、孙琪童、孙玉、孟德坤、张鹏、于永辉等研究生和本科生。

尽管本书的编写历时一年多，反复斟酌，数易其稿，但由于知识更新速度快及编者水平所限，书中难免有错误和不妥之处，敬请读者批评指正。批评和建议请致信 liulin2009@126.com。也欢迎高校学子、业界同行致信，共同探讨 GIS 开发和应用的相关问题。本书是《GIS 开发与实践丛书》中的一本，以后将陆续推出 GIS 竞赛相关编著，欢迎广大读者、同行予以关注。

柳林

2015 年 7 月 28 日，于青岛山东科技大学

目　录

第一编　ArcGIS Engine 开发

第三编　ArcGIS API for JavaScript

第四编 ArcGIS for Android 移动开发

第五编　MapGIS IGServer for Flex 开发

第六编　SuperMap for JavaScript 开发

第一编　ArcGIS Engine 开发

第1章　ArcGIS Engine开发基础

1.1　对象模型技术

对象模型技术(Object Modeling Technique，OMT)是美国通用电气提出的一套系统开发技术。它以面向对象的思想为基础，通过对问题进行抽象，构造出一组相关的模型，这些模型描述现实世界中“类与对象”以及它们之间的关系，表示目标系统的静态数据结构，从而能够全面地捕捉问题空间的信息。

对象模型图(OMD)是基于对象模型技术(OMT)的表示方法，它可以帮助编程人员快速了解该类支持哪些接口、完成任务需要哪些对象、如何使用该类的对象、是否可以直接实例化类、接口有哪些方法和属性、是否有其他类也支持该接口，以及对象间的关系。图1.1介绍了UML关系符号，图1.2以ArcMap为例介绍了ARCOBJECT对象组成。

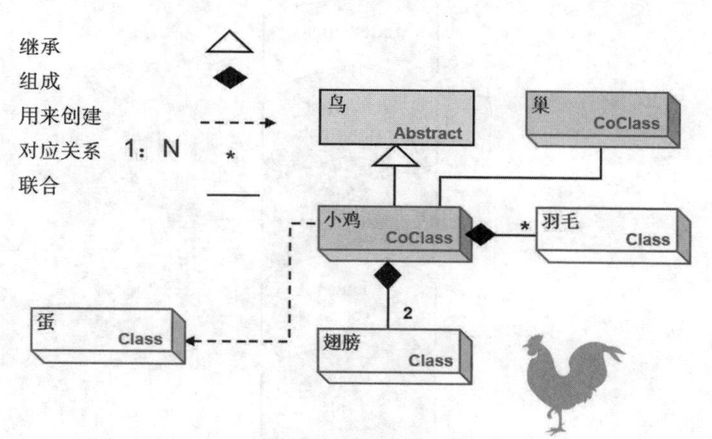

图1.1　ARCOBJECT_UML关系符号例子

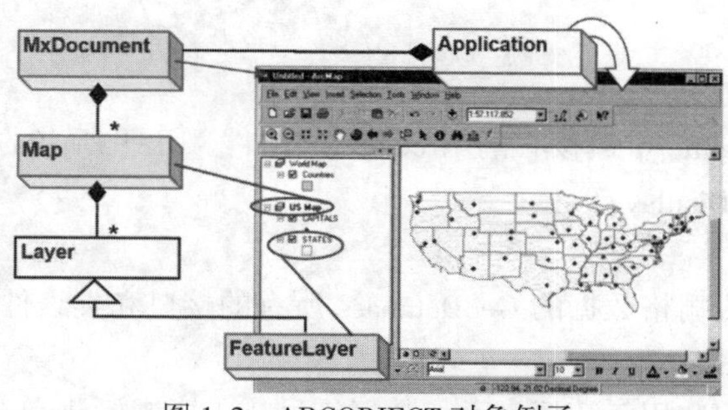

图1.2　ARCOBJECT对象例子

1.2 ArcGIS Engine 简介

1.2.1 ArcGIS Engine 体系结构

ArcGIS Engine 是一组完备的并且打包过的嵌入式 GIS 组件库和工具库，开发人员可用以创建新的或扩展已有的桌面应用程序。使用 ArcGIS Engine，开发人员可以将 GIS 功能嵌入到已有的应用软件中，如自定义行业专用产品；或嵌入到商业生产应用软件中，如 Word 和 Excel；还可以创建集中式自定义应用软件，并将其发送给机构内的多个用户。

ArcGIS Engine 由两个产品组成：构建软件所用的开发工具包和能够运行的可再发布的 Runtime(运行时环境)。ArcGIS Engine 开发工具包是一个基于组件的软件开发产品，可用于构建自定义 GIS 和制图应用软件。它并不是一个终端用户产品，而是软件开发人员的工具包，适于为 Windows、UNIX 或 Linux 用户构建基础制图和综合动态 GIS 应用软件。ArcGIS Engine Runtime 是一个使终端用户软件能够运行的核心 ArcObjects 组件产品，其将被安装在每一台运行 ArcGIS Engine 应用程序的计算机上。ArcGIS Engine 逻辑体系结构如图 1.3 所示，包括以下 5 个部分：

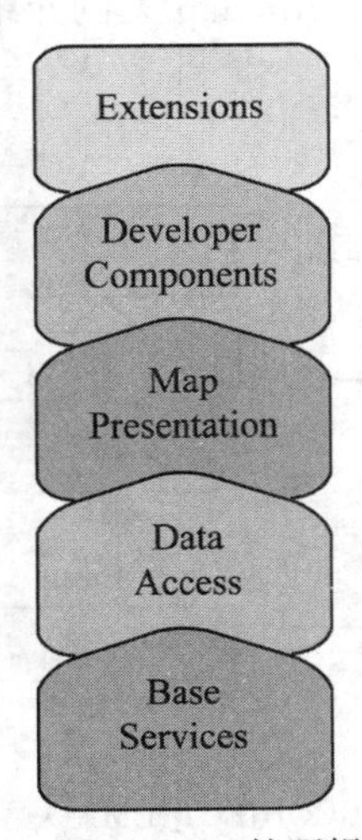

图 1.3 ArcGIS Engine 的逻辑体系结构

1. Base Services

包含了 ArcGIS Engine 中最核心的 ArcObjects 组件，几乎所有的 GIS 组件都需要调用它们，如 Geometry 和 Display 等。

2. Data Access

包含了访问矢量或栅格数据的 GeoDatabase 所有的接口和类组件。

3. Map Presentation

包含了 GIS 应用程序用于数据显示、数据符号化、要素标注和专题图制作等需要的接口和类组件。

4. Developer Components

包含了进行快速开发所需要的全部可视化控件，如 MapControl、PageLayoutControl、SceneControl、GlobeControl、TOCControl、ToolbarControl、SymbologyControl 和 LicensenControl 控件等，除了这些，该库还包括大量可以用 ToolbarControl 调用的内置 commands、tools、Menus，它们可以极大地简化二次开发工作。

5. Extensions

由图 1. 3 中可看出，ArcGIS Engine 的开发体系是一条纵线，功能丰富、层次清晰。最上层的 Extensions 包含了许多高级开发功能，如 GeoDatabase Update、空间分析、三维分析、网络分析、Schematics 逻辑示意图以及数据互操作等。ArcGIS Engine 标准版并不包含这些 ArcObjects 许可，它们只能作为扩展功能存在，需要特定的 License 才能运行。

1. 2. 2　ArcGIS Engine 类库简介

ArcGIS Engine 开发中，为了更好地管理这些 COM 对象，ESRI 将这些 COM 对象放在不同的组件库中，下面对一些常用类库进行介绍：

①Version 库是 ArcGIS10 新增的一个类库，该类库包含了将独立应用程序绑定到特定的 ArcGIS 系列产品的函数和方法。

②System 库是 ArcGIS 架构中最底层的库。该库包含了组成 ArcGIS 其他库所使用的服务组件。System 库中定义了许多接口，它们可以由开发者来实现。开发者不扩展该库，但可以通过实现其中的接口来扩展 ArcGIS 系统。

③SystemUI 库中包含了可在 ArcGIS Engine 中扩展的用户界面组件的接口定义，包括 ICommand、ITool 和 IToolControl 接口。开发者使用这些接口来扩展 UI 组件。该库所包含的对象是 Utility 对象，开发者可用于简化某些用户界面的开发。

④Geometry 库用于处理存储在特征类（feature classes）或其他图形要素（graphical elements）中的 geometry 或 shape。投影和地理坐标系统的空间参考对象都包含在 Geometry 库中，开发者可以通过添加新的空间参考和投影来扩展空间参考系统。

⑤Display 库包含了用于 GIS 数据显示的对象。除了负责实际图像输出的主要显示对象，该库中还包含了表示颜色和符号的对象，这些颜色和符号用于控制显示上所绘制实体的属性。该库中也包含了为用户在与显示交互时提供可视化反馈的对象。该库的所有对象都可以被扩展，常被扩展的内容有符号、颜色和显示反馈。

⑥Output 库用于将创建图形输出到设备，如打印机、绘图仪和硬拷贝格式，如增强型图元文件（enhancedmetafiles）和栅格影像格式（JPG、BMP 等）。开发者使用该库和 ArcGIS 系统其他部分中的对象来创建图形输出。通常这些是 Display 和 Carto 库中的对象。开发者可以扩展 output 库用于定制的设备和输出格式。

⑦GeoDatabase 库提供了用于 GeoDatabase 的编程 API。GeoDatabase 是一个构建在标准工业关系和对象数据库技术基础上的地理数据储存库。库中的对象为 ArcGIS 支持的所有数据源提供了统一的编程模型。GeoDatabase 库定义了许多由架构中较高层次数据源提供者实现的接口。开发者可以扩展 GeoDatabase 来支持特殊的数据对象（Features、Classes

等)类型。此外，还可以使用 PlugInDataSource 对象添加自定义的矢量数据源。

⑧DataSourcesFile 库包含用于基于文件数据源的 GeoDatabase API 的实现。这些基于文件的数据源包括 Shapefile、Coverage、TIN、CAD、SDC、ArcGIS StreetMap™ 和 VPF。开发者不能扩展 DataSourcesFile 库。

⑨Carto 库支持地图的创建和显示。支持的地图可以包含一幅地图或具有多幅地图和相关旁注的复合数据。PageLayout 对象是宿主一幅或多幅地图和相关旁注(指北针、图例、比例尺条等)的容器。Map 对象是图层的容器。Map 对象可以操作地图中所有图层的属性、空间参考、地图比例尺等，也可以操作地图的相关图层。多种不同类型的图层都可以被加载到地图中。不同数据源通常有一个相关图层负责在地图上显示数据。矢量数据由 FeatureLayer 对象处理，栅格数据由 RasterLayer 对象处理，TIN 数据由 TinLayer 对象处理。Renderer 对象的相关属性控制数据在地图中怎样显示，Renderers 一般使用 Display 库中的 Symbols 进行实际绘图。一个 Map 和一个 PageLayout 可以包含要素(elements)，要素利用位置属性来定义它在地图或页面上的位置。Carto 库也包含对地图注记和动态标注的支持。

⑩Location 库包含支持地理编码和与 Route 事件一起工作的对象。可通过 Full 控件的 Finegrained 对象实现访问地理编码的功能，开发者可以创建他们自己的地理编码对象。线性参考功能允许用户添加事件到线性特征的对象，并使用多种绘制方法渲染这些事件，开发者可以扩展线性参考功能。

⑪NetworkAnalysis 库提供的对象允许网络数据加载到 GeoDatabase 中时，对数据进行网络分析。开发者可以扩展该库来支持定制的网络跟踪。

⑫Controls 库允许开发者使用 Controls 库来构建或扩展具有 ArcGIS 功能的应用程序。ArcGIS Controls 通过封装 ArcObjects 和提供一个 Coarser-grained API 简化开发过程。

⑬GeoAnalyst 库包含核心空间分析功能的对象。这些功能在 ArcGIS Spatial Analyst 和 ArcGIS 3DAnalyst 库中使用。开发者可以通过创建一个新的栅格操作类型来扩展该库。

⑭3DAnalyst 库包含在三维场景中使用的对象，它们的工作方式类似于 Carto 库中包含的对象在二维地图中工作。3DAnalyst 库拥有一个开发控件和一组与该控件协同工作的命令和工具，该控件可以与 Controls 库中的对象联合使用。

⑮GlobeCore 库包含处理 Globe 数据的对象，实现方式类似于 Carto 库中包含的对象在二维地图中实现模式。

1.3 .NET 平台概述

.NET 是 Microsoft XML Web Services 平台。XML Web Services 允许应用程序通过 Internet 进行通信和共享数据，而不管所采用的是哪种操作系统、设备或编程语言。Microsoft.NET 平台提供创建 XML Web Services，并将这些服务集成在一起。

微软总裁兼首席执行官 Steve Ballmer 给 .NET 下的定义为：“.NET 代表一个集合，一个环境，一个可以作为平台支持下一代 Internet 的可编程结构。”即 .NET=新平台+标准协议+统一开发工具。

.NET Framework 具有两个主要组件：公共语言运行库和 .NET Framework 类库。公共

语言运行库是 . NET Framework 的基础。. NET 框架是一个多语言组件开发和执行环境，它提供了一个跨语言的统一编程环境。. NET 框架的设计目标是让开发人员更容易地建立 Web 应用程序和 Web 服务，使得 Internet 上的各应用程序之间，可以使用 Web 服务进行沟通。. NET 框架由以下 5 部分组成：

①程序设计语言及公共语言规范(CLS)；

②应用程序平台(ASP. ENT 及 Windows)；

③ADO. NET 及类库；

④公共语言运行库(CLR)；

⑤程序开发环境(VS2010)。

. NET 标准控件根据其应用环境分为两类：

①Windows Form 控件：主要用于 Windows 应用程序的开发。所有的 Windows 控件都是从 Control 类中派生来的，该类包含了所有用户界面的 Windows Form 组件，其中也包括 Form 类。Control 类中包括了很多控件所共享的属性、事件和方法。

②Web 窗体控件：主要用于 Web 应用程序的开发。它是专门针对 ASP. NET Web 窗体设计的服务器控件。

. NET Compact Framework 提供了可以满足大多数设备项目需要的 Windows Form 控件。若要使用这些控件没有的功能，则可以从公共控件派生特定功能的自定义控件。最简单的控件自定义是重写公共控件的方法，例如，可以重写 TextBox 控件的 OnKeyPress 继承方法，提供将输入限制为数字字符的代码。

第2章　ArcGIS Engine 二次开发控件

ArcGIS Engine 为开发者提供了多个可视化的控件，方便用户能够快速建构一个具有 GIS 功能的独立应用程序。ArcGIS Engine 的每种控件都有属性可以设置，均可以在可视化环境中进行编辑。将控件拖放到窗体上后，用户可以右键点击，选择属性菜单来编辑这些属性，它们可以让用户在无需编写代码的情况下快速构建一个 GIS 应用程序。

在 VS .NET 中使用 ArcGIS Engine，需要使用 ESRI. interop 程序集，这些程序集在托管的 .NET 代码和非托管的 COM 代码之间起到桥梁作用。ESRI. interop 程序集为 ArcGIS 控件提供了能够位于 .NET 窗体上的控件，这些控件名前缀为"Ax"，如 AxMapControl、AxPageLayoutControl、AxTOCControl 和 AxToolbarControl。例如，新建一个 AxMapControl 类的对象，再新建一个 MapControlClass 对象，会发现它们拥有的方法和属性事件等基本上差不多，但是后者还是要多几个，这就体现了 COM 组件与 .NET 托管间的区别。.NET 是托管环境，而 ArcGIS 则是基于 COM 技术实现的，也就是说，COM 组件和托管的代码不能直接访问，必须经过转换，或叫做包装。MapControlClass 在 COM 组件库中就是 MapControl，在 .NET 中给所有的类加上了 Class 的后缀，AxMapControl 就是 MapControl 经过包装后的 ActiveX 控件，由于所有的在 Windows. Form 环境下显示的控件都继承 AxHost 类，在工具条上拖过来的就是 .NET 包装的 MapControl，它可以在 Winform 编程中可视化显示。如果想通过 AxMapControl 包装类实现 COM 类的所有方法属性可以通过以下两个方法：一是通过 Object 属性，二是利用 GetOcx 方法。

2.1　制图控件

制图控件包括 MapControl 和 PageLayoutControl。MapControl 控件封装了 MapControl 对象，主要实现各类数据的显示和分析操作功能。MapControl 对象 OMD 图如图 2.1 所示，包含了 MapControl 对象的所有方法属性及其实现的接口类型。PageLayoutControl 控件封装了 PageLayoutControl 对象用于生成标准地图布局以输出。

2.1.1　MapControl 控件

MapControl 控件封装了 Map 对象，在实现 Map 对象所有的属性、方法、事件基础上，提供了额外的属性、方法、事件，其功能包括：

①管理控件的外观、显示属性和地图属性。如 Rotation 属性，其可以用于更改 MapControl 控件显示区域内地图的旋转角度。

②添加并管理控件中的数据层。如 AddShapeFile 方法，允许用户添加指定路径的

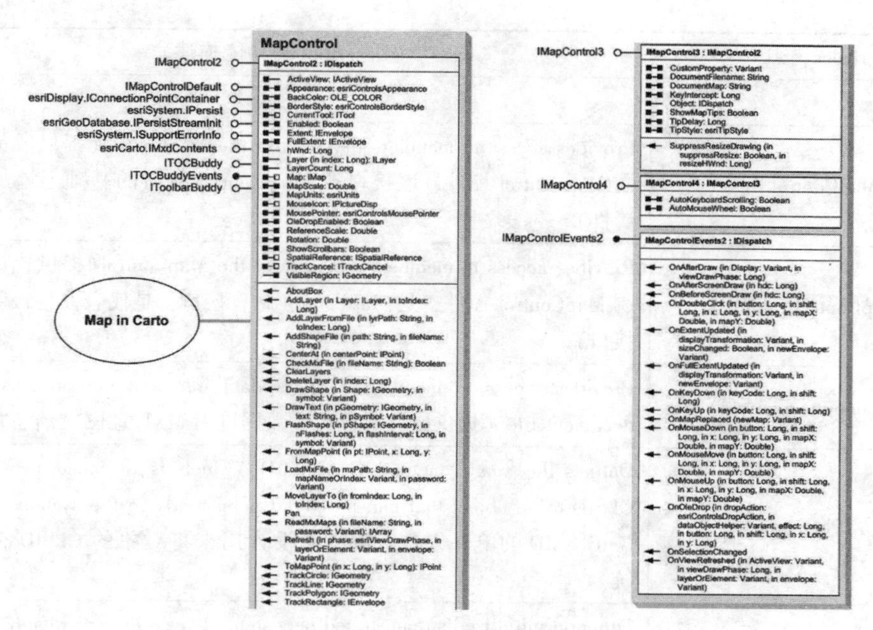

图 2.1　MapControl 对象 OMD 图

ShapeFile 文件到 MapControl 控件中。

③装载 Map 文档(.mxd)到控件中。MapControl 控件允许用户利用 LoadMxFile 方法添加指定路径的 .mxd 地图文件到 MapControl 控件中。

④从其他应用程序拖放数据到控件中。

⑤跟踪鼠标所画形状并在控件中绘图显示。

MapControl 控件实现的主要接口有 IMapControl 类、IMapControlEvents 类等接口类型，见表 2.1。

表 2.1　**MapControl 控件实现的接口及说明**

Interfaces(接口)	Description(功能描述)
Interfaces	
IConnectoinPointContainer	Supports connection points for connectable objects(对于可连接对象支持点连接)
IMapControl 2	Provides access to members that control the MapControl(提供成员对象控制 MapControl 控件，该接口是所有与 MapControl 相关任务的出发点)
IMapControl 3	Provides access to members that control the MapControl(提供成员对象控制 MapControl 控件，该接口在 IMapControl 2 基础上增加了部分属性和方法)

续表

Interfaces(接口)	Description(功能描述)
Interfaces	
IMapControl 4	Provides access to members that control the MapControl(提供成员对象控制 MapControl 控件，该接口在 IMapControl 接口的基础上增加了两个属性)
IMapControlDafault	Provides access to members that control the MapControl(提供成员对象控制 MapControl 控件，是 MapControl 缺省接口，是代表控件最新版本的接口)
IMxdContent(esriCato)	Provides access to members to pass data into and out off a MXD map document file.(提供成员对象传递数据进出 MXD 地图文档。)
IPersist	Defines the single method GetClassID, which is designed to supply the CLSID of an object that can be stored persistently in the system.(定义了 GetClassID 的单一方法，该方法设计用来将对象的 CLSID 存储在系统。)
IPesistSteamInit	Supports initialized stream-based persistence, regardless of whatever else the object does(对于任何对象支持基于流的初始化)
ISupportErrorInfo	Indicates whether a specific interface can return Automation error objects(表明特定接口是否可以自动返回错误对象)
ITOCBuddy	Provides access to members that control the TOC Buddy(提供成员对象控制 TOCBuddy 控件)
IToolbarBuddy	Provides access to members that control the ToolbarControl Buddy(提供成员对象控制 ToolbarBuddy 控件)
Event Interfaces	
IMapControlEvents2(default)	Provides access to events that occur with interaction to the MapControl(定义与 MapControl 控件进行交互的事件)
ITOCBuddyEvents	Provides access to events that notify the TOC of a change(定义 TOC 控件改变所触发的事件)

IMapControl 2、IMapControl 3 等是接口，AxMapControl 是 .NET 控件，并实现了以上接口，AxMapControl1 是 AxMapControl 实例化的对象。三者关系如下所示：

this. AxMapControl1 =new AxMapControl()；//实例化一个 Mapcontrol 对象

IMapControl2 pMapControl2=AxMapControl1. Object；//实例化的 Mapcontrol 对象实现了 IMapControl2 接口的全部方法

随着 ArcGIS Engine 的更新升级，组件会增加一些新的功能，接口的数字越靠后说明这个接口越新，ESRI 建议新开发的程序使用新接口。但考虑到和以前老版本写代码的兼容性，原来的旧接口不能删除。

1. IMapControl 2 接口

IMapControl 2 接口是任何一个与 MapControl 相关的任务的出发点，如设置控件外观，设置 Map 对象或控件的显示属性，添加或者管理数据图层、地图文档，在控件上绘制图形和返回 Geometry 等。IMapControl 2 接口提供的属性和方法见表 2.2。

表 2.2　**IMapControl 2 接口提供的属性和方法**

Members(成员)	Description(功能描述)
AboutBox	Displays a dialog of information about the MapControl(弹出关于 MapControl 控件信息的对话框)
ActiveView	The activeview of the Map contained by the MapControl(MapControl 控件包含在 ActiveView 当前活动视图)
AddLayer	Adds a layer to the Map's collection of layers at the specified index position(在地图指定的索引位置添加图层)
AddLayerFormFile	Loads a layer file and adds it to the Map's collection of layers at the specified index position(加载图层并添加在地图指定的索引位置)
AddShapeFile	Adds a shapefile as a layer to the Map(在地图中添加作为图层的 shapefile 文件)
Appearance	The appearance of the MapControl(MapControl 控件是否显示)
BackColor	Background color of the MapControl(MapControl 控件的背景色)
BordeStyle	The border style of the MapControl(MapControl 控件的边界样式)
CenterAt	Moves the center of the MapControl to the specified location(将 MapControl 控件的中心移到指定位置)
CheckMxFile	Checks the shapefile filename to see if it is a map document that can be loaded into the MapControl(检查 MXD 文件是否可以加载到 MapControl 控件中)
ClearLayers	Removes all layers from the Map(移除所有图层)
CurrentTool	Current active tool for MapControl(MapControl 控件中的当前活动工具)
DeleteLayer	Removes a layer from the Map's collection of layers at the specified index position(在地图指定的索引位置删除图层)
DrawShap	Draw a geometry shape on the MapControl(在 MapControl 控件绘制几何形状)
Drawtext	Draws text along the supplied geometry(沿着给定几何形状绘制文本)
Enable	Indicates whether the MapControl can respond to user generated events(指示 MapControl 控件是否可以响应用户生成的事件)
Extent	Current extent of the Map in map units(以地图单位放大当前地图)
FlashShape	Flashes a shape on the MapControl, duration is in milliseconds(闪烁 MapControl 控件中的形状，持续时间以毫秒为单位)

续表

Members(成员)	Description(功能描述)
FromMapPoint	Converts a point on Map(in map units) to device co-ordinates(typically pixels)(转换地图点坐标到设备坐标)
FullExtent	Rectangular shape that encloses all features of all layers in the Map(放大地图矩形包含的所有图层的要素)
hWnd	Handle to the window associated with the MapControl(获取与 MapControl 控件相关窗口的句柄)
Layer	Layer at the supplied index(给定索引代表的层)
LayerCount	Number of layers in the Map(地图中图层的数目)
LoadMxFile	Loads the specified Map from the map document into the MapControl.(加载地图文档到指定 MapControl 控件。)
Map	The Map contained by the MapControl(包含 MapControl 控件的 Map 对象)
MapUnits	The geographical units of the map(地图坐标单位)
MouseIcon	Custom mouse icon used if MousePointer is 99(自定义鼠标图标)
MousePointer	The mouse pointer displayed over the MapControl(鼠标指针在 MapControl 控件中的显示样式)
MoveLayerTo	Moves a layer within the Map's collection from its current index position to a new index position(移动指定索引位置图层到新的索引位置)
OleDropEnabled	Indicates if the MapControl will fire events when data is dragged over the control's window(判断 MapControl 控件是否响应将数据拖入窗口触发的事件)
Pan	Tracks the mouse while panning the MapControl(跟踪鼠标平移 MapControl 控件内容)
ReadMxMaps	Opens a map document specified by the supplied filename and reads the maps into an array object(根据指定的文件名打开地图文档并映射成一个数组对象)
ReferenceScale	Reference scale of the Map as a representative fraction(地图投影数字比例尺)
Refresh	Redraws the Map, optionally just redraw specified phases or envelope(重绘地图中的形状和包络线)
Rotation	Determines how many degrees the map display is rotated(设定地图显示旋转度)
ShowScrollbars	Indicates whether or not the Map's scrollbars are visible(设定地图滚动条是否可见)
SpatialReference	Spatial reference of the Map(地图的空间坐标)
ToMapPoint	Converts a point in device co-ordinates(typically pixels) to a point on the Map(in map units)(转换点的机器坐标到地图坐标)

续表

Members(成员)	Description(功能描述)
TrackCancel	The object used by the MapControl to check if drawing has been aborted(判断 MapControl 控件中绘制是否取消)
TrackCircle	Rubber-bands a circle on the MapControl(在 MapControl 控件边缘设置圆形)
TrackLine	Rubber-bands a polyline on the MapControl(在 MapControl 控件边缘设置多线)
TrackPolgon	Rubber-bands a polygon on the MapControl(在 MapControl 控件边缘设置多边形)
TrackRectangle	Rubber-bands a rectangle on the MapControl(在 MapControl 控件边缘设置矩形)
VisibleRegion	The geometry specifying the visible region of the Map(设置地图的指定几何可见区域)

2. IMapControl 3 接口

IMapControl 3 接口继承于 IMapControl 2，并增加了以下 8 个属性和一个方法：

①CustomProperty：设置自定义控件属性；

②DocumentFilename：返回 MapControl 装入的地图文档的文件名；

③DocumentMap：返回 MapControl 最后装入的地图(Map)名称；

④KeyIntercept：返回或设置 MapControl 截取键盘按键信息；

⑤Object：返回潜在的 MapControl 控件；

⑥ShowMapTips：确定是否显示地图的 Map Tips；

⑦TipDelay：设置 Map Tips 的延迟时间；

⑧TipStyle：设置 Map Tips 的显示样式；

⑨SuppressResizeDrawing ( )：在控件尺寸发生变化的过程中，阻止数据实时重绘。

3. IMapControl 4 接口

与 IMapControl 3 相比，IMapControl 4 多了以下两个可读写属性：

①public bool AutoKeyboardScrolling {get; set;}: Indicates whether keyboard scrolling is enabled. 判读键盘滚动是否激活。

② public bool AutoMouseWheel {get; set;}: Indicates whether the mouse wheel is enabled. 判读鼠标滚轮是否激活。

4. IMapControlDefault 接口

IMapControlDefault 接口是地图控件缺省接口，多数开发环境自动使用这个接口定义的属性、方法。由于 MapControl 是一个自动化控件，当它被放到一个容器，如窗体上后，它会自动产生一个被称为 AxMapControl1 的对象，这个对象可以直接使用缺省接口定义的属性和方法。这个接口也代表了控件最新版本的接口，MapControl 当前最新版本接口为 IMapControl 4。

5. IMapControlEvents 2 接口

IMapControlEvents 2 是一个事件接口，它定义了 MapControl 能够处理的全部事件，如 OnMouseDown、OnAfterDraw、OnMouseMove 等，这些事件在建构独立程序过程中经常用到。如 OnExtentUpdated 是地图的 Extent 属性发生变化时触发的事件。MapControl 控件提供的事件处理方法见表 2. 3。

表 2. 3　　**MapControl 控件提供的事件处理方法**

Members(成员)	Description(功能描述)
OnAfterDraw	Fires after the Map draws a specified view phase(地图上绘制图形后所触发的事件)
OnAfterScreenDraw	Fires after the Map contained by the MapControl has finished drawing(绘屏完成后所触发的事件)
OnBeforeScreenDraw	Fires before the Map contained by the MapControl starts to draw(绘屏开始前所触发的事件)
OnDoubleClick	Fires when the user presses and releases the mouse button twice in quick succession(双击触发的事件)
OnExtentUpDated	Fires after the extent(Visible bounds) of the MapControl is changed(地图可见区域缩放后触发的事件)
OnFullExtentUpDated	Fires after the full extent(bounds) of the MapControl has changed(地图缩放后触发的事件)
OnKeyDown	Fires after a key is pressed on the keyboard(按下一个按键后触发的事件)
OnKeyUp	Fires after a pressed key is released(释放一个按键后触发的事件)
OnMapReplaced	Fires after the Map contained by the MapControl has been replaced(地图控件中的地图改变后所触发的事件)
OnMouseDown	Fires when the user presses any mouse button while over the MapControl(当鼠标按下后触发的事件)
OnMouseMove	Fires when the user moves the mouse over the MapControl(当鼠标移动后触发的事件)
OnMouseUp	Fires when the user releases a mouse button while over the MapControl(当鼠标弹起后触发的事件)
OnOleDrop	Fires when an OLE drop action occurs on the MapControl(在 MapControl 上拖放一个对象后所触发的事件)
OnSelectionChange	Fires when the current selection changes(当选择改变后触发的事件)
OnViewRefreshed	Fires when the view is refreshed before drawing occurs(当视图刷新后触发的事件)

2.1.2　PageLayoutControl 控件

PageLayoutControl 控件用于制图，可以方便地操作各种元素对象，以产生一幅制作精美的地图对象。PageLayoutControl 控件封装了 PageLayout 对象，PageLayoutControl 控件除了提供 PageLayout 对象在布局视图中控制元素的属性和方法外，还拥有许多附加的事件、属性和方法。

①Printer：提供地图打印相关设置。

②Page：提供控件页面相关效果。

③Element：用于管理控件中的地图元素。

PageLayoutControl 控件只允许通过 .mxd 文件来加载数据。PageLayoutControl 控件实现的主要接口有 IPageLayoutControl 类、IPageLayoutControlEvents 等接口类型，见表 2.4。

表 2.4　**PageLayoutControl 控件实现的接口及说明**

Interfaces(接口)	Description(描述)
Interfaces	
IconnectionPointConstainer	Supports connection points for connectable objects.(对于可连接对象支持点连接)
IMxdContents(esriCarto)	Provides access to pass data into and out off a MXD map document file.(提供成员对象传递数据进出 MXD 地图文档)
IpageLyoutControl	Provides access to members that control the PageLayoutControl.(提供成员对象控制 PageLayoutControl 控件，定义了管理布局视图、输出、打印等功能的属性和方法)
IpageLyoutControl 2	Provides access to members that control the PageLayoutControl.(提供成员对象控制 PageLayoutControl 控件，该接口在 IpageLyoutControl 接口的基础上增加了部分属性和方法)
IpageLyoutControl 3	Provides access to members that control the PageLayoutControl.(提供成员对象控制 PageLayoutControl 控件，该接口在 IpageLyoutControl 2 接口的基础上增加了部分属性和方法，是当前版本的最新功能接口)
IpageLyoutControlDefault	Provides access to members that control the PageLayoutControl.(提供成员对象控制 PageLayoutControl 控件，是 PageLyoutControl 缺省接口，代表控件最新版本的接口，多数开发环境自动使用这个接口定义的属性、方法，PageLayoutControl 当前最新版本接口为 IPageLayoutControl 3)
Ipersist	Defines the single method GetClassID, which is designed to supply the CLSID of an object that can be stored persistently in the system.(定义了 GetClassID 的单一方法，该方法可以将对象的 CLSID 存储到系统)
IPersistStreamInit	Supports initialized stream-based persistencely, regardless of whatever else the object does.(对于任何对象支持基于流的初始化)

续表

Interfaces(接口)	Description(描述)
Interfaces	
ISupportErrorInfo	Indicates whether a specific interface can return Automation error objects.(表明特定接口是否可以自动返回错误对象)
ITOCBuddy	Provides access to members that control the TOC buddy.(提供成员对象控制 TOCBuddy 控件)
IToolbarBuddy	Provides access to members that control the ToolbarControl buddy.(提供成员对象控制 ToolbarBuddy 控件)
Event Interface	
IPageLayoutControlEvents(default)	Provides access to events that occur with user interaction to the PageLayoutControl.(定义当用户与 PageLayoutControl 控件进行交互时触发的事件)
ITOCBuddyEvents	Provides access to events that notify the TOC of a change.(定义当 TOC 控件改变触发的事件)

1. IPageLayoutControlDefault 接口

IPageLayoutControlDefault 接口是页面布局控件缺省接口，多数开发环境自动使用这个接口定义属性、方法。IPageLayoutControlDefault 接口也代表了控件最新版本的接口，PageLayoutControl 当前最新版本接口为 IPageLayoutControl 3。

2. IPageLayoutControlEvents 接口

IPageLayoutControlEvents 是事件接口，它定义了 PageLayoutControl 能够处理的全部事件，如 OnMouseDown、OnAfterDraw、OnMouseMove 等，这些事件在建构独立程序过程中经常用到。如 OnExtentUpdated 是地图的 Extent 属性发生变化时触发的事件。MapControl 控件提供的事件处理方法见表 2.5。

表 2.5　　**MapControl 控件提供的事件处理方法**

Members(成员)	Description(描述)
OnAfterDraw	Fires after the Map draws a specified view phase(地图上绘制图形后所触发的事件)
OnAfterScreenDraw	Fires after the Map contained by the MapControl has finished drawing(绘屏完成后所触发的事件)
OnBeforeScreenDraw	Fires before the Map contained by the MapControl starts to draw(绘屏开始前所触发的事件)
OnDoubleClick	Fires when the user presses and releases the mouse button twice in quick succession(双击触发的事件)

续表

Members(成员)	Description(描述)
OnExtentUpDated	Fires after the extent(Visible bounds) of the MapControl is changed(地图控件的显示范围(可视边界)改变时所触发的事件)
OnFullExtentUpDated	Fires after the full extent(bounds) of the MapControl has changed(当地图控件的最大范围(边界)改变时所触发的事件)
OnKeyDown	Fires after a key is pressed on the keyboard(按下键盘上的某个按键后所触发的事件)
OnKeyUp	Fires after a pressed key is released(释放键盘上某个按键后所触发的事件)
OnMapReplaced	Fires after the Map contained by the MapControl has been replaced(地图控件所包含的地图被替代时所触发的事件)
OnMouseDown	Fires when the user presses any mouse button while over the MapControl(用户在地图控件上按下鼠标所触发的事件)
OnMouseMove	Fires when the user moves the mouse over the MapControl(用户在地图控件上移动鼠标所触发的事件)
OnMouseUp	Fires when the user releases a mouse button while over the MapControl(用户在地图控件上释放鼠标后所触发的事件)
OnOleDrop	Fires when an OLE drop action occurs on the MapControl(在 MapControl 控件上拖放对象后所触发的事件)
OnSelectionChange	Fires when the current selection changes(当前选择改变后触发的事件)
OnViewRefreshed	Fires when the view is refreshed before drawing occurs(绘图发生前，视图刷新所触发的事件)

2.2　框架控件

框架控件包括 TOCControl 和 ToolbarControl。TOCControl 控件主要实现图层管理相关操作。ToolbarControl 控件主要实现添加命令、工具和菜单，如何自定义工具、命令和菜单。

2.2.1　TOCControl 控件

TOCControl 控件是用来对图层数据进行相关管理、操作的控件，其需要同一个如 MapControl、PageLayoutControl、SceneControl 或 GlobeControl 等实现了 IActiveView 接口类的“伙伴控件”协同工作。“伙伴控件”可以在设计时利用 TOCControl 属性页设置或用 SetBuddyControl 方法，通过编程设置。TOCControl 控件可以通过“伙伴控件”的 ITOCBuddy 接口实现显示“伙伴控件”图层数据的交互视图，并保持内容与“伙伴控件”同步。

TOCControl 的 OMD 图如图 2.2 所示，其主要实现了 ITOCControl 和 ITOCControlEvents

接口。ITOCControl 接口定义了 TOCControl 控件的相关属性和方法，如管理图层的顺序可见性等。ITOCControlEvents 接口定义了 TOCControl 控件所能处理的全部事件，如 OnBeginLabelEdit 是处理标签开始编辑的事件。

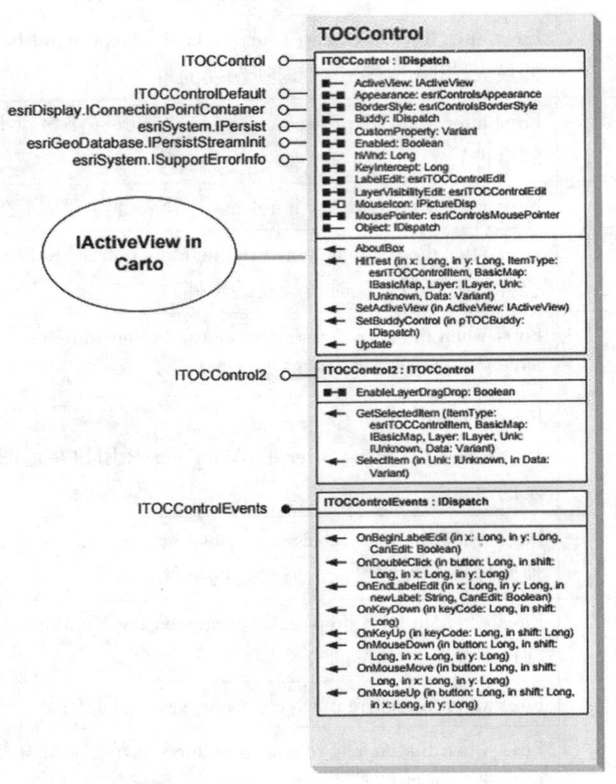

图 2.2　TOCControl 对象的 OMD 图

2.2.2　ToolbarControl 控件

ToolbarItem 是一个不可创建的对象。引用不可创建的对象必须通过其他对象的获取，ToolbarControl 要与一个“伙伴控件”协同工作。ToolbarControl 使用钩子(hook)来联系命令对象和 MapControl 或 PageLayoutControl 等控件，并提供属性、方法、事件用于管理控件的外观，设置伙伴控件，添加、删除命令项，设置当前工具，定制工具。

ToolbarControl 的主要接口有：IToolbarControl、IToolbarControl 2、IToolbarControlDefault、IToolbarControlEvents (default)。

IToolbarControl 接口是任何与 ToolbarControl 有关的任务的出发点，如设置控件的外观，设置伙伴控件，添加或去除命令、工具、菜单，定制 ToolbarControl 的内容。IToolbarControl 接口提供的主要属性有：Buddy、CommandPool、CurrentTool、Customize、CustomProperty、Enabled、Object、OperationStack、ToolTips、TextAlignment、UpdateInterval 等。IToolbarControl 接口提供的主要方法：AddItem、AddMenuItem、AddToolbarDef、Find、

GetItem、GetItemRect、HitTest、MoveItem、Remove、RemoveAll、SetBuddyControl、Update 等。

与 IToolbarControl 相比，IToolbarControl 2 增加了以下属性：

①public bool AlignLeft {get; set;};

②public uint BackColor {get; set;};

③public uint FadeColor {get; set;};

④public esriToolbarFillDirection FillDirection {get; set;};

⑤public int IconSize {get; set;};

⑥public void LoadItems (IStream pStream);

⑦public esriToolbarOrientation Orientation {get; set;};

⑧public void SaveItems (IStream pStream);

⑨public bool ShowHiddenItems {get; set;};

⑩public bool ThemedDrawing {get; set;};

⑪public bool Transparent {get; set;}。

2.3　桌面应用示例构建

本示例主要是使用 MapControl、PageLayoutControl、ToolbarControl、TOCControl 4 个控件建立基本的桌面 GIS 应用程序框架，运行结果如图 2.3 所示。

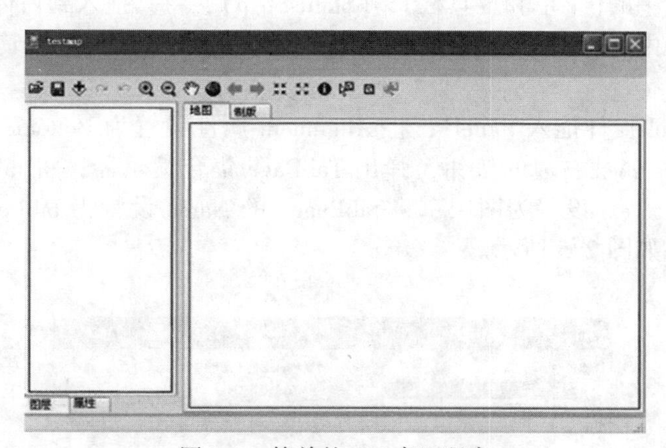

图 2.3　简单的 GIS 桌面程序

2.3.1　新建项目

①启动 VS2010，选择【文件】≫【新建】≫【项目】，在项目类型中选择【Visual C#】，再选择“Windows 应用程序”模板，输入名称“testMap”，点击【确定】，如图 2.4 所示。

②在解决方案管理器中将“Form1. cs”重命名为“testMap. cs”，在设计视图中，选中窗体，将其属性中的“Text”改为“testMap”。

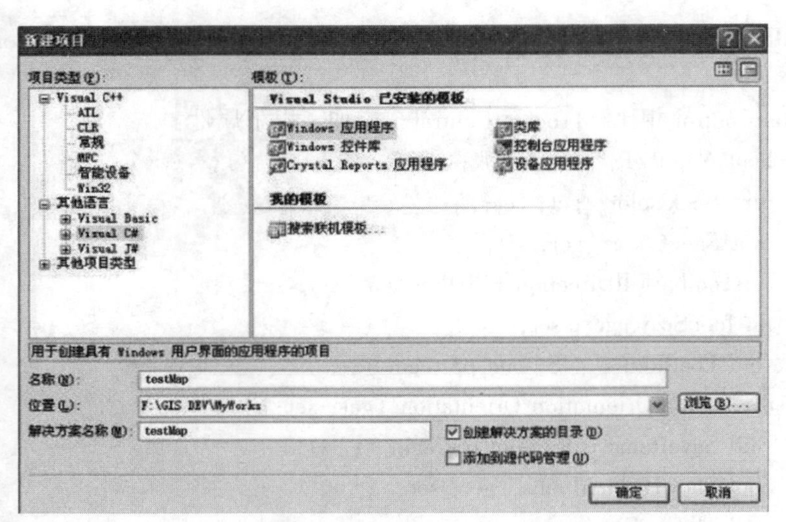

图 2.4　新建项目

2.3.2　添加控件

①选择工具箱中的【菜单和工具栏】≫【MenuStrip】，将其拖入窗体。

②选择工具箱中的【ArcGIS Windows Forms】，将 ToolbarControl 控件拖入窗体，并将其属性中的 Dock 设置为 Top。

③选择工具箱中的【菜单和工具栏】≫【StatusStrip】，将其拖入到窗体。

④选择工具箱中的【容器】≫【SplitContainer】容器拖入窗体，并将其属性中的 Dock 设置为 Fill。

⑤将 TabControl 控件拖入 Panel1，将 Alignment 属性设置为 Bottom，Dock 属性设置为 Fill。点击 TabPages 属性右边的按钮，弹出 TabPage 集合编辑器，将 tabPage1 的 Name 设置为 tabPageLayer，Text 设置为图层，将 tabPage2 的 Name 设置为 tabPageProp，Text 设置为属性，设置结果如图 2.5 所示。

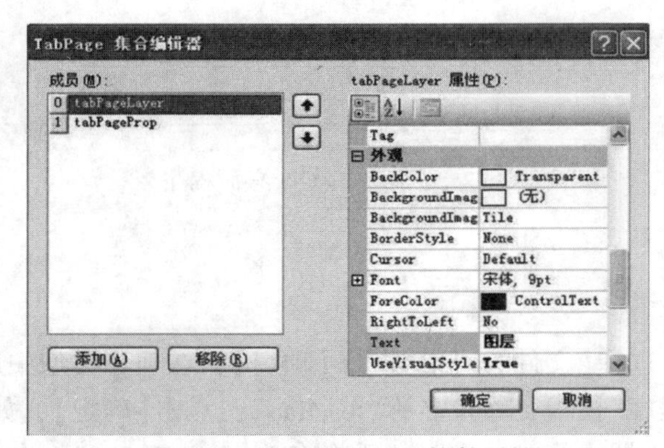

图 2.5　设置 TabControl 控件属性

⑥选择【图层】选项卡，拖入 TOCControl 控件，设置 Dock 属性为 Fill。

⑦选择【属性】选项卡，拖入 DataGridView 控件，设置 Dock 属性为 Fill。

⑧拖入 TabControl 控件到 Panel2，设置 Dock 属性为 Fill。并按上述类似的方法，将两个选项卡的 Name 和 Text 分别设置为：(tabPageMap，地图)，(tabPageLayout，制版)。

⑨选择【地图】选项卡，拖入 MapControl 控件，设置 Dock 属性为 Fill。

⑩选择【制版】选项卡，拖入 PageLayoutControl 控件，设置 Dock 属性为 Fill。

⑪最后将 LicenseControl 控件拖入到窗体的任意地方。

⑫单击 F5 键编译运行，可以看到布局好的程序界面。

2.3.3　控件绑定

通过以上步骤添加的控件还只是单独存在，而程序需要各控件间协同工作，因此要进行控件绑定。分别右击 ToolbarControl、TOCControl 控件，将 Buddy 设置为 axMapControl1，如图 2.6 所示。

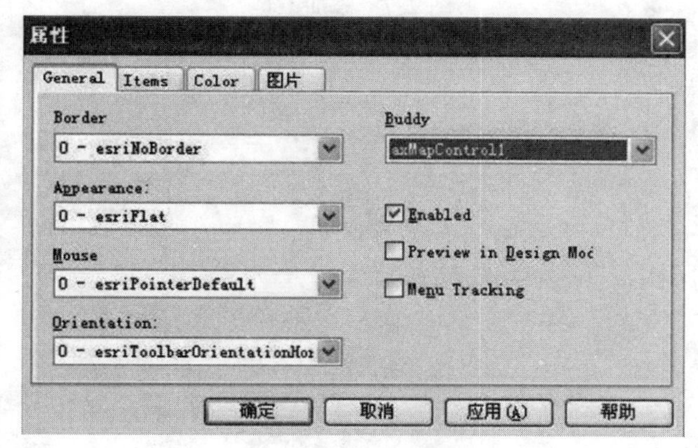

图 2.6　控件绑定

2.3.4　添加工具

此时，工具条中还没有任何工具，添加的方法也很简单，如图 2.7 所示。右击 ToolbarControl，选择【属性】≫【Items】，点击【Add】，选择【Commands】选项卡中的【Generic】，双击【Open】、【SaveAs】、【Redo】、【Undo】即可将相应的工具添加到工具条中。

常见的工具有：Map Navigation 中的导航工具，Map Inquiry 中的查询工具，Feature Selection 中的选择工具，可以根据需要酌情添加工具。

2.3.5　编译运行

单击 F5 键即可编译运行程序，至此，桌面 GIS 应用程序框架基本框架已经搭建好，可以通过工具条的工具打开地图文档，浏览地图。

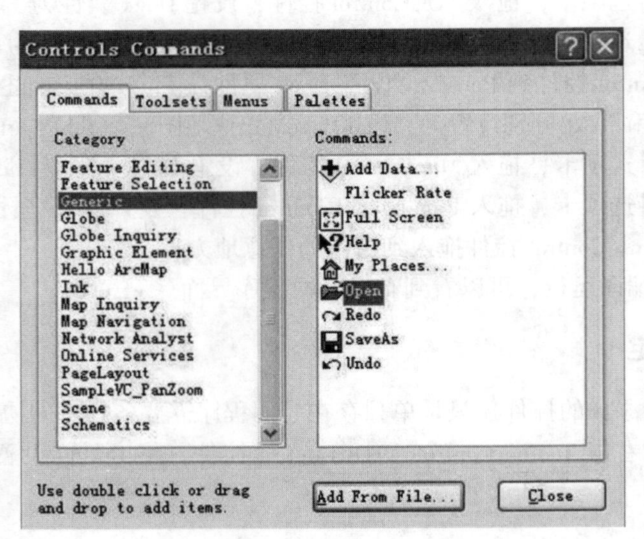
Controls Commands
Commands
Toolsets
Menus
Palettes
Category
Feature Editing
Feature Selection
Generic
Globe
Globe Inquiry
Graphic Element
Hello ArcMap
Ink
Map Inquiry
Map Navigation
Network Analyst
Online Services
PageLayout
SampleVC_PanZoom
Scene
Schematics
Commands:
Add Data...
Flicker Rate
Full Screen
Help
My Places...
Open
Redo
SaveAs
Undo
Use double click or drag and drop to add items.
Add From File...
Close

图 2.7　添加工具

第3章　坐标系与几何对象

3.1　空间坐标变换

空间数据与其他数据的最大区别在于空间数据具有空间位置信息。空间位置可以用不同的坐标系统来描述，同一实体，在不同坐标系统中具有不同坐标值。坐标系统分为地理坐标系统（Geographic Coordinate System，GCS）和投影坐标系统（Projected Coordinate System，PCS）两大类型。

地理坐标系统是使用三维球面来定义地球表面位置，以实现通过经纬度对地球表面点位引用的坐标系。一个地理坐标系包括角度测量单位、本初子午线和参考椭球体三部分。投影坐标系统使用基于 *X*，*Y* 值的坐标系统来描述地球上某个点所处的位置。这个坐标系是从地球的近似椭球体投影得到的，它对应于某个地理坐标系。

ArcGIS Engine 提供了 3 种空间参考对象模型：GeographicCoordinate Systems、ProjectedCoordinate Systems、UnknownCoordinate Systems，3 种空间参考对象模型都实现了 ISpatialRefrence 接口，该接口定义了与设置数据集空间参考的相关属性方法，见表 3.1。

表 3.1　**ISpatialRefrence 接口**

Properties(属性)	Description(描述)
Abbreviation	The abbreviated name of this spatial reference component.(空间坐标系缩写名称)
Alias	The alias of this spatial reference component.(空间坐标系的别名)
Changed	Notify this object that of its parts have changed (parameter values, zunit, etc.).(通知对象的部分坐标系改变)
FactoryCode	The factory code(WKID) of the spatial reference.(空间坐标的工厂代码)
GetDomain	The XY domain extent.(获取 XY 域范围)
GetFalseOriginAndUnits	Get the false origin and units.(获取错误的数据源和单位)
GetMDomain	The measure domain extent.(获取测量域范围)
GetMFalseOriginAndUnits	Get the measure false origin and units.(获取测量错误的数据源和单位)
GetZDomain	The z domain extent.(获取高程范围)
GetZFalseOriginAndUnits	Get the z false origin and units.(获取高程的错误数据源和单位)

续表

Properties(属性)	Description(描述)
HasMPrecision	Returns true when m-value precision information has been defined.(当存在 M 值信息时返回真)
HasXYPrecision	Returns true when (x, y) precision information has been defined.(当存在 XY 值信息时返回真)
HasZPrecision	Returns true when z-value precision information has been defined.(当存在 Z 值信息时返回真)
IsPrecisionEqual	Returns TRUE when the precision information for two spatial references is the same.(两个空间坐标的精确信息相同时返回真)
Name	The name of this spatial reference component.(获取空间坐标系的名称)
Remarks	The comment string of this spatial reference component.(获取空间坐标系的注释字符串)
SetDomain	The XY domain extent.(设置 XY 域范围)
SetFalseOriginAndUnits	Set the false origin and units.(重新设置数据源和单位)
SetMDomain	The measure domain extent.(设置测量域范围)
SetMFalseOriginAndUnits	Set the measure false origin and units.(重新设置测量数据源和单位)
SetZDomain	The z domain extent.(设置高程范围)
SetZFalseOriginAndUnits	Set the z false origin and units.(重新设置高程数据源和单位)
ZCoordinateUnit	The unit for the z coordinate.(高程投影单位)

功能调用代码示例如下:

```
//使用空间参考对话框设置一个地图的空间参考
{
IProjectedCoordinateSystem pSpatialreference;
ISpatialreferenceDialog pDialog;
pDialog=new SpatialreferenceDialog pDialogClass();
pSpatialreference= pDialog.DoModalCreate(true,false,false,0)as
IProjectedCoordinateSystem;
pMap.Spatialreference= pSpatialreference;
pActiveView.Refresh();
}
```

3.2 Geometry 对象集

3.2.1 Envelope 对象

Envelope 包络线，是每个几何形体的最小外接矩形，其示意图如图 3.1 所示。AE 中 IEnvelope 接口定义了包络线对象，包含用于获取设置包络线对象空间信息的 XMax、XMin、YMax、YMin、Height、Width 等属性。IEnvelope 接口定义了一些方法，如 expand 用于按比例缩放包络线。IEnvelope 接口还提供了两个拓扑运算方法 Intersect 和 Union。IEnvelope 接口定义的方法属性见表 3.2。

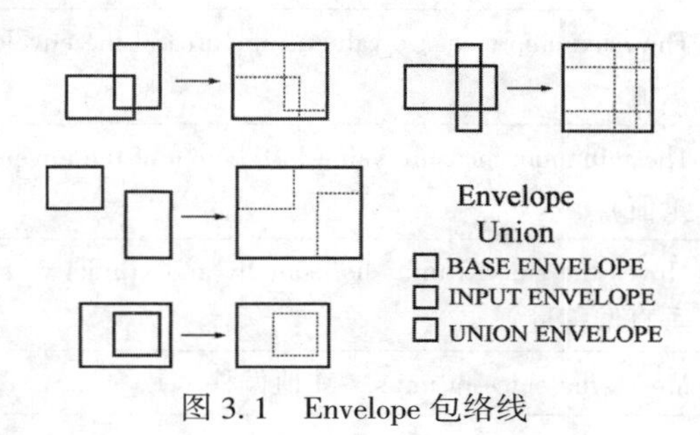

图 3.1　Envelope 包络线

表 3.2　**IEnvelope 接口**

Properties(属性)	Description(描述)
CenterAt	Moves this envelope so it centered at p.(移动包络线使其集中在 p 点)
DefineFromPoint	Defines the envelope to cover all the points.(定义覆盖所有点的包络线)
Depth	The depth of the envelope.(包络线的深度)
Dimension	The topological dimension of this geometry.(几何对象的拓扑维数)
Envelope	Creates a copy of this geometry's envelope and returns it.(创建几何对象包络线的副本并返回)
Expand	Moves the X and Y coordinates of the sides toward or away from each other.(移动边界 XY 坐标靠近或远离)
ExpandM	Moves the measure of the sides toward or away from each other.(移动边界 M 值靠近或远离)
ExpandZ	Moves the Z attribute of the sides toward or away from each other(移动边界 Z 值靠近或远离)
GeometryType	The type of this geometry.(几何对象的类型)
GeoNormalize	Shifts longitudes, if need be, into a continuous range of 360 degrees.(可连续 360 度的经度变化)

续表

Properties(属性)	Description(描述)
GeoNormalizeFromLongitude	Normalizes longitudes into z continuous range containing the longitude.(在经度范围标准化经度为连续 z 值。)
Height	The height of the envelope.(包络线高度)
Intersect	Adjusts to include only the area also include by inEnvelope. (调整对象到指定包络线包含的区域)
IsEmpty	Indicates whether this geometry contains any points.(判断几何体是否包含点)
LowerLeft	The lower left corner. (左下角)
LowerRight	The lower right corner. (右下角)
MMax	The maximum measure value in the area of the envelope.(包络线区域的最大测量值)
MMin	The minimum measure value in the area of the envelope.(包络线区域的最小测量值)
Offset	Moves the sides x units horizontally and y units vertically.(设置边界的水平和垂直偏移量)
OffsetM	Moves the sides m units.(M 值偏移量)
OffsetZ	Moves the sides z units.(z 值偏移量)
Project	Projects this geometry into a new spatial reference.(将几何对象投影到新的空间坐标系)
PutCoords	Constructs an envelope from the coordinate values of lower, left and upper, right corners.(用左下角和右上角的坐标值构造包络线)
QueryCoords	Returns the coordinates of lower, left and upper, right corners.(返回左下角和右上角的坐标)
QueryEnvelope	Copies this geometry's envelope properties into specified envelope.(复制几何对象包络线属性赋给指定的包络线)
SetEmpty	Removes all points from this geometry.(删除几何对象的所有点)
SnapToSpatialReference	Moves points of this geometry so that they can be represented in the precision of the geometry's associated spatial reference system. (移动几何对象的点使其在一定的几何精度内匹配指定的空间坐标系)
SpatialReference	The spatial reference associated with this geometry.(几何对象的空间坐标系)
Union	Adjusts to overlap inEnvelope.(调整包络线的重叠关系)
UpperLeft	The upper left corner.(左上角)
UpperRight	The upper right corner.(右上角)
Width	The width of the envelope.(包络线的宽度)

续表

Properties(属性)	Description(描述)
XMax	The position of the right side. (最大 X 值)
XMin	The position of the left side. (最小 X 值)
YMax	The position of the top. (最大 Y 值)
YMin	The position of the bottom. (最小 Y 值)
ZMax	The maximum Zvalue in the area of the envelope. (包络线区域中 Z 的最大值)
ZMin	The minimum Zvalue in the area of the envelope. (包络线区域中 Z 的最小值)

3.2.2　Curve 对象

几何对象中除了点、点集、包络线外，其他类型都可归为曲线，即 Curve 对象。如线、多边形等都实现了 ICurve 接口。该接口提供了操作这些对象的属性和方法，但不能产生一个新的曲线对象。ICurve 接口定义的属性方法见表 3.3。

表 3.3　**ICurve 接口**

Properties(属性)	Description(描述)
Dimension	The topological dimension of this geometry. (几何对象的拓扑维度)
Envelope	Creates a copy of this geometry's envelope and returns it. (创建几何体包络线副本并返回)
FromPoint	The 'from' point of the curve. (曲线的起始点)
GeometryType	The type of this geometry. (几何体类型)
GeoNormalize	Shifts longitudes, if need be, into a continuous range of 360 degrees. (可连续 360 度的经度变化)
GeoNormalizeFromLongitude	Normalizes longitudes into z continuous range containing the longitude, this method is obsolete. (在经度范围标准化经度为连续 z 值)
GetSubcurve	Extracts a portion of this curve into a new curve. (提取部分曲线构成子曲线)
IsClosed	Indicates if 'from' and 'to' point(of each part) are identical(判断曲线的起点和终点是否闭合)
IsEmpty	Indicates whether this geometry contains any points. (判断几何对象是否包含点)
Length	The length of the curve. (曲线长度)
Project	Projects this geometry into a new spatial reference. (将几何体投影到新空间坐标系)

续表

Properties(属性)	Description(描述)
QueryEnvelope	Copies this geometry's envelope properties into the specified envelope.(复制几何体包络线属性赋给指定包络线)
QueryFromPoint	Copies this curve's 'from' point to the input point.(复制曲线起始点到输入点)
QueryNormal	Constructs aline normal to a curve from a point at a specified distance along the curve.(沿曲线以指定距离沿点构造曲线的法线)
QueryPoint	Copies to outpoint the properties of a point on the curve at a specified distance from the beginning of the curve.(从开始以特定距离复制曲线上点的属性到输出点)
QueryPointAndDistance	Finds the point on the curve closest to inpoint, then copies that point to outpoint.(查到曲线上离输入点最近的点并将其复制到输出点)
QueryTangent	Constructs a line tangent to a curve from a point at a specified distance along the curve.(给定曲线上的一点查询其正切值)
QueryToPoint	Copies the curve's 'to' point into the input point.(复制曲线的终点到输入点)
ReverseOrientation	Reverses the parameterization of the curve.(改变曲线参数使其反转)
SetEmpty	Removes all points from this geometry.(删除几何对象的所有点)
SnapToSpatialReference	Moves points of this geometry so that they can be represented in the precision of the geometry's associated spatial reference system.(移动几何对象的点使其在一定的几何精度内匹配指定的空间坐标系统)
SpatialReference	The spatial reference associated with this geometry.(几何对象的空间坐标)
ToPoint	The'to'point of the curve.(曲线的终点)

3.2.3 Point 及 Multipoint 对象

Point 是一维具有 *X*、*Y* 坐标的几何对象，Geometry 中的任何类型都是用点生成的。IPoint 接口定义了 Point 对象的属性和方法。IConstructPoint 接口提供了 10 种创建点的方法，见表 3.4。

表 3.4　**IConstructPoint 接口**

Properties(属性)	Description(描述)
ConstructAlong	Constructs a point distance units along the input curve.(沿曲线一定距离构造点)

续表

Properties(属性)	Description(描述)
ConstructAnqleBisector	Constructs a point on the bisector of the angle (from, through, to), when Acute Angle is false, the sign of distance will select a point right or left of (from, through, to).(在角的平分线上构造点，当是钝角时，标记需选择左或右)
ConstructAnqleDistance	Constructs a point at a specified angle (in radians) from the horizontal axis and a specified distance away from the input point.(和水平轴成指定角(弧度)距离指定输入点一定距离来构造点)
ConstructAnqleIntersection	Constructs the point of intersection between two lines defined by the input points and angles (in radians).(构造由输入点和角度(弧度)所定义的两条线的交叉点)
ConstructDeflection	Constructs a point in the polar coordinate system defined by baseLine and its'from'point.(在极坐标系下构造一个点)
ConstructDeflectionIntersection	Constructs the point of intersection of two rays with origins at the endpoints of the base line and specified angles (in radians).(构建由基线和特定角度所确定的两条射线的交点)
ConstructOffset	Constructs a point distance units along the input curve and offset units perpendicularly away from it.(沿着输入曲线和垂直方向偏移，构建一个单位距离的点)
ConstructParallel	Constructs a point distance units from start, parallel to the tangent at the point nearest to start on the (extended) segment.(从给定起点构建一个距其所在的直线段或曲线的切线一定距离的点)
ConstructPerpendicular	Constructs a point'distance'units from p and lying along the line normal to base and passing through p.(在参考点与路径垂直的方向上，构建给定距离点)
ConstructThreePointResection	Constructs the point of observation from which two signed angles between three points were measured; returns an angle which can help establish the confidence of the observation location; A small angle indicates greater uncertainty in the location.(按后方交会方法构建观测点，对三个已知点进行观测，确定两个角度；返回可以建立观察位置置信度的角度；角度越小表明位置的不确定性越大)

第4章　地图对象与图层控制

4.1　Map 对象

Map 对象是 ArcGIS Engine 的重要组成部分，主要实现数据管理和数据显示的功能。Map 对象一共实现了 35 个接口，本节介绍 IMap、IGraphicsContainer 与 IActiveView 3 个基本接口。

4.1.1　IMap 接口

IMap 接口定义的方法主要通过对图层相关操作实现得 Map 对象中相关对象的操作功能。IMap 接口定义的属性和方法见表 4.1。

表 4.1　**IMap 接口定义的属性和方法**

Properties(属性)	Description(描述)
ActiveGraphicsLayer	The active graphics layer.(活动图形图层)
AddLayer	Adds a layer to the map.(添加图层到地图中)
AddLayers	Adds multiple layers to the map, arranging them nicely if specified.(将多图层添加到地图，如果有指定的顺序则按顺序添加)
AddMapSurround	Adds a map surround to the map.(添加辅助地图到地图)
AnnotationEngine	The annotation (label) engine the map will use.(将要用到的地图标注引擎)
AreaOfInterest	Area of interest for the map.(地图的兴趣区域)
Barriers	The list of barriers and their weight for labeling.(标签的边界宽度列表)
BasicGraphicsLayer	The basic graphics layer.(基础图形图层)
ClearLayers	Removes all layers from the map.(移除地图中的所有图层)
ClearMapSurrounds	Removes all map surrounds from the map.(移除地图中所有辅助地图)
ClearSelection	Clears the map selection.(清除图层选项)
ClipBorder	An optional border drawn around ClipGeometry.(裁剪几何体的可选边界)
ClipGeometry	A shape that layers in the map are clipped to.(地图中需要裁剪的形状)

续表

Properties(属性)	Description(描述)
ComputeDistance	Computers the distance between two points on the map and returns the result.(计算地图上两点的距离并返回结果)
CreateMapSurround	Create and initialize a map surround.(创建并初始化地图)
DelayDrawing	Suspends drawing.(暂停绘制)
DelayEvents	Used to batch operations together to minimize notifications.(用于批处理操作并以最小化通知)
DeleteLayer	Deletes a layer from the map.(从地图中删除一个图层)
DeleteMapSurround	Deletes a map surround from the map.(从地图中删除辅助地图)
Description	Description of the map.(地图描述)
DistanceUnits	The distance units for the map.(地图距离单位)
Expanded	Indicates if the Map is expanded.(指示地图是否放大)
FeatureSelection	The feature selection for the map.(地图中选中的要素)
GetPageSize	Gets the page size for the map.(获取地图页面的大小)
IsFramed	Indicates if map is drawn in a frame rather than on the whole window.(指示地图绘制是在框架下还是在整个窗口中)
Layer	The layer at the given index.(指定索引图层)
LaterCount	Number of layers in the map.(地图中图层数目)
Layers	The layers in the map of the type specified in the uid.(地图图层中指定类型的uid)
MapScale	The scale of the map.(地图比例尺)
MapSurround	The map surround at the given index.(指定索引的辅助地图)
MapSurroundCount	Number of map surrounds associated with the map.(与当前地图相关的辅助地图数目)
MapUnits	The units for the map.(地图单位)
MoveLayer	Moves a layer to another position.(移动图层到指定位置)
Name	Name of the map.(地图名称)
RecalcFullExtent	Forces the full extent to be recalculated.(重新计算地图全图显示)
ReferenceScale	The reference scale of the map. (地图基准比例尺)
SelectByShape	Selects features in the map given a shape and a selection environment (optional).(通过给定形状和选择环境(可选)选择地图中的要素)
SelectFeature	Selects a feature.(选中一个要素)
SelectionCount	Number of selected features.(选中要素的数目)

续表

Properties(属性)	Description(描述)
SetPageSize	Sets the page size for the map (optional).(设置地图页面的尺寸(可选))
SpatialReference	The spatial reference of the map.(地图的空间参考系)
SpatialReferenceLocked	Indicates whether the spatial reference is prevented from being changed.(指明空间参考系是否被锁定)
UseSymbollevels	Indicates if the Map draws using symbol levels.(指明地图绘图是否使用了符号级别)

①public void AddLayer (ILayer Layer):添加一个图层到 Map 对象。

②public void ClearSelection ():清除选择区域。

③ public void SelectByShape (IGeometry Shape, ISelectionEnvironment env, bool justOne):通过图形选择实体。

④public int SelectionCount {get;}:获取选择实体的个数。

下面通过典型代码示例说明 IMap 接口的一些用法。

1. 添加 Shp 图层实现示例(AddLayer 方法)

具体实现步骤如下:

①利用 OpenFileDialog 找到 Shp 图层;

②创建工作空间工厂;

③打开要素类;

④创建要素图层;

⑤关联图层和要素类;

⑥添加到地图控件中。

示例代码如下:

```
openFileDialog1=new OpenFileDialog();
openFileDialog1.Title ="打开图层文件";
openFileDialog1.Filter ="map documents(*.shp)|*.shp";
openFileDialog1.ShowDialog();
FileInfo fileInfo=new FileInfo(openFileDialog1.FileName);
string pPath = openFileDialog1.FileName;
string pFolder = System.IO.Path.GetDirectoryName(pPath);
string pFileName = System.IO.Path.GetFileName(pPath);
IWorkspaceFactory pWorkspaceFactory=new ShapefileWorkspace Fac-
tory();
IWorkspace pWorkspace = pWorkspaceFactory.OpenFromFile(pFolder,
0);
FeatureWorkspace pFeatureWorkspace= pWorkspace as IFeatureWorkspace;
```

```
IFeatureClass pFC = pFeatureWorkspace.OpenFeatureClass(pFileName);
IFeatureLayer pFLayer =new FeatureLayer();
pFLayer.FeatureClass = pFC;
pFLayer.Name = pFC.AliasName;
ILayer pLayer = pFLayer as ILayer;
IMap pMap = axMapControl1.Map;
pMap.AddLayer(pLayer);
axMapControl1.ActiveView.Refresh();
```

2. 删除 Shp 图层实现示例(DeleteLayer 方法)

具体实现步骤如下:

①获得当前地图控件中的地图;

②在地图中获得最上面的一个数据层;

③删除数据层。

示例代码如下:

```
IMap pMap = axMapControl1.Map;
ILayer pLayer = pMap.get_Layer(0);
pMap.DeleteLayer(pLayer);
//被高亮显示的要素称为"FeatureSelection"(要素选择集),程序有多种方式可以用于选择要素,比如 SelectByShape 方法就是其中之一
IMap pMap = axMapControl1.Map;
pActiveView = pMap as IActiveView;
objEnvelope = axMapControl1.TrackRectangle();
ISelectionEnvironment pSelectionEnvironment =new SelectionEnvironment
();
pMap.SelectByShape(objEnvelope,pSelectionEnvironment,false);
//枚举在 MapDocument 地图和将自己的名字打印到控制台上
IMapDocument pMapDocument =new MapDocumentClass();
if(pMapDocument.get_IsMapDocument(path))
{
   pMapDocument.Open(path,null);
   IMap pMap;
for(int i =0; i <= pMapDocument.MapCount -1; i++)
{
       pMap = pMapDocument.get_Map(i);
       Console.WriteLine(pMap.Name);
       IEnumLayer pEnumLayer = pMap.get_Layers(null,true);
       pEnumLayer.Reset();
       ILayer pLayer = pEnumLayer.Next();
```

```
while(pLayer ! =null)
{
        Console.WriteLine(pLayer.Name);
        pLayer= pEnumLayer.Next();
}
}
}
```

3. 添加点和清空点示例

代码示例如下：

```
//添加
IFeatureLayer pFeatLayer = axMapControl1.get_Layer(0)as IFeatureLayer;
IFeatureClass pFeatCla= pFeatLayer.FeatureClass;
IPoint pnt=new PointClass();
pnt.X=113.301;
pnt.Y=23.113;
GISTool.CreateTimePoint(pFeatCla, pnt,100.1);
axMapControl1.ActiveView.Refresh();
staticpublic void  CreateTimePoint(IFeatureClass fClass,IPoint pt,double time)
{
    IFeature featCreate= fClass.CreateFeature();
    featCreate.Shape = pt;
int lFld;
    lFld= fClass.FindField("timevalue");
    featCreate.set_Value(lFld, time);
    featCreate.Store();
}
//清空
if(axMapControl1.get_Layer(0).Name ! ="TimePoints")
{
    MessageBox.Show("图层不对!");
return;
}
IFeatureLayer pFeatLayer = axMapControl1.get_Layer(0)as IFeatureLayer;
IFeatureCursor pFeatureCursor= pFeatLayer.Search(null,false);
IFeature pFeature= pFeatureCursor.NextFeature();
```

```
while(pFeature ! =null)
{
    pFeature.Delete();
    pFeature= pFeatureCursor.NextFeature();
}
//重新刷新
axMapControl1.ActiveView.Refresh();
MessageBox.Show("清除完毕!");
```

4. 通过鼠标在屏幕上绘制多边形示例

代码示例如下:

```
Private void axMapControl1_OnMouseDown
(objectsender,IMapControlEvents2_OnMouseDownEvent e)
{
//选定要素
    IMap pMap=this.axMapControl1.Map;
    IFeatureLayer pFeaturelayer=null;
    IFeatureClass pFeatureClass;
for(int i =0; i < axMapControl1.LayerCount;i++)
{
if(pMap.get_Layer(i).Name= =this.comboBox1.Text)
{
            pFeaturelayer= pMap.get_Layer(i)as IFeatureLayer;
break;
}
}
    pFeatureClass= pFeaturelayer.FeatureClass;
    IPolygon pPolygon;
    ITopologicalOperator pTopologicaloperator;
    ISpatialFilter pSpatialFilter;
//得到鼠标交互所产生的多边形
    pPolygon= axMapControl1.TrackPolygon()as IPolygon;
//使该多边形有效
    pTopologicaloperator=(ITopologicalOperator)pPolygon;
    pTopologicaloperator.Simplify();
    pSpatialFilter=new SpatialFilterClass();
//设置查询几何体
    pSpatialFilter.Geometry =(IGeometry)pPolygon;
//根据被查询要素类几何类型决定空间查询关系
```

```
switch(pFeatureClass.ShapeType)
{
case esriGeometryType.esriGeometryPoint:
        pSpatialFilter.SpatialRel = esriSpatialRelEnum.esriSpa-
tialRelContains;
break;
case esriGeometryType.esriGeometryLine:
        pSpatialFilter.SpatialRel = esriSpatialRelEnum.esriSpa-
tialRelIntersects;
break;
case esriGeometryType.esriGeometryPolygon:
        pSpatialFilter.SpatialRel = esriSpatialRelEnum.esriSpa-
tialRelIntersects;
break;
}
//设置过滤器几何字段名称
    pSpatialFilter.GeometryField = pFeatureClass.ShapeFieldName;
//QI
     IFeatureSelection pFeatureSelection = (IFeatureSelection)
pFeaturelayer;
//判断该图层是否已经存在选择集
if(pFeatureSelection.SelectionSet.Count>0)
{
        pFeatureSelection.Clear();
}
//添加要素到选择集
    pFeatureSelection.SelectFeatures((IQueryFilter)pSpatialFil-
ter, esriSelectionResultEnum.esriSelectionResultNew,false);
//刷新视图
    IActiveView pActiveView=(IActiveView)pMap;
    pActiveView.PartialRefresh
(esriViewDrawPhase.esriViewGeoSelection, null, pActiveView.Ex-
tent);
}
```

4.1.2 IGraphicsContainer 接口

Map 对象通过 IGraphicsContainer 接口来管理图形元素。IGraphicsContainer 接口定义的属性和方法见表 4.2。

表 4.2　　**IGraphics Container 接口定义的属性和方法**

Properties(属性)	Description(描述)
AddElenment	Add a new graphic element to the layer.(添加一个新图形元素到图层)
AddElements	Add new graphic elements to the layer.(添加新图形元素到图层)
BringForward	Move the specified elements one step closer to the top of the stack of elements.(单步移动指定元素到上一层)
BringToFront	Make the specified elements draw in front of all other elements.(在所有元素的上层绘制指定元素)
DeleteAllElements	Delete all the elements.(删除所有元素)
DeleteElement	Delete the given element.(删除指定元素)
FindFrame	Find the frame that contains the specified object.(查找包含指定对象的框架)
GetElementOrder	Get the ordering of the elements.(获得元素的排序)
LocateElements	Returns the elements at the given coordinate.(返回指定坐标的元素)
LocateElementsByEnvelope	Returns the elements inside the given envelope.(返回指定包络线内的元素)
MoveElementFromGroup	Move the element from the group to the container.(从组移动元素到容器内)
MoveElementToGroup	Move the element from the container to the group.(从容器内移动元素到组)
Next	Returns the next graphic in the container.(返回容器的下一个图形元素)
PutElementOrder	Used to do ordering operations.(进行排序操作)
Reset	Reset internal cursor so that Next returns the first element.(重置内部指针以便返回第一个元素)
SendBackward	Move the specified elements one step closer to the bottom of the stack of elements.(单步将指定元素移至下一层)
SendToBack	Make the specified elements draw behind all other elements.(将指定元素移至最后层)
UpdateElement	The graphic element's properties have changed.(标志图形元素的属性已经改变)

①public void AddElement (IElement Element, int zorder):添加一个新的元素到图层。

②public IEnumElement LocateElementsByEnvelope (IEnvelope envelope):返回在包络线中的元素。

通过下面典型示例代码,说明 IGraphicsContaine 接口的一些用法:

```
//添加元素是一个往其中一个图形图层上添加元素的对象的过程
IGraphicsContainer graphicsContainer;
IMap map=this.axMapControl1.Map;
ILineElement lineElement=new LineElementClass();
```

```
IElement element;
IPolyline polyline=new PolylineClass();
IPoint point=new PointClass();
point.PutCoords(1,5);
polyline.FromPoint = point;
point.PutCoords(80,5);
polyline.ToPoint = point;
IElement element= lineElement as IElement;//接口查询
element.Geometry = polyline as IGeometry;
graphicsContainer= map as IGraphicsContainer;
graphicsContainer.AddElement(element,0);
this.axMapControl1.ActiveView.PartialRefresh
(esriViewDrawPhase.esriViewGraphics,null,null);//添加元素后一定要刷新,不然无法显示
//更新元素:用户改变的是 Map 中一个元素的形状或者符号,用户一定希望它能够及时更新后在地图上显示出来,用到 UpdateElement
IGraphicsContainer graphicsContainer;
IPolyline polyline=new PolylineClass();
IPoint point=new PointClass();
point.PutCoords(1,5);
polyline.FromPoint = point;
point.PutCoords(80,20);
polyline.ToPoint = point;
IElement el;
graphicsContainer=this.axMapControl1.Map as IGraphicsContainer;
graphicsContainer.Reset();
el= graphicsContainer.Next();
if(el ! =null)
{
    el.Geometry = polyline;
    graphicsContainer.UpdateElement(el);
}
this.axMapControl1.ActiveView.PartialRefresh ( esriViewDrawPhase.
esriViewGraphics,null,null);
//删除元素:DeleteElement 用于删除 Map 对象中的一个给定元素
IGraphicsContainer graphicsContainer;
IElement el;
graphicsContainer=this.axMapControl1.Map as IGraphicsContainer;
```

```
graphicsContainer.Reset();
el= graphicsContainer.Next();
while(el ! =null)
{
    graphicsContainer.DeleteElement(el);
    el= graphicsContainer.Next();
}
this.axMapControl1.ActiveView.Refresh ();
```

4.1.3 IActiveView 接口

前面的两个接口都是管理图形的，而 IActiveView 接口定义了 Map 对象的另一个功能——数据显示功能。使用该接口可以改变视图的范围，刷新视图。

在 ArcMap 中，有两个对象实现了这个接口，分别是 Map 和 PageLayout。这两个对象分别代表了 ArcMap 中两种不同的视图，数据视图和版式视图，在任何一个时刻仅仅只能有一个视图处于活跃状态。

如果 ArcMap 处于版式视图状态，则 ActiveView 返回一个 IActiveView 对象指向 PageLayout 对象的，反之，处于视图状态，则指向 Map 对象。

IActiveView 接口定义了 Map 对象的数据显示功能。使用该接口可以改变视图的范围，刷新视图。

①IActiveView 的 PartialRefresh(esriViewGeography, pLayer, null)用于刷新指定图层。

②IActiveView 的 PartialRefresh(esriViewGeography, null, null) 用于刷新所有图层。

③IActiveView 的 PartialRefresh(esriViewGeoSelection, null, null) 用于刷新所选择的对象。

④IActiveView 的 PartialRefresh(esriViewGraphics, null, null) 用于刷新图形元素。

⑤IActiveView 的 PartialRefresh(esriViewGraphics, pElement, null) 用于刷新指定图形元素。

⑥IActiveView 的 PartialRefresh(esriViewGraphics, null, null) 用于刷新所有图形元素。

⑦IActiveView 的 PartialRefresh(esriViewGraphicSelection, null, null)用于刷新所选择的图元。

4.2 Layer 对象

Map 对象可以装载地理数据，这些数据是以图层的形式加载到地图对象上的，图层对象 Layer 作为一个数据的“中介”存在，它本身没有转载地理数据，而仅仅是获得了数据的引用，用于管理数据源的连接。地理数据始终保存在 GeoDatabase 或者地理数据文件中。

4.2.1 ILayer 接口

所有图层类都实现这一接口，它定义了所有图层的公共方法和属性，如 Name 属性可以返回图层名称，MaxmunScale 和 MinimunScale 两个可写属性，用于显示和设置图层可以出现的最大尺寸和最小尺寸。

Showtips 属性用于指示当鼠标放在图层某个要素上的时候，是否会出现提示(Tips)，TipText 确定图层提示的字段。SpatialReference 属性用于设置图层的空间参考，这个对象是从 Map 对象中传入，对于地理数据极其重要。

4.2.2 IFeatureLayer 接口

IFeatureLayer 接口主要用于设置要素图层的数据源(FeatureClass)，IFeatureLayer 的 DataSourceType 获取 FeatureLayer 对象的数据源类型。此外，通过 IFeatureLayer 的 Search 方法可以查询要素图层上符合某一条件的要素集。

4.2.3 图层操作

1. 添加矢量数据示例

示例代码如下：

```
Public static void addFeatLayerToMapFromGDB(IMap map,string gdbPath,string name)
{
    IWorkspaceFactory toWsf=new FileGDBWorkspaceFactoryClass();
    IFeatureWorkspace toFeatWs = (IFeatureWorkspace)toWsf.OpenFromFile(gdbPath,0);
    IFeatureClass featCls= toFeatWs.OpenFeatureClass(name);
    ILayer layer=new FeatureLayerClass();
    layer.Name = name;
((IGeoFeatureLayer)layer).FeatureClass=featCls;
if(featCls！=null)
{
map.AddLayer(layer);
}
}
```

2. 添加栅格数据示例

示例代码如下：

```
//直接用 IRasterLayer 接口打开一个栅格文件并加载到地图控件
IRasterLayer rasterLayer=new RasterLayerClass();
rasterLayer.CreateFromFilePath(fileName);// fileName 指存本地的栅格文件路径
axMapControl1.AddLayer(rasterLayer,0);
//用 IRasterDataset 接口打开一个栅格数据集
IWorkspaceFactory workspaceFactory = new RasterWorkspaceFactory
```

```
();
    IWorkspace workspace;
    workspace= workspaceFactory.OpenFromFile(inPath,0);//inPath 栅格数据存储路径
    if(workspace ==null)
    {
        Console.WriteLine("Could not open the workspace.");
    return;
    }
    IRasterWorkspace rastWork=(IRasterWorkspace)workspace;
    IRasterDataset rastDataset;
    rastDataset= rastWork.OpenRasterDataset(inName);//inName 栅格文件名
    if(rastDataset ==null)
    {
        Console.WriteLine("Could not open the raster dataset.");
    return;
    }
```

3. 加载 CAD 图层示例

CAD 图层的加载可以分为：分图层加载和整幅图加载。

(1)分图层加载示例

示例代码如下：

```
Public void AddCADByLayer()
{
    IWorkspaceFactory Fact=new CadWorkspaceFactoryClass();//定义工作空间,并用 CadWorkspaceFactoryClass()实例化它;
    //打开相应的工作空间,并赋值给要素空间,OpenFromFile()中的参数为 CAD 文件夹的路径
    IFeatureWorkspace Workspace= Fact.OpenFromFile(@ "I:\test\",0)as IFeatureWorkspace;
    /* 打开线要素类,如果要打开点类型的要素,需要把下边的代码改成:
    IFeatureClass Fcls = Workspace.OpenFeatureClass ( " modle.dwg:point");
    由此可见 modle.dwg 为 CAD 图的名字,后边加上要打开的要素类的类型,中间用冒号隔开,大家可以想想多边形和标注是怎么打开的。*/
    IFeatureClass Fcls = Workspace.OpenFeatureClass ( " modle.dwg:polyline");
    IFeatureLayer Fly=new FeatureLayerClass ();
```

```
Fly.FeatureClass = Fcls;
MapCtr.Map.AddLayer (Fly);
MapCtr.ActiveView.Refresh ();
}
```

(2)整幅 CAD 图的加载示例

示例代码如下:

```
Public void AddWholeCAD()
{
    IWorkspaceFactory Fact =new CadWorkspaceFactoryClass();
    IWorkspace Workspace= Fact.OpenFromFile(@ "I:\test\",0);
//定义一个 CAD 画图空间,并把上边打开的工作空间赋给它
ICadDrawingWorkspace dw=  Workspace as ICadDrawingWorkspace;
//定义一个 CAD 的画图数据集,并且打开上边指定的工作空间中一幅 CAD 图
//然后赋值给 CAD 数据集
//通过 ICadLayer 类,把上边得到的 CAD 数据集赋值给 ICadLayer 类对象的 Cad-
DrawingDataset 属性
ICadLayer CadLayer=new  CadLayerClass();
CadLayer.CadDrawingDataset = ds;
//利用 MapControl 加载 CAD 层
MapCtr.ActiveView.Refresh ();
}
```

第5章　空间数据模型及数据库

5.1 ArcSDE 简介

ArcSDE(SDE 即 Spatial Database Engine，空间数据库引擎)是 ArcGIS 与关系数据库之间的 GIS 通道。它允许用户在多种数据管理系统中管理地理信息，并使所有的 ArcGIS 应用程序都能够使用这些数据。ArcSDE 是多用户 ArcGIS 系统的一个关键部件。它为 DBMS 提供了一个开放的接口，允许 ArcGIS 在多种数据库平台上管理地理信息。这些平台包括 Oracle、Oracle with Spatial/Locator、Microsoft SQL Server、IBM DB2 和 Informix。通过 ArcSDE，ArcGIS 可以在 DBMS 中轻而易举地管理一个共享的、多用户的空间数据库。ArcSDE 具体功能如下：

1. 高性能的 DBMS 通道

ArcSDE 是多种 DBMS 的通道。它本身并非一个关系数据库或数据存储模型。它是一个能在多种 DBMS 平台上提供高级的、高性能的 GIS 数据管理的接口。

2. 开放的 DBMS 支持

ArcSDE 允许用户在多种 DBMS 中管理地理信息，包括 Oracle、Oracle with Spatial or Locator、Microsoft SQL Server，Informix，以及 IBM DB2。

3. 多用户

ArcSDE 为用户提供大型空间数据库支持，并且支持多用户编辑。

4. 连续、可伸缩的数据库

ArcSDE 可以支持海量的空间数据库和任意数量的用户，直至 DBMS 的上限。

5. GIS 工作流和长事务处理

GIS 中的数据管理工作流，例如，多用户编辑、历史数据管理、check-out/check-in 以及松散耦合的数据复制等，都依赖于长事务处理和版本管理。ArcSDE 为 DBMS 提供了这种支持。

6. 丰富的地理信息数据模型

ArcSDE 保证了存储于 DBMS 中的矢量和栅格几何数据的高度完整性。这些数据包括矢量和栅格几何图形，支持 x、y、z 和 x、y、z、m 的坐标，曲线，立体，多行栅格，拓扑，网络，注记，元数据，空间处理模型，地图，图层等。

7. 灵活的配置

ArcSDE 允许用户在客户端应用程序内或跨网络、跨计算机的对应用服务器设置多种多层结构的配置方案。ArcSDE 支持 Windows、UNIX、Linux 等多种操作系统。

ArcSDE 是为了解决 DBMS 的多样性和复杂性而存在的。ArcSDE 的体系结构给用户提供了巨大的灵活性，允许用户能够自由地选择 DBMS 来存储空间数据。

5.2 GeoDatabase 对象模型

5.2.1 GeoDatabase 对象模型简介

GeoDatabase 是第三代地理数据模型，每个要素不再仅仅是一条有几何字段的记录，而是一个拥有属性和行为的对象。即它是一个基于面向对象模型的关系数据库，而不仅仅是一个普通关系数据库。GeoDatabase 模型支持要素间的拓扑关系，并且扩展了基于要素和其他面向对象类型数据的复杂网络和关系功能。由于 GeoDatabase 模型使用了一般性数据框架描述，从这个框架的视角上看，同类型的不同格式数据之间除了存储的物理文件有所不同外，在操作上使用的代码没有差别。

GeoDatabase 的每一种数据集都是对世界某一方面的空间描述，它包括：

①基于矢量的离散要素(点、线和面)的有序集合。

②基于数字高程模型 DEM 和影像的栅格数据集合，描述连续的地理现象。

③模拟物流、人流等形式的拓扑网络。

④地球表面的三维模拟，如 TIN 等数据集。

⑤地址和定位器等其他数据集。

GeoDatabase 实际上是一个关系数据库基础上扩展的"面向对象数据库"，由于将数据看成不同类型的对象而不是简单的一行数据，从而可以给这些对象更加准确的行为和关系。对于一个对象而言，它具备多态性、封装性和继承性三大基本特点。地理数据库实现了数据的统一存储，即可以将所有的地理数据库保存在一个数据库中，GeoDatabase 将地理数据作为一个对象而保存，这些对象以行为形式被存储在要素类、对象或要素数据集中。对象类和要素类的区别在于前者存储的是非空间数据，而后者保存的是具有相同字段的几何形体对象及其属性信息。

GeoDatabase 核心结构模型图如图 5.1 所示。

1. WorkspaceFactory 对象

WorkspaceFactory 是 GeoDatabase 的入口，是一个抽象类，拥有很多子类，如 SdeWorkspaceFactory、AccessWorkspaceFactory、ShapfileWorkspaceFactory、CadWorkspaceFactory 等。IWorkspaceFactory 接口定义了所有工作空间对象的一般属性和方法。

① Public IWorkspaceName Create (string parentDirectory, string Name, IPropertySet ConnectionProperties, int hWnd)：用于产生一个新的工作空间名称对象，前两个参数是数据库的路径和名称，第三个参数是一个属性集(ProperSet)对象。

②public IWorkspace Open (IPropertySet ConnectionProperties, int hWnd)：打开一个已经存在的工作空间，如 SDE 数据库等。

③public IWorkspace OpenFromFile (string fileName, int hWnd)：打开一个文件类型的数据。

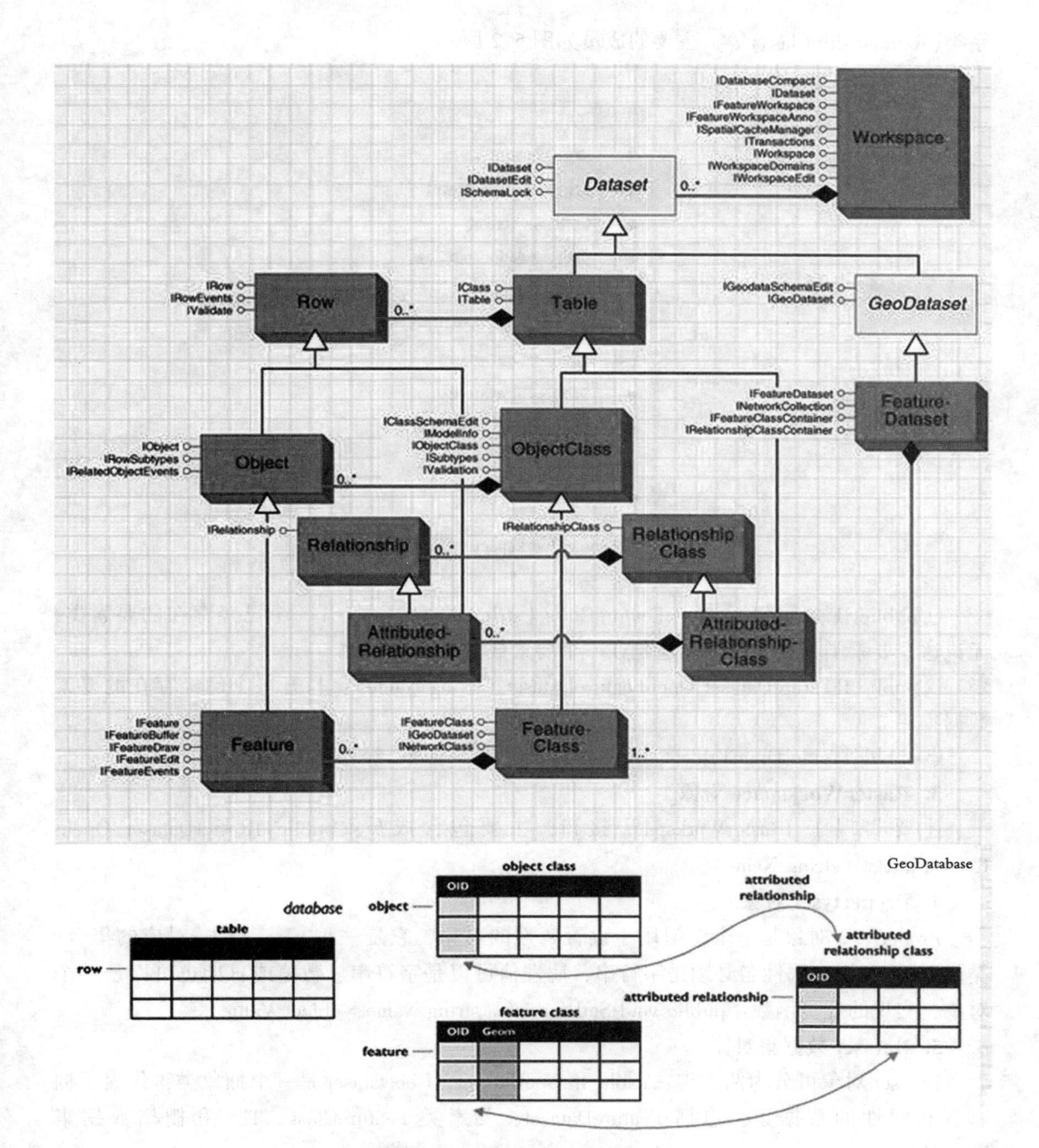

图 5.1　GeoDatabase 核心结构模型图

2. WorkSpace 对象

WorkSpace 在逻辑上是一个包含空间数据集和非空间数据集的数据容器，数据包括要素类、栅格数据集、表等对象。IFeatureWorkspace 接口主要用于管理要素的数据集，如表(Table)、对象类(ObjectClass)、要素类(FeatureClass)、要素数据集(FeatureDataset)和关系类(RelationshipClass)等，主要的成员如图 5.2 所示。

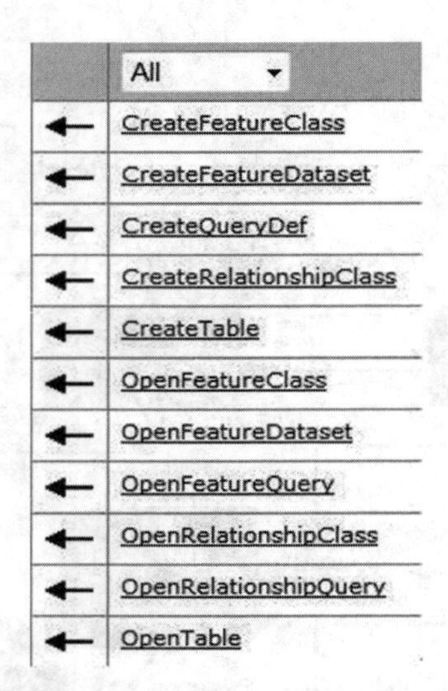

图 5.2　IFeatureWorkspace 接口成员方法

①public IFeatureClass OpenFeatureClass (string Name)：打开一个已经存在的要素类，无论这个要素类是在工作空间还是在一个要素数据集中。

②public IFeatureDataset OpenFeatureDataset (string Name)：打开一个已经存在的要素数据集。

③public ITable OpenTable (string Name)：打开一个已经存在的表。

3. RasterWorkspace 对象

主要实现了 IRasterWorkspace 接口，主要包含成员：public IRasterDataset Open-RasterDataset (string Name)。

4. Propertyset 对象

Propertyset 对象是一个专门用于设置属性的对象，它是一种 name-value 对应的集合，类似于 Hash 表。属性名必须是字符串，属性值可以是字符串、数值或日期也可以是一个对象，实现的成员函数：public void SetProperty (string Name, object Value)。

5. Dataset 数据集对象

Dataset 对象可分为两大类：Table 和 GeoDataset，GeoDataset 是一个抽象类，代表了拥有空间属性的数据集，包括 FeatureDataset、要素类 FeatureClass、TIN 和栅格数据集 RasterDataset。

6. FeatureDataset 对象

实现的主要接口：

①IFeatureDataset 接口。Public IEnumDataset Subsets {get}：连接数据集。

②IFeatureClassContainer 接口。用于管理要素数据集里面的要素类，该接口的 ClassByName 和 Class(index)等属性可以用来获取数据集中的特定的要素类。

5.2.2 GeoDatabase 加载数据示例

1. 加载 Shapefiles 文件示例

示例代码如下：

```
OpenFileDialog dlg=new OpenFileDialog();
dlg.Title="打开 shapefile 文件";
dlg.Filter="(*.shp)|*.shp";
dlg.ShowDialog();
string filename=dlg.FileName;
int index=filename.LastIndexOf(@"\");
string path=filename.Substring(0,index);
string name=filename.Substring(index+1);
IWorkspaceFactory pWsFactory=new ShapefileWorkspaceFactoryClass();
IFeatureWorkspace pWorkSpace= pWsFactory.OpenFromFile(path,0)as
IFeatureWorkspace;
IFeatureClass pFeatureClass= pWorkSpace.OpenFeatureClass(name);
IFeatureLayer pFeatureLayer=new FeatureLayerClass();
pFeatureLayer.Name = pFeatureClass.AliasName;
pFeatureLayer.FeatureClass = pFeatureClass;
axMapControl1.Map.AddLayer(pFeatureLayer);
axMapControl1.ActiveView.PartialRefresh(esriViewDrawPhase.esri
ViewGraphics,null,null);
```

2. 加载栅格数据示例

示例代码如下：

```
OpenFileDialog dlg=new OpenFileDialog();
dlg.Title ="打开 Raster 文件";
dlg.Filter =" Layer File(*.lyr)|*.jpg;*.bmp;*.tiff ";
dlg.ShowDialog();
string filename = dlg.FileName;
int index = filename.LastIndexOf(@"\");
string path = filename.Substring(0, index);
string name = filename.Substring(index +1);
IWorkspaceFactory pWsFactory=new RasterWorkspaceFactoryClass();
IRasterWorkspace pRasterWorkspace= pWsFactory.OpenFromFile(path,0)
as IRasterWorkspace;
IRasterDataset pRasterDataset = pRasterWorkspace.OpenRasterDataset
(name);
IRasterLayer pRasterLayer=new RasterLayerClass();
```

```
pRasterLayer.CreateFromDataset(pRasterDataset);
axMapControl1.Map.AddLayer(pRasterLayer);
axMapControl1.ActiveView.PartialRefresh(esriViewDrawPhase.esri-
ViewGraphics,null,null);
```

3. 加载 CAD 数据示例

示例代码如下:

```
OpenFileDialog dlg=new OpenFileDialog();
dlg.Title ="打开 CAD 文件";
dlg.Filter =" CAD( *.dwg) |*.dwg |All Files( *.*) |*.* ";
dlg.ShowDialog();
string filename = dlg.FileName;
int index = filename.LastIndexOf(@ "\");
string path = filename.Substring(0, index);
string name = filename.Substring(index +1);
IWorkspaceFactory pWsFactory=new CadWorkspaceFactoryClass();
IFeatureWorkspace pFeatureWorkspace = pWsFactory.OpenFromFile
(path,0)as IFeatureWorkspace;
IFeatureDataset pFeatureDataset = pFeatureWorkspace.OpenFeature
Dataset(name);
IFeatureClassContainer pFCContainer = pFeatureDataset as IFea-
tureClassContainer;
for(int i =0; i < pFCContainer.ClassCount; i++)
{
    IFeatureClass pFeatureClass= pFCContainer.get_Class(i);
if(pFeatureClass.FeatureType == esriFeatureType.esriFTCoverage-
Annotation)
{
        IFeatureLayer pFeatureLayer=new CadAnnotationLayerClass();
}
else
{
        IFeatureLayer pFeatureLayer=new FeatureLayerClass();
        pFeatureLayer.FeatureClass = pFeatureClass;
        pFeatureLayer.Name = pFeatureClass.AliasName;
        axMapControl1.Map.AddLayer(pFeatureLayer);
}
}
axMapControl1.ActiveView.PartialRefresh
```

```
(esriViewDrawPhase.esriViewGraphics,null,null);
```

4. 加载 PersonGeoDatabase 数据示例

示例代码如下：

```
    OpenFileDialog dlg=new OpenFileDialog();
    dlg.Title ="打开 Geodatabase 文件";
    dlg.Filter =" Personal Geodatabase( * .mdb) |* .mdb |All Files( * .
*)|*.* ";
    dlg.ShowDialog();
    string path = dlg.FileName;
    IWorkspaceFactory pAccessWorkspaceFactory;
    IFeatureWorkspace pFeatureWorkspace;
    IFeatureLayer pFeatureLayer;
    IFeatureDataset pFeatureDataset;
    pAccessWorkspaceFactory=new AccessWorkspaceFactoryClass();
    IWorkspace pWorkspace= pAccessWorkspaceFactory.OpenFromFile
(path,0);
    IEnumDataset pEnumDataset= pWorkspace.get_Datasets(ESRI.ArcGIS.
Geodatabase.esriDatasetType.esriDTAny);
    pEnumDataset.Reset();
    IDataset pDataset= pEnumDataset.Next();
    if(pDataset is IFeatureDataset)
    {
        pFeatureWorkspace
         =(IFeatureWorkspace)pAccessWorkspaceFactory.OpenFromFile
(path,0);
        pFeatureDataset= pFeatureWorkspace.OpenFeatureDataset(pDataset.
Name);
        IEnumDataset pEnumDataset1= pFeatureDataset.Subsets;
        pEnumDataset1.Reset();
        IDataset pDataset1= pEnumDataset1.Next();
    if(pDataset1 is IFeatureClass)
    {
            pFeatureLayer=new FeatureLayerClass();
    pFeatureLayer.FeatureClass = pFeatureWorkspace.OpenFeatureClass
(pDataset1.Name);
            pFeatureLayer.Name = pFeatureLayer.FeatureClass AliasName;
            axMapControl1.Map.AddLayer(pFeatureLayer);
            axMapControl1.ActiveView.Refresh();
```

```
        }
    }
    else
    {
        pFeatureWorkspace=(IFeatureWorkspace)pWorkspace;
        pFeatureLayer=new FeatureLayerClass();
        pFeatureLayer.FeatureClass = pFeatureWorkspace.OpenFeature-
Class(pDataset.Name);
        pFeatureLayer.Name=pFeatureLayer.FeatureClass.AliasName;
        axMapControl1.Map.AddLayer(pFeatureLayer);
        axMapControl1.ActiveView.Refresh();
    };
```

5.2.3 在 AE 中使用数据库

个人数据库是保存在 Access 中的数据库，其加载方式有两种：通过名字和通过属性加载。如果通过名称加载，首先通过 IPropertySet 接口定义要连接数据库的一些相关属性，在个人数据库中为数据库的路径，例如：

```
IPropertySet Propset=new PropertySetClass();
Propset.SetProperty("DATABASE", @"D:\test\MapData.mdb");
```

当定义完属性并设置属性后，就可以进行打开数据库的操作，在 AE 开发中存在 IWorkspaceFactory、IFeatureWorkspace、IFeatureClass、IFeatureLayer 等几个常用的用于打开和操作数据空间地物的接口。例如，IWorkspaceFactory 是一个用于创建和打开工作空间的接口，它是一个抽象的接口，在具体应用时，要用对应的工作空间对其进行实例化，实例化过程如下：

```
IWorkspaceFactory Fact = new AccessWorkspaceFactoryClass ();
```

如果打开的是 SDE 数据库，就要用 SdeWorkspaceFactoryClass 实例化 Fact。当完成了工作空间的实例化后，就可以根据以上设置的属性打开对应的 Access 数据库了。打开方式如下：

```
IFeatureWorkspace Workspace = Fact.Open(Propset, 0) as IFeatureWorkspace;
```

AE 中加载数据库的相关典型代码如下：

1. 通过数据库名字加载个人数据库示例

示例代码如下：

```
publicvoid AddAccessDBByName()
{
    IWorkspaceName pWorkspaceName=new WorkspaceNameClass();
    pWorkspaceName.WorkspaceFactoryProgID ="esriDataSourcesGDB.
AccessWorkspaceFactory";
    pWorkspaceName.PathName = @"D:\test\Ao\data\sh\MapData.mdb";
```

```
IName n= pWorkspaceName as IName ;
IFeatureWorkspace Workspace= n.Open()as IFeatureWorkspace;
IFeatureClass Fcls= Workspace.OpenFeatureClass ("District");
IFeatureLayer Fly=new FeatureLayerClass();
Fly.FeatureClass = Fcls;
MapCtr.Map.AddLayer (Fly);
MapCtr.ActiveView.Refresh();
```

2. 加载 SDE 数据库示例

示例代码如下：

```
Public void AddSDELayer(bool ChkSdeLinkModle)
{
    IPropertySet  Propset=new PropertySetClass();//定义一个属性
if(ChkSdeLinkModle==true)//采用 SDE 连接
{
        Propset.SetProperty ("SERVER","192.168.1.1");//设置数据库服务器名,服务器所在的 IP 地址
        Propset.SetProperty ("INSTANCE","port:5151");//设置 SDE 的端口,这是安装时指定的,默认安装时"port:5151"
        Propset.SetProperty ("USER","sa");//SDE 的用户名
        Propset.SetProperty ("PASSWORD","sa");//密码
        Propset.SetProperty ("DATABASE","sde");//设置数据库名称
        Propset.SetProperty ("VERSION","SDE.DEFAULT");//SDE 的版本,在这为默认版本
}
else//直接连接
{
          Propset.SetProperty ( " INSTANCE "," sde: sqlserver: zh-pzh");//设置数据库服务器名,如果是本机,可以用"sde:sqlserver:."
        Propset.SetProperty ("USER","sa");//SDE 的用户名
        Propset.SetProperty ("PASSWORD","sa");//密码
        Propset.SetProperty ("VERSION","SDE.DEFAULT");//SDE 的版本,在这为默认版本
}
    IWorkspaceFactory Fact=new SdeWorkspaceFactoryClass();//定义一个工作空间,并实例化为 SDE 的工作空间
    IFeatureWorkspace Workspace=(IFeatureWorkspace )Fact.Open(Propset,0);//打开 SDE 工作空间,并转化为地物工作空间
    IFeatureClass Fcls= Workspace.OpenFeatureClass ("sde.dbo.管
```

```
点");
        IFeatureLayer Fly=new FeatureLayerClass();
        Fly.FeatureClass = Fcls;
        MapCtr.Map.AddLayer(Fly);
        MapCtr.ActiveView.Refresh();
    }
```

第6章　栅格数据处理

6.1 栅格数据简介

栅格数据是按网格单元的行与列排列、具有不同灰度或颜色的阵列数据。栅格结构是用大小相等、分布均匀、紧密相连的像元(网格单元)阵列来表示空间地物或现象分布的数据组织。栅格结构是最简单、最直观的空间数据结构，它将地球表面划分为大小均匀、紧密相邻的网格阵列。每一个单元(像素)的位置由它的行列号定义，所表示的实体位置隐含在栅格行列位置中，数据组织中的每个数据表示地物或现象的非几何属性或指向其属性的指针。对于栅格结构：点实体由一个栅格像元来表示；线实体由一定方向上连接成串的相邻栅格像元表示；面实体(区域)由具有相同属性的相邻栅格像元的块集合来表示。

ArcGIS 中所支持的栅格数据有：GRID、TIFF、ERDAS、IMAGE、JPEG 等。其主要用于空间分析功能和叠加功能。

6.2 栅格数据加载

访问栅格数据时，必须打开一个工作空间。在 ArcGIS 中，RasterWorkspace 对象是通过 RasterWorkspaceFactory 来创建的。RasterWorkspace 实现了 IRasterWorkspace 类的接口，其定义了有关打开、设置数据集的相关属性和方法，如图 6.1 所示。

All	Description
CanCopy	Indicates if this dataset can be copied.
Copy	Copies this workspace to a new workspace with the specified name.
CreateRasterDataset	Creates a RasterDataset in the workspace given its name.
IsWorkspace	Indicates if the file path specified is a raster workspace.
OpenRasterDataset	Opens a RasterDataset in the workspace given its name.

图 6.1　IRasterWorkspace 接口

加载栅格数据集示例代码如下：

```
static IRasterDataset OpenFileRasterDataset(string folderName,
string datasetName)
{
    IWorkspaceFactory workspaceFactory=new RasterWorkspaceFacto-
ryClass();
     IRasterWorkspace rasterWorkspace = (IRasterWorkspace) work-
```

```
spaceFactory.OpenFromFile(folderName,0);
        IRasterDataset rasterDataset = rasterWorkspace.OpenRaster-
Dataset(datasetName);
    return rasterDataset;
    }
```

6.3　栅格数据配准

AE 中 RasterLayer 类实现了 IGeoReference 接口定义的与栅格数据配准的相关属性方法，如图 6.2 所示。

All	Description
CanCopy	Indicates if this dataset can be copied.
Copy	Copies this workspace to a new workspace with the specified name.
CreateRasterDataset	Creates a RasterDataset in the workspace given its name.
IsWorkspace	Indicates if the file path specified is a raster workspace.
OpenRasterDataset	Opens a RasterDataset in the workspace given its name.

图 6.2　IGeoReference 接口

6.4　栅格数据处理

栅格数据一般可以存储为 Grid(由一系列文件组成)、Tiff(包括一个 Tif 文件和一个 Aux 文件)、Image 等文件格式。本节将给出栅格数据处理的典型示例代码。

6.4.1　栅矢转换

1. 栅格转矢量示例

示例代码如下：

```
    ILayer pLayer = pAxMapControl.get_Layer(this.cmbLayer.Selecte-
dIndex);
    IRasterLayer pRasterLayer= pLayer as IRasterLayer;
    IWorkspaceFactory2 pWorkspaceFactory2 =new RasterWorkspaceFacto-
ryClass();
    IRasterWorkspace pRasterWorkspace;
    int Index = pRasterLayer.FilePath.LastIndexOf("\\");
    string filePath = pRasterLayer.FilePath.Substring(0, Index);
    string fileName = pRasterLayer.FilePath.Substring(Index +1);
    pRasterWorkspace = (IRasterWorkspace) pWorkspaceFactory2.Open-
FromFile(filePath,0);
    IRasterDataset pRasterDataset = pRasterWorkspace.OpenRasterData-
set(fileName);
```

```
IGeoDataset pRasterGeoDataset= pRasterDataset as IGeoDataset;
IWorkspaceFactory2 pWorkspaceFactoryShp=new ShapefileWorkspace-
FactoryClass();
IWorkspace pWorkspace = pWorkspaceFactoryShp.OpenFromFile(this.
tbFolderPath.Text,0);
IConversionOp pConversionOp=new RasterConversionOpClass();
ISpatialReference pSpatialReference= pRasterGeoDataset.Spatial-
Reference;
IGeoDataset pGeoDataset;
pGeoDataset = pConversionOp.RasterDataToPolygonFeatureData
(pRasterGeoDataset, pWorkspace,this.tbName.Text,true);
IDataset pDataset1= pGeoDataset as IDataset;
IFeatureClass pFeatureClass= pDataset1 as IFeatureClass;
IFeatureLayer pFeatureLayer=new FeatureLayerClass();
pFeatureLayer.FeatureClass = pFeatureClass;
pAxMapControl.ClearLayers();
pAxMapControl.AddLayer(pFeatureLayer);
pAxMapControl.Refresh();
```

2. 矢量转栅格示例

示例代码如下:

```
ILayer pLayer= pAxMapControl.get_Layer(this.cmbLayer.SelectedIn-
dex);
IRasterLayer pRasterLayer= pLayer as IRasterLayer;
IWorkspaceFactory2 pWorkspaceFactory2=new RasterWorkspaceFacto-
ryClass();
IRasterWorkspace pRasterWorkspace;
int Index = pRasterLayer.FilePath.LastIndexOf("\\");
string filePath = pRasterLayer.FilePath.Substring(0,Index);
string fileName = pRasterLayer.FilePath.Substring(Index +1);
pRasterWorkspace = (IRasterWorkspacep) WorkspaceFactory2.Open-
FromFile(filePath,0);
IRasterDataset pRasterDataset = pRasterWorkspace.OpenRasterData-
set(fileName);
IGeoDataset pRasterGeoDataset= pRasterDataset as IGeoDataset;
WorkspaceFactory2 pWorkspaceFactoryShp=new ShapefileWorkspace-
FactoryClass();
IWorkspace pWorkspace = pWorkspaceFactoryShp.OpenFromFile(this.
tbFolderPath.Text,0);
```

```
IConversionOp pConversionOp=new RasterConversionOpClass();
ISpatialReference pSpatialReference = pRasterGeoDataset.Spatial-
Reference;
IGeoDataset pGeoDataset;
pGeoDataset = pConversionOp.RasterDataToPolygonFeatureData
(pRasterGeoDataset,pWorkspace,this.tbName.Text,true);
IDataset pDataset1= pGeoDataset as IDataset;
IFeatureClass pFeatureClass= pDataset1 as IFeatureClass;
IFeatureLayer pFeatureLayer=new FeatureLayerClass();
pFeatureLayer.FeatureClass = pFeatureClass;
pAxMapControl.ClearLayers();
pAxMapControl.AddLayer(pFeatureLayer);
pAxMapControl.Refresh();
```

6.4.2 叠加分析

栅格图层叠加分析主要运用的接口有 IMathOp、ILogicalOp、ITrigOp 等，组件类 RasterMathOps 封装了 IMathOp 接口，实现了栅格图层相关数学运算。IMathOp 定义的方法属性见表 6.1。

表 6.1　**IMathOp 接口**

Properties(属性)	Description(描述)
Abs	Calculates the absolute value of cells in a raster.(计算栅格元像素的绝对值。)
Divide	Divides the values of two inputs.(计算两个输入参数相除数的商)
Exp	Calculates the base e exponential of cells in a raster.(计算栅格元像素的以 e 为底的指数)
Exp10	Calculates the base 10 exponential of cells in a raster.(计算栅格元像素的以 10 为底的指数)
Exp2	Calculates the base 2 exponential of cells in a raster.(计算栅格元像素的以 2 为底的指数)
Float	Converts a raster into floating point representation.(转换栅格元像素为浮点数)
Int	Converts a raster to integer by truncation.(转换栅格元像素为整数)
Ln	Calculates the natural logarithm (base e) of cells in a raster.(计算栅格元像素的自然对数)
Loq10	Calculates the base 10 logarithm of cells in a raster.(计算栅格元像素以 10 为底的对数)

续表

Properties(属性)	Description(描述)
Loq2	Calculates the base 2 logarithm of cells in a raster.(计算栅格元像素以 2 为底的对数)
Minus	Subtracts the values of two inputs.(对输入的两个值进行减法计算)
Mod	Finds the remainder of the first input when divided by the second.(计算余数)
Negate	Changes the sign of the input raster (multiplies by-1).(改变输入栅格的正负(乘以-1))
Plus	Adds the values of two inputs.(对两个输入值进行加法计算)
Power	Raises the cells in a raster to the power.(对输入栅格元像素进行加权运算)
PowerByCellValue	Raises the cells in a raster to the power of values found in another raster.(将其他栅格元像素的权值加权给栅格元像素)
RoundDown	Returns the next lower whole number for each cell in a raster.(返回栅格元像素的最小整数)
RoundUp	Returns the next higher whole number for each cell in a raster.(返回栅格元像素的最大整数)
Square	Calculates the square of cells in a raster.(计算栅格元像素的平方)
SquareRoot	Calculates the square root of cells in a raster.(计算栅格元像素的开方)
Times	Multiplies the value of two inputs.(计算两个输入值的乘积)

叠加分析示例代码如下：

```
public IFeatureClass Intsect(IFeatureClass _pFtClass,IFeature-
Class_pFtOverlay,string_FilePath,string_pFileName)
{
    IFeatureClassName pOutPut =new FeatureClassNameClass();
    pOutPut.ShapeType =_pFtClass.ShapeType;
    pOutPut.ShapeFieldName =_pFtClass.ShapeFieldName;
    pOutPut.FeatureType = esriFeatureType.esriFTSimple;
    IWorkspaceName pWsN =new WorkspaceNameClass();
    pWsN.WorkspaceFactoryProgID ="esriDataSourcesFile.Shapefile
WorkspaceFactory";
    pWsN.PathName = _FilePath;
//也可以用这种方法,IName 和 IDataset 的用法
/* IWorkspaceFactory pWsFc = new ShapefileWorkspaceFactoryClass
();
IWorkspace pWs = pWsFc.OpenFromFile(_FilePath, 0);
```

```
    IDataset pDataset = pWs as IDataset;
    IWorkspaceName pWsN = pDataset.FullName as IWorkspaceName;
    */
        IDatasetName pDatasetName= pOutPut as IDatasetName;
        pDatasetName.Name =_pFileName;
        pDatasetName.WorkspaceName =pWsN;
        IBasicGeoprocessor pBasicGeo=new BasicGeoprocessorClass();
        IFeatureClass pFeatureClass= pBasicGeo.Intersect(_pFtClass
as ITable ,false,_pFtOverlay as ITable ,false,0.1, pOutPut);
    return pFeatureClass;
    }
```

第7章 数据编辑

空间数据编辑是GIS最基本功能之一，ArcEngine中通过IWorkspaceEdit等接口提供了丰富的编辑功能，例如，IWorkspaceEdit接口定义的UndoEditoperation方法用于编辑操作的撤销。

7.1 捕捉功能

捕捉功能是新建地理空间对象与已有地理空间对象的一种关联。其在矢量化过程中非常有用，比如在矢量化线要素时，要让新建道路与已有道路连接起来，这时候系统就需要设定一定的容差，鼠标坐标与已有端点坐标小于一定距离时，就认为两个端点是重合的，系统自动将线要素的端点与已有要素的端点连接起来。

捕捉功能实现示例代码如下：

```
bool bCreateElement =true;
int internalTime =5;//时间间隔
int snapTime =10;//初始值
IElement m_element=null;//界面绘制点元素
IPoint currentPoint=new PointClass();//当前鼠标点
IPoint snapPoint=null;//捕捉到的点
IMovePointFeedback pFeedback=new MovePointFeedbackClass();
publicvoid DoMouseMove (object  sender, IMapControlEvents2 _On-
MouseMoveEvent e)
{
    AxMapControl axMapControl1= sender as AxMapControl;
    currentPoint.PutCoords(e.mapX, e.mapY);
if(action == ActionType.CreateFeature)
{
        snapTime++;
        snapTime= snapTime% internalTime ;
        ILayer layer= GetLayerByName(snapLayer, axMapControl1);
        IFeatureLayer m_iFeatureLyr= layer as IFeatureLayer;
if(bCreateElement)
{
```

```
            CreateMarkerElement(currentPoint,axMapControl1);
            bCreateElement = false;
    }
    if(snapPoint ==null)
        ElementMoveTo(currentPoint, axMapControl1);
    //鼠标自动捕获顶点
    if(snapTime ==0)
        snapPoint= Snapping(e.mapX, e.mapY, m_iFeatureLyr, axMap-
Control1);
    if(snapPoint != null&& snapTime ==0)
        ElementMoveTo(snapPoint, axMapControl1);
    }
    }

    ///<summary>
    ///捕捉
    ///</summary>
    ///<param name="x"></param>
    ///<param name="y"></param>
    ///<param name="iFeatureLyr"></param>
    ///<param name="axMapControl1"></param>
    ///<returns></returns>
    public IPoint Snapping(double x,double y, IFeatureLayer iFeature-
Lyr, AxMapControl axMapControl1)
    {
        IPoint iHitPoint=null;
        IMap iMap= axMapControl1.Map;
        IActiveView iView= axMapControl1.ActiveView;
        IFeatureClass iFClss= iFeatureLyr.FeatureClass;
        IPoint point=new PointClass();
        point.PutCoords(x,y);
    double length = ConvertPixelsToMapUnits(axMapControl1.ActiveView,8);
        ITopologicalOperator pTopo= point as ITopologicalOperator;
        IGeometry pGeometry= pTopo.Buffer(length).Envelope as IGeom-
etry;
        ISpatialFilter spatialFilter=new SpatialFilterClass();
        spatialFilter.GeometryField = iFeatureLyr.FeatureClass.Shape
FieldName;
```

```
        spatialFilter.SpatialRel = esriSpatialRelEnum.esriSpatial-
RelCrosses;
        spatialFilter.Geometry = pGeometry;
        IFeatureCursor cursor= iFClss.Search(spatialFilter,false);
        IFeature iF= cursor.NextFeature();
    if(iF = =null)returnnull;
        IPoint iHitPt =new ESRI.ArcGIS.Geometry.Point();
        IHitTest iHitTest= iF.Shape as IHitTest;
    double hitDist =0;
    int partIndex =0;
    int vertexIndex =0;
    bool bVertexHit =false;
    //Tolerance in pixels for line hits
    double tol = ConvertPixelsToMapUnits(iView,8);
    if(iHitTest.HitTest(point, tol, esriGeometryHitPartType.esriGe-
ometryPartBoundary,iHitPt, ref hitDist, ref partIndex, ref vertexIn-
dex,ref bVertexHit))
    {
            iHitPoint= iHitPt;
    }
        axMapControl1.ActiveView.Refresh();
    return iHitPoint;
    }

  ◦ ///<summary>
    ///创建新的 element 用于显示
    ///</summary>
    ///<param name="point"></param>
    ///<param name="axMapControl1"></param>
    publicvoid CreateMarkerElement(IPoint point, AxMapControl axMap-
Control1)
    {
        IActiveView iView= axMapControl1.ActiveView;
         IGraphicsContainer iGraphContainer = axMapControl1.Map as
IGraphicsContainer;
    //建立一个 marker 元素
        IMarkerElement iMarkerElement =new MarkerElement()as IMark-
erElement;
```

```
        ISimpleMarkerSymbol iSym=new SimpleMarkerSymbol();
    //符号化元素
        IRgbColor iColor=new RgbColor();
        iColor.Red =0;
        iColor.Blue =100;
        iColor.Green =255;
        iSym.Color = iColor;
        IRgbColor iColor2=new RgbColor();
        iColor2.Red =0;
        iColor2.Blue =0;
        iColor2.Green =0;
        iSym.Outline =true;
        iSym.OutlineColor = iColor2 as IColor;
        iSym.OutlineSize =1;
        iSym.Size =5;
        iSym.Style = esriSimpleMarkerStyle.esriSMSCircle;
        ISymbol symbol= iSym as ISymbol;
        symbol.ROP2 = esriRasterOpCode.esriROPNotXOrPen;
        iMarkerElement.Symbol = iSym;
        m_element= iMarkerElement as IElement;
        m_element.Geometry = point as IGeometry;
        iGraphContainer.AddElement(m_element,0);
        iView.PartialRefresh(esriViewDrawPhase.esriViewGraphics, m_
element,null);
        IGeometry iGeo= m_element.Geometry;
        pFeedback.Display = iView.ScreenDisplay;
        pFeedback.Symbol = iSym as ISymbol;
        pFeedback.Start(iGeo as IPoint, point);
    }
    ///<summary>
    ///移动元素
    ///</summary>
    ///<param name="iPt"></param>
    ///<param name="axMapControl1"></param>
    publicvoid ElementMoveTo(IPoint iPt, AxMapControl axMapControl1)
    {
    //移动元素
        pFeedback.MoveTo(iPt);
```

```
        IGeometry iGeo1=null;
        IGeometry iGeoResult;
    if(m_element ! =null)
    {
            iGeo1= m_element.Geometry;
            iGeoResult= pFeedback.Stop();
    //map.ActiveView.Refresh();
            m_element.Geometry = iGeoResult;
    //更新该元素的位置
            axMapControl1.ActiveView.GraphicsContainer.UpdateElement
(m_element);
    //重新移动元素
            pFeedback.Start(iGeo1 as IPoint, iPt);
    //map.ActiveView.Refresh();
    axMapControl1.ActiveView.PartialRefresh(esriViewDrawPhase.esri
ViewGraphics,null,null);
    }
    }

    public ILayer GetLayerByName(string layerName, AxMapControl ax-
MapControl1)
    {
    for(int i =0; i < axMapControl1.LayerCount; i++)
    {
    if(axMapControl1.get_Layer(i).Name.Equals(layerName))
    return axMapControl1.get_Layer(i);
    }
    Return null;
    }
    publicdouble ConvertPixelsToMapUnits (IActiveView pActiveView,
double pixelUnits)
    {
    double realWorldDisplayExtent;
    int pixelExtent;
    double sizeOfOnePixel;
    pixelExtent = pActiveView.ScreenDisplay.DisplayTransformation.
get_DeviceFrame().right -pActiveView.ScreenDisplay.DisplayTransfor-
mation.get_DeviceFrame().left;
```

```
    realWorldDisplayExtent = pActiveView.ScreenDisplay.Display
Transformation.VisibleBounds.Width;
    sizeOfOnePixel = realWorldDisplayExtent /pixelExtent;
    return pixelUnits * sizeOfOnePixel;
    }
```

示例功能演示如图 7.1 所示。

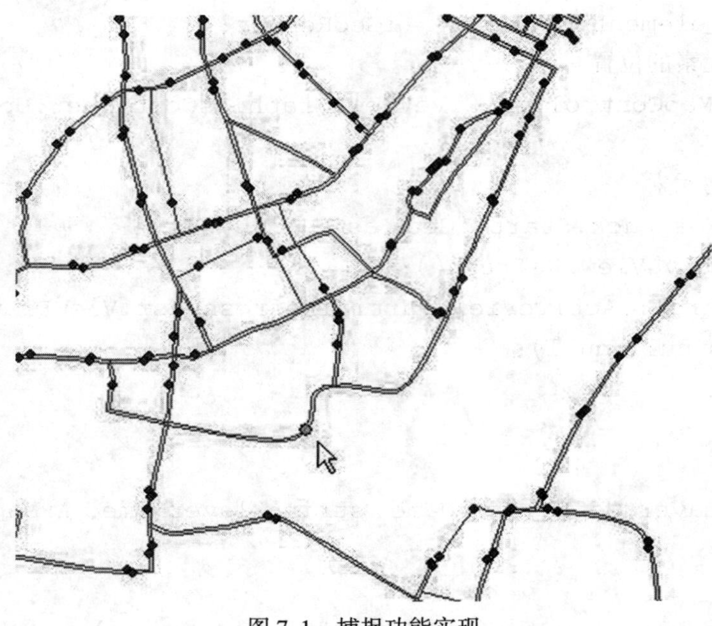

图 7.1　捕捉功能实现

7.2　要素编辑

7.2.1　开始/结束编辑

开始编辑使用的是 IWorkspaceEdit 接口定义的 StartEditing 方法，该方法的 withUndo Redo 参数用来决定工作空间是否支持“恢复/取消恢复”操作。用户完成编辑后，可以使用 IWorkspaceEdit 接口定义的 StopEditOperation 方法来保存结束已完成的编辑。

7.2.2　图形编辑

图形编辑功能主要用到 IFeature、IFeatureEdit、IWorkspaceEdit、IFeatureClass 接口实现。IFeatureEdit 中定义的 MoveSet、RotateSet 等方法，分别是移动、旋转有一个多个要素组成的要素集。IFeatureEdit 接口定义的属性方法见表 7.1。

表 7.1　**IFeatureEdit 接口**

Properties(属性)	Description(描述)
BeqinMoveSet	Prepares the set of features for a move operation.(为移动操作准备要素集)
DeleteSet	Deletes the set of rows.(删除多个要素集)
MoveSet	Moves the set of features through a distance and direction specified by move Vector.(按指定距离和方向通过移动向量移动要素集)
RotateSet	Rotates the set of features according to the specified origin and angle.(根据指定的起点和角度旋转要素集)
Split	Split the feature.(分离要素)
SplitAttributes	Split the feature attributes.(分离要素属性)
SplitWithUpdate	Split the feature by updating the split feature and creating new feature(s) for the small portion.(通过更新分离要素和生成新的要素来分离要素)

下面给出了添加点要素、线要素实现的相关代码示例。

1. 添加点示例

AE 提供了多种方法添加点，但基本的思路一样，只是少量的接口有变化。通过 FeatrueClass 的 CreateFeature 方法添加地物的示例代码如下：

```
Public void AddPointByStore()
{
//得到要添加地物的图层
    IFeatureLayer l= MapCtr.Map.get_Layer(0)as IFeatureLayer;
//定义一个地物类,把要编辑的图层转化为定义的地物类
    IFeatureClass fc= l.FeatureClass ;
//先定义一个编辑的工作空间,然后转化为数据集,最后转化为编辑工作空间
    IWorkspaceEdit w =(fc as IDataset).Workspace as IWorkspace-
Edit;
    IFeature f;
    IPoint p;
//开始事务操作
    w.StartEditing(false);
//开始编辑
    w.StartEditOperation();
for(int i =0; i<100; i++)
{
//创建一个地物
        f= fc.CreateFeature();
        p=new PointClass();
```

```
//设置点的坐标
        p.PutCoords (i,i);
//确定图形类型
        f.Shape = p;
//保存地物
        f.Store();
}
//结束编辑
    w.StopEditOperation();
//结束事务操作
    w.StopEditing(true);
}
```

2. 添加线示例

添加线示例代码如下:

```
Public void AddLineByWrite()
{
    IFeatureLayer l = MapCtr.Map.get_Layer(0)as IFeatureLayer;
    IFeatureClass fc= l.FeatureClass ;
    IFeatureClassWrite fr= fc as IFeatureClassWrite ;
    IWorkspaceEdit w = ( fc as IDataset ) .Workspace as IWorkspace-
Edit;
    IFeature f;
//可选参数的设置
object Missing = Type.Missing;
    IPoint p=new PointClass();
    w.StartEditing (true);
    w.StartEditOperation();
for(int i =0; i<100; i++)
{
        f= fc.CreateFeature();
//定义一个多义线对象
        IPolyline PlyLine=new PolylineClass();
//定义一个点的集合
        IPointCollection ptclo= PlyLine as IPointCollection;
//定义一系列要添加到多义线上的点对象,并赋初始值
for(int j=0;j<4;j++)
{
            p.PutCoords(j,j);
```

```
            ptclo.AddPoint(p,ref Missing,ref Missing);
    }
        f.Shape = PlyLine;
        fr.WriteFeature (f);
    }
    w.StopEditOperation();
    w.StopEditing(true);
}
```

第8章　空 间 分 析

8.1　空间分析简介

空间分析是为了解决地理空间问题而进行的数据分析与数据挖掘，是从 GIS 目标之间的空间关系中获取派生的信息和新的知识，是从一个或多个空间数据图层中获取信息的过程。空间分析通过地理计算和空间表达挖掘潜在的空间信息，其本质包括探测空间数据中的模式；研究数据间的关系并建立空间数据模型；使得空间数据更为直观地表达出其潜在含义；改进地理空间事件的预测和控制能力。

空间分析主要通过空间数据和空间模型的联合分析来挖掘空间目标的潜在信息，而这些空间目标的基本信息，无非是其空间位置、分布、形态、距离、方位、拓扑关系等，其中距离、方位、拓扑关系组成了空间目标的空间关系，它是地理实体之间的空间特性，可以作为数据组织、查询、分析和推理的基础。通过将地理空间目标划分为点、线、面不同的类型，可以获得这些不同类型目标的形态结构。将空间目标的空间数据和属性数据结合起来，可以进行许多特定任务的空间计算与分析。

8.2　空间查询

8.2.1　基于属性的查询

属性查询是在当前 map 的 layer 中获取符合条件的 feature 的集合。实现过程是获取 featurelayer 的 featureclass，然后定义过滤条件，在 featureclass 中执行 search 函数，获取查询结果。主要示例代码如下：

```
ILayerlayer=axMapControl1.get_Layer(0);
IFeatureLayerfeatureLayer=layerasIFeatureLayer;
//获取 featureLayer 的 featureClass
IFeatureClassfeatureClass=featureLayer.FeatureClass;
IFeaturefeature=null;
IQueryFilterqueryFilter=newQueryFilterClass();
IFeatureCursorfeatureCusor;
queryFilter.WhereClause="name='"+searchName+"'";
featureCusor=featureClass.Search(queryFilter,true);
```

```
//search 的参数第一个为过滤条件,第二个为是否重复执行。
feature=featureCusor.NextFeature();
if(feature! =null)
{
    axMapControl1.Map.SelectFeature(axMapControl1.get_Layer(0),
feature);
    axMapControl1.Refresh(esriViewDrawPhase.esriViewGeoSelection,
null,null);
}
```

8.2.2 基于位置的查询

空间查询实现起来比较简单，使用 ArcGIS 封装好的函数即可，主要是获取选取的 Geometry 对象，相关示例代码如下：

```
axMapControl1.MousePointer=ESRI.ArcGIS.Controls.esriControl-
sMousePointer.esriPointerCrosshair;
IGeometrygeometry=null;
switch(actionFlag)
{
    caseflag.POINTSELECT://点选
    ESRI.ArcGIS.Geometry.Pointpt=newESRI.ArcGIS.Geometry.Point
();
    pt.X=e.mapX;
    pt.Y=e.mapY;
    geometry=ptasIGeometry;
break;
    caseflag.CIRCLESELECT://圆选
    geometry=axMapControl1.TrackCircle();
break;
    caseflag.RECTSELECT://长方形选取
    geometry=axMapControl1.TrackRectangle();
break;
    caseflag.POLYGONSELECT://多边形选取
    geometry=axMapControl1.TrackPolygon();
break;
}
axMapControl1.Map.SelectByShape(geometry,null,false);
axMapControl1.Refresh ( esriViewDrawPhase.esriViewGeoSelection,
null,null);
```

8.3 空间插值

本节主要介绍IDW(反距离)插值。IDW插值的基本思想是目标离观察点越近则权重越大，受该观察点的影响则越大。IDW插值的优点是观察点本身是绝对准确的，而且可以限制插值点的个数。通过power属性可以确定最近原则对于结果影响的程度，Searchradius属性可以控制插值点的个数。相关示例代码如下：

```
    Public class IDW
    {
    publicIGeoDatasetGetIDW(IFeatureClass _pFeatureClass, string _
pFieldName,double_pDistance,double_pCell,int_pPower)
    {
            IGeoDatasetGeo=_pFeatureClassasIGeoDataset;
            objectpExtent=Geo.Extent;
            objecto=Type.Missing;
            IFeatureClassDescriptorpFeatureClassDes=newFeatureClass
DescriptorClass();
             pFeatureClassDes.Create(_pFeatureClass,null,_pField-
Name);
            IInterpolationOppInterOp=newRasterInterpolationOpClass
();
             IRasterAnalysisEnvironmentpRasterAEnv=pInterOpasIRas-
terAnalysisEnvironment;
    //pRasterAEnv.Mask=Geo;
    pRasterAEnv.SetExtent(esriRasterEnvSettingEnum.esriRasterEnv
Value,refpExtent,refo);
            objectpCellSize=_pCell;//可以根据不同的点图层进行设置
    pRasterAEnv.SetCellSize(esriRasterEnvSettingEnum.esriRasterEnv
Value,refpCellSize);
            IRasterRadiuspRasterrad=newRasterRadiusClass();
            objectobj=Type.Missing;
    //pRasterrad.SetFixed(_pDistance,refobj);
            pRasterrad.SetVariable(12);
            objectpBar=Type.Missing;
    IGeoDatasetpGeoIDW=pInterOp.IDW(pFeatureClassDesasIGeoDataset,
_pPower,pRasterrad,refpBar);
            returnpGeoIDW;
    }
```

```
    publicvoidCreateRasterFromPoints  ( IMappMap,  IFeatureLayerpp,
StringPath,StringName,StringField)
    {
    //1.将 Shape 文件读取成 FeatureClass
    //2.根据 FeatureClass 生成 IFeatureClassDescriptor
    //3.创建 IRasterRaduis 对象
    //设置 Cell
    //4.插值并生成表面
            objectobj=null;
            IWorkspaceFactorypShapSpace;
            pShapSpace=newShapefileWorkspaceFactory();
            IWorkspacepW;
            pW=pShapSpace.OpenFromFile(Path,0);
            IFeatureWorkspacepFeatureWork;
            pFeatureWork=pWasIFeatureWorkspace;
           IFeatureClasspFeatureClass=pFeatureWork.OpenFeatureClass
("Name");
            IGeoDatasetGeo=pFeatureClassasIGeoDataset;
            objectextend=Geo.Extent;
            objecto=null;
            IFeatureClassDescriptorpFeatureClassDes=newFeatureClass
DescriptorClass();
            pFeatureClassDes.Create(pFeatureClass,null,Field);
            IRasterRadiuspRasterrad=newRasterRadiusClass();
            pRasterrad.SetVariable(10,refobj);
            objectdCell=0.5;//可以根据不同的点图层进行设置
            IInterpolationOpPinterpla=newRasterInterpolationOpClass
();
    IRasterAnalysisEnvironmentpRasterAnaEn = PinterplaasIRasterAnal-
ysisEnvironment;
    pRasterAnaEn.SetCellSize(esriRasterEnvSettingEnum.esriRasterEnv
Value,refdCell);
    pRasterAnaEn.SetExtent(esriRasterEnvSettingEnum.esriRasterEnv
Value,refextend,refo);
            IGeoDatasetpGRaster;
            objecthh=3;
            pGRaster=Pinterpla.IDW((IGeoDataset)pFeatureClassDes,2,
pRasterrad,refhh);
```

```
        ISurfaceOppSurF;
        pSurF=newRasterSurfaceOpClass();
        IGeoDatasetpCour;
        objecto1=0;
        pCour=pSurF.Contour(pGRaster,5,refo1);
        IFeatureLayerpLa;
        pLa=newFeatureLayerClass();
        IFeatureClasspClass=pCourasIFeatureClass;
        pLa.FeatureClass=pClass;
        pLa.Name=pClass.AliasName;
        pMap.AddLayer(pLaasILayer);
    }
}
```

8.4 缓冲区分析

缓冲区分析是指以点、线、面实体为基础，自动建立其周围一定宽度范围内的缓冲区多边形图层，然后将该图层与目标图层叠加，进行分析而得到所需结果。它是用来解决邻近度问题的空间分析工具之一。邻近度描述了地理空间中两个地物距离相近的程度。

下面给出了缓冲区分析实例代码，其用户交互界面如图 8.1 所示。

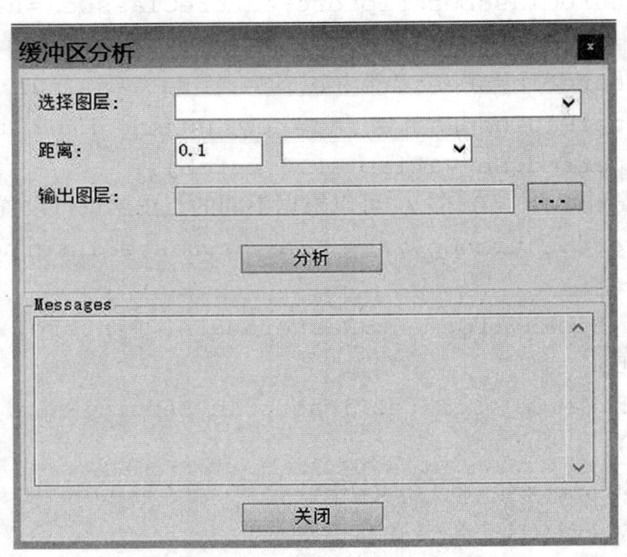

图 8.1　缓冲区分析

具体实例代码如下：

```
Namespace GISDesign
```

```
{
publicpartialclassBufferDlg:Form
{
//inordertoscrollthemessagestextboxtothebottomwemustimportthis
Win32call
[DllImport("user32.dll")]
privatestaticexternintPostMessage(IntPtrwnd, uintMsg, IntPtrw-
Param,IntPtrlParam);
        privateIHookHelperm_hookHelper=null;
        privateconstuintWM_VSCROLL=0x0115;
        privateconstuintSB_BOTTOM=7;
        publicBufferDlg(IHookHelperhookHelper)
{
            InitializeComponent();
            m_hookHelper=hookHelper;
}
        privatevoidbufferDlg_Load(objectsender,EventArgse)
{
if(null==m_hookHelper || null==m_hookHelper.Hook || 0==m_
hookHelper.FocusMap.LayerCount)
return;
//loadallthefeaturelayersinthemaptothelayerscombo
            IEnumLayerlayers=GetLayers();
            layers.Reset();
            ILayerlayer=null;
while((layer=layers.Next())!=null)
{
                cboLayers.Items.Add(layer.Name);
}
//selectthefirstlayer
if(cboLayers.Items.Count>0)
            cboLayers.SelectedIndex=0;
            stringtempDir=System.IO.Path.GetTempPath();
txtOutputPath.Text = System.IO.Path.Combine(tempDir, ((string)
cboLayers.SelectedItem+"_buffer.shp"));
//setthedefaultunitsofthebuffer
            intunits=Convert.ToInt32(m_hookHelper.FocusMap.MapU-
nits);
```

```
            cboUnits.SelectedIndex=units;
    }
            privatevoidbtnOutputLayer _ Click ( objectsender, Even-
tArgse)
    {
    //settheoutputlayer
            SaveFileDialogsaveDlg=newSaveFileDialog();
            saveDlg.CheckPathExists=true;
            saveDlg.Filter="Shapefile(*.shp)|*.shp";
            saveDlg.OverwritePrompt=true;
            saveDlg.Title="OutputLayer";
            saveDlg.RestoreDirectory=true;
            saveDlg.FileName=(string)cboLayers.SelectedItem+"_
buffer.shp";
            DialogResultdr=saveDlg.ShowDialog();
    if(dr==DialogResult.OK)
            txtOutputPath.Text=saveDlg.FileName;
    }
          privatevoidbtnBuffer_Click(objectsender,EventArgse)
    {
    //makesurethatallparametersareokay
            doublebufferDistance;
    //转换 distance 为 double 类型
    double.TryParse(txtBufferDistance.Text,outbufferDistance);
    if(0.0==bufferDistance)
    {
              MessageBox.Show("Badbufferdistance!");
    return;
    }
    //判断输出路径是否合法
    if (! System.IO.Directory.Exists ( System.IO.Path.GetDirectory
Name ( txtOutputPath.Text )) || ".shp"! = System.IO.Path.GetExtension
(txtOutputPath.Text))
    {
              MessageBox.Show("Badoutputfilename!");
    return;
    }
    //判断图层个数
```

```
    if(m_hookHelper.FocusMap.LayerCount==0)
    return;
    //getthelayerfromthemap
                IFeatureLayerlayer=GetFeatureLayer((string)cboLay-
ers.SelectedItem);
    if(null==layer)
    {
    txtMessages.Text+="Layer"+(string)cboLayers.SelectedItem+"can-
notbefound! \r\n";
    return;
    }
    //scrollthetextboxtothebottom
               ScrollToBottom();
               txtMessages.Text+="\r\n分析开始,这可能需要几分钟时间,请稍
候...\r\n";
               txtMessages.Update();
    //getaninstanceofthegeoprocessor
               Geoprocessorgp=newGeoprocessor();
               gp.OverwriteOutput=true;
    //createanewinstanceofabuffertool
    ESRI.ArcGIS.AnalysisTools.Bufferbuffer = newESRI.ArcGIS.Analys-
isTools.Buffer ( layer, txtOutputPath.Text, Convert.ToString ( buff-
erDistance)+""+(string)cboUnits.SelectedItem);
               buffer.dissolve_option="ALL";//这个要设成ALL,否则相交
部分不会融合
    //buffer.line_side="FULL";//默认是"FULL",最好不要改,否则出错
    //buffer.line_end_type="ROUND";//默认是"ROUND",最好不要改,否则出错
               IGeoProcessorResultresults=null;
    try
    {
                   results=(IGeoProcessorResult)gp.Execute(buffer,
null);
    }
    catch(Exceptionex)
    {
                     txtMessages.Text+="Failedtobufferlayer:"+layer.
Name+"\r\n";
    }
```

```
    if(results.Status! =esriJobStatus.esriJobSucceeded)
    {
                        txtMessages.Text + = " Failedtobufferlayer:" +
layer.Name+" \r \n";
    }
    //scrollthetextboxtothebottom
             ScrollToBottom();
             txtMessages.Text+=" \r \n 分析完成 . \r \n";
    txtMessages.Text+="------------------------------------------------------------------------
-------------- \r \n";
    //scrollthetextboxtothebottom
             ScrollToBottom();
    }
         privatestringReturnMessages(Geoprocessorgp)
    {
             StringBuildersb=newStringBuilder();
    if(gp.MessageCount>0)
    {
    for(intCount=0;Count<=gp.MessageCount-1;Count++)
    {
                                 System.Diagnostics.Trace.WriteLine
(gp.GetMessage(Count));
                           sb.AppendFormat ( "{0} \n", gp.GetMessage
(Count));
    }
    }
             returnsb.ToString();
    }
         privateIFeatureLayerGetFeatureLayer(stringlayerName)
    {
    //getthelayersfromthemaps
             IEnumLayerlayers=GetLayers();
             layers.Reset();
             ILayerlayer=null;
    while((layer=layers.Next())! =null)
    {
    if(layer.Name==layerName)
                 returnlayerasIFeatureLayer;
```

```
    }
            returnnull;
    }
        privateIEnumLayerGetLayers()
    {
            UIDuid=newUIDClass();
            uid.Value="{40A9E885-5533-11d0-98BE-00805F7CED21}";
            IEnumLayerlayers=m_hookHelper.FocusMap.get_Layers
(uid,true);
            returnlayers;
    }
        privatevoidScrollToBottom()
    {
    PostMessage((IntPtr)txtMessages.Handle,WM_VSCROLL,(IntPtr)SB_
BOTTOM,(IntPtr)IntPtr.Zero);
    }
        privatevoidbtnCancel_Click(objectsender,EventArgse)
    {
    this.Close();
    }
    }
    }
```

8.5　叠加运算

叠加分析是 GIS 中的一项非常重要的空间分析功能，它是指在统一空间参考系统下，通过对两个数据进行一系列集合运算，产生新数据的过程。这里提到的数据可以是图层对应的数据集，也可以是地物对象。其中，叠置分析的目标是分析在空间位置上有一定关联的空间对象的空间特征和专属属性之间的相互关系。多层数据的叠置分析，不仅仅产生了新的空间关系，还可以产生新的属性特征关系，能够帮助用户发现多层数据间的相互差异、联系和变化等特征。

在叠加分析中至少涉及三种数据，其中一种数据的类型可以是点、线、面等，被称为输入数据，另一种数据是面数据，被称为叠加数据，还有一种数据就是叠加结果数据，包含叠加后数据的几何信息和属性信息。

叠加运算示例代码如下：

```
publicIFeatureClassIntsect ( IFeatureClass _ pFtClass, IFeature-
Class_ pFtOverlay,string_FilePath,string_ pFileName)
    {
```

```
        IFeatureClassNamepOutPut=newFeatureClassNameClass();
        pOutPut.ShapeType=_pFtClass.ShapeType;
        pOutPut.ShapeFieldName=_pFtClass.ShapeFieldName;
        pOutPut.FeatureType=esriFeatureType.esriFTSimple;
    //setoutputlocationandfeatureclassname
        IWorkspaceNamepWsN=newWorkspaceNameClass();
        pWsN.WorkspaceFactoryProgID="esriDataSourcesFile.Shapefile-
WorkspaceFactory";
        pWsN.PathName=_FilePath;
    //也可以用这种方法,IName 和 IDataset 的用法
    /*IWorkspaceFactorypWsFc=newShapefileWorkspaceFactoryClass();
    IWorkspacepWs=pWsFc.OpenFromFile(_FilePath,0);
    IDatasetpDataset=pWsasIDataset;
    IWorkspaceNamepWsN=pDataset.FullNameasIWorkspaceName;
    */
        IDatasetNamepDatasetName=pOutPutasIDatasetName;
        pDatasetName.Name=_pFileName;
        pDatasetName.WorkspaceName=pWsN;
        IBasicGeoprocessorpBasicGeo=newBasicGeoprocessorClass();
        IFeatureClasspFeatureClass=pBasicGeo.Intersect(_pFtClassa-
sITable,false,_pFtOverlayasITable,false,0.1,pOutPut);
        returnpFeatureClass;
    }
```

第9章　地图制图

9.1　地图图例

图例是集中于地图一角或一侧的地图上各种符号和颜色所代表内容与指标的说明，有助于使用者更好地认识地图。它具有双重任务，在编图时作为图解，表示地图内容的准绳，在用图时作为必不可少的阅读指南。图例应符合完备性和一致性的原则。

地图符号一般包括各种大小、粗细、颜色不同的点、线、图形等。符号的设计要能表达出地面景物的形状、大小和位置，而且还能反映出各种景物的质和量的特征，以及相互关系。因此，图例常设计成与实地景物轮廓相似的几何图形。

下面给出利用自定义工具类，实现添加图名、比例尺、指北针功能的示例代码：

9.1.1　添加图名示例

示例代码具体如下：

```
//自定义工具类,实现添加图名功能
namespaceWindowsApplication1
{
classaddPageLayoutName:BaseTool
{
        publicForm1formTemp;
        TextBoxtextbox;
        AxPageLayoutControlaxLayoutControl;
        IPointpPoint;
//doublexMap,yMap;
        publicstaticdoublexMap;
        publicstaticdoubleyMap;
        publicoverridevoidOnMouseDown(intButton,intShift,intX,
intY)
{
if(Button==1)
{
pPoint=formTemp.returnPageLayoutControl().ActiveView.ScreenD-
```

```
isplay.DisplayTransformation.ToMapPoint(X,Y);
                xMap=pPoint.X;
                yMap=pPoint.Y;
                formTemp.returnTextbox1().Location=newSystem.
Drawing.Point(X,Y);
                formTemp.returnTextbox1().Visible=true;
                formTemp.returnTextbox1().Focus();
                formTemp.returnTextbox1().Text="请在此输入图名";
    }
    }
        publicoverridevoidOnCreate(objecthook)
    {
            axLayoutControl=hookasAxPageLayoutControl;
    }
    publicvoidAddTextElement(AxPageLayoutControlPageLayoutControl,
doublex,doubley,stringtextName)
    {
            IPageLayoutpPageLayout;
            IActiveViewpAV;
            IGraphicsContainerpGraphicsContainer;
            IPointpPoint;
            ITextElementpTextElement;
            IElementpElement;
            ITextSymbolpTextSymbol;
            IRgbColorpColor;
            pPageLayout=PageLayoutControl.PageLayout;
            pAV=(IActiveView)pPageLayout;
            pGraphicsContainer=(IGraphicsContainer)pPageLayout;
            pTextElement=newTextElementClass();
            IFontDisppFont=newStdFontClass()asIFontDisp;
            pFont.Bold=true;
            pFont.Name="宋体";
            pFont.Size=13;
            pColor=newRgbColorClass();
            pColor.Red=255;
            pTextSymbol=newTextSymbolClass();
            pTextSymbol.Color=(IColor)pColor;
            pTextSymbol.Font=pFont;
```

```
            pTextElement.Text=textName;
            pTextElement.Symbol=pTextSymbol;
            pPoint=newPointClass();
            pPoint.X=x;
            pPoint.Y=y;
            pElement=(IElement)pTextElement;
            pElement.Geometry=(IGeometry)pPoint;
            pGraphicsContainer.AddElement(pElement,0);
            pAV.PartialRefresh(esriViewDrawPhase.esriViewGraphics,
null,null);
    }
    }
    }
```

9.1.2 添加指北针示例

示例具体代码如下：

```
Namespace WindowsApplication1
{
sealedclassaddNorthArrow:BaseTool
{
        AxPageLayoutControlaxPageLayout=null;
        IPointpPoint;
        boolbInuse;
        INewEnvelopeFeedbackpNewEnvelopeFeedback=null;
        publicaddNorthArrow()
{
base.m_caption="添加指北针";
base.m_toolTip="添加指北针";
base.m_category="customCommands";
base.m_message="添加指北针";
base.m_deactivate=true;
}
        publicoverridevoidOnCreate(objecthook)
{
            axPageLayout=(AxPageLayoutControl)hook;
}
         publicoverridevoidOnMouseDown(intButton,intShift,intX,
intY)
```

```
{
pPoint=axPageLayout.ActiveView.ScreenDisplay.DisplayTransform-
ation.ToMapPoint(X,Y);
            bInuse=true;
}
        publicoverridevoidOnMouseMove(intButton,intShift,intX,
intY)
{
if(bInuse==false)
{
return;
}

if(pNewEnvelopeFeedback==null)
{
                pNewEnvelopeFeedback=newNewEnvelopeFeedbackClass
();
                pNewEnvelopeFeedback.Display=axPageLayout.Active-
View.ScreenDisplay;
                pNewEnvelopeFeedback.Start(pPoint);
}
pNewEnvelopeFeedback.MoveTo(axPageLayout.ActiveView.ScreenDis-
play.DisplayTransformation.ToMapPoint(X,Y));
}
        publicoverridevoidOnMouseUp(intButton,intShift,intX,in-
tY)
{
if(bInuse==false)
{
return;
}
if(pNewEnvelopeFeedback==null)
{
                pNewEnvelopeFeedback=null;
                bInuse=false;
return;
}
            IEnvelopepEnvelope=pNewEnvelopeFeedback.Stop();
```

```
    if((pEnvelope.IsEmpty) ‖ (pEnvelope.Width = = 0) ‖ (pEnvelope.
Height = =0))
    {
                pNewEnvelopeFeedback=null;
                bInuse=false;
    return;
    }
            addNorthArrowFormnorthArrow=newaddNorthArrowForm();
            IStyleGalleryItempStyleGalleryItemTemp=Form1.pStyle-
GalleryItem;
    if(pStyleGalleryItemTemp==null)
    {
    return;
    }
    IMapFramepMapframe = axPageLayout.ActiveView. GraphicsContain-
er.FindFrame(axPageLayout.ActiveView.FocusMap)asIMapFrame;
            IMapSurroundFramepMapSurroundFrame=newMapSurroundFr-
ameClass();
            pMapSurroundFrame.MapFrame=pMapframe;
            pMapSurroundFrame.MapSurround=(IMapSurround)pStyle-
GalleryItemTemp.Item;
    //在 pageLayout 中根据名称查要素 Element,找到之后删除已经存在的指北针
            IElementpElement=axPageLayout.FindElementByName
("NorthArrows");
    if(pElement! =null)
    {
                axPageLayout.ActiveView.GraphicsContainer.Delete
Element(pElement);
    //删除已经存在的指北针
    }
            pElement=(IElement)pMapSurroundFrame;
            pElement.Geometry=(IGeometry)pEnvelope;
            axPageLayout.ActiveView.GraphicsContainer.AddElement
(pElement,0);
axPageLayout.ActiveView.PartialRefresh(esriViewDrawPhase.esriView
Graphics,null,null);

            pNewEnvelopeFeedback=null;
```

```
            bInuse=false;
    }
    }
    }
```

9.1.3 添加比例尺示例

示例具体代码如下：

```
Namespace WindowsApplication1
{
sealedclassaddScaleBar:BaseTool
{
//privateIHookHelperpHookHelper=null;
        privateAxPageLayoutControlaxPagelayoutControl=null;
        privateIPointpPoint;
        privateINewEnvelopeFeedbackpNewEnvelopeFeedback;
        privateboolbInuse;
        publicaddScaleBar()
{
base.m_caption="ScaleBar";
base.m_category="myCustomCommands(C#)";
base.m_message="Addascalebarmapsurround";
base.m_name="myCustomCommands(C#)_ScaleBar";
base.m_toolTip="Addascalebar";
base.m_deactivate=true;
}
        publicoverridevoidOnCreate(objecthook)
{
//pHookHelper.Hook=hook;
            axPagelayoutControl=hookasAxPageLayoutControl;
}
          publicoverridevoidOnMouseDown(intButton,intShift,intX,
intY)
{
pPoint = axPagelayoutControl.ActiveView.ScreenDisplay.Display-
Transformation.ToMapPoint(X,Y);
            bInuse=true;
}
          publicoverridevoidOnMouseMove(intButton,intShift,intX,
```

```
intY)
    {
    if(bInuse==false)
    {
    return;
    }
    if(pNewEnvelopeFeedback==null)
    {
                pNewEnvelopeFeedback=newNewEnvelopeFeedbackClass
();
    pNewEnvelopeFeedback.Display=axPagelayoutControl.ActiveView.
ScreenDisplay;
                pNewEnvelopeFeedback.Start(pPoint);
    }
    pNewEnvelopeFeedback.MoveTo(axPagelayoutControl.ActiveView.
ScreenDisplay.DisplayTransformation.ToMapPoint(X,Y));
    }
          publicoverridevoidOnMouseUp(intButton,intShift,intX,in-
tY)
    {
    if(bInuse==false)
    {
    return;
    }
    if(pNewEnvelopeFeedback==null)
    {
                pNewEnvelopeFeedback=null;
                bInuse=false;
    return;
    }
            IEnvelopepEnvelope=pNewEnvelopeFeedback.Stop();
    if((pEnvelope.IsEmpty)||(pEnvelope.Width==0)||(pEnvelope.Height==
0))
    {
                pNewEnvelopeFeedback=null;
                bInuse=false;
    return;
    }
```

```
            AddScaleBarFormscaleBarForm=newAddScaleBarForm();
    //scaleBarForm.Show();
            IStyleGalleryItempStyleItem=Form1.pStyleGalleryItem;
    if(pStyleItem==null)
    {
    return;
    }
    IMapFramepMapframe = axPagelayoutControl.ActiveView.GraphicsCon-
tainer.FindFrame ( axPagelayoutControl.ActiveView.FocusMap ) asIMap-
Frame;
            IMapSurroundFramepSurroundFrame=newMapSurroundFrame-
Class();
            pSurroundFrame.MapFrame=pMapframe;
            pSurroundFrame.MapSurround=(IMapSurround)pStyleItem.
Item;
    //在 pageLayout 中根据名称查要素 Element,找到之后删除已经存在的比例尺
    IElementpelement=axPagelayoutControl.FindElementByName("Scale-
Bars");
    if(pelement! =null)
    {
    axPagelayoutControl.ActiveView.GraphicsContainer.DeleteElement
(pelement);//删除已经存在的指北针
    }
            pelement=(IElement)pSurroundFrame;
            pelement.Geometry=(IGeometry)pEnvelope;
    axPagelayoutControl.ActiveView.GraphicsContainer.AddElement
(pelement,0);
    axPagelayoutControl.ActiveView.PartialRefresh(esriViewDrawPhase.
esriViewGraphics,null,null);
            pNewEnvelopeFeedback=null;
            bInuse=false;
    }
    }
    }
```

9.2 要素渲染

Feature 常用的渲染方法包括简单绘制、唯一值绘制/多字段唯一值绘制、点密度/多

字段点密度绘制、数据分级绘制、质量图(饼图/直方图)、按比例尺渲染、比例符号渲染。

9.2.1　简单渲染

简单渲染是 ArcEngine 的默认渲染，当系统打开一个 FeatureClass，建立一个 FeatureLayer 的时候，如果没有给 FeatureLayer 设置 Renderer，所使用的就是简单渲染。简单渲染对整个图层中的所有 Feature 使用同一种方式显示。简单渲染在 ArcEngine 中用 ISimpleRenderer 来表示，ISimpleRenderer 的示例代码如下：

```
//假设 layer 是一个 IFeatureLayer,获取 IGeoFeatureLayer
IGeoFeatureLayergeoLayer=layerasIGeoFeatureLayer;
//构造 SimpleRenderer
ISimpleRendererrenderer=newSimpleRendererClass();
renderer.description="简单地渲染一下";
renderer.Label="符号的标签";
//假设 sym 是一个和该图层中 Geometry 类型对应的符号
renderer.Symbol=sym;
//为图层设置渲染,注意需要刷新该图层
geoLayer.Renderer=renderer;
```

9.2.2　独立值渲染

独立值(多字段)渲染是根据 Feature 的某一个字段的数据或某几个字段的组合结果来确定符号。具有相同值或相同组合值的 Feature，使用同一符号。字段的取值顺序和在 Renderer 中设置的一样。示例代码如下：

```
//假设 layer 是一个 IFeatureLayer,获取 IGeoFeatureLayer
IGeoFeatureLayergeoLayer=layerasIGeoFeatureLayer;
//构造一个 UniqueValueRenderer
IUniqueValueRendererrenderer=newUniqueValueRendererClass();
//假设使用两个字段来渲染
renderer.FieldCount=2;
//假设 YSLX 字段表示要素类型
//假设 YSYT 字段表示要素用途
renderer.set_Field(0,"YSLX");
renderer.set_Field(1,"YSYT");
//字段之间使用"|"来连接(默认取值)
renderer.FieldDelimiter="|";
//设置默认符号
renderer.DefaultSymbol=defaultSymbol;
renderer.DefaultLabel="默认 Label";
```

```
//添加值
renderer.addValue("房屋|民居","民居房屋",MJSymbol);
renderer.addValue("房屋|商业用地","商业用地",SYSymbol);
//还可以通过 set_Symbol,set_Heading、set_Value 来修改上述设置
geoLayer.Renderer=renderer
```

9.2.3 点密度渲染

点密度图通过在 Feature 的图形上打点来表示数据的数量级，点越密集表示数据量越大。还可以使用多字段的点密度图即可在同一个 Feature 上显示几种不同的点。需要注意的是，点密度图有一个特殊的地方——点密度图使用的符号是面状符号。示例代码如下：

```
IDotDensityRendererrenderer=newDotDensityRendererClass();
IRendererFieldsflds=(IRendererFields)renderer;
flds.AddField("MJ","面积");
flds.AddField("RK","人口");
IDotDensityFillSymbolddSym=newDotDensityFillSymbolClass();
ISymbolArraysymArray=(ISymbolArray)ddSym;
symArray.AddSymbol(mjSymbol);
symArray.AddSymbol(rkSymbol);
ddSym.Outline=(ILineSymbol)outlineSymbol;
ddSym.DotSize=10;
ddSym.FixedPlacement=true;
renderer.DotDensitySymbol=ddSym;
renderer.DotValue=20;
renderer.MaintainSize=this.m_dotdensityParam.MaintainSize;
IGeoFeatureLayergeoLayer=(IGeoFeatureLayer)layer;
geoLayer.Renderer=(IFeatureRenderer)renderer;
```

9.3 专题图制作

9.3.1 外表关联示例

本节设计了一个外表关联函数示例，通过此示例来演示外表关联的方法，设计代码如下：

```
ITabledispTable=((IDisplayTable)feaLayer).DisplayTable;//图层
ITableattTable;//外表
IMemoryRelationshipClassFactoryfac = newMemoryRelationshipClass-
FactoryClass();
IRelationshipClassrelClass = fac.Open ("JZMJ", ( IObjectClass )
```

```
dispTable,"ZDDJH",
    IObjectClass)attTable,"G03",
    "Forward","Backward",
    esriRelCardinality.esriRelCardinalityOneToOne);
    IDisplayRelationshipClassdispRelClass = feaLayerasIDisplayRela-
tionshipClass;
    dispRelClass.DisplayRelationshipClass(relClass,esriJoinType.es-
riLeftInnerJoin);
```

9.3.2 统计分析示例

本节设计了一个统计分析功能函数示例，通过此示例来演示统计分析的方法，设计代码如下：

```
    ITableHistogramtableHistogram=newBasicTableHistogramClass();
    tableHistogram.Table=((IDisplayTable)layer).DisplayTable;
    tableHistogram.Field=fieldName;
    objectvalueArray=null,freq=null;
    IBasicHistogrambasicHistogram=(IBasicHistogram)tableHistogram;
    basicHistogram.GetHistogram(outvalueArray,outfreq);
    IClassifyclassify=null;
    intbreakNum=6;
    //分类方法
    switch(ClassifyMethod)
    {
        caseClassifyMethodName.lsClassifyMethodEqualInterval:
    {
            EqualIntervalClasseq=newEqualIntervalClass();
            eq.Classify(valueArray,freq,refbreakNum);
            classify=(IClassify)eq;
    break;
    }
        caseClassifyMethodName.lsClassifyMethodStandardDeviation:
    {
            StandardDeviationClasssd=newStandardDeviationClass();
            IStatisticsResultsstat=histogramasIStatisticsResults;
            classify=sdasIClassify;
            classify.SetHistogramData(valueArray,freq);
            IDeviationIntervaldi=sdasIDeviationInterval;
            di.DeviationInterval=1;
```

```
            di.Mean=stat.Mean;
            di.StandardDev=stat.StandardDeviation;
            classify.Classify(refbreakNum);
    break;
    }
        caseClassifyMethodName.lsClassifyMethodQuantile:
    {
            Quantileqc=newQuantileClass();
            qc.Classify(valueArray,freq,refbreakNum);
            classify=qcasIClassify;
    break;
    }
        caseClassifyMethodName.lsClassifyMethodNaturalBreaks:
    {
            NaturalBreaksClassnb=newNaturalBreaksClass();
            nb.Classify(valueArray,freq,refbreakNum);
            classify=nbasIClassify;
    break;
    }
        caseClassifyMethodName.lsClassifyMethodDefinedInterval:
    {
            DefinedIntervalClassdi=newDefinedIntervalClass();
            di.IntervalRange=this.m_classBreaksParam.Interval;
            di.Classify(valueArray,freq,refbreakNum);
            classify=diasIClassify;
    break;
    }
    default:
    {
            EqualIntervalClasseq=newEqualIntervalClass();
            eq.Classify(valueArray,freq,refbreakNum);
            classify=(IClassify)eq;
    break;
    }
    }
    objecto=classify.ClassBreaks;
    System.ArraybreakArray=oasSystem.Array;
```

9.4　打印输出

9.4.1　剪贴板方式输出

实际上 AE 并没有提供实现这一功能的方法，但可以调用 Windows 32 API 函数来实现。基本思路是：获取 MapControl 句柄，然后实现一个一般的剪贴板拷贝程序，将 MapControl 范围内的视图以位图的形式复制过去。示例代码如下：

```
CWnd * pWndCal =GetDlgItem(IDC_MAPCONTROL1)//获取指 MapControl 的指针
CBitmap bitmap
CClientDC dc(pWndCal)//获得 MapControl 客户区的设备环境
memdc.CreateCompatibleDC(&dc)//创建和指定 dc 兼容的内存 dc
pWndCal->GetClientRect(rect)//获得 MapControl 的范围(Client Rectangle)//创建一个和指定设备环境兼容的位图
bitmap.CreateCompatibleBitmap(&dc,rect.Width(),rect.Height())
CBitmap* poldbitmap = memdc.SelectObject(&bitmap)
memdc.BitBlt(0,0,rect.Width(),rect.Height(),&dc,0,0,SRCCOPY)
pWndCal->OpenClipboard()//打开剪贴板
SetClipboardData(CF_BITMAP,bitmap.GetSafeHandle())//向剪贴板增加地图
```

9.4.2　图片方式输出

采用 IActiveView 接口下的 OutPut 方法，可以将地图输出为数十种格式，具体的格式受 IExport 类型限制，如 ExportBMP，ExportPNG、ExportJPEG 等，下面以输出为 JPEG 格式来说明。

首先定义 ExportJPEG 的实例 pExport，然后设置其相关的参数，过程比较简单，这里重点描述一下相关的参数设置。

方法：OutPut(hdc，Dpi，pixelBounds，VisibleBounds，TrackCancel)

①hdc 是输出设备，由 pExport. StartExporting 指定。

②Dpi 是输出图片的精度，但是 resolution 属性并不能改变图片的精度，无论设置多大的 Dpi，输出同一范围图片的大小、精度都是一样的。要想改变精度，需要指定参数 IOutputRasterSettings:: ResampleRatio，可以设置 1~5 个级别的采样率，在输出图片很大的时候这个参数能提高图片的质量。

③PixelBounds 设置输出像素所占的范围。

④VisibleBounds 指定地图可视的范围，这个范围是以地图坐标为单位。

⑤PixelBounds 定义输出图片的大小，即图片尺寸，相当于画布大小。当输出像素的范围大于图片大小的时候就会裁切图片，只输出部分地图，当它小于图片尺寸时，地图会缩小到画布的一角。

第10章 “噪音污染分析与决策系统”案例

“噪音污染分析与决策系统”是“ESRI杯”中国大学生GIS软件开发竞赛获奖作品(研发者：孙群、谢中凯、任沂斌，指导教师：柳林老师)。本章以此获奖作品为案例，介绍了基于ArcEngine的GIS系统二次开发过程，并给出了核心代码。“噪音污染分析与决策系统”案例更多内容参见随书光盘。

10.1 设计思想

系统运用英国CRTN88噪声预测模型，结合国内实际研究情况，对研究范围内监测点的交通噪声值进行实时动态预测，再运用GIS分析的插值思想，得到整个分析区域的任意时刻的噪音状况。通过对噪音状况的全面综合分析，得到各种成果数据——表格、文档、图表等。并最终结合分析结果，进行降噪效果模拟，通过内嵌的飞信终端和电子邮件功能，把噪音的相关信息发布给当地居民和当地的交通规划部门，以供交通管理部门制定相关降噪措施达到对区域减噪的目的。

10.2 功能及实现效果

“噪音污染分析与决策系统”案例系统结构如图10.1所示，涉及的主要功能包括：

①静态污染预测：预测出某一时刻的噪音情况。

②静态污染分析：对噪音进行各种时空对比分析。

③动态污染模拟：动态模拟一辆车在行驶的过程中对监测点的噪音影响情况。

④防治与治理：根据噪音状况提出合理的降噪措施。

“噪音污染分析与决策系统”案例实现了以上功能，获得2010年“EIRI杯”中国大学生GIS软件开发竞赛桌面组奖项，部分功能的实现效果如图10.2~图10.6所示。

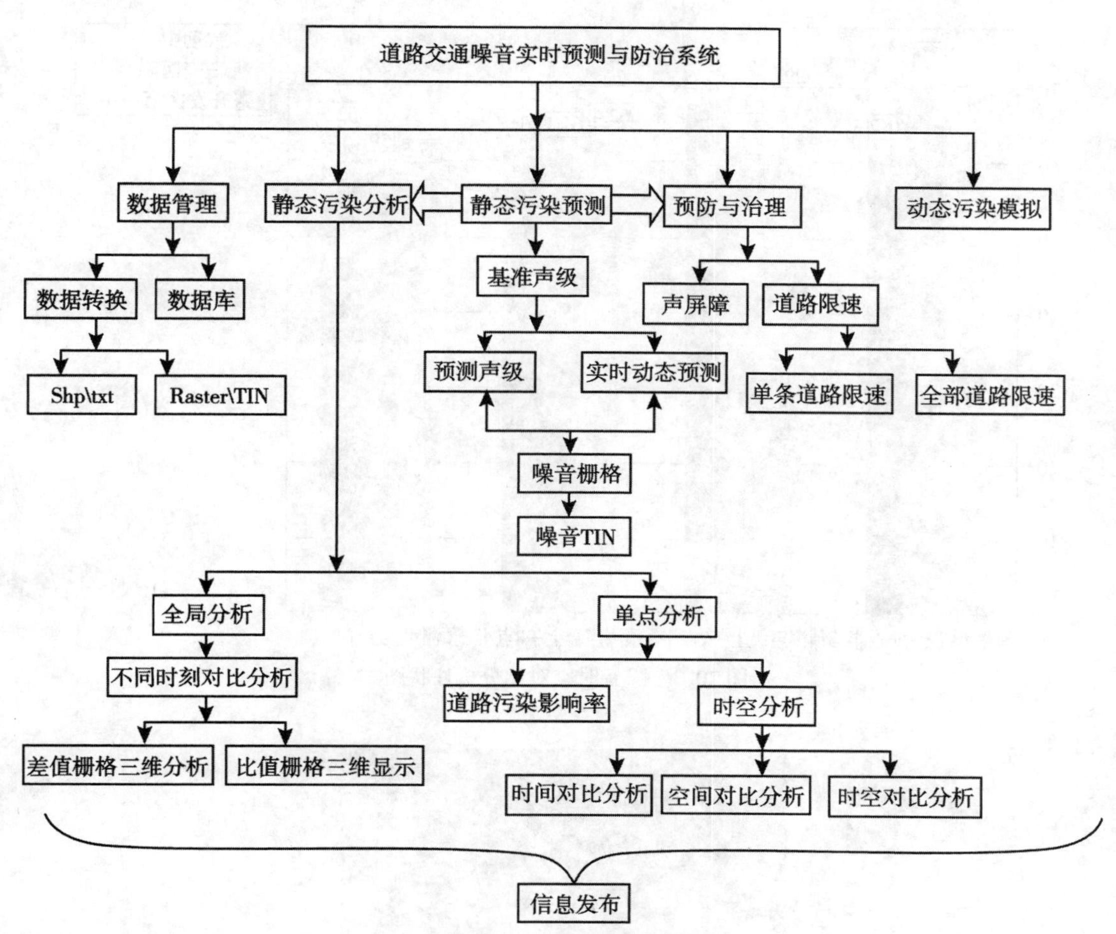

图 10.1　案例系统结构图

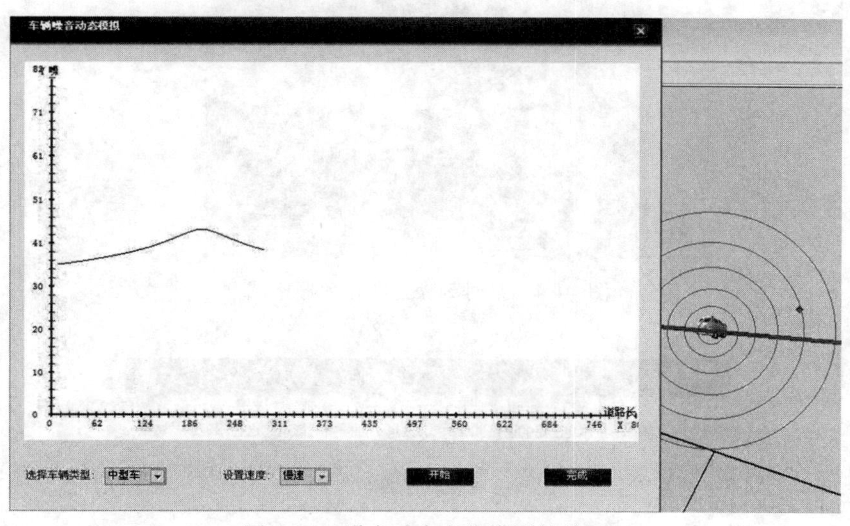

图 10.2　噪音动态变化模拟效果图

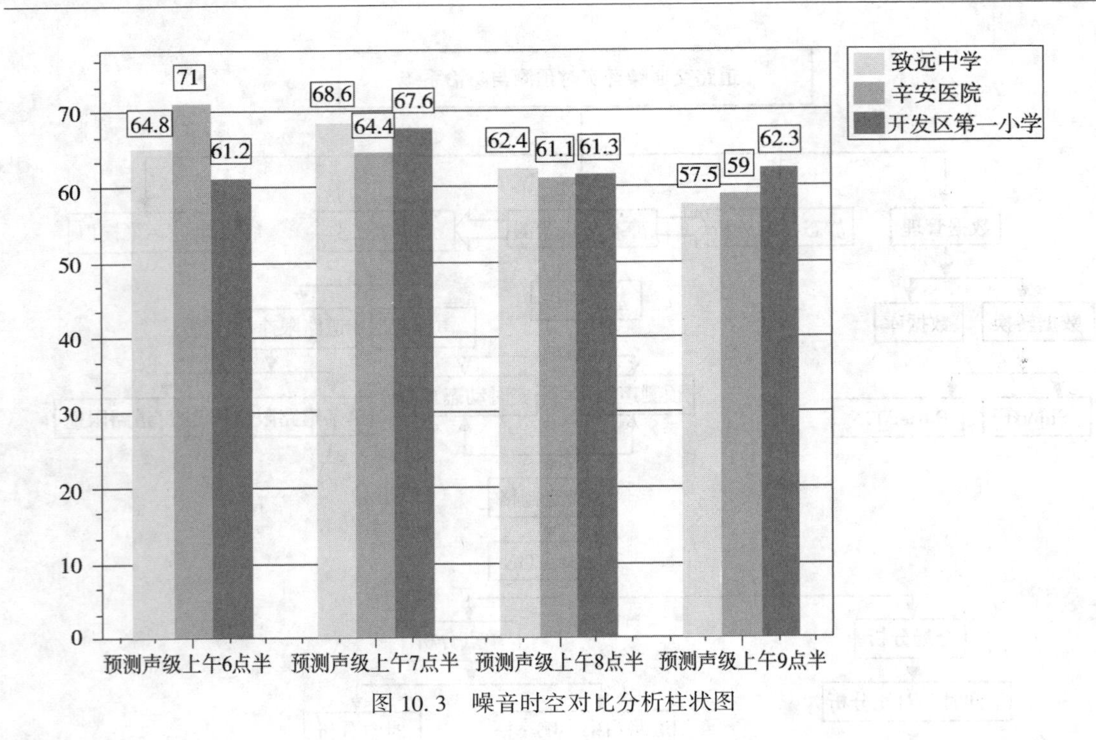

图 10.3　噪音时空对比分析柱状图

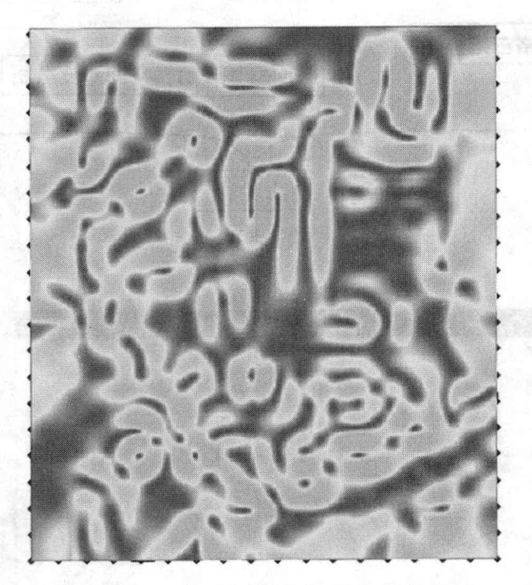

图 10.4　某时刻噪音变化率栅格图

图 10.5　案例道路限速功能

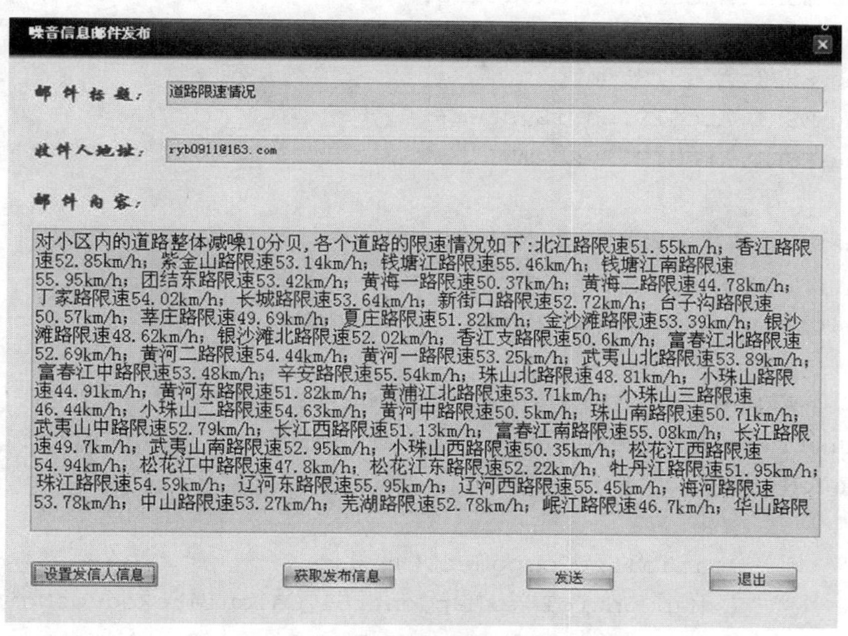

图 10.6 道路限速邮箱发布

10.3 核心代码

“噪音污染分析与决策系统”案例原来是基于 ArcGIS 9 版本开发的，本节将代码已转换为 ArcGIS 10 新版本。

10.3.1 计算噪音减小量代码

具体代码如下：

```
Using System;
Using System.Collections.Generic;
Using System.ComponentModel;
Using System.Data;
Using System.Drawing;
Using System.Text;
Using System.Windows.Forms;

Using ESRI.ArcGIS.Carto;
Using ESRI.ArcGIS.Controls;
Using ESRI.ArcGIS.Geometry;
Using ESRI.ArcGIS.GeoDatabaseUI;
```

```
Using ESRI.ArcGIS.Geodatabase;
Using ESRI.ArcGIS.Display;
Using ESRI.ArcGIS.Output;
Using ESRI.ArcGIS.esriSystem;
Using ESRI.ArcGIS.SystemUI;

Namespace NoisePredict.NoiseHandle
{
publicpartialclassFrmCalculateReduceNoise:Form
{
        AxMapControlaxMapControl=null;
publicFrmCalculateReduceNoise(AxMapControlaxMapControl_Calcu-
lateReduceNoise)
{
            InitializeComponent();
            axMapControl=axMapControl_CalculateReduceNoise;
this.skinEngine1.SkinFile=QuanJuBianLiang.SkinName;
}

        privatevoidBtnCalculate_Click(objectsender,EventArgse)
{
try
{
if(QuanJuBianLiang.pNoiseHandleSensitivePoint==null)
{
                    MessageBox.Show("未设置敏感点!","警告",Message-
BoxButtons.OK,MessageBoxIcon.Information);
return;
}
if(QuanJuBianLiang.pNoiseHandleRoad==null)
{
                    MessageBox.Show("未设置要进行降噪的道路!","警告",
MessageBoxButtons.OK,MessageBoxIcon.Information);
return;
}
if(QuanJuBianLiang.pNoiseBarriePolyline==null)
{
                    MessageBox.Show("未设置声屏障!","警告",Message-
```

```
BoxButtons.OK,MessageBoxIcon.Information);
    return;
    }
    doubleSensitinvPoint_Road=GetDistance(QuanJuBianLiang.pNoise-
HandleSensitivePoint,QuanJuBianLiang.pNoiseHandleRoad);//敏感点和道
路之间的距离
    doubleSensitivePoint_NoiseBarrier=GetDistance(QuanJuBianLiang.
pNoiseHandleSensitivePoint,QuanJuBianLiang.pNoiseBarriePolyline);
//敏感点和声屏障之间的距离
    if(SensitinvPoint_Road<SensitivePoint_NoiseBarrier)
    {
                    MessageBox.Show("声屏障位置设置错误,声屏障应设置在
敏感点和要进行降噪的道路之间,否则无效,请重新设置正确位置的声屏障!","警告",
MessageBoxButtons.OK,MessageBoxIcon.Information);
                    QuanJuBianLiang.pNoiseBarriePolyline=null;
    return;
    }
    doubleNoiseBarrier_Road=SensitinvPoint_Road-SensitivePoint_
NoiseBarrier;//声屏障和道路之间的距离
    doubleNoiseBarrierHeight=Convert.ToDouble(this.numNoiseBarri-
erHeight.Value);//声屏障高度
    doubleXieJu_SensitivePoint_NoiseBarrier=Math.Sqrt(Math.Pow
(SensitivePoint_NoiseBarrier,2)+Math.Pow(NoiseBarrierHeight,2));
    doubleXieJu_NoiseBarrier_Road=Math.Sqrt(Math.Pow(NoiseBarrier_
Road,2)+Math.Pow(NoiseBarrierHeight,2));
    doublePathDifference=XieJu_SensitivePoint_NoiseBarrier+XieJu_
NoiseBarrier_Road-SensitinvPoint_Road;//计算声称差
                PathDifference=Math.Round(PathDifference,2);
                doublet=19.6*PathDifference;
    //MessageBox.Show("声称差为"+PathDifference.ToString()+"米","反
馈信息",MessageBoxButtons.OK,MessageBoxIcon.Information);
                doubleb=Math.Sqrt(Math.Pow(t,2)-1);
                doublefenzi=3*Math.PI*b;
                doublefenmu=2*Math.Log(t+b);
                doubleL1=10*Math.Log10(fenzi/fenmu);
    //MessageBox.Show("L1="+L1.ToString(),"反馈信息",MessageBoxBut-
tons.OK,MessageBoxIcon.Information);
    doubleSensitivePoint_FromPointA=GetPointsDistance(QuanJuBian-
```

```
Liang.pNoiseHandleSensitivePoint,QuanJuBianLiang.pNoiseBarriePoly-
line.FromPoint);
doubleSensitivePoint_ToPointA=GetPointsDistance(QuanJuBianLiang.
pNoiseHandleSensitivePoint,QuanJuBianLiang.pNoiseBarriePolyline.
ToPoint);
doubleA1 = Math.Acos ( SensitivePoint _NoiseBarrier/SensitivePoint_
FromPointA);
doubleA2=Math.Acos(SensitivePoint_NoiseBarrier/SensitivePoint_To
PointA);
doubleSensitivePoint_FromPointB=GetPointsDistance(QuanJuBianLia-
ng.pNoiseHandleSensitivePoint,QuanJuBianLiang.pNoiseHandleRoad.
FromPoint);
doubleSensitivePoint_ToPointB=GetPointsDistance(QuanJuBianLiang.
pNoiseHandleSensitivePoint,QuanJuBianLiang.pNoiseHandleRoad.To
Point);
                doubleB1=Math.Acos(SensitinvPoint_Road/Sensitive-
Point_FromPointB);
                doubleB2=Math.Acos(SensitinvPoint_Road/Sensitive-
Point_ToPointB);
                doubleA=B1+B2;
                doubleL2=10*Math.Log10(A/(B1+B2));
                doubleL=L1-L2;
                L=Math.Round(L,2);
                stringsComment=string.Empty;
    if(L<7)
                sComment="减噪效果较差";
                elseif(L<10)
                sComment="减噪效果一般";
    else
                sComment="减噪效果较好!";
                MessageBox.Show("声称差为"+PathDifference.ToString
()+"米,声屏障减噪"+L.ToString()+"分贝,"+sComment,"反馈信息",Message-
BoxButtons.OK,MessageBoxIcon.Information);
    }
    catch(Exceptionex)
    {
                MessageBox.Show(ex.Message.ToString(),"警告",Mes-
sageBoxButtons.OK,MessageBoxIcon.Warning);
```

```
        }
    }
            privatedoubleGetDistance(IPointpPoint,IPolylinepPoly-
line)
    {
            IProximityOperatorpProximityOperator=pPointasIProx-
imityOperator;
            doubledistance=pProximityOperator.ReturnDistance
(pPolyline);
            returndistance;
    }
        privatedoubleGetPointsDistance(IPointPoint1,IPointPoint2)
    {
            IProximityOperatorpProximityOperator=Point1asIProximity
Operator;
            doubledistance=pProximityOperator.ReturnDistance
(Point2);
            returndistance;
    }
    privatevoidFrmCalculateReduceNoise_FormClosing(objectsender,
FormClosingEventArgse)
    {
            QuanJuBianLiang.pNoiseHandleSensitivePoint=null;
            QuanJuBianLiang.pNoiseHandleRoad=null;
            QuanJuBianLiang.pNoiseBarriePolyline=null;
    axMapControl.ActiveView.PartialRefresh(esriViewDrawPhase.esriView
Graphics,null,null);
    }
    }
    }
```

10.3.2 限定道路速度代码

具体代码如下:

```
Using System;
Using System.Collections.Generic;
Using System.ComponentModel;
Using System.Data;
Using System.Drawing;
```

```
Using System.Text;
Using System.Windows.Forms;

Using ESRI.ArcGIS.Carto;
Using ESRI.ArcGIS.Controls;
Using ESRI.ArcGIS.Geometry;
Using ESRI.ArcGIS.GeoDatabaseUI;
Using ESRI.ArcGIS.Geodatabase;
Using ESRI.ArcGIS.Display;
Using ESRI.ArcGIS.Output;
Using ESRI.ArcGIS.esriSystem;
Using ESRI.ArcGIS.SystemUI;
Namespace NoisePredict.NoiseHandle
{
publicpartialclassFrmRestrictRoadSpeed:Form
{
        AxMapControlaxMapControl=null;
        intIndex;
        stringPublishMessage=string.Empty;
        publicFrmRestrictRoadSpeed(AxMapControlaxMapControl_Re-
strictRoadSpeed)
{
            InitializeComponent();
            axMapControl=axMapControl_RestrictRoadSpeed;
this.skinEngine1.SkinFile=QuanJuBianLiang.SkinName;
}
        privatevoidAddRoadLayer()
{
for(inti=0;i<axMapControl.LayerCount;i++)
{
                ILayerpLayer=axMapControl.get_Layer(i);
if(pLayerisIFeatureLayer)
{
                    IFeatureLayerpFeatureLayer=pLayerasIFeature-
Layer;
if(pFeatureLayer.FeatureClass.ShapeType==esriGeometryType.es-
riGeometryPolyline)
{
```

```
this.comRoadLayer.Items.Add(pFeatureLayer.Name.ToString());
}
}
}
if(this.comRoadLayer.Items.Count>0)
this.comRoadLayer.SelectedIndex=0;
}
    privateILayerGetSelectedLayer()
{
for(inti=0;i<axMapControl.LayerCount;i++)
{
        ILayerpLayer=axMapControl.get_Layer(i);
if(pLayer.Name.ToString()==(string)this.comRoadLayer.Selecte-
dItem)
{
            returnpLayer;
}
}
        returnnull;
}
    privatevoidFrmRestrictRoadSpeed_Load(objectsender,Even-
tArgse)
{
this.txtNewFieldName.Text = " 降噪 " + this.numericUpDown1.Val-
ue.ToString()+"分贝限速";
        AddRoadLayer();
}
    privatevoidcomRoadLayer_SelectedIndexChanged(objectsend-
er,EventArgse)
{
this.comTrafficVolume.Items.Clear();
this.comRoadFields.Items.Clear();
        IFeatureLayerpFeatureLayer=GetSelectedLayer()asIFea-
tureLayer;
        IFieldspFields=pFeatureLayer.FeatureClass.Fields;
for(inti=0;i<pFields.FieldCount;i++)
{
                stringsFieldName = pFields.get _Field ( i ).
```

```
Name.ToString();
   if(sFieldName.Contains("车流量"))
   this.comTrafficVolume.Items.Add(sFieldName);
   if(sFieldName=="OBJECTID"‖sFieldName=="道路名")
   this.comRoadFields.Items.Add(sFieldName);
   }
   if(this.comTrafficVolume.Items.Count>0)
   this.comTrafficVolume.SelectedIndex=0;
   if(this.comRoadFields.Items.Count>0)
   this.comRoadFields.SelectedIndex=0;
   }
           privatevoidcomRoadFields_SelectedIndexChanged(object-
sender,EventArgse)
   {
   this.comRoadName.Items.Clear();
              IFeatureLayerpFeatureLayer=GetSelectedLayer()asIFea-
tureLayer;
   Index = pFeatureLayer.FeatureClass.Fields.FindField((string)
this.comRoadFields.SelectedItem);
              IQueryFilterpQueryFilter=newQueryFilterClass();
              pQueryFilter.WhereClause=null;
              IFeaturepRoadFeature=null;
   IFeatureCursorpFeatureCursor=pFeatureLayer.FeatureClass.Search
(pQueryFilter,true);
   while((pRoadFeature=pFeatureCursor.NextFeature())! =null)
   {
                     stringvalue = pRoadFeature.get_Value(Index)
.ToString();
   this.comRoadName.Items.Add(value);
   }
   if(this.comRoadName.Items.Count>0)
   this.comRoadName.SelectedIndex=0;
   }
          privatevoidBtnCalculate_Click(objectsender,EventArgse)
   {
   try
   {
   if(this.radSingleRoad.Checked)
```

```
{
                    axMapControl.Refresh();
doubleDecreaseNoise = Convert.ToDouble(this.numericUpDown1.Value);//获取道路降噪分贝
                    ILayerpLayer=GetSelectedLayer();
                    IFeatureLayerpFeatureLayer=pLayerasIFeatureLayer;
intIndex _ Q = pFeatureLayer.FeatureClass.FindField((string)this.comTrafficVolume.SelectedItem);//获取所选择的某一时间段的车流量的索引值
                    IQueryFilterpQueryFilter=newQueryFilterClass();
                    pQueryFilter.WhereClause=null;
                    IFeaturepRoadFeature=null;
IFeatureCursorpFeatureCursor=pFeatureLayer.FeatureClass.Search(pQueryFilter,true);
                    IFeaturepSelectedRoadFeature=null;
while((pRoadFeature=pFeatureCursor.NextFeature())! =null)
{
                        stringvalue=pRoadFeature.get_Value(Index).ToString();
if(value==(string)this.comRoadName.SelectedItem)
{
                            pSelectedRoadFeature=pRoadFeature;
break;
}
}
                    System.GC.Collect();//强制对所有代码进行垃圾回收
//ESRI.ArcGIS.ADF.ComReleaser.ReleaseCOMObject(pFeatureCursor);//此处不能加此行代码,否则会报错“指针无效”
QuanJuBianLiang.Road _ RestricteSpeed = pSelectedRoadFeature.ShapeasIPolyline;
                    ISimpleLineSymbollineSymbol=newSimpleLineSymbolClass();
                    IRgbColorrgbColor=newRgbColorClass();
                    rgbColor.Red=150;
                    rgbColor.Green=50;
                    rgbColor.Blue=25;
```

```
                    lineSymbol.Color=rgbColor;
                    lineSymbol.Width=3;
                    objectsymbol=lineSymbolasobject;
    axMapControl.DrawShape ( QuanJuBianLiang.Road _ RestricteSpeed,
refsymbol);
                      doubleQ1=(double)pSelectedRoadFeature.get_
Value(Index_Q);
                    doublev=16*Math.Log10(Q1)-1.6*DecreaseNoise+
18.68;
                    v=Math.Round(v,2);
                    MessageBox.Show("对"+(string)this.comRoad-
Fields.SelectedItem+"为"+(string)this.comRoadName.SelectedItem+"的
道路减噪大约"+DecreaseNoise.ToString()+"分贝,则需要对该道路进行限速"+
v.ToString()+"km/h","信息反馈",MessageBoxButtons.OK,MessageBoxIcon.
Information);
    }
                 elseif(this.radAllRoads.Checked)
    {
    doubleDecreaseNoise = Convert.ToDouble(this.numericUpDown1.Val-
ue);//获取道路降噪分贝
                    ILayerpLayer=GetSelectedLayer();
                    IFeatureLayerpFeatureLayer=pLayerasIFeature-
Layer;
                    IDatasetpDataset=pLayerasIDataset;
                    IWorkspacepWorkspace=pDataset.Workspace;
                    IWorkspaceEditpWorkspaceEdit=pWorkspaceasI-
WorkspaceEdit;
                    ITablepTable=pFeatureLayer.FeatureClassasI-
Table;
    intIndex _ Q = pFeatureLayer.FeatureClass.FindField ((string)
this.comTrafficVolume.SelectedItem);//获取所选择的某一时间段的车流量的
索引值
                    IFieldpNewField=newFieldClass();
                    IFieldEditpNewFieldEdit=pNewFieldasIFieldEdit;
                    pNewFieldEdit.Name_2=this.txtNewFieldName.
Text.ToString();
                    pNewFieldEdit.Type_2=esriFieldType.esriField
TypeDouble;
```

```
                    pTable.AddField(pNewField);
                    IQueryFilterpQueryFilter=newQueryFilterClass();
                    pQueryFilter.WhereClause=null;
                    IFeaturepRoadFeature=null;
   IFeatureCursorpFeatureCursor=pFeatureLayer.FeatureClass.Search
(pQueryFilter,true);
   this.progressBarControl1.Visible=true;
   this.progressBarControl1.Properties.Maximum=pFeatureLayer.Fea-
tureClass.FeatureCount(pQueryFilter);// 设置 progressBarControl1.
Properties.Maximum
   while((pRoadFeature=pFeatureCursor.NextFeature())! =null)
   {
                       pWorkspaceEdit.StartEditing(false);
                       pWorkspaceEdit.StartEditOperation();
                       doubleQ1=(double)pRoadFeature.get_Value
(Index_Q);
                       doublev=16*Math.Log10(Q1)-1.6*Decrease-
Noise+18.68;
                       v=Math.Round(v,2);
   intIndex_NewFied = pRoadFeature.Fields.FindField(this.txtNew-
FieldName.Text.ToString());
                       pRoadFeature.set_Value(Index_NewFied,v);
                       pRoadFeature.Store();
                       pWorkspaceEdit.StopEditOperation();
                       pWorkspaceEdit.StopEditing(true);
                       progressBarControl1.PerformStep();
                       progressBarControl1.Update();
   }
   this.progressBarControl1.Visible=false;
                    MessageBox.Show("计算完成,请查看属性表","提示",
MessageBoxButtons.OK,MessageBoxIcon.Information);
   this.BtnCalculate.Enabled=false;
   this.BtnFetionPublish.Enabled=true;
   this.BtnEmailPublish.Enabled=true;
   this.BtnExportToWord.Enabled=true;
   }
   }
   catch(Exceptionex)
```

```
{
if(ex.Message.ToString().Contains("数据库引擎无法锁定"))
{
this.BtnCalculate_Click(sender,e);
}
}
}
        privatevoidBtnClose_Click(objectsender,EventArgse)
{
this.Close();
        QuanJuBianLiang.Road_RestricteSpeed=null;
}
privatevoidnumericUpDown1 _ ValueChanged ( objectsender, Even-
tArgse)
{
this.txtNewFieldName.Text = " 降 噪 " + this.numericUpDown1.Val-
ue.ToString()+"分贝限速";
}
        privatevoidradSingleRoad _CheckedChanged ( objectsender,
EventArgse)
{
if(this.radSingleRoad.Checked)
{
this.comRoadFields.Enabled=true;
this.comRoadName.Enabled=true;
}
        elseif(this.radAllRoads.Checked)
{
this.comRoadFields.Enabled=false;
this.comRoadName.Enabled=false;
}
}
        privatevoidBtnFetionPublish _Click ( objectsender, Even-
tArgse)
{
        stringAllMessageFetion = " 对 小 区 内 的 道 路 整 体 减 噪 " +
this.numericUpDown1.Value.ToString()+"分贝,各个道路的限速情况如下:";
if(this.PublishMessage= =string.Empty)
```

```
{
                        stringsFieldName_Speed = this.txtNewFieldName.
Text.ToString();
                    ILayerpLayer=GetSelectedLayer();
                    IFeatureLayerpFeatureLayer=pLayerasIFeatureLayer;
                    IFeatureClasspFeatureClass=pFeatureLayer.Feature-
Class;
                    intIndex_Speed=pFeatureClass.FindField(sField-
Name_Speed);
                    intIndex_RoadName=pFeatureClass.FindField("道路
名");
                    IQueryFilterpQueryFilter=newQueryFilterClass();
                    pQueryFilter.WhereClause=null;
    IFeatureCursorpFeatureCursor=pFeatureClass.Search(pQueryFilter,
true);
                    IFeaturepFeature=null;
    while((pFeature=pFeatureCursor.NextFeature())! =null)
    {
    PublishMessage = PublishMessage+pFeature.get_Value(Index_Road-
Name).ToString()+"限速"+pFeature.get_Value(Index_Speed).ToString
()+"km/h;";
    }
    }
                AllMessageFetion=AllMessageFetion+PublishMessage;
                QuanJuBianLiang.FetionPublishMessage=AllMessageFe-
tion;
    if(! QuanJuBianLiang.frmFetion.Created)
                QuanJuBianLiang.frmFetion=newNoisePredict.Message-
Publish.FrmFetion();
                QuanJuBianLiang.frmFetion.Show();
    }
            privatevoidBtnEmailPublish_Click(objectsender,Event-
Argse)
    {
                    stringAllMessageEmail = "对小区内的道路整体减噪" +
this.numericUpDown1.Value.ToString()+"分贝,各个道路的限速情况如下:";
if(this.PublishMessage==string.Empty)
    {
```

```
                    stringsFieldName_Speed=this.txtNewFieldName.Text.
ToString();
                    ILayerpLayer=GetSelectedLayer();
                    IFeatureLayerpFeatureLayer=pLayerasIFeatureLayer;
                    IFeatureClasspFeatureClass=pFeatureLayer.Feature-
Class;
                    intIndex_Speed=pFeatureClass.FindField(sField-
Name_Speed);
                    intIndex_RoadName=pFeatureClass.FindField("道路
名");
                    IQueryFilterpQueryFilter=newQueryFilterClass();
                    pQueryFilter.WhereClause=null;
IFeatureCursorpFeatureCursor = pFeatureClass.Search ( pQueryFilter,
true);
                    IFeaturepFeature=null;
while((pFeature=pFeatureCursor.NextFeature())! =null)
    {
PublishMessage=PublishMessage+pFeature.get_Value(Index_RoadName)
.ToString()+"限速"+pFeature.get_Value(Index_Speed).ToString()+"km/
h;";
    }
    }
                AllMessageEmail=AllMessageEmail+PublishMessage;
                QuanJuBianLiang.EmailPublishMessage=AllMessageEmail;
NoisePredict.MessagePublish.FrmEmailfrmEmail = newNoisePredict.Mes-
sagePublish.FrmEmail();
                frmEmail.Show();
    }
              privatevoidBtnExportToWord _ Click ( objectsender, Even-
tArgse)
    {
    try
    {
    if(this.PublishMessage==string.Empty)
    {
                        stringsFieldName_Speed=this.txtNewFieldName.
Text.ToString();
                        ILayerpLayer=GetSelectedLayer();
```

```
                    IFeatureLayerpFeatureLayer=pLayerasIFeature-
Layer;
                    IFeatureClasspFeatureClass=pFeatureLayer.Fea-
tureClass;
                    intIndex_Speed=pFeatureClass.FindField
(sFieldName_Speed);
                    intIndex_RoadName=pFeatureClass.FindField("道
路名");
                    IQueryFilterpQueryFilter=newQueryFilterClass();
                    pQueryFilter.WhereClause=null;
IFeatureCursorpFeatureCursor = pFeatureClass.Search ( pQueryFilter,
true);
                    IFeaturepFeature=null;
while((pFeature=pFeatureCursor.NextFeature())! =null)
    {
PublishMessage=PublishMessage+pFeature.get_Value(Index_RoadName)
.ToString()+"限速"+pFeature.get_Value(Index_Speed).ToString()+"km/
h;";
    }
    }

                stringMessageWord=this.PublishMessage;
                SaveFileDialogsaveFileDialog=newSaveFileDialog();
                saveFileDialog.Title="道路限速Word输出";
                saveFileDialog.Filter="Word文档(*.doc)|*.doc";
                stringsaveFileName=string.Empty;
   if(DialogResult.OK! =saveFileDialog.ShowDialog())
   return;
                saveFileName=saveFileDialog.FileName;
                objectnone=System.Reflection.Missing.Value;
                Word.Applicationword=newWord.Application();
Word.Documentdocument=word.Documents.Add(refnone,refnone,refnone,
refnone);
Word.Tabletable = document.Tables.Add ( document.Paragraphs.Last.
Range,2,1,refnone,refnone);
                table.Cell(1,1).Range.Text="辛安街办整体减噪"+this.
numericUpDown1.Value.ToString()+"分贝,各道路限速情况汇总\n";
                table.Cell(2,1).Range.Text=MessageWord;
table.Cell(1,1).Range.ParagraphFormat.Alignment=Word.WdParagraph-
```

```
Alignment.wdAlignParagraphCenter;
                table.Cell(1,1).Range.Font.Color=Word.WdColor.wd-
ColorDarkBlue;
                table.Cell(1,1).Range.Font.Bold=16;
                table.Cell(1,1).Range.Font.Size=16;
table.Cell(2,1).Range.ParagraphFormat.Alignment=Word.WdParagraph
Alignment.wdAlignParagraphCenter;
                objectob=@ saveFileName;
document.SaveAs(refob, refnone, refnone, refnone, refnone, refnone,
refnone,refnone,refnone,refnone,refnone);
                document.Close(refnone,refnone,refnone);
                MessageBox.Show("成功导出 Word 文档!","提示",Message-
BoxButtons.OK,MessageBoxIcon.Information);
   }
   catch
   {
                MessageBox.Show("未能成功导出 Word 文档!","信息反馈",
MessageBoxButtons.OK,MessageBoxIcon.Warning);
   }
   }
   }
   }
```

第二编　ArcGIS API for Flex

第 11 章 相关技术

11. 1 ArcGIS for Server 架构

11. 1. 1 ArcGIS for Server 架构概述

ArcGIS for Server 是一个用于构建集中管理、支持多用户的企业级 GIS 应用的平台，能将地图、地理处理等资源作为服务发布出去，让用户可以通过浏览器、移动端等对这些资源进行访问，同时可以使用 ArcGIS for Server 为用户提供 GIS 功能。

ArcGIS 10. 1 版本中 ArcGIS for Server 采用了全新架构，所以了解它的结构是很有必要的。对于系统管理者而言，知道了其本质，就能更好地去管理 Server；而对于开发者而言，原则上只需要知道这个 Server 暴露的 API 就可以了。但开发人员还要兼顾管理者的角色，因此不论管理人员还是开发人员，知晓 Server 内部结构都是基本的要求，可以为以后研发奠定良好的基础。ArcGIS 10. 1 for Server 是一个纯 64 位的应用程序，不能在 32 位的计算机上安装，这一点需要注意。

ArcGIS 10.1 for Server 构架如图 11.1 所示，包括服务管理员、服务发布者、数据服务器、GIS 服务器、Web 服务器，以及使用 ArcGIS Server 服务的各种终端(桌面端、移动端、浏览器等)。

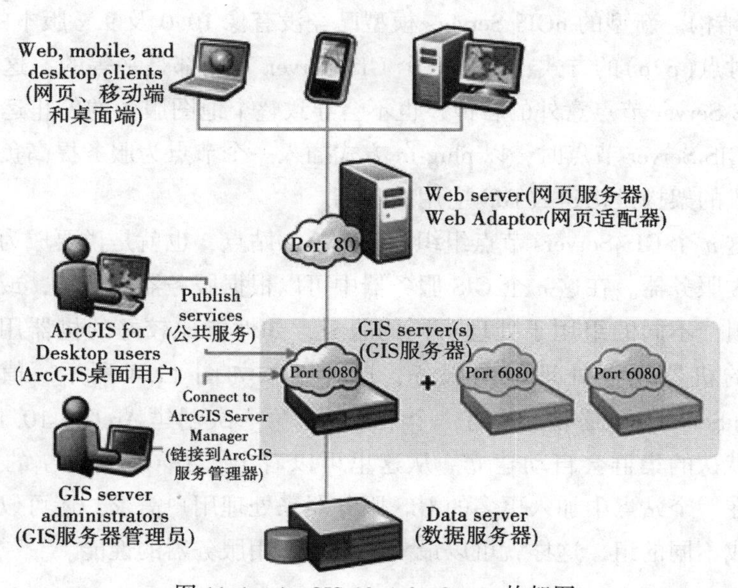

图 11. 1 ArcGIS 10. 1 for Server 构架图

ArcGIS Server 站点是由 4 个部分构成：Web 服务器、Web Adaptor、GIS 服务器和数据库服务器，如图 11.2 所示。

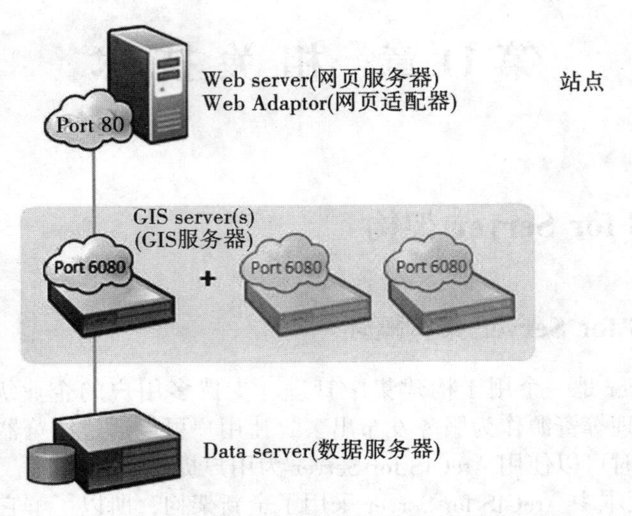

图 11.2 ArcGIS Server 站点的组成

数据服务器存储了 GIS 服务所需要的数据，Web 服务器是可以向发出请求的浏览器提供文档的程序，Web Adaptor 实际上是安装在 Web Server 机器上的一个 Web 应用程序，负责将 Web Server 接收到的 GIS 请求转发到 ArcGIS Server Site 内的 GIS Server 机器上去。

如图 11.3 所示，阴影部分的矩形框中就是 ArcGIS Server 站点的核心——GIS 服务器。这里将它称为 nGIS Servers，即多节点 GIS Servers。这种模型架构取代了 10.0 以前的基于 SOM——SOCs 结构。新型的 nGIS Servers 模型已经没有像 10.0 及 9.x 版本的 SOM 主控制节点，采用点对点(p2p)的方式，即每一个 GIS Server 节点都是平等的。这样，新模型即使是某一个 GIS Server 节点意外的宕掉，也不会导致整个地图服务的停止运行；同样，当需要增加一个 GIS Server 节点时，以 plug-in 方式插入一个节点为服务提高负载能力。这种松散的、热插拔的架构是构建云 GIS 应用的基础。

逻辑上，这 n 个 GIS Servers 节点组织为一个 Site 站点，也就是说要成为一个站点至少应该有一台 GIS 服务器。在这 n 个 GIS 服务器中可以根据服务器的性能，或者根据应用的不同而进行分组，不同的组用于处理不同的服务，如性能比较好的机器用于处理 GP 服务，性能一般的机器用于处理地图服务，这种结构如图 11.4 所示。值得注意的是，ArcGIS 10.1 for Server site 必须至少有一个集群，当第一次创建 ArcGIS 10.1 for Server site 的时候，一个默认的集群会自动建立。从这里可以看出，ArcGIS Server 的站点架构有优势，不仅可以在一个站点中加入更多的 GIS 服务器来处理用户需求，还可以根据需要将这些服务器细分成不同的组，这样就可以最大限度地利用服务器的性能。

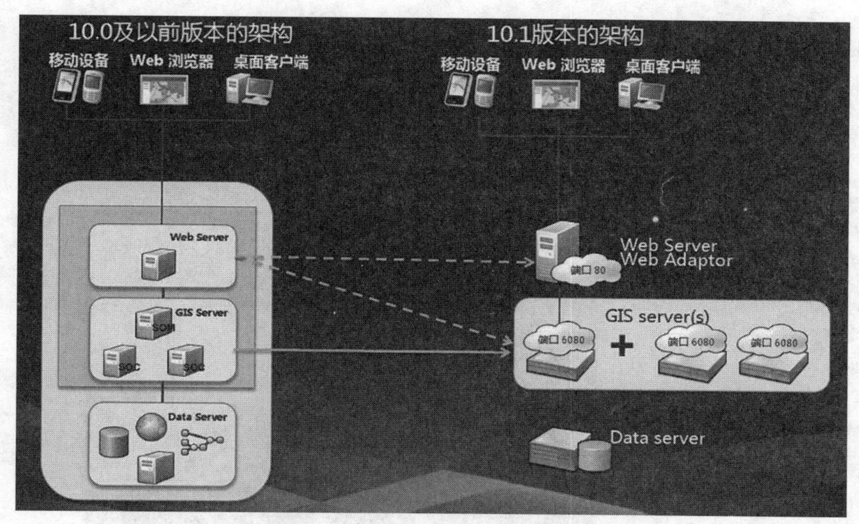

图 11.3　ArcGIS 10.1 for Server 节点结构

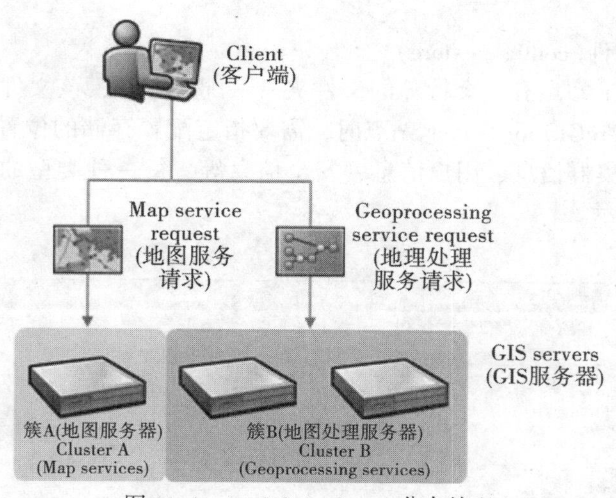

图 11.4　ArcGIS for Server 分布处理

11.1.2　ArcGIS for Server 逻辑构成

ArcGIS for Server 10.1 新架构模型的逻辑关系可简单概括为：以 Site 为架构单位、Cluster 为 GIS 服务的逻辑单位、GIS Server 为实际处理单位、GIS Instance 实例为每个 GIS 功能的处理容器。

1. Site 为架构单位

ArcGIS for Server 10.1 在安装完成以后，需要确定创建一个新的 Site 站点还是添加到已经存在的 Site 站点。如果创建一个新的 ArcGIS for Server 环境，就需要选择 New Site 操作，包括创建站点管理员账户、配置 Directories 和 Configuration Store 路径等信息。图 11.5 为在已经安装的 ArcGIS for Server 中启动 Manager 页面时的界面图，这时候还不能称之为一个站点，因为站点还没有创建。

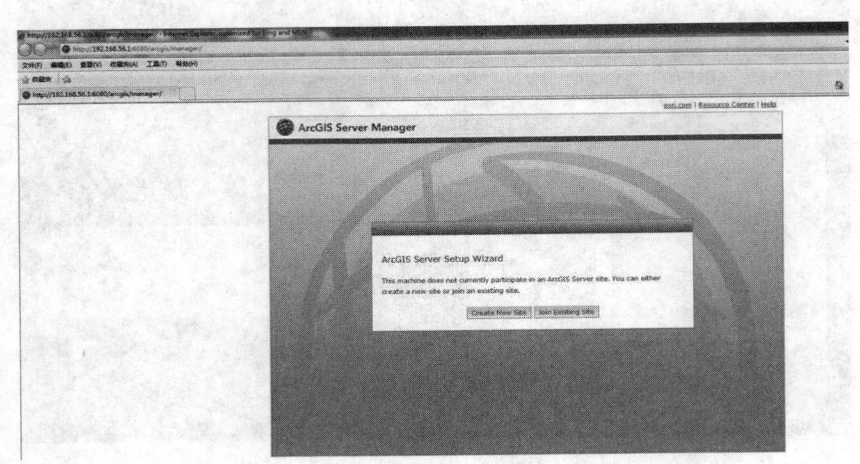

图 11.5 ArcGIS 10.1 for Server 管理界面

(1)配置存储文件(config - store)。

ArcGIS for Server 站点有一个特殊的文件夹——配置存储，该文件夹中包含了站点的重要信息。在创建 ArcGIS for Server 站点时，需要指定配置存储的位置，该目录包含整个 ArcGIS for Server 的集群信息、用户信息、安全信息等，每一种类型的信息都有相应的文件夹，如图 11.6 所示。

config-store ▸

名称	修改日期	类型	大小
.site	2012-7-23 19:30	文件夹	
clusters	2012-7-23 19:30	文件夹	
data	2012-7-23 19:34	文件夹	
events	2012-7-23 19:30	文件夹	
extns	2012-7-23 19:30	文件夹	
locks	2012-7-23 19:30	文件夹	
machines	2012-7-23 19:30	文件夹	
security	2012-7-23 19:30	文件夹	
serverdirs	2012-7-23 19:30	文件夹	
services	2012-7-23 19:30	文件夹	
system	2012-7-23 19:30	文件夹	
uploads	2012-7-23 19:30	文件夹	
web-adaptors	2012-7-23 19:30	文件夹	
arcgis-logsettings.json	2012-7-10 10:25	JSON 文件	1 KB
arcgis-logsettings.json.rlock	2012-7-23 11:34	RLOCK 文件	0 KB
version.json	2012-7-9 9:46	JSON 文件	1 KB

图 11.6 ArcGIS 10.1 for Server 存储结构图

(2)服务器目录(directories)。

一个服务器目录是计算机上的一个物理目录，ArcGIS for Server 站点将某些特定的信息写入和存储在该目录中，该目录包含有缓存文件夹、输出文件夹、系统文件夹、KML 文件夹等。图 11.7 为创建的 directories 目录的截图。

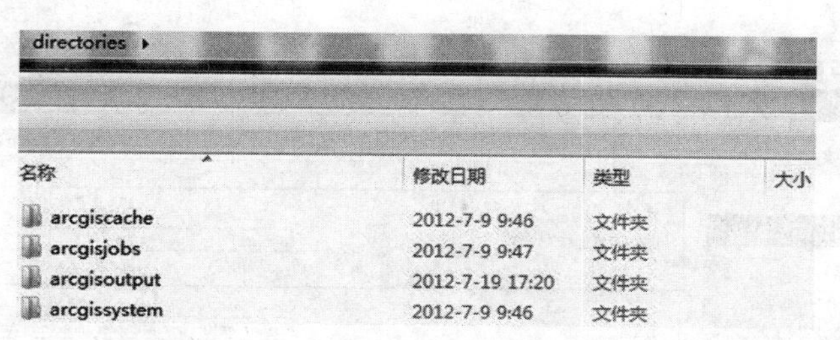

图 11.7 ArcGIS 10.1 for Server 服务器目录图

只有添加到Site站点的GIS Server，才能称为Siteful的GIS Server节点，要不就为孤立的节点，是不属于架构之内。每个Runnable的GIS Server所需的一系列数据，它们都被保存到Site的相关属性里。例如，所属的集群信息、服务信息、服务所依赖的数据信息、目录信息以及日志信息，等等。GIS Server也是基于这些信息才能提供具体服务的。一个具体的应用GIS环境只有一个Site站点。GIS服务器可以认为是构成ArcGIS Server站点中安装了ArcGIS Server的机器。GIS服务器可以用来绘制地图、提供服务、同步数据库、投影几何对象、搜索数据、并执行许多大量由ArcGIS提供的操作。所以，可以说GIS服务器是ArcGIS Server站点中的工作中心。在ArcGIS Server 10.1中GIS服务器内嵌了一个Web服务器，但是可以提供一个自己的专有Web服务器，以便提供更高级的功能，如使用本地Web服务器的安全功能等。

2. Cluster为GIS服务的逻辑单位

安装完成GIS Server节点，创建一个新的Site站点后，ArcGIS Server默认会产生一个名为“default”的默认集群。以后创建的Runnable GIS Server节点都可以添加到这个集群内，某一个Site站点可以创建多个集群。对于某个特定的Cluster，它是某个具体服务的逻辑容器，承载的具体服务如Map Service、GP Service，等等。例如，现在需要发布某区域的基础地形的地图服务，就需要选择是由哪个Cluster承载这个地图服务。到此为止，用户发布地图服务的过程就完成了。当然，具体的服务能力是由下面的GIS Server提供，但并不是一个Cluster只承载某一个服务或者某一类服务，每一个Cluster可以为不同类型、多个服务提供容器。

ArcGIS Server为Cluster内的GIS Server通信提供了完善的协同保障，如TCP轮询、UDP广播、心跳感应等。

ArcGIS Server的服务是位于一个集群中，而这个集群至少包含一台GIS服务器，创建站点默认的集群如图11.8所示。

3. GIS Server为实际处理单位

每一个安装ArcGIS Server的机器为一个GIS Server节点，这里的机器可以是物理机，也可以是虚拟机，当然，这样的每个机器内只能有一个GIS Server节点。上述的GIS Server节点，其实也是Siteless的节点。要转成为Runnable的GIS Server节点，首先需要添加到Site站点内，转为Siteful的GIS Server节点，然后添加到Cluster内，就成为Runnable的GIS Server节点。在每一个Cluster逻辑内可以存在多个GIS Server节点，这些

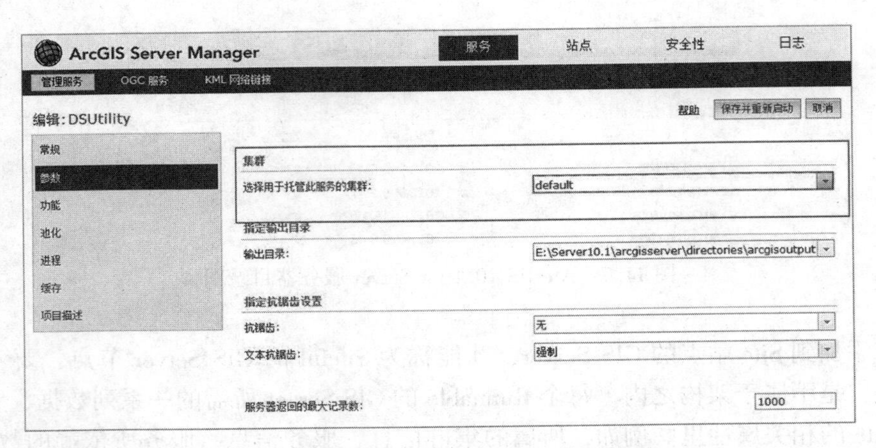

图 11.8　ArcGIS 10.1 for Server 集群管理

GIS Server 节点负载均衡上层的逻辑功能。ArcGIS Server 提供了多种负载均衡的算法，对于不同的请求情况，如密集 I/O 型、长事务型、高 CPU 型等，会自动配置到不同的负载算法。在新模式下，GIS Server 是全缓存模式的，这样性能将得到提升。

服务虽然位于集群中，但是处理服务的请求操作都是这个集群中的 GIS 服务器，一个集群至少包含一个 GIS 服务器，如图 11.9 所示。

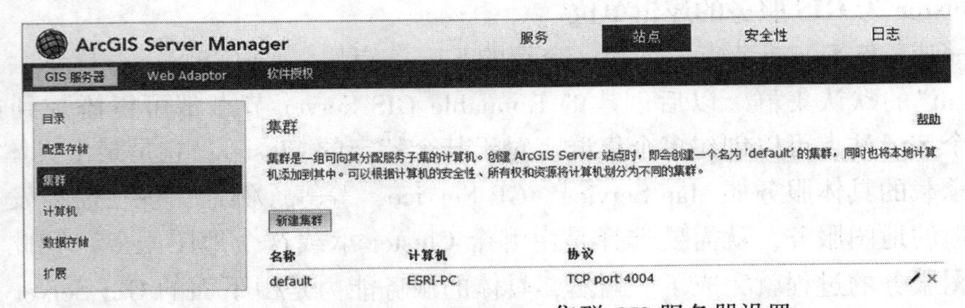

图 11.9　ArcGIS 10.9 for Server 集群 GIS 服务器设置

4. GIS Instance 实例为每个 GIS 功能的处理容器

GIS Instance 为 GIS Server 的处理实例。默认情况下，一个 GIS Server 节点自动设置最大实例数为 2 个。对于 ArcGIS Server for windows 版本，如果这个节点运行饱和下就是产生两个进程，这些就是处理具体功能的实例进程。对于某个负载较重的 GIS Server 节点，通过相关接口可以调整最大实例数，以满足处理量的需求。

11.2　ArcGIS API for Flex

11.2.1　Adobe Flash Builder

该软件可以帮助软件开发人员使用开放源 Flex 框架快速开发跨平台富 Internet 应用程

序(RIA)和内容。它包含对智能编码、调试及可视设计的支持，提供功能强大的测试工具，这些工具可以提高开发速度并创建出性能更高的应用程序。

Flex 是一个高效、免费的开放源框架，可用于构建具有表现力的 Web 应用程序，这些应用程序利用 Adobe Flash Player 和 Adobe AIR 运行时跨浏览器、桌面和操作系统实现一致的部署。2009 年 5 月 16 日，Adobe 公司宣布“下一版本的 Flex Builder 将被命名为 Flash Builder”。官方的解释是，“这样可以使 Flash 家族工具的命名具有更好的一致性，并借此将 Flash Builder 定位为开发工具”。

在 Flex Builder 中构建应用程序时，实际上需要用到开源的 Flex SDK 及 ActionScript，应用程序最终会被编译为运行在 Flash Player 中的 SWF 文件。图 11.10 展示了 Flash 家族主要工具与技术之间的关系：

①用 Flash Catalyst 生成 SWF 文件，并运行在 Flash Player 中。

②用 Flash Catalyst 创建的项目可以导入 Flash Builder 中。

③用 Flash CS4 IDE 和 AS3 生成 SWF 文件，并运行在 Flash Player 中。

④用 Flash Builder 和 Flex SDK 及 AS3 生成 SWF 文件，并运行在 Flash Player 中。

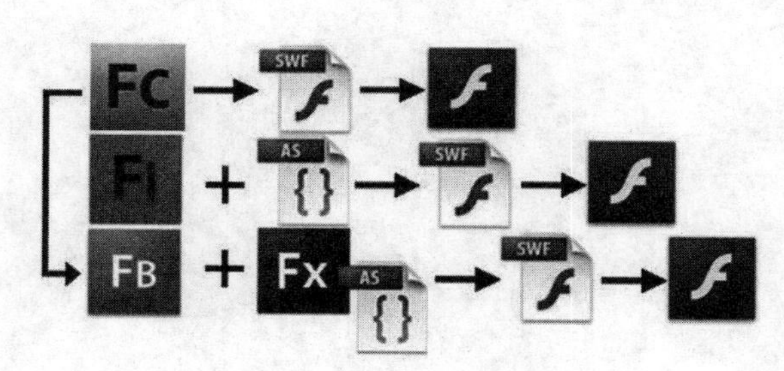

图 11.10　Flash 产品结构

11.2.2　ArcGIS API for Flex

ArcGIS API for Flex 是建立在 ArcGIS Server 之上的动态的富互联网应用(RIAs)。用户可以利用 ArcGIS 服务器资源创建交互式的表达 Web 应用程序(如地图、定位、特色服务和地理处理模型)和 Flex 组件(如网格、树、图)。

使用 ArcGIS API for Flex 可以实现如下效果：

①空间数据展示：加载地图服务、影像服务、WMS 等。

②客户端 Mashup：将来自不同服务器、不同类型的服务在客户端聚合后统一呈现给客户。

③图形绘制：在地图上交互式地绘制查询范围或地理标记等。

④符号渲染：提供对图形进行符号化、要素图层生成专题图和服务器端渲染等功能。

⑤查询检索：基于属性和空间位置进行查询，支持关联查询、对查询结果的排序、分组以及对属性数据的统计。

⑥地理处理：调用 ArcGIS for Server 发布的地理处理服务(GP 服务)，执行空间分析、地理处理或其他需要服务器端执行的工具、模型、运算等。

⑦网络分析：计算最优路径、临近设施和服务区域。

⑧在线编辑：通过要素服务编辑要素的图形、属性、附件，进行编辑追踪。

⑨时态感知：展示、查询具有时间特征的地图服务或影像服务数据。

⑩影像处理：提供动态镶嵌、实时栅格函数处理等功能。

⑪地图输出：提供多种地图图片导出和服务器端打印等功能。

第12章　环境搭建

12.1　Flash Builder

安装软件官方下载地址：

32bit：http://trials3. adobe. com/AdobeProducts/FLBR/4_7/win32/FlashBuilder_4_7_LS10. exe.

64bit：http://trials3. adobe. com/AdobeProducts/FLBR/4 _ 7/win64/FlashBuilder _ 4 _ 7 _ LS10_win64. exe.

需要先在此网址（http://www. adobe. com/downloads/）登录，才能下载成功。需要使用Keygen 下载地址：http://pan. baidu. com/share/link? shareid=2422372807 &uk=2769186556.

其破解过程为：

①拔掉网线断掉网络，清理 HOST 文件，确保没有诸如"127. 0. 0. 1 lmlicenses. wip4. adobe. com"、"127. 0. 0. 1 lm. licenses. adobe. com"，如果存在，则删除掉。

②使用 xf-mccs6. exe 生成序列号。

③将生成的序列号前 4 位替换为"1424"。

④使用此序列号安装 flash builder 4. 7（不要关掉 keygen），当出现 "Please connect to the internet and retry"（请链接网络并重试）的时候，点击 "connect later"（稍后链接）。

⑤启动 Flash Builder 4. 7。

⑥确认有 "connection problem" 网络链接问题，希望 "activate offline" 离线激活。

⑦这时会出现一个 request code（邀请码），使用该邀请码和之前生成的序列号再次生成 "activation code" 激活码，如图 12. 1 所示。

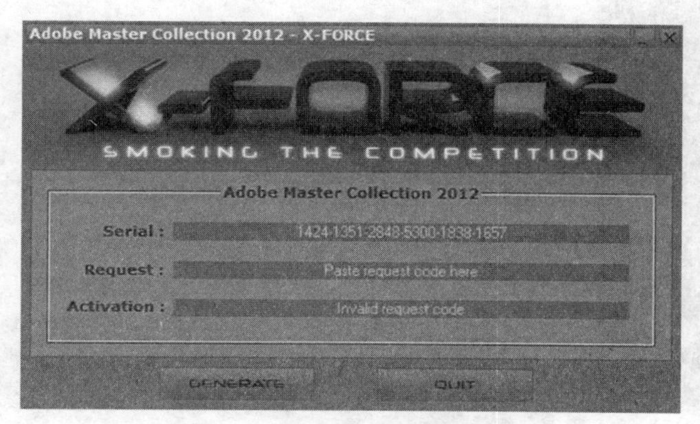

图 12. 1　Flash Builder 破解过程

⑧将其激活。

⑨安装结束后，双击运行【disable_activation. cmd】，完成。

此方法原理在于激活 Flash Builder 4. 7 而不是破解(靠修改文件，或改变版本号)，所以比破解更加稳定。

12. 2 ArcGIS for Server

12. 2. 1 ArcGIS Server 安装

受支持的操作系统如图 12. 2 所示。

操作系统	最低版本	最高版本
Red Hat Enterprise Linux Server 5 (64-bit)	Update 7 + libX11 patch*	
Red Hat Enterprise Linux Server 6 (64-bit)		
SUSE Linux Enterprise Server 11 (64-bit)	Update 1	
Windows Server 2003 Standard, Enterprise, and Datacenter (64-bit [EM64T])	SP2	SP2
Windows Server 2008 Standard, Enterprise, and Datacenter (64-bit [EM64T])	SP2	SP2
Windows Server 2008 R2 Standard, Enterprise, and Datacenter (64-bit [EM64T])	SP1	
Windows 7 Ultimate, Enterprise, Professional, Home Premium (64-bit [EM64T])	SP1	
Windows Vista Ultimate, Enterprise, Business, Home Premium (64-bit [EM64T])	SP2	SP2
Windows XP Professional Edition, Home Edition (64-bit [EM64T])	SP2	SP2

图 12. 2 操作系统支持

1. 安装

安装步骤如下：

①加载光盘运行后界面如图 12. 3 所示。

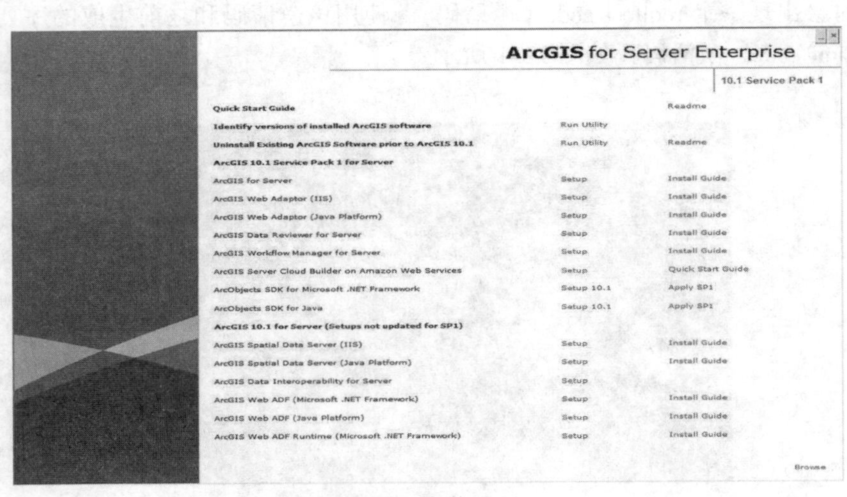

图 12. 3 安装过程 1

②选择安装 ArcGIS for Server，点击 ArcGIS for Server 后面的【Setup】。在弹出的窗口中(图 12.4)点击【Next】。

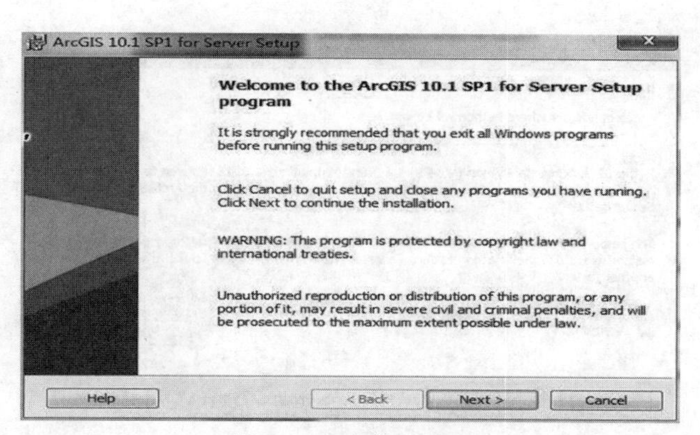

图 12.4　安装过程 2

③勾选【I accept the license agreement】(图 12.5)。点击【Change】可以修改安装路径，点击【Next】继续(图 12.6)。

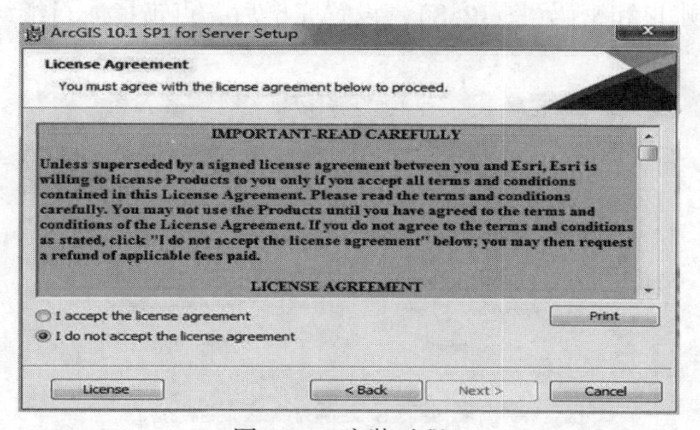

图 12.5　安装过程 3

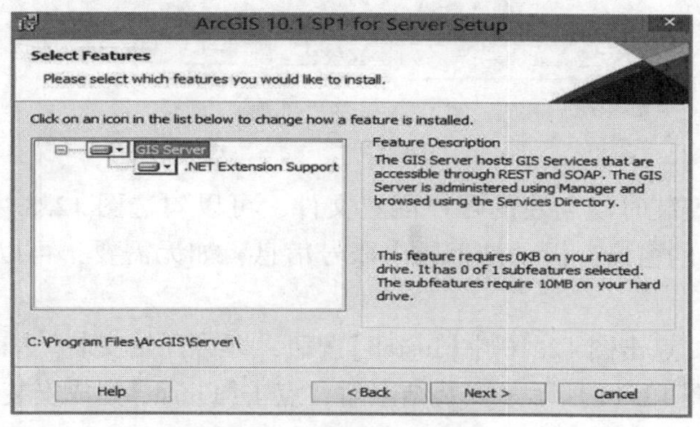

图 12.6　安装过程 4

④点击【Browse】，可以选择 Python 的安装目录(图 12.7)，选择好后，点击【Next】继续。

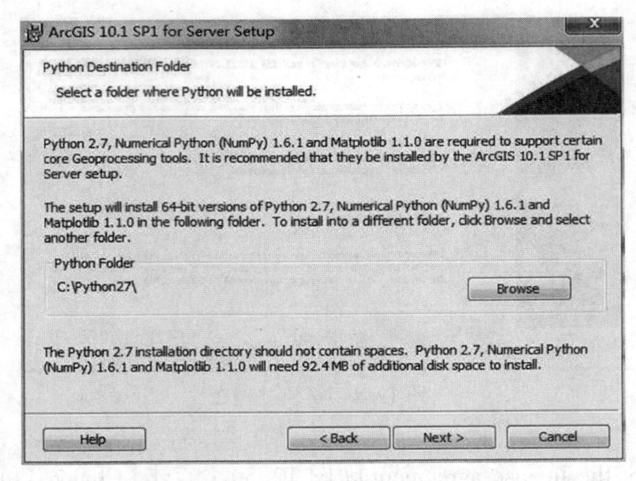

图 12.7　安装过程 5

⑤创建 Server 账户和密码(图 12.8)，设置完毕后，点击【Next】继续。

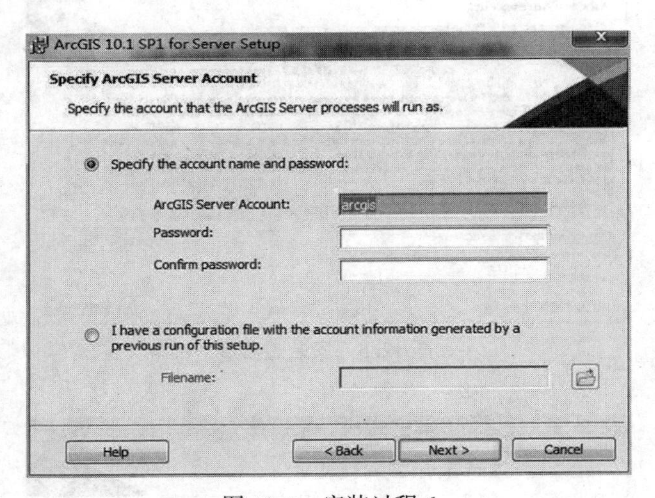

图 12.8　安装过程 6

⑥如果有之前安装时候创建的账户配置文件，可以勾选图 12.8 中下面的那个选项，弹出如图 12.9 所示的窗口，提示是否导出账号信息，如无需要，可以选择不导出，点击【Next】继续。

⑦设置完成后，点击图 12.10 中【Install】按钮，即可开始安装。

⑧安装过程如图 12.11 所示。安装结束后，点击【Finish】完成安装。

图 12.9　安装过程 7

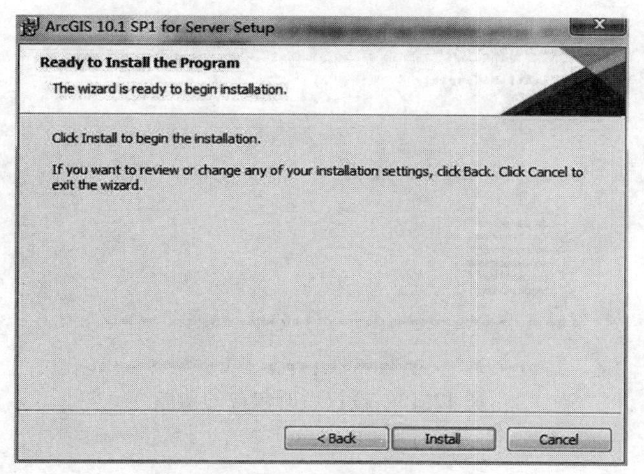

图 12.10　安装过程 8

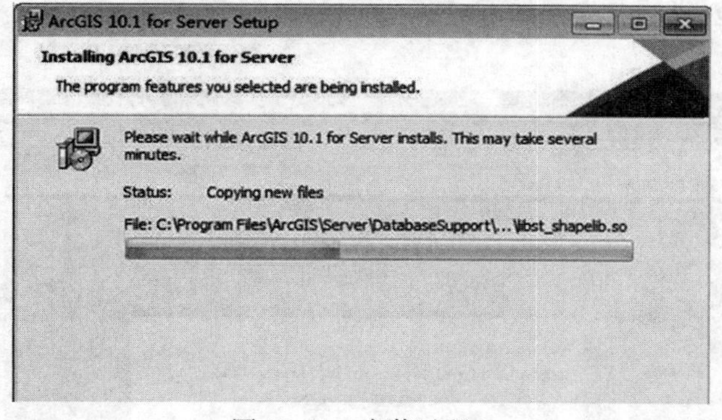

图 12.11　安装过程 9

2. 授权

选择相应的授权方式进行授权(图 12.12)，授权完成后，点击【完成】结束授权，不同的授权包含不同的功能权限，如图 12.13 所示。

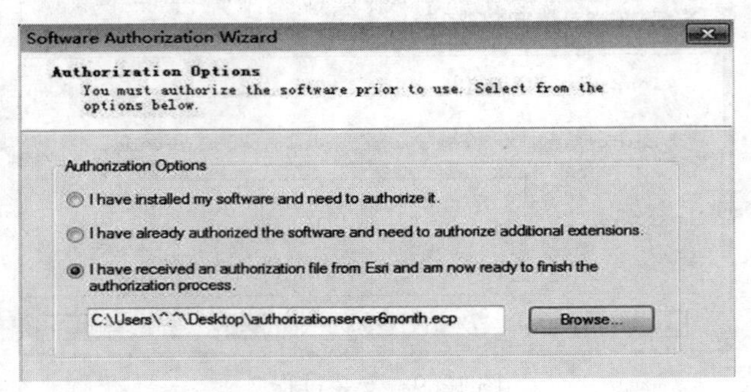

图 12.12　软件授权

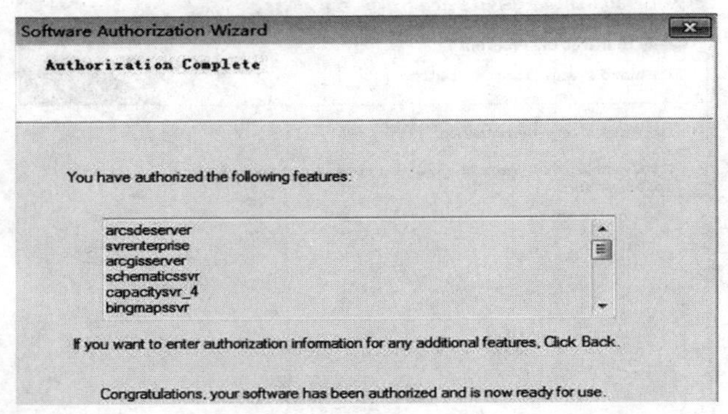

图 12.13　授权后所包含的功能

3. 创建站点

安装完成后，会自动弹出 Manager 页面，可以选择创建站点，也可以加入现有站点，如图 12.14 所示。创建站点步骤如下：

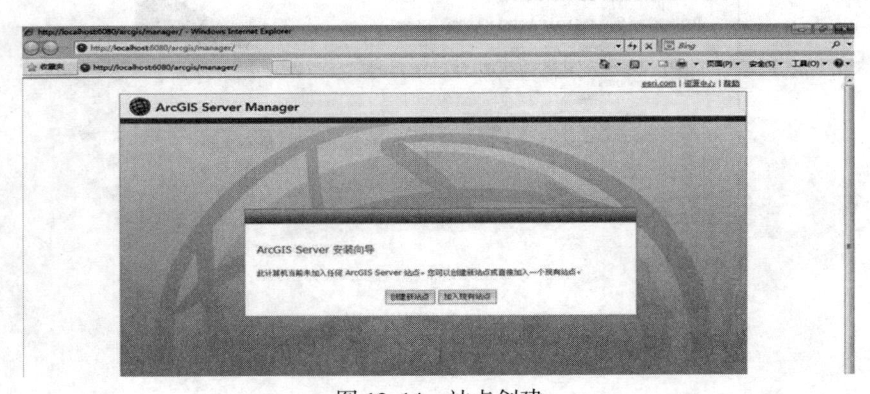

图 12.14　站点创建

①输入主站点管理员账户，点击“下一步”(图 12.15)。这里需要注意，这个账户和刚刚安装 ArcGIS Server 步骤中创建的账户的关系。上一节中创建的 ArcGIS Server 账户，是为 ArcGIS Server 创建的操作系统账户(安装完成后可以切换操作系统账户看一下，是不是多了一个 ArcGIS Server)，该账户用来管理 ArcGIS Server 的操作系统进程(启动、停止 ArcGIS Server 服务)；而此时创建的站点账户，主要是为了管理站点中的各种服务形式的 GIS 资源(各种类型服务的发布、启动、停止、删除；站点安全；集群管理，等等)。所以，这两个账户的设置完全没有关系。

图 12.15　设置账户密码

②指定相关目录，然后点击【下一步】，如图 12.16 所示。

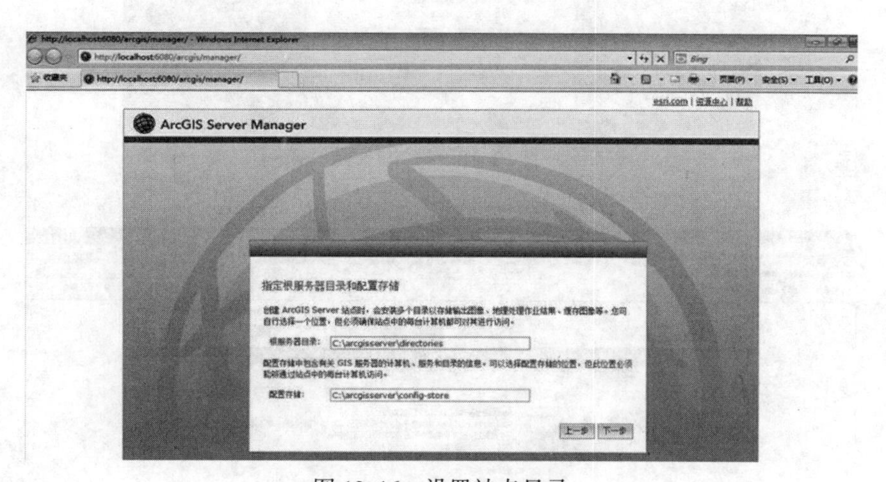

图 12.16　设置站点目录

③点击【完成】，完成创建，如图 12.17 所示。

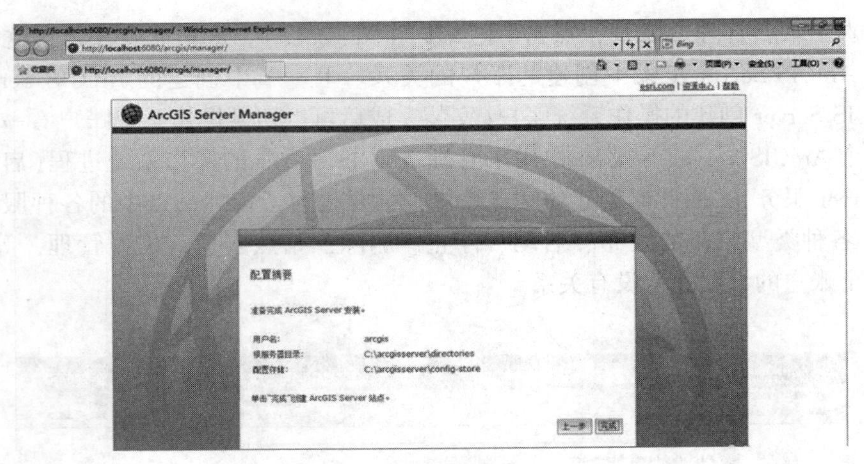

图 12.17　完成创建

④输入之前的站点管理员账户(图 12.18)，点击【登录】，便可登录所创建的站点，如图 12.19 所示。

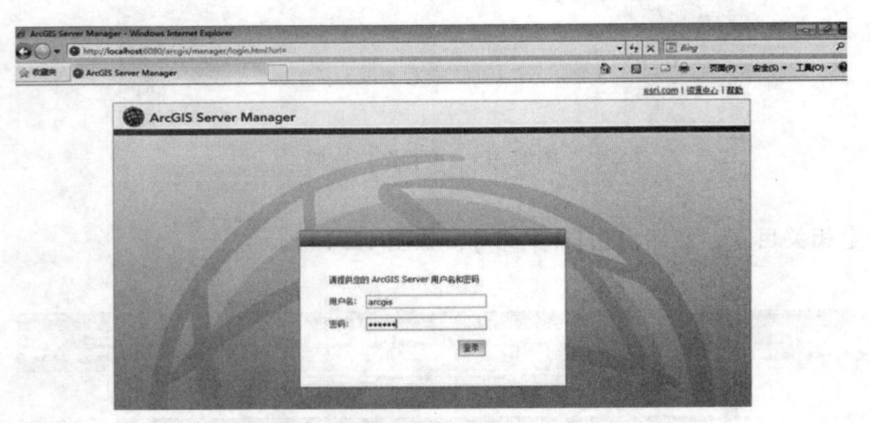

图 12.18　账户登录

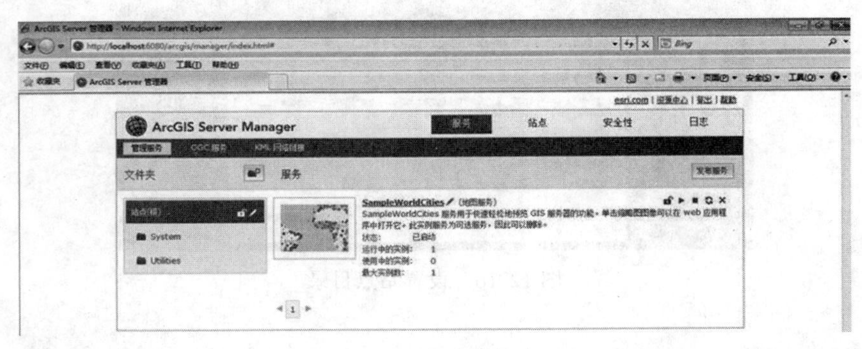

图 12.19　登录站点

至此，已经成功安装了 ArcGIS 10.1 for Server。

12.2.2　地图发布

要将地图作为服务发布，需执行以下步骤：

①在 ArcMap 中打开地图文档，从主菜单中选择【文件】≫【共享为】≫ 服务。

②在“共享为”服务窗口中，选择【发布服务】，单击【下一步】。

③在“发布服务”对话框中，单击【连接到 ArcGIS Server】，以创建到服务器的新连接。

④在添加 ArcGIS Server 窗口中，选择【发布 GIS 服务】，单击【下一步】。

⑤对于服务器 URL，输入要连接的 ArcGIS Server 站点的 URL。例如，http：//myserver：6080/arcgis。

⑥在服务器类型下拉列表中选择【 ArcGIS Server】。

⑦发布过程中，将创建服务定义文件并将其临时存储到本地磁盘上；发布过程完成后，服务定义将上传到服务器并删除本地文件。考虑到本教程的目的，接受默认的过渡文件夹并继续。

⑧如果服务器管理员已为站点启用了安全功能，输入用户名和密码，单击【完成】。

⑨还可以在发布服务窗口中，输入新的服务名称，单击【下一步】。名称长度不能超过 120 个字符，并且只能包含字母数字字符和下画线。

⑩默认情况下，服务会发布到 ArcGIS Server 的根文件夹下；也可将服务组织到根文件夹下的子文件夹中。选择要将服务发布到其中的目标文件夹或创建一个新的文件夹来存储服务，单击【继续】。

⑪使用【服务编辑器】选择用户可对地图服务执行的操作，可以对服务器显示服务的方式进行精细的控制。单击【服务能力】选项卡，默认情况下，自动启用了地图和 KML，单击【地图】可以查看以下属性：

a. URL 是客户端用来访问地图服务的统一资源定位器，其格式为：http：//<服务器名称>：6080/arcgis/rest/services/<文件夹名称(如果适用)>/<服务名称>/MapServer。

b. 数据选项允许客户端应用程序对地图服务中的要素执行属性搜索。

c. 地图选项允许客户端应用程序查看地图服务中的地图图层。

d. 查询选项允许客户端应用程序查询地图服务中的要素。

⑫点击【分析】，可用于对地图文档进行检查，看其是否能够发布到服务器。配置地图服务时为获得更多视图区域，可以单击服务编辑器顶部的【折叠】按钮。

⑬在准备窗口，对地图文档中的错误进行修复，也可以修复警告和通知消息，以进一步完善地图服务的性能和外观。

⑭在服务编辑器中，单击【预览】，以了解在 Web 上查看地图时地图的外观。

⑮单击【发布 】，将地图服务发布到 ArcGIS Server 上，便可以在 Web 应用程序中使用地图服务。

注意：可将文件夹和地理数据库注册到 ArcGIS Server 站点，从而确保服务器可识别并使用数据。如果继续以下步骤，那么地图文档中所引用的来自取消注册的文件夹或地理数据库的任何数据都将在发布时复制到服务器。这是一种预防性措施，可确保服务器能够访问服务所使用的所有数据。

12.2.3　使用服务

可以使用 ArcGIS. com 地图查看器通过 Web 查看地图服务并与其进行交互。该地图查看器托管在 ArcGIS. com 中，并可通过 ArcGIS Server 服务目录直接部署。启动和运行地图查看器无需进行任何配置或编程操作。执行以下步骤即可：

①打开 Web 浏览器并导航至 ArcGIS Server 服务目录。通常，此目录位于 http：//<服务器名称>：6080/arcgis/rest/services。还可以通过打开随 ArcGIS Server 一起安装的快捷方式来访问“服务目录”。

②在服务列表中，单击地图服务的名称。通常，地图服务的名称显示为<地图服务名称>(MapServer)。如果看不到地图服务，它可能位于服务目录中列出的一个文件夹中。

③在描述地图服务的页面中，单击在下列程序中查看地图：ArcGIS. com 地图。此时会打开一个显示 ArcGIS. com 地图查看器的新窗口(或选项卡)。将地图查看器显示画面缩放至地图服务的范围。

④围绕地图服务进行平移和缩放，也可以切换底图图层以更准确地补充地图的上下文，也可以选择保存 Web 地图。

12.3　ArcGIS API for Flex

12.3.1　环境配置

Flash Builder 的环境配置如下：

①打开 Flex Builder，创建一个 Flex 项目，如图 12.20 所示。

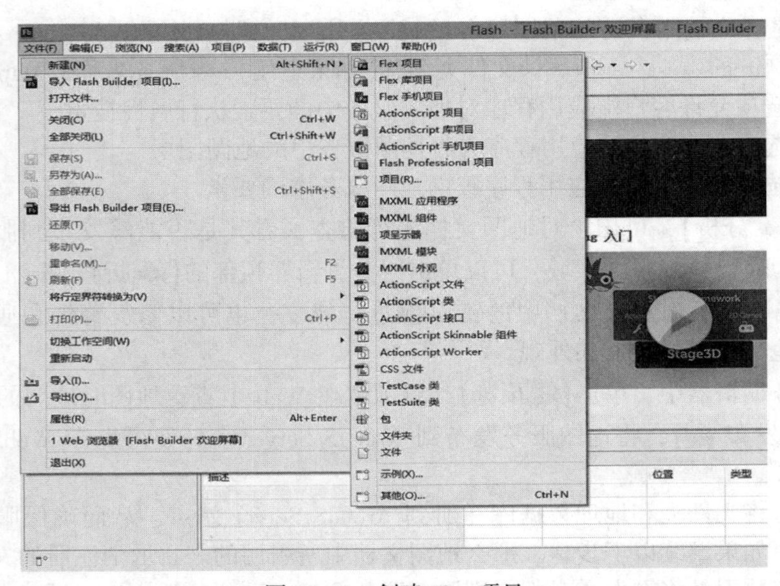

图 12.20　创建 Flex 项目

②在图 12.21 中进行项目名称和位置等设置，点击【下一步】进入服务器设置，弹出如图 12.22 所示的窗口。

图 12.21　设置项目信息

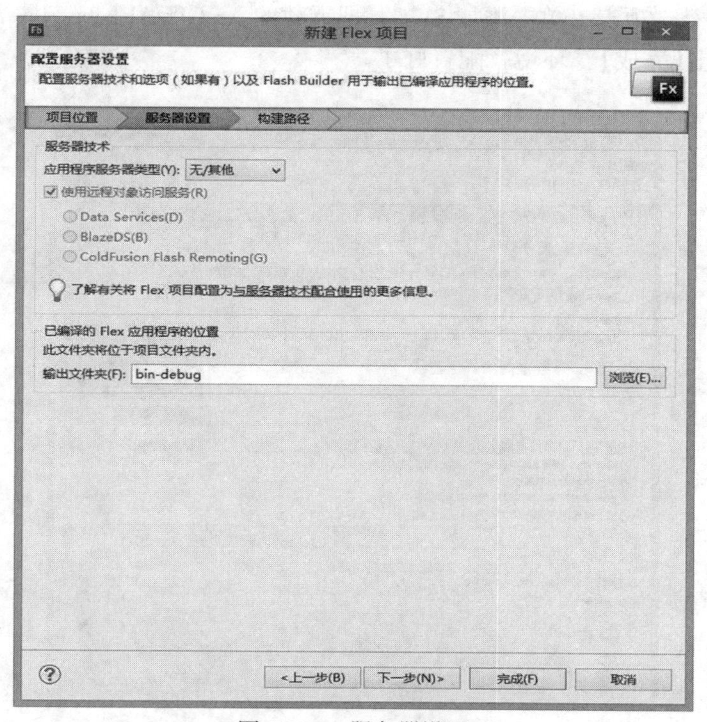

图 12.22　服务器设置

③进行服务器相关设置或默认设置，点击【下一步】，进入构建路径窗口，如图 12.23 所示。

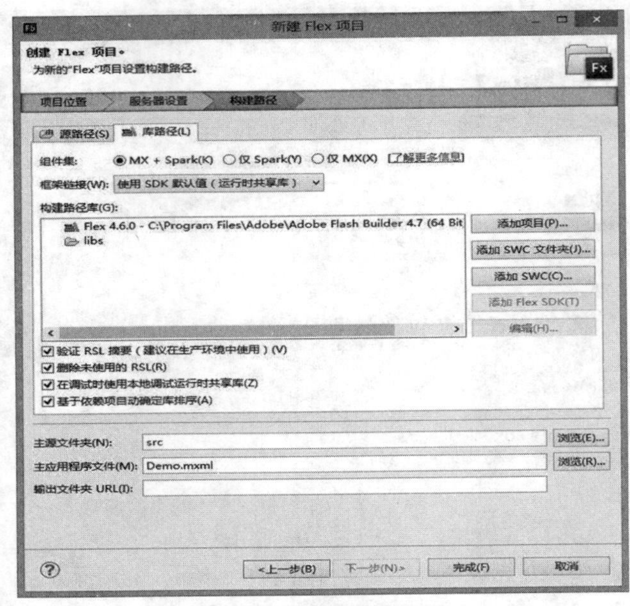

图 12.23　构建路径设置

④选择库路径选项卡，单击添加 SWC，把下载的 ArcGIS API for Flex 添加进去，如图 12.24 所示，点击【完成】，环境就配置好了。

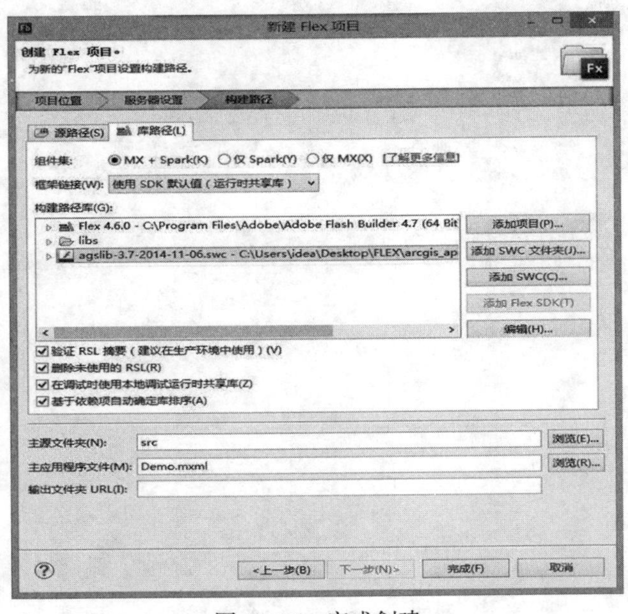

图 12.24　完成创建

12.3.2　环境测试

在 Demo. mxml 文件中输入以下代码进行环境部署测试：

```
<? xml version="1.0" encoding="utf-8"? >
<s:Application xmlns:fx="http://ns.adobe.com/mxml/2009"
               xmlns:s="library://ns.adobe.com/flex/spark"
xmlns:esri="http://www.esri.com/2008/ags"
               pageTitle="World Topographic Map">
<fx:Declarations>
</fx:Declarations>
<esri:Map >
<esri:ArcGISTiledMapServiceLayer
url="http://server.arcgisonline.com/ArcGIS/rest/services/World_
Topo_Map/MapServer"/>
<! --可以替换为本地发布的地图,如:<esri:ArcGISDynamicMapServiceLayer
url = " http://localhost: 6080/arcgis/rest/services/qingdao/MapS-
erver"/>-->
</esri:Map>
</s:Application>
```

操作界面如图 12.25 所示。

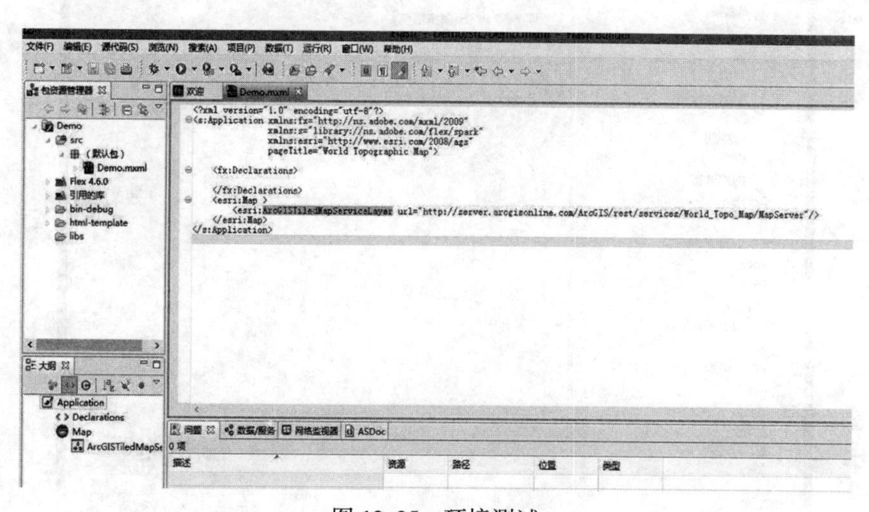

图 12.25　环境测试

点击【调试】或【运行】按钮运行程序，如果得到如图 12.26 所示的结果，即表示 ArcGIS API For Flex 环境部署成功。

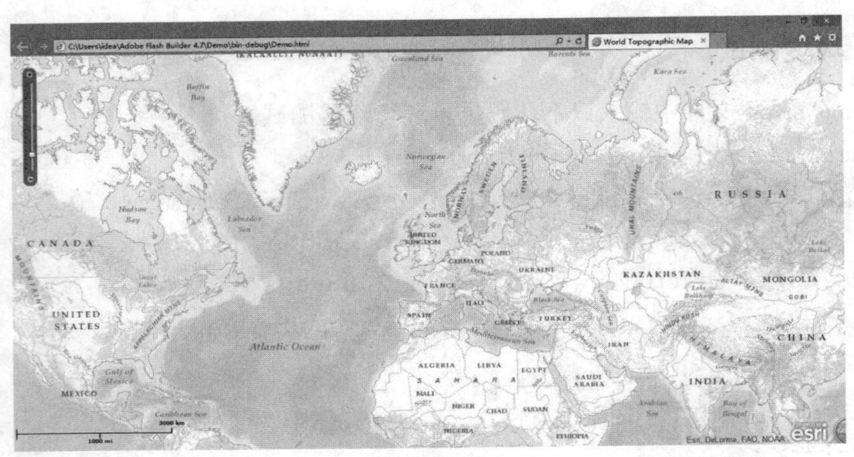

图 12.26　测试结果

注意：第一次使用 Flex 在运行程序时，若默认浏览器为非 IE 内核的浏览器，则 Flex 会提示安装最新的 Adobe Flash Player，此时可根据提示一步步安装更新。若是想启用 IE 为默认程序运行的浏览器，可在【菜单】≫【窗口】≫【首选项】中设置，如图 12.27 所示。

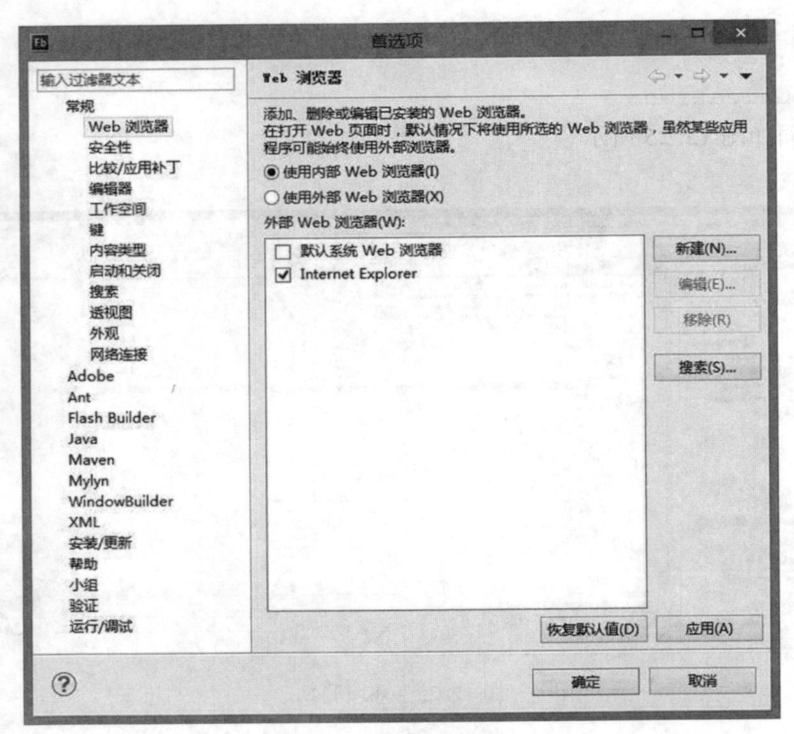

图 12.27　设置默认浏览器

第13章 应用接口

13.1 接口概述

表13.1展示了在ArcGIS API for Flex的所有包的逻辑分组。

表13.1　　**ArcGIS API for Flex 逻辑分组**

程序集	描　述
com. esri. ags	本程序包包含了主要的地图和其他的一些常用类：FeatureSet、Graphic、SpatialReference、TimeExtent and Units
com. esri. ags. clusterers	主类集群[2.0版本以后添加]
com. esri. ags. clusterers. supportClasses	聚类器中的集群符号和其他配套类[2.0版本以后添加]
com. esri. ags. components	这个软件包包含了属性检查器(AttributeInspector)，编辑器(Editor)，信息窗口(InfoWindow)，时间滑(TimeSlider)等
com. esri. ags. components. geoEnrichmentClasses	GeoEnrichment 组件的支持类[3.5版本以后添加]
com. esri. ags. components. supportClasses	不同组件的支持类[2.0版本以后添加]
com. esri. ags. events	地图导航、加载图层、地理处理任务和工具栏的时间类
com. esri. ags. geometry	地图点(MapPoint)、多点(Multipoint)、折线(Polyline)和多边形(Polygon)的主要几何类
com. esri. ags. layers	地图的图层功能接口，包含ArcGIS (dynamic、tiled or image)，ArcIMS，Graphics and Feature layers

续表

程序集	描　述
com. esri. ags. layers. supportClasses	ArcGIS 图层相关的支持类[2.0 版本以后添加]
com. esri. ags. portal	ArcGIS 门户网站的相关接口类[3.0 版本以后添加]
com. esri. ags. portal. supportClasses	支持 ArcGIS 门户网站工作的相关接口类[3.0 版本以后添加]
com. esri. ags. renderers	渲染器可以根据图形的属性来渲染图形，多用于专题图[1.2 版本以后添加]
com. esri. ags. renderers. supportClasses	渲染器的支持类 [2.0 版本以后添加]
com. esri. ags. skins	皮肤包，包含各种皮肤类[2.0 版本以后添加]
com. esri. ags. symbols	符号用于表示在地图上的“几何形状”；标记符号可用于点和多点的几何形状；线符号可用于折线几何图形；填充符号可用于多边形的几何形状
com. esri. ags. tasks	ArcGIS Server 任务的主要类，包含：最近设施、发现、几何、地理处理器、识别、定位、打印、查询、路线、服务区、与地理富集服务
com. esri. ags. tasks. geoEnrichmentClasses	支持 GeoEnrichment 的服务和相关任务[3.5 版本以后添加]
com. esri. ags. tasks. supportClasses	ArcGIS Server 任务的所有支持类[2.0 版本以后添加]
com. esri. ags. tools	这个软件包包含非 UI 工具类，用它们来创建自己的工具栏
com. esri. ags. utils	ArcGIS API for Flex 的实用工具类，如获得几何图形的外边框范文或者转换地理坐标与墨卡托投影数据[1.2 版本以后添加]
com. esri. ags. virtualearth	有关 Bing 地图和 Bing 地图地理编码(原微软虚拟地球)[1.2 版本以后添加]

13.2　接口图解

图 13.1(a)和图 13.1(b)、图 13.1(c)给出了 ArcGIS API for Flex 的所有类和包的逻辑分组。

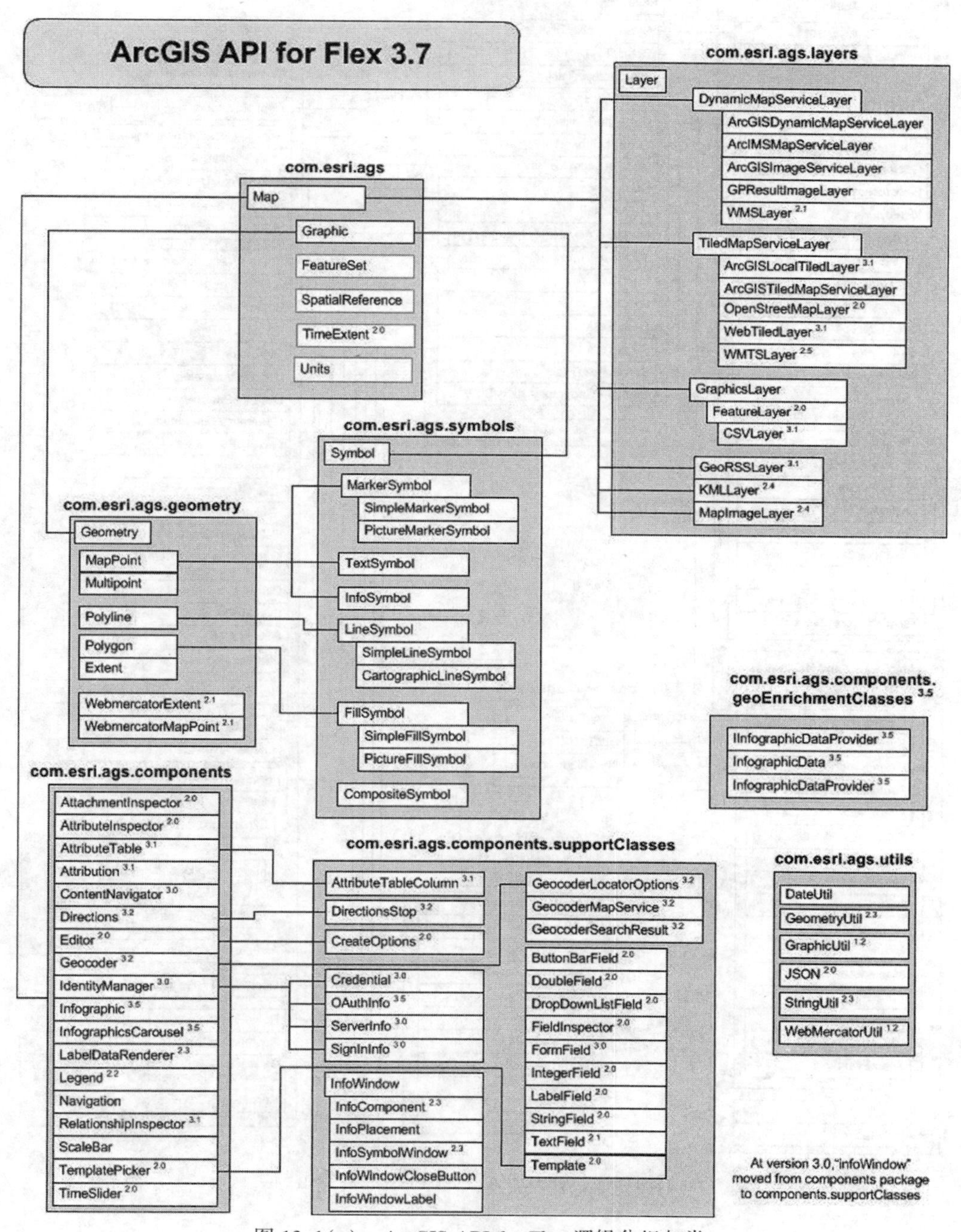

图 13.1(a)　ArcGIS API for Flex 逻辑分组与类

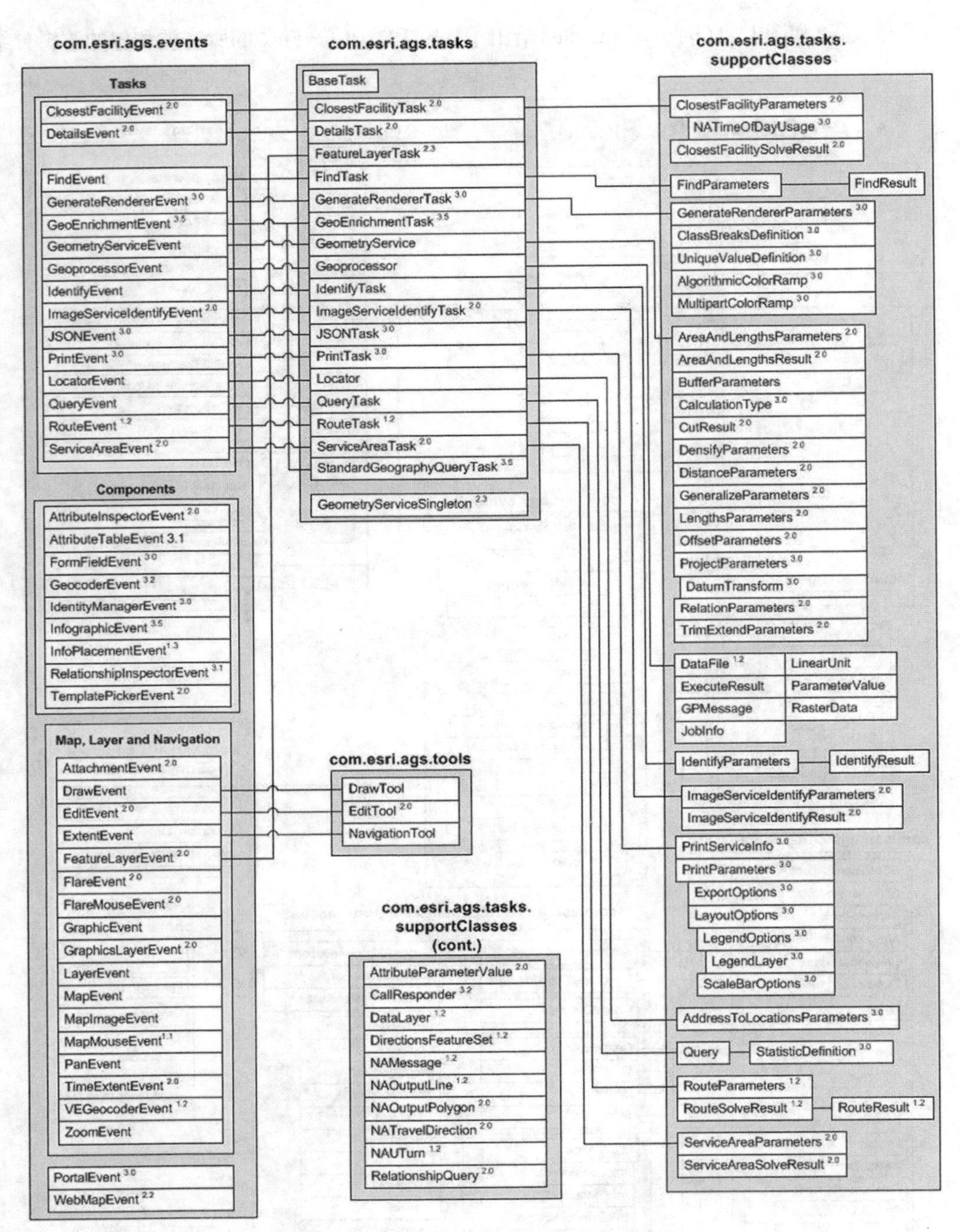

图 13.1(b)　ArcGIS API for Flex 逻辑分组与类

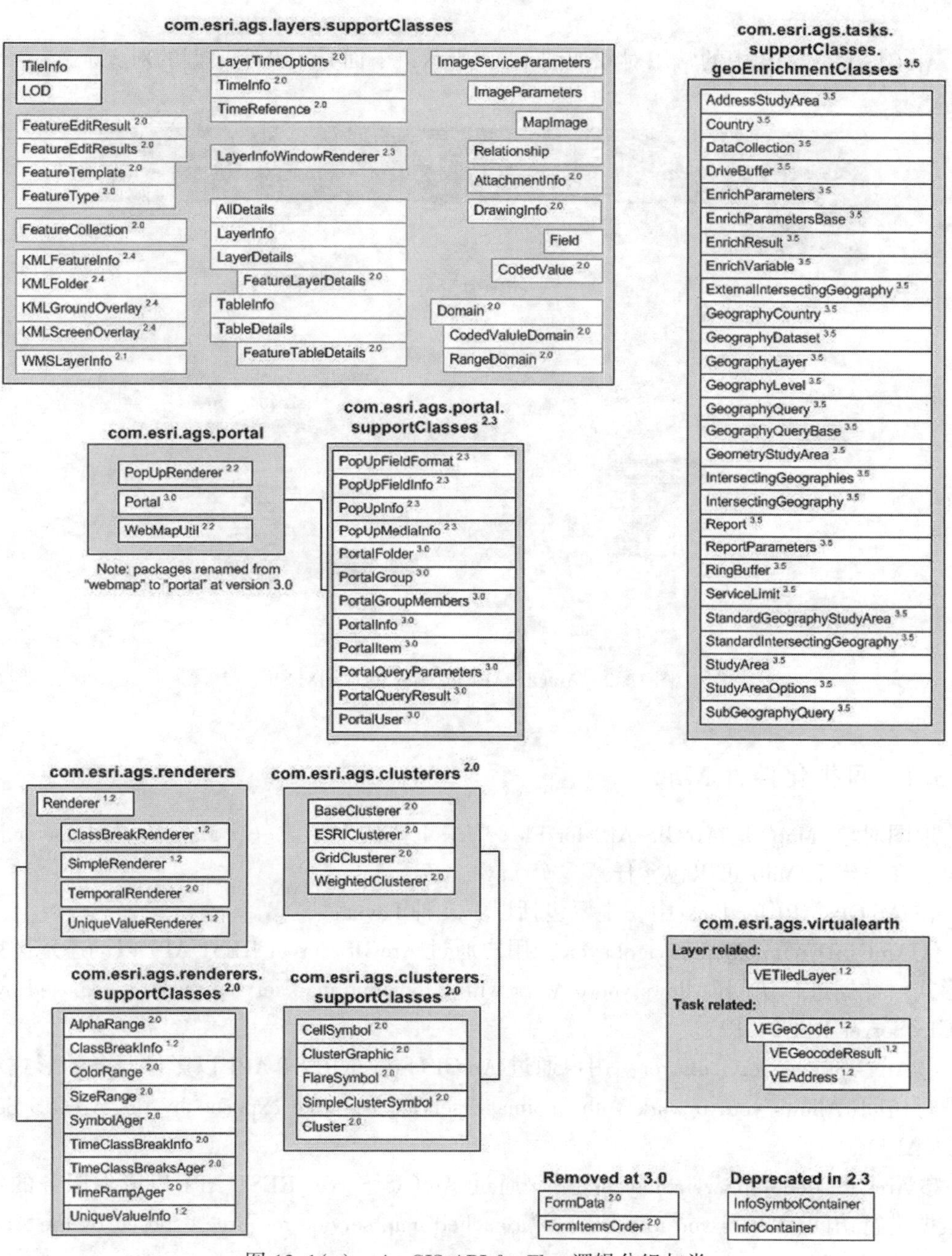

图 13.1(c)　ArcGIS API for Flex 逻辑分组与类

13.3 常用对象

ArcGIS Flex API 中的常用对象如图 13.2 所示，下面介绍几个典型的对象。

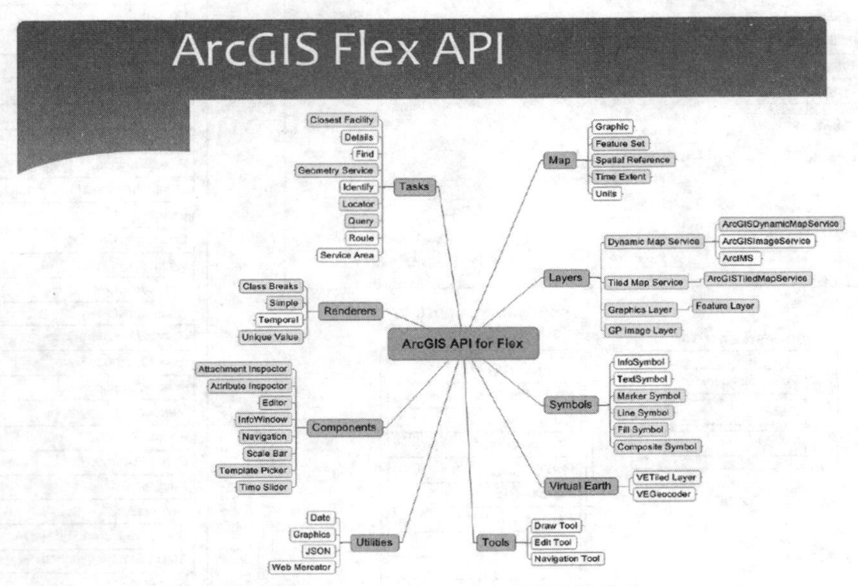

图 13.2 ArcGIS API for Flex 的常用对象

13.3.1 可视化控件 Map

地图对象(Map)是 ArcGIS API for Flex 中最重要的对象之一，Map 负责对 Layer 进行组织，在逻辑上 Map 可以包含任意多个 Layer。

在 ArcGIS API for Flex 中，主要包括以下几种 Layer：

①ArcGISDynamicMapServiceLayer：用户通过 ArcGIS Server REST API 以动态地图服务的形式对数据进行访问(Allows you to work with a dynamic map service resource exposed by the ArcGIS Server REST API)。

②ArcGISImageServiceLayer：用户通过 ArcGIS Server REST API 以影像服务的形式对数据进行访问(Allows you to work with an image service resource exposed by the ArcGIS Server REST API)。

③ArcGISTiledMapServiceLayer：用户通过 ArcGIS Server REST API 以影像服务的形式对数据进行访问(Allows you to work with a cached map service resource exposed by the ArcGIS Server REST API)。

④ArcIMSMapServiceLayer：用户可以访问 ArcIMS image service 提供的数据(Allows you to work with an ArcIMS image service)。

⑤GraphicsLayer：用于显示用户在客户端绘制的图形要素的图层，该图层可以包含一

个或多个图形要素(a layer that contains one or more graphic features)。

13.3.2　图形对象 Graphics

使用图形对象(Graphics)，当 Flex 的 ArcGIS API 第一次发布时，显示信息的唯一选择客户端是通过使用一个图形层。图形层允许在地图上动态显示图形。图形可以由用户标记或输入一个任务，或者他们可以由应用程序来响应一个任务。例如，应用程序可能会将查询的结果添加到地图图形。

一个图形由以下要素组成：

①Geometry(几何)：几何确定图形的位置，可以是点、多点、多段线、多边形或范围(extent)。

②Symbol(符号)：符号决定了图形的外观，可以是一个标志符号(和多点几何)、线符号(折线)、填充符号(多边形)。

③Attributes(属性)：属性是描述图形的名称——值对。如果是新创建的图形，需要分配的属性；如果图形是响应创建一个图层执行的任务，则自动包含图层的字段属性。Query.outFields 是限制任务返回的属性的一个例子。

注意：图形(Graphics)对象不需要所有上面列出的三个项目。例如，许多任务结果作为 FeatureSet 图形对象返回，这些图形具有 Geometry 和 Attributes。但如果想向地图添加图形，则必须定义一个 Symbol。

DrawTool 是一个类，它可以获取用户在屏幕上绘制的几何，然后定义一个符号和应用于一个新的图形对象的几何形状。当添加一个 task 任务时，根据 task 的类型到地图上作为图形方式，如果 task 返回一个 feature 集合，用户会得到一个数组的图形(FeatureSet.features)，可以指定 graphicProvider GraphicsLayer。

代码示例：定义一个符号的图形层，分配给所有图形并添加到这一层：

```
private function doQuery():void
{
queryTask.execute(query,new AsyncResponder(onResult, onFault));
function onResult(featureSet:FeatureSet, token:Object = null):
void
{
    myGraphicsLayer.graphicProvider = featureSet.features;
}
function onFault(info:Object,token:Object =null):void
{
    Alert.show( info.toString());
}
}
<esri:SimpleMarkerSymbol id = "sms" size = "10" color = "0xFF0000"
style="circle"/>
```

```
<esri:GraphicsLayer id="myGraphicsLayer" symbol="{sms}"/>
```

其他任务没有提供 Feature 结果，但提供 Geometry(几何)，可以使用它来创建一个图形。例如，查询一个位置定位，其属性是一个 MapPoint 几何位置。可以使用此几何创建一个图形显示的地址位置，实现代码如下：

```
for each (var candidate:AddressCandidate in candidates)
{
if(candidate.score >80){
        var myGraphic:Graphic =new Graphic();
        myGraphic.geometry = candidate.location;
        myGraphic.symbol = mySymbol;
        myGraphic.attributes ={ address: candidate.address,score:
candidate.score, locatorName: candidate.attributes.Loc_name };
        myGraphicsLayer.add(myGraphic);
}
}
```

13.3.3 图形样式 Symbol

Symbol 是用来在地图上重新显示地理图元渲染方式的类(Symbols are used to represent "geometries" on the map)，在实际应用中非常重要。使用美观的样式，无疑可以为所研发的系统增添很多亮点，用户看着也会觉得舒服。详细介绍见第 14 章第 2 小节。

13.3.4 查询分析 QueryTask

QueryTask 是一个进行空间和属性查询的功能类，它可以在某个地图服务的某个子图层内进行查询。QueryTask 进行查询的地图服务并不必须加载到 Map 中进行显示。QueryTask 的执行需要两个先决条件：一是需要查询的图层 URL，二是进行查询的过滤条件。

FindTask 是在某个地图服务中进行属性查询的功能类，可查询单图层和跨图层数据，输入查询的图层 Layers、查询的字段名 SearchFields、查询的条件 Search Text，可以查询非唯一的数据。FindTask 与 QueryTask 的使用类似，QueryTask 在执行时需要给出 Query 对象作为参数，FindTask 则是给出 FindParameters 对象作为参数；FindTask 的 URL 属性需要指向所查询的地图服务的 REST URL，而 QueryTask 需要指定子图层的 URL。

IdentifyTask 是一个在地图服务中识别要素的功能类。当用户在客户端使用 Draw 工具绘制一个几何对象后，就可以作为 IdentifyTask 的参数发送到服务器进行识别，满足条件的要素将会被输出到 ArcGIS Flex API 中，这些要素也可以作为 Graphic 被添加到地图上。

第14章　地图功能开发

14.1　地图控件

14.1.1　Map 控件属性

1. Map 控件常用属性

Map 控件的常用属性见表 14.1。

表 14.1　**Map 控件的常用属性**

属性名称	数据类型	说　明
extent	Extent	地图的当前视图对应的地理坐标范围
infoWindow	InfoWindow	气泡窗口(只读属性)
infoWindowContent	UIComponent	气泡窗口中的内容
layers	Object	地图中包含的图层数组
loaded	Boolean	标识地图是否已经加载完成。当地图中包含多个图层时，第一个图层加载完成后，该属性即变成 true
lods	Array	地图的缩放级别
panEasingFactor	Number	平移地图时惯性大小，取值区间 0~1，1 表示完全没有惯性
scale	Number	地图的当前比例尺
spatialReference	SpatialReference	地图的坐标系，只读属性
staticLayer	Group	静态图层，用于添加图例、logo 等元素，可以固定在地图控件的指定位置，不随地图缩放、平移而发生变化
units	String	地图单位

2. Map 控件导航功能开关属性

Map 控件的导航功能开关属性见表 14.2。

表 14.2 **Map 控件的导航功能开关属性**

属性名称	数据类型	说　明
clickRecenterEnabled	Boolean	Shift+点击中心定位功能开关
doubleClickZoomEnabled	Boolean	鼠标双击放大功能开关
keyboardNavigationEnabled	Boolean	键盘导航开关
rubberbandZoomEnabled	Boolean	Shift+左键拉框放大开关
panEnabled	Boolean	鼠标漫游开关
scrollWheelZoomEnabled	Boolean	鼠标滚轮缩放开关
mapNavigationEnabled	Boolean	地图导航开关。设置为 false，相当于锁定地图，鼠标、键盘都无法导航
keyboardNavigationEnabled	Boolean	键盘导航开关
openHandCursorVisible	Boolean	地图上鼠标的样式是否为张开的小手，为 false 时就是普通箭头

3. Map 控件辅助性静态对象显示属性

Map 控件辅助性静态对象显示属性见表 14.3。

表 14.3 **Map 控件辅助性的静态对象显示属性**

属性名称	数据类型	说　明
crosshairVisible	Boolean	控制地图中心的十字符号是否显示
scaleBarVisible	Boolean	控制是否有比例尺条
panArrowsVisible	Boolean	控制控件边缘上的 8 个漫游按钮是否显示
zoomSliderVisible	Boolean	控制是否显示地图缩放滑动条
logoVisible	Boolean	控制是否显示 ESRI 公司的 logo

14.1.2 Map 控件方法

1. Map 地图控件的导航方法

Map 地图控件的导航方法见表 14.4。

表 14.4　　**Map 地图控件的导航方法**

方法签名	说　明
centerAt(mapPoint)：void	把地图中心定位到指定点
panDown( )：void	向下平移地图
panLeft( )：void	向左平移地图
panLowerLeft( )：void	向左下平移地图
panLowerRight( )：void	向右下平移地图
panRight( )：void	向右平移地图
panUp( )：void	向上平移地图
panUpperLeft( )：void	向左上平移地图
panUpperRight( )：void	向右上平移地图
zoomIn( )：void	放大地图
zoomOut( )：void	缩小地图

2. Map 地图控件的图层控制方法

Map 地图控件的图层控制方法见表 14.5。

表 14.5　　**Map 地图控件的图层控制方法**

方法签名	说　明
addLayer(layer：Layer，index：int=-1)：String	添加图层；并且可以通过参数指定新图层所在的上下位置，这会影响图层叠加后显示的效果
getLayer(layerId：String)：Layer	根据图层 ID，获取图层对象
removeAllLayers( )：void	删除所有图层
removeLayer(layer：Layer)：void	删除指定地图
reorderLayer(layerId：String，index：int)：void	改变地图中图层的顺序

3. Map 地图控件坐标转换方法

Map 地图控件的坐标转换方法见表 14.6。

表 14.6　　**Map 地图控件的坐标转换方法**

方法签名	说　明
toMap(screenPoint：Point)：MapPoint	从屏幕坐标(相对于 Map 控件左上角)转换为地理坐标
toMapFromStage (stageX：Number，stageY：Number)：MapPoint	从屏幕坐标(相对于整个 Flex 程序界面的左上角)转换为地理坐标
toScreen(mapPoint：MapPoint)：Point	从地理坐标转换为屏幕坐标

14.1.3 Map 控件事件

Map 控件的常用事件见表 14.7。

表 14.7 **Map 控件的常用事件**

事件名称	说明
extentChange	地图的当前可视区域对应的地理范围发生改变时触发，即 Map 控件的 extent 属性发生变化时触发
layerAdd	往 Map 中添加图层时触发
layerRemove	从 Map 中删除图层时触发
layerRemoveAll	删除所有图层时触发
layerReorder	地图中的图层上下叠加顺序发生变化时触发
load	当地图中有图层加载成功时触发，不管共有多少图层，只要有一个图层加载成功，即触发
mapClick	鼠标点击地图时
panEnd	地图平移完成时
panStart	地图开始平移时
zoomEnd	地图缩放完成时，一般通过监听此事件来获取比例尺信息
zoomStart	地图开始缩放时
mapMouseDown	鼠标左键在地图上按下时

14.1.4 Map 控件实例

1. 地图操作示例

具体示例代码如下：

```
<? xml version="1.0" encoding="utf-8"? >
<s:Application xmlns:fx="http://ns.adobe.com/mxml/2009"
           xmlns:s="library://ns.adobe.com/flex/spark"
           xmlns:esri="http://www.esri.com/2008/ags"
           pageTitle="World Topographic Map">
<! --添加 Map 控件-->
<! -- Map 控件的属性
    clickRecenterEnabled:true/false 是否开启按 Shift 键点击地图定位中
心的功能
    crosshairVisible:true/false 是否在地图的中心显示“十”字
doubleClickZoomEnabled:true/false 是否开启双击地图放大的功能
```

```
        keyboardNavigationEnabled:true/false 是否开启键盘进行地图导航放大缩小的功能
        mapNavigationEnabled:true/false 是否开启地图导航的功能
        panArrowsVisible:true/false 是否在地图的四边以及 4 个角显示移动地图箭头按钮的功能
        panEnabled:true/false 是否开启拖拽地图移动的功能
        rubberbandZoomEnabled:true/false 是否开启按 Shift+拉框放大地图的功能
        scaleBarVisible:true/false 是否显示放大缩小功能按钮条
        scrollWheelZoomEnabled:true/false 是否开启鼠标滚轮放大缩小的功能
        logoVisible:true/false 是否地图上显示 logo 图标
    -->
    <esri:Map horizontalCenter="-34" verticalCenter="1"
                crosshairVisible="true" panArrowsVisible="true"
logoVisible="false">
    <!--url 发布的 Map Service 的 rest 服务地址-->
    <esri:ArcGISTiledMapServiceLayer
    url="http://server.arcgisonline.com/ArcGIS/rest/services/World_
Topo_Map/MapServer"/>
    <!--可以替换为本地发布的地图,如:<esri:ArcGISDynamicMapServiceLayer
    url = " http://localhost: 6080/arcgis/rest/services/qingdao/MapS-
erver"/>-->
    </esri:Map>
    </s:Application>
```

提示：在页面中加入 Map 控件后，可以通过设置 Map 中的属性来添加用户需要的功能。例如：

```
    <esri:Map horizontalCenter="-34" verticalCenter="1"
    crosshairVisible="true" panArrowsVisible="true" logoVisible="
false"
```

此示例运行结果如图 14.1 所示。

2. 图层切换示例

此示例演示如何让用户改变不同的底图。有不同的方式完成这一功能，本示例是其中的一种。点击不同的按钮下面的栏切换底图。具体的示例代码如下：

```
<?xml version="1.0" encoding="utf-8"?>
<s:Application xmlns:fx="http://ns.adobe.com/mxml/2009"
               xmlns:s="library://ns.adobe.com/flex/spark"
               xmlns:esri="http://www.esri.com/2008/ags"
               pageTitle="Toggle between Map Services">
```

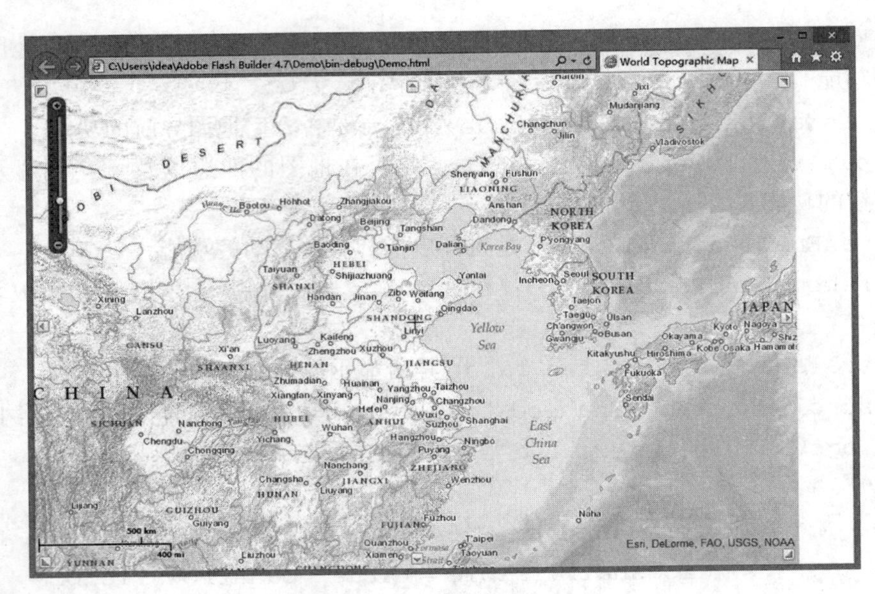

图 14.1　示例运行结果图

```
<fx:Script>
<![CDATA[
    import com.esri.ags.geometry.MapPoint;
    import com.esri.ags.layers.TiledMapServiceLayer;
    import mx.events.FlexEvent;

    privatefunction layerShowHandler(event:FlexEvent):void
    {
        //更新图层
        var tiledLayer:TiledMapServiceLayer = event.target
as TiledMapServiceLayer;
        myMap.lods = tiledLayer.tileInfo.lods;//显示级别
    }
]]>
</fx:Script>

<s:controlBarContent>
    <s:RichText width="100% ">
        此示例演示如何让用户改变不同的底图。有不同的方式直观地完成这一
行为,这仅仅是一个例子。点击不同的按钮下面的栏切换底图
    </s:RichText>
</s:controlBarContent>
```

```
        <esri:Map id="myMap"
              level="4"
              load="myMap.centerAt(new MapPoint(-11713000, 4822000))">
            <esri:ArcGISTiledMapServiceLayer show="layerShowHandler
(event)"
    url="http://server.arcgisonline.com/ArcGIS/rest/services/World
_Street_Map/MapServer"
                                           visible="{bb.selectedIndex == 0}"/>
            <esri:ArcGISTiledMapServiceLayer show="layerShowHandler
(event)"
               url="http://server.arcgisonline.com/ArcGIS/rest/serv-
ices/World_Topo_Map/MapServer"
                                           visible="{bb.selectedIndex == 1}"/>
            <esri:ArcGISTiledMapServiceLayer show="layerShowHandler
(event)"
    url="http://server.arcgisonline.com/ArcGIS/rest/services/World
_Imagery/MapServer"
                                           visible="{bb.selectedIndex == 2}"/>
            <esri:ArcGISTiledMapServiceLayer show="layerShowHandler
(event)"
    url="http://server.arcgisonline.com/ArcGIS/rest/services/Ocean
_Basemap/MapServer"
                                           visible="{bb.selectedIndex == 3}"/>
          <esri:ArcGISTiledMapServiceLayer show="layerShowHandler(e-
vent)"
    url = " http://server.arcgisonline.com/ArcGIS/rest/services/Nat-
Geo_World_Map/MapServer"
                                           visible="{bb.selectedIndex == 4}"/>
    </esri:Map>
    <s:ButtonBar id="bb"
               right="5" top="5"
               requireSelection="true">
          <s:dataProvider>
            <s:ArrayList>
               <fx:String>Streets</fx:String>
               <fx:String>Topographic</fx:String>
               <fx:String>Imagery</fx:String>
             <fx:String>Oceans</fx:String>
```

```
        <fx:String>National Geographic</fx:String>
      </s:ArrayList>
    </s:dataProvider>
  </s:ButtonBar>
</s:Application>
```

提示：上面代码中，对每一个图层绑定一个 layerShowHandler(event)事件，同时将本图层的可见性绑定到 Button 上，当点击 Button 后，函数 layerShowHandler(event)被激活，然后将相应的图层变为可见。

图层切换示例运行结果如图 14.2、图 14.3、图 14.4 所示。

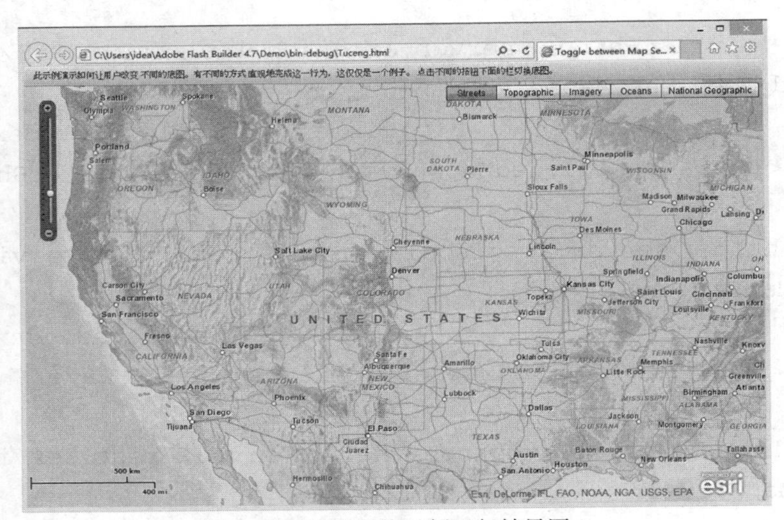

图 14.2　图层切换示例运行结果图 1

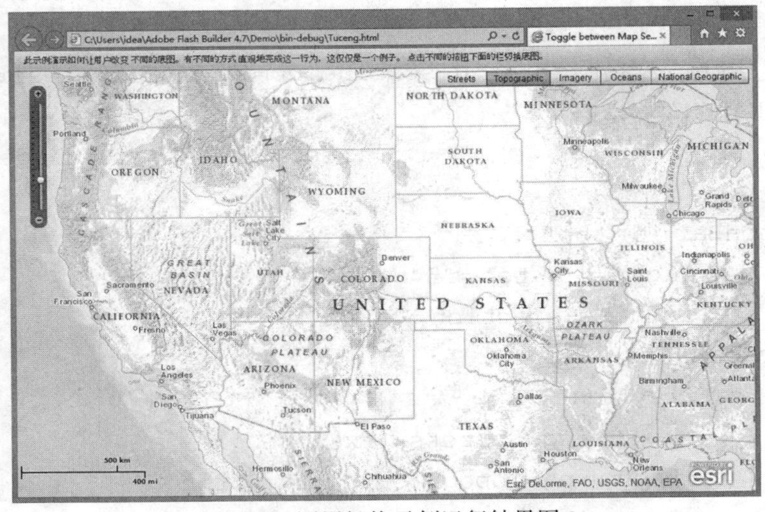

图 14.3　图层切换示例运行结果图 2

图 14.4　图层切换示例运行结果图 3

3. 地图可视区域和鼠标事件示例

此示例演示如何使用事件侦听器监听鼠标，显示鼠标所在位置、地图的范围和缩放等级，通过导航地图的变化程度(平移/缩放)或使用导航滑块来放大/缩小。

具体的示例代码如下：

```
<? xml version="1.0" encoding="utf-8"? >
<s:Application xmlns:fx="http://ns.adobe.com/mxml/2009"
               xmlns:s="library://ns.adobe.com/flex/spark"
               xmlns:esri="http://www.esri.com/2008/ags"
               pageTitle="Map Extent and Mouse Coordinates">

    <fx:Script>
        <! [CDATA[
            import com.esri.ags.geometry.Extent;
            import com.esri.ags.geometry.MapPoint;
            import com.esri.ags.utils.WebMercatorUtil;

            //监听鼠标移动事件
            privatefunction loadHandler():void
            {
                myMap.addEventListener(MouseEvent.MOUSE_MOVE,
mouseMoveHandler);
            }
```

```
            //显示鼠标所在位置
             privatefunction mouseMoveHandler(event:MouseEvent):
void
            {
            const mapPoint:MapPoint = myMap.toMapFromStage(event.
stageX, event.stageY);
            const latlong:MapPoint =
WebMercatorUtil.webMercatorTo-Geographic(mapPoint) as MapPoint;
            mousecoords.text =
                "x,y is " + mapPoint.x.toFixed(0) + "," + mapPoint.
y.toFixed(0)
                + " and Lat/Long is: " + latlong.y.toFixed(6)
                + " / " + latlong.x.toFixed(6);
        }

        //转换坐标系
        protectedfunctionshowExtentInGeographic(extent:Extent):
String
        {
            const geoExtent:Extent =
WebMercatorUtil.webMercatorToGeographic(myMap.extent) as Extent;
            //return geoExtent.toString() + ".." ;
            return" " + geoExtent.xmin.toFixed(6)
                + ", " + geoExtent.ymin.toFixed(6)
                + ", " + geoExtent.xmax.toFixed(6)
                + ", " + geoExtent.ymax.toFixed(6)
                + " (wkid: " + geoExtent.spatialReference.wkid + ")";
        }
    ]]>
</fx:Script>

<s:controlBarLayout>
    <s:VerticalLayout gap="10"
                      paddingBottom="7"
                      paddingLeft="10"
                      paddingRight="10"
                      paddingTop="7"/>
</s:controlBarLayout>
```

```
<s:controlBarContent>
    <s:RichText width="100%">
```

此示例演示如何使用事件侦听器监听鼠标，显示鼠标所在位置的信息。地图的范围和缩放等级也显示出来。通过导航地图的变化程度(平移/缩放)或使用导航滑块来放大/缩小。

```
    </s:RichText>
    <s:HGroup>
        <s:Label fontWeight="bold" text="Current map extent:"/>
        <s:RichEditableText editable="false"
text='xmin="{myMap.extent.xmin.toFixed(0)}" ymin="{myMap.extent.
ymin.toFixed(0)}" xmax="{myMap.extent.xmax.toFixed(0)}" ymax="
{myMap.extent.ymax.toFixed(0)}"
(wkid="{myMap.spatialReference.wkid}")'/>
    </s:HGroup>
    <s:HGroup>
        <s:Label fontWeight="bold" text="Current map extent (in
geographic):"/>
        <s:RichEditableText editable="false"
text="{showExtentInGeographic(myMap.extent)}"/>
    </s:HGroup>
    <s:HGroup>
        <s:Label fontWeight="bold" text="Current Mouse Coordi-
nates:"/>
        <s:RichEditableText id="mousecoords"
                            editable="false"
                            text="Move the mouse over the map to
see its current coordinates..."/>
    </s:HGroup>
    <s:HGroup>
        <s:Label fontWeight="bold" text="Current map scale is"/>
        <s:RichEditableText editable="false" text="1:
{myMap.scale.toFixed(0)} (level {myMap.level})"/>
    </s:HGroup>
</s:controlBarContent>

<esri:Map id="myMap" load="loadHandler()">
```

```
        <esri:extent>
            <esri: Extent xmin = " 3035000 " ymin = " 4305000 " xmax = "
3475000" ymax="10125000">
              <esri:SpatialReference wkid="102100"/>
              <! —跟上面的地图实例相同 -->
            </esri:Extent>
        </esri:extent>
        <esri:ArcGISTiledMapServiceLayer
url = " http://server.arcgisonline.com/ArcGIS/rest/services/World _
Street_Map/MapServer"/>
    </esri:Map>

    </s:Application>
```

提示：当程序加载 Map 控件时，激活 load 事件函数 loadHandler( )，在 loadHandler( ) 函数中监听鼠标移动事件；当鼠标移动时激活 mouseMoveHandler( ) 函数，将当前的鼠标位置显示在地图上。与此同时定义了一个 RichEditableText 绑定 showExtentInGeographic (myMap. extent) 函数，实时地显示当前地图的可视域。其运行结果如图 14.5 所示。

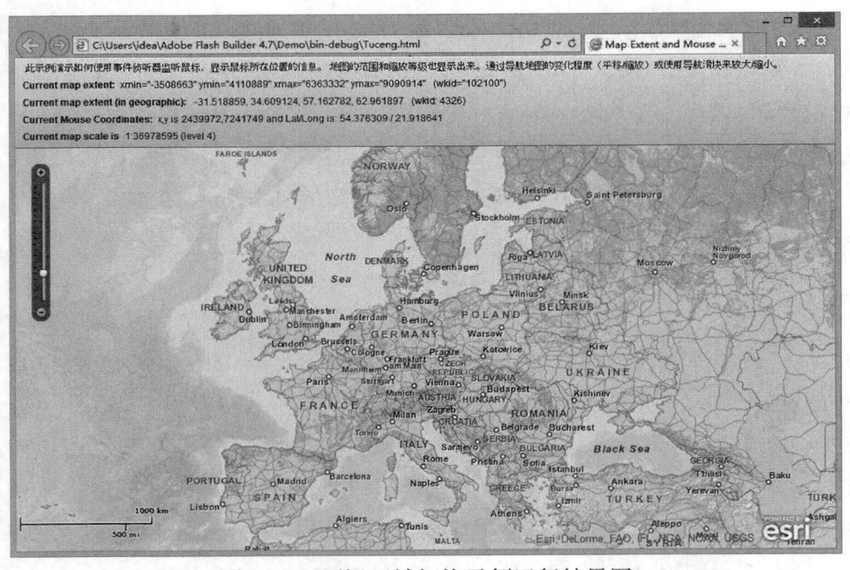

图 14.5　可视区域切换示例运行结果图

注意：ArcGIS Server 发布的地图(无论是在线地图还是本地地图)，可以将地址直接粘贴到浏览器查看相应的参数(如坐标系、可视域、图层信息等)，如图 14.6 和图 14.7 所示。

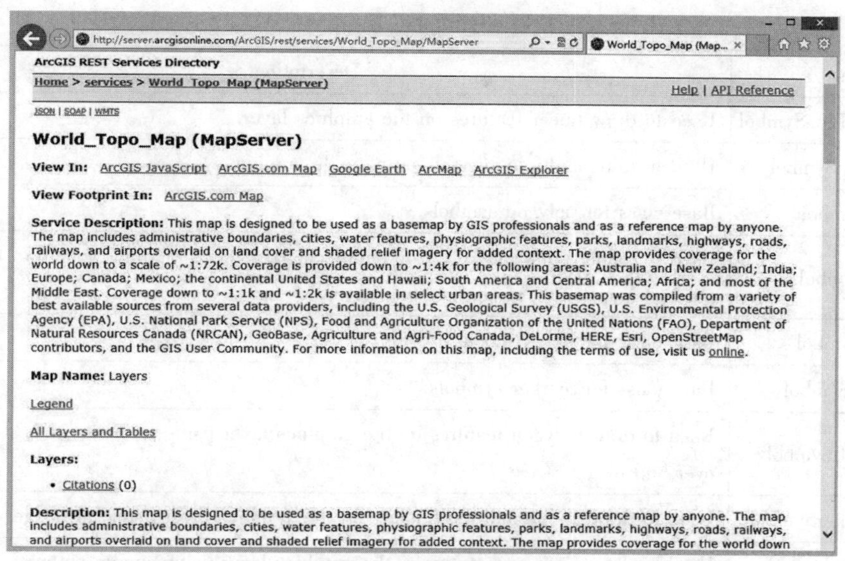

图 14.6　地图参数查看 1

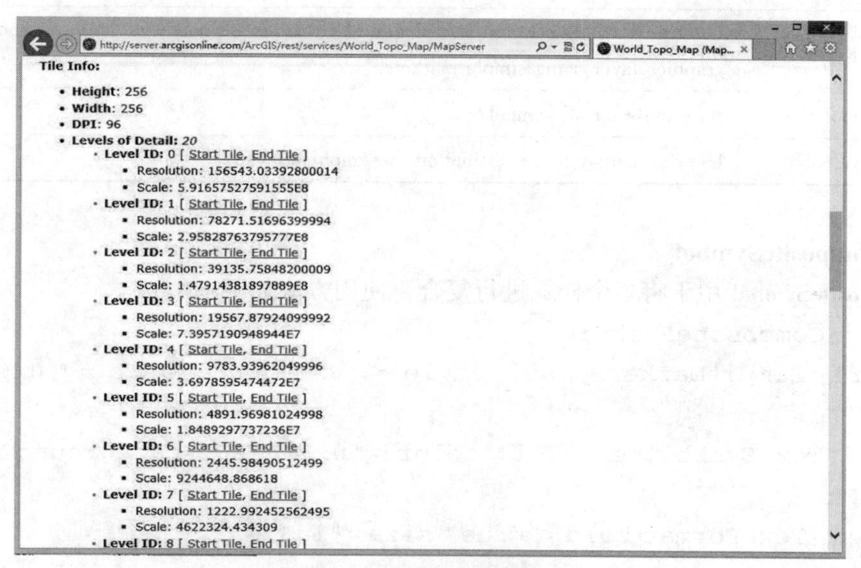

图 14.7　地图参数查看 2

14.2　地图样式

14.2.1　Symbol 介绍

Symbol 是用来在地图上重新显示地理图元渲染方式的类库，包括以下对象，见表 14.8。

表 14.8 **Symbol 对象**

Class	Description
CartographicLineSymbol	Used to draw linear features on the graphics layer.
CompositeSymbol	Used to draw multiple symbols on a single graphic.
FillSymbol	Base class for polygon symbols.
InfoSymbol	Used to display info markers (or bubble markers), containing your own content, at points (MapPoint) on the graphics layer.
LineSymbol	Base class for line symbols.
MarkerSymbol	Base class for marker symbols.
PictureFillSymbol	Used to draw polygon features on the graphics layer using an image that gets repeated over and over.
PictureMarkerSymbol	Used to draw points and multipoints on the graphics layer using an image.
SimpleFillSymbol	Used to draw polygon features on the graphics layer using simple patterns.
SimpleLineSymbol	Used to draw linear features on the graphics layer using simple line patterns.
SimpleMarkerSymbol	Used to draw points and multipoints (or nodes of polylines and polygons) on the graphics layer using simple markers.
Symbol	Base class for all symbols.
TextSymbol	Used to display text at points on the graphics layer.

1. CompositeSymbol

CompositeSymbol 用于将多个样式进行复合，使用方法如下：

```
<esri:CompositeSymbol>
<esri:SimpleMarkerSymbol style="circle" color="0x0000FF" size="20"/>
<esri:TextSymbol text="i" color="0xFFFFFF" backgroundColor="0x0000FF">
<flash:TextFormat bold="true" size="16"/>
</esri:TextSymbol>
</esri:CompositeSymbol>
```

上述代码的效果就是在圆里面写个字母 i。这个效果可以用来进行标注的渲染，使标注多样化。

2. SimpleMarkerSymbol

SimpleMarkerSymbol 指定预先定义好的一些 Marker，来渲染地图图元。ESRI 提供的样式有以下几种：

```
public static const STYLE_CIRCLE:String = "circle"—圆形
public static const STYLE_CROSS:String = "cross"—十字叉
```

public static const STYLE_DIAMOND:String = "diamond"—菱形

public static const STYLE_SQUARE:String = "square"—矩形框

public static const STYLE_TRIANGLE:String = "triangle"—三角形

public static const STYLE_X:String = "x"—x 形叉

3. PictureMarkerSymbol

使用一张图片定义样式，这是最常用的渲染方式之一。通常的应用都会使用这一样式，使地图图元的意义更加明朗，地图也更加美观。

其定义方法为：

```
PictureMarkerSymbol(source:Object = null, width:Number = 0, height:Number = 0, xoffset:Number = 0, yoffset:Number = 0, angle:Number = 0)
```

Source:图片路径,可以为网络路径或本地地址。

Width:图片宽度;Height:图片高度;Xoffset:x 方向上的偏移;Angle:图片的旋转角度。

例如：

```
Var graphicPointSym: PictureMarkerSymbol = new PictureMarkerSymbol(widgetIcon, 30, 30)
```

其中的值都是有默认值的。

4. SimpleFillSymbol

SimpleFillSymbol 用于对多边形类型的 gra 进行样式填充。需要使用 SimpleLineSymbol 设置边界的样式。

5. PictureFillSymbol

使用图片对多边形进行填充。

6. SimpleLineSymbol

SimpleLineSymbol 可以定义线的简单样式。

7. InfoSymbol

InfoSymbol 可以嵌入 Flex 的控件进行渲染，功能强大，例如饼图等都可以完成。用好此样式可以使自己的系统功能强大，事半功倍。

8. 绘制饼图示例

按照如下方法定义一个 infosymbol：

```
<esri:InfoSymbol id="PointSym" infoPlacement="center">
<esri:infoRenderer>
<mx:Component>
<mx:VBox width="100%" height="100%">
<mx:PieChart id="pieChart" dataProvider="{data}" width="90" height="90" showDataTips="true">
<mx:series>
<mx:PieSeries field="num" labelField="name">
```

```
<mx:fills>
<mx:Array>
<mx:RadialGradient>
<mx:entries>
<mx:Array>
<mx:GradientEntry color="#FF0000" ratio="0"/>
</mx:Array>
</mx:entries>
</mx:RadialGradient>
<mx:RadialGradient>
<mx:entries>
<mx:Array>
<mx:GradientEntry color="#00ff00" ratio="0"/>
</mx:Array>
</mx:entries>
</mx:RadialGradient>
</mx:Array>
</mx:fills>
</mx:PieSeries>
</mx:series>
</mx:PieChart>
</mx:VBox>
</mx:Component>
</esri:infoRenderer>
</esri:InfoSymbol>
```

将此样式赋给要显示饼图的 gra，注意 PieChart 的 dataprivide 为{data}这是固定写法，使用它可以将 gra 的 attribute 的信息内容赋给 data。

Dataprivide = data = gra. attribute;

data 的定义结构：

```
var GraInfo: Object =
{
  num: CurrentPec,
name: gra. attributes. FREQ_ LC+" MHz: /n" +CurrentPec+" %"
};
```

将 num 值赋予饼图当做饼图的值，name 值当做鼠标移动的提示。

至此就可以完成一个 infosymbol 的使用。

14.2.2 Symbol 应用示例

本示例代码包含多种类型的 Symbol，演示如何添加图形到 MXML 或 ActionScript 地图中，并使简单标记符号、图片标记符号、简单填充符号和图片填充符号显示。

具体示例代码如下：

```
<?xml version="1.0" encoding="utf-8"?>
<s:Application xmlns:fx="http://ns.adobe.com/mxml/2009"
               xmlns:s="library://ns.adobe.com/flex/spark"
               xmlns:esri="http://www.esri.com/2008/ags"
               pageTitle="Adding graphics using MXML and/or Ac-
tionScript">
    <s:layout>
        <s:VerticalLayout horizontalAlign="center"/>
    </s:layout>

    <fx:Script>
        <![CDATA[
            import com.esri.ags.SpatialReference;
            import com.esri.ags.symbols.PictureMarkerSymbol;

            privatefunction addSomeMarkers():void
            {
                var myGraphicMarker:Graphic = new Graphic(new Map-
Point(1447100, 7477200, new SpatialReference(102100)),
                        new SimpleMarkerSymbol(SimpleMarkerSymbol.
STYLE_DIAMOND, 22, 0x009933));
                myGraphicMarker.toolTip = "Marker added with Ac-
tionScript";
                myGraphicsLayer.add(myGraphicMarker);

                //PictureMarker-embedded image
                [Embed(source='assets/blue_globe.png')]
                var picEmbeddedClass:Class;
                var pictureMarker:PictureMarkerSymbol = newPicture-
MarkerSymbol(picEmbeddedClass);

                var myGraphicPic:Graphic = new Graphic(new MapPoint
(-411000, 4924000, new SpatialReference(102100)));
```

```
            myGraphicPic.symbol = pictureMarker;
            myGraphicsLayer.add(myGraphicPic);

            var myPolyline:Polyline = new Polyline(
                [[
                    new MapPoint(-1726185, 9543036),
                    new MapPoint(34923, 6920940),
                    new MapPoint(1874303, 6255632),
                    new MapPoint(1835168, 6255632),
                    new MapPoint(1913439, 6138225)
                ]], new SpatialReference(102100));
            var myGraphicLine:Graphic = new Graphic(myPolyline);
             myGraphicLine.symbol = new SimpleLineSymbol(Sim-
pleLineSymbol.STYLE_DASH, 0xDD2222, 1.0, 4);
            myGraphicsLayer.add(myGraphicLine);

            var myPolygon:Polygon = new Polygon(
                [[
                    new MapPoint(2352491, -1992338),
                    new MapPoint(2332923, -2461967),
                    new MapPoint(2646009, -2266288),
                    new MapPoint(3076503, -2324992),
                    new MapPoint(3272181, -2520670),
                    new MapPoint(3506996, -2559806),
                    new MapPoint(3702675, -3049003),
                    new MapPoint(3370021, -3675175),
                    new MapPoint(2763416, -4046965),
                    new MapPoint(2117676, -4144804),
                    new MapPoint(1961133, -3890422),
                    new MapPoint(2000269, -3655607),
                    new MapPoint(1667615, -3185978),
                    new MapPoint(1550208, -2422831),
                    new MapPoint(1334961, -1953202),
                    new MapPoint(2352491, -1992338)
                ]], new SpatialReference(102100));
            var myGraphicPolygon:Graphic = new Graphic();
            myGraphicPolygon.geometry = myPolygon;
            myGraphicPolygon.symbol = new SimpleFillSymbol(
```

```
                    SimpleFillSymbol.STYLE_SOLID, //fill style 填充
样式
                    0xFF0000, //fill color 填充色
                    0.7 //fill alpha
                );
                myGraphicPolygon.toolTip = "Polygon added with Ac-
tionScript";
                myGraphicsLayer.add(myGraphicPolygon);
                btn.enabled = false;
            }
        ]]>
    </fx:Script>

    <s:controlBarLayout>
        <s:VerticalLayout gap="10"
                    paddingBottom="7"
                    paddingLeft="10"
                    paddingRight="10"
                    paddingTop="7"/>
    </s:controlBarLayout>
    <s:controlBarContent>
        <s:RichText width="100%">
    此示例演示如何添加图形到 MXML 或 ActionScript 地图中，简单标记符号、图片标记
符号、简单填充符号和图片填充符号显示。
    </s:RichText>
        <s:Button id="btn"
                click="addSomeMarkers()"
                label="通过代码添加更多样例"/>
    </s:controlBarContent>

    <esri:Map id="myMap"
            level="2"
            wrapAround180="true">
        < esri: ArcGISTiledMapServiceLayer  url = " http://server.
arcgisonline.com/ArcGIS/rest/services/World_Street_Map/MapServer"/>
        <esri:GraphicsLayer id="myGraphicsLayer">
            <esri:Graphic toolTip = "California MapPoint with a Sim-
pleMarkerSymbol">
```

```
            <esri:geometry>
                <esri:MapPoint x="-13163000" y="4035000"
                                    spatialReference="{new SpatialRef-
erence(102100)}"/>
                </esri:geometry>
                <esri:symbol>
                    <esri:SimpleMarkerSymbol color="0x0033DD"
size="18"/>
                </esri:symbol>
            </esri:Graphic>
            <esri:Graphic toolTip="Hurricane polyline with a Sim-
pleLineSymbol">
                <esri:geometry>
                    <esri:Polyline spatialReference="{new Spatial-
Reference(102100)}">
                        <fx:Array>
                          <fx:Array>
                            <esri:MapPoint x="-4700503" y="1128848"/>
                            <esri:MapPoint x="-7909635" y="2819513"/>
                            <esri:MapPoint x="-8144450" y="4199048"/>
                            <esri:MapPoint x="-7244327" y="5261584"/>
                          </fx:Array>
                        </fx:Array>
                    </esri:Polyline>
                </esri:geometry>
                <esri:symbol>
                  <esri:SimpleLineSymbol width="6" color="0xFF0000"/>
                </esri:symbol>
            </esri:Graphic>
            <esri:Graphic toolTip="Brazilian polygon with a Sim-
pleFillSymbol">
                <esri:geometry>
                    <esri:Polygon spatialReference="{new Spatial-
Reference(102100)}">
                        <fx:Array>
                          <fx:Array>
                            <esri:MapPoint x="-3867905" y="-671044"/>
                            <esri:MapPoint x="-4533702" y="-2578326"/>
```

```
                                    <esri:MapPoint x="-5316417" y="-2832708"/>
                                    <esri:MapPoint x="-5844750" y="-3869806"/>
                                    <esri:MapPoint x="-6333947" y="-3498016"/>
                                    <esri:MapPoint x="-6412218" y="-1942370"/>
                                    <esri:MapPoint x="-8211974" y="-954779"/>
                                    <esri:MapPoint x="-7703209" y="229077"/>
                                    <esri:MapPoint x="-5736637" y="454597"/>
                                    <esri:MapPoint x="-3867905" y="-671044"/>
                                </fx:Array>
                            </fx:Array>
                        </esri:Polygon>
                    </esri:geometry>
                    <esri:symbol>
                        <esri:SimpleFillSymbol alpha="0.7" color="
0x009933"/>
                    </esri:symbol>
                </esri:Graphic>
            </esri:GraphicsLayer>
        </esri:Map>
    </s:Application>
```

提示：本示例代码包含两部分，一部分是在设计代码中添加地图点、线、面以及相应的图例 Symbol，另一部分是在实现代码中添加点、线、面及相应的图例。通过两个部分完整地展现了 Symbol 的添加形式。其运行结果如图 14.8 所示。

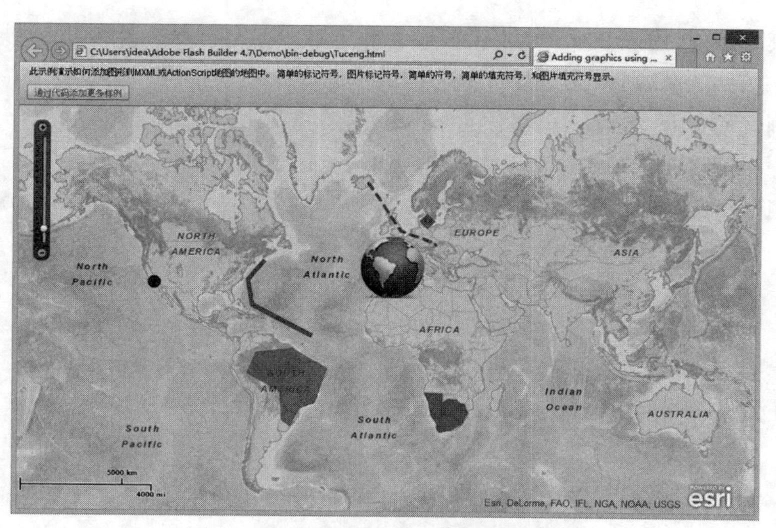

图 14.8　添加 Symbol 示例运行结果图

注意：代码中 source='assets/i_draw_point.png'为引用的图片地址，可在 Flex 项目中添加自己的图片以便于调用，如图 14.9 所示。

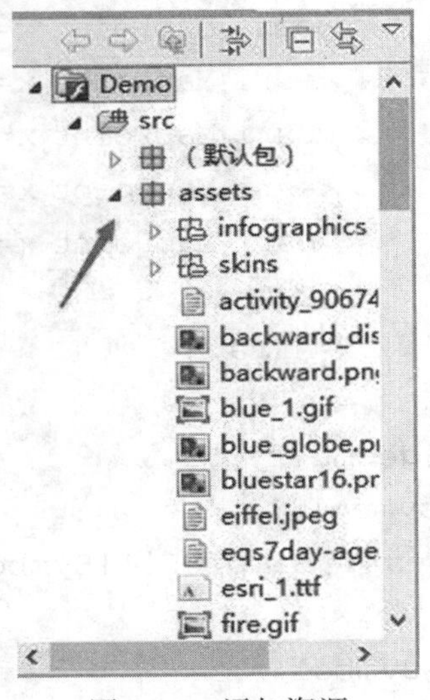

图 14.9　添加资源

14.3　常用工具

ArcGIS API For Flex 中的工具主要包含三类：绘图工具、编辑工具、浏览工具。

14.3.1　绘图工具

绘图工具支持创建新的几何形状：点(单个或多点)、线(直线、折线或 freehand_polyline)、多边形(freehand_polygon 或多边形)、矩形(程度)和圆(圆)。

绘图示例代码如下：

```
<? xml version="1.0" encoding="utf-8"? >
<s:Application xmlns:fx="http://ns.adobe.com/mxml/2009"
               xmlns:s="library://ns.adobe.com/flex/spark"
               xmlns:mx="library://ns.adobe.com/flex/mx"
               xmlns:esri="http://www.esri.com/2008/ags"
               pageTitle="DrawTool">
       <fx:Style>
       @ namespace mx "library://ns.adobe.com/flex/mx";
       mx|ToolTip
       {
           font-size: 14;
```

```
        }
    </fx:Style>
        <fx:Script>
        <![CDATA[
            import com.esri.ags.events.DrawEvent;

            import mx.events.ItemClickEvent;

             protectedfunction tbb_itemClickHandler(event:Item-
ClickEvent):void
             {
                if(tbb.selectedIndex < 0)
                {
                    //切换工具时关闭,停用上一个画笔
                    myDrawTool.deactivate();
                }
                else
                {
                    switch(event.item.label)
                    {
                        case"MAPPOINT":
                        {
                          myDrawTool.activate(DrawTool.MAPPOINT);
                          break;
                        }

                        case"POLYLINE":
                        {
                          myDrawTool.activate(DrawTool.POLYLINE);
                          break;
                        }
                        case"FREEHAND_POLYLINE":
                        {
                        myDrawTool.activate(DrawTool.FREEHAND_
POLYLINE);
                          break;
                        }
                        case"POLYGON":
```

```
                {
                    myDrawTool.activate(DrawTool.POLYGON);
                    break;
                }
                case"FREEHAND_POLYGON":
                {
                    myDrawTool.activate(DrawTool.FREEHAND_
POLYGON);
                    break;
                }
                case"EXTENT":
                {
                    myDrawTool.activate(DrawTool.EXTENT);
                    break;
                }
                case"CIRCLE":
                {
                    myDrawTool.activate(DrawTool.CIRCLE);
                    break;
                }
                case"ELLIPSE":
                {
                    myDrawTool.activate(DrawTool.ELLIPSE);
                    break;
                }
            }
        }
    }
    //绘制完成函数
    protectedfunction drawTool_drawEndHandler(event:Draw-
Event):void
    {

        myDrawTool.deactivate();
        tbb.selectedIndex = -1;
    }
  ]]>
</fx:Script>
```

```
<fx:Declarations>
    <!--所有点的样式 Symbol-->
    <esri:SimpleMarkerSymbol id="sms"
                             color="0x00FF00"
                             size="12"
                             style="square"/>

    <!--所有线的样式 Symbol-->
    <esri:SimpleLineSymbol id="sls"
                           width="3"
                           color="0x00FF00"/>

    <!--所有多边形的样式 Symbol-->
    <esri:SimpleFillSymbol id="sfs"
                           color="0xFFFFFF"
                           style="diagonalcross">
        <esri:outline>
            <esri:SimpleLineSymbol width="2" color="0x00FF00"/>
        </esri:outline>
    </esri:SimpleFillSymbol>

    <esri:DrawTool id="myDrawTool"
                   drawEnd="drawTool_drawEndHandler(event)"
                   fillSymbol="{sfs}"
                   graphicsLayer="{myGraphicsLayer}"
                   lineSymbol="{sls}"
                   map="{myMap}"
                   markerSymbol="{sms}"/>
</fx:Declarations>

<s:controlBarLayout>
    <s:VerticalLayout gap="10"
                      paddingBottom="7"
                      paddingLeft="10"
                      paddingRight="10"
                      paddingTop="7"/>
</s:controlBarLayout>
```

```
    <s:controlBarContent>
        <s:RichText width="100%">
    工具的支持功能，可以创建新的几何形状：点(单个或多点)、线(直线、折线或
freehand_polyline)、多边形(freehand_polygon 或多边形)、矩形(程度)和圆(圆)。
    </s:RichText>
    <s:HGroup width="100%" horizontalAlign="center">
      <mx:ToggleButtonBar id="tbb"
                          itemClick="tbb_itemClickHandler(event)"
                          labelField="null"
                          selectedIndex="-1"
                          toggleOnClick="true">
            <fx:Object icon="@Embed(source='assets/i_draw_
point.png')" label="MAPPOINT"/>
            <fx:Object icon="@Embed(source='assets/i_draw_
line.png')" label="POLYLINE"/>
            <fx:Object icon="@Embed(source='assets/i_draw_free-
line.png')" label="FREEHAND_POLYLINE"/>
            <fx:Object icon="@Embed(source='assets/i_draw_po-
ly.png')" label="POLYGON"/>
            <fx:Object icon="@Embed(source='assets/i_draw_freep-
oly.png')" label="FREEHAND_POLYGON"/>
            <fx:Object icon="@Embed(source='assets/i_draw_re-
ct.png')" label="EXTENT"/>
            <fx:Object icon="@Embed(source='assets/i_draw_cir-
cle.png')" label="CIRCLE"/>
            <fx:Object icon="@Embed(source='assets/i_draw_el-
lipse.png')" label="ELLIPSE"/>
        </mx:ToggleButtonBar>
      </s:HGroup>
    </s:controlBarContent>

    <esri:Map id="myMap"
              level="3"
              wrapAround180="true">
        <esri:ArcGISTiledMapServiceLayer
url="http://server.arcgisonline.com/ArcGIS/rest/services/World_Im-
agery/MapServer"/>
        <esri:GraphicsLayer id="myGraphicsLayer"/>
```

```
    </esri:Map>
</s:Application>
```

提示：首先在加载页面时，设置 DrawTool 的参数：

```
<esri:DrawTool id="myDrawTool"
    drawEnd="drawTool_drawEndHandler(event)"//绘图完成后运行的函数
    fillSymbol="{sfs}" //画多边形时的 Symbol
    graphicsLayer="{myGraphicsLayer}"//画出的图形所在的图层(承载层)
    lineSymbol="{sls}"  //画线时的 Symbol
    map="{myMap}"  //绑定绘制图形的地图
    markerSymbol="{sms}"/>  //画点时的 Symbol
```

当 ToggleButtonBar 被点击时，激活 tbb_itemClickHandler(event)函数，开始绘制图形；当双击完成图形的绘制后，激活 drawTool_drawEndHandler(event)函数，将画笔工具变为不可用状态，同时 Button 按钮的选中状态设置为“无”，方便下次使用。示例运行结果如图 14.10 所示。

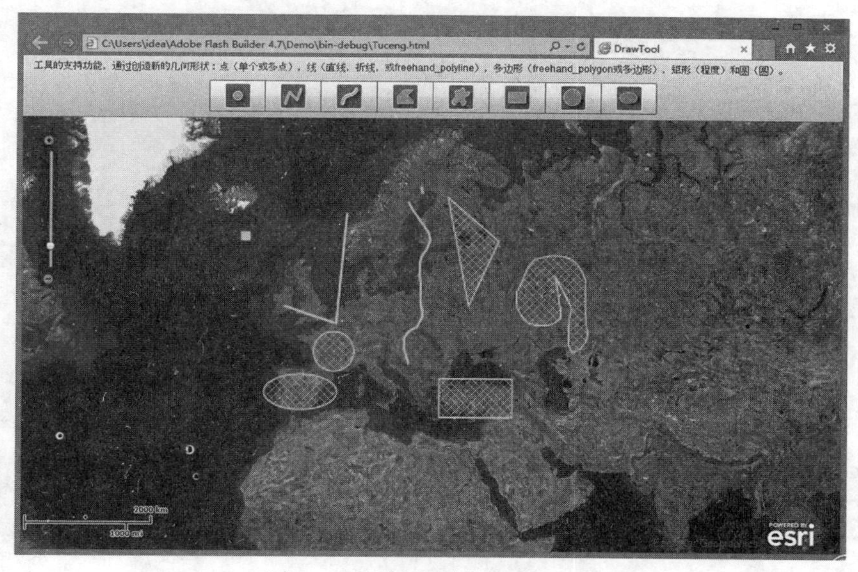

图 14.10　绘图示例运行结果图

14.3.2　编辑工具

编辑工具的功能是移动图形或修改单个顶点，即编辑现有的几何图形。

编辑工具示例代码如下：

```
<? xml version="1.0" encoding="utf-8"? >
<s:Application xmlns:fx="http://ns.adobe.com/mxml/2009"
               xmlns:s="library://ns.adobe.com/flex/spark"
```

```
            xmlns:mx="library://ns.adobe.com/flex/mx"
            xmlns:esri="http://www.esri.com/2008/ags"
            pageTitle="Edit graphics with the EditTool">

    <fx:Script>
        <![CDATA[
            import com.esri.ags.Graphic;
            import com.esri.ags.events.DrawEvent;
            import com.esri.ags.events.MapMouseEvent;
            import com.esri.ags.geometry.Extent;
            import com.esri.ags.geometry.Polygon;
            import com.esri.ags.geometry.Polyline;
            import com.esri.ags.symbols.SimpleFillSymbol;
            import com.esri.ags.symbols.SimpleLineSymbol;
            import com.esri.ags.symbols.SimpleMarkerSymbol;

            importmx.events.FlexEvent;
            import mx.events.ItemClickEvent;

            importspark.events.IndexChangeEvent;

            privatevar graphic:Graphic;
            privatevar lastEditGraphic:Graphic;
            privatevar lastActiveEditTypes:String;

             protectedfunction tbb_itemClickHandler(event:Item-
ClickEvent):void
            {
                if (tbb.selectedIndex < 0)
                {
                    //when toggling a tool off, deactivate it
                    myDrawTool.deactivate();
                }
                else
                {
                    switch (event.item.label)
                    {
                        case"MAPPOINT":
```

```
{
  myDrawTool.activate(DrawTool.MAPPOINT);
  break;
}
case"POLYLINE":
{
  myDrawTool.activate(DrawTool.POLYLINE);
  break;
}
case"FREEHAND_POLYLINE":
{
  myDrawTool.activate(DrawTool.FREEHAND_POLY-
LINE);
  break;
}
case"POLYGON":
{
    myDrawTool.activate(DrawTool.POLYGON);
    break;
}
case"FREEHAND_POLYGON":
{
    myDrawTool.activate(DrawTool.FREEHAND_
POLYGON);
    break;
}
case"EXTENT":
{
    myDrawTool.activate(DrawTool.EXTENT);
    break;
}
case"CIRCLE":
{
    myDrawTool.activate(DrawTool.CIRCLE);
    break;
}
case"ELLIPSE":
{
```

```
                        myDrawTool.activate(DrawTool.ELLIPSE);
                        break;
                    }
                }
            }
        }

        protectedfunction drawTool_drawEndHandler(event:Draw-
Event):void
        {

            myDrawTool.deactivate();
            tbb.selectedIndex = -1;
        }

        privatefunction myMap_mapMouseDownHandler(event:Map-
MouseEvent):void
        {
            event.currentTarget.addEventListener(MouseEvent.
MOUSE_MOVE, map_mouseMoveHandler);
            event.currentTarget.addEventListener(MouseEvent.
MOUSE_UP, map_mouseUpHandler);
        }

            privatefunction map_mouseMoveHandler(event:Mou-
seEvent):void
        {
                event.currentTarget.removeEventListener(Mou-
seEvent.MOUSE_MOVE, map_mouseMoveHandler);
                event.currentTarget.removeEventListener(Mou-
seEvent.MOUSE_UP, map_mouseUpHandler);
        }

        privatefunction map_mouseUpHandler(event:MouseEvent):
void
        {
                event.currentTarget.removeEventListener(Mou-
seEvent.MOUSE_MOVE, map_mouseMoveHandler);
```

```
                event.currentTarget.removeEventListener ( Mou-
seEvent.MOUSE_UP, map_mouseUpHandler);

            if (event.target is Graphic || event.target.parent
is Graphic)
            {
                if (event.target is Graphic)
                {
                    graphic = Graphic(event.target);
                }
                elseif (event.target.parent is Graphic) //check
for PictureMarkerSymbol
                {
                    graphic = Graphic(event.target.parent);
                }

                if (lastEditGraphic ! == graphic)
                {
                    lastEditGraphic = graphic;
                    lastActiveEditTypes = "moveRotateScale"; //
make sure move and edit vertices is the 1st mode
                }
                    if (graphic.geometry is Polyline || graph-
ic.geometry is Polygon)
                {
                  if (lastActiveEditTypes == "moveEditVertices")
                  {
                      lastActiveEditTypes = "moveRotateScale";
                      myEditTool.activate(EditTool.MOVE | Edit-
Tool.SCALE | EditTool.ROTATE, [ graphic ]);
                    }
                    else
                    {
                      lastActiveEditTypes = "moveEditVertices";
                      myEditTool.activate(EditTool.MOVE | Edit-
Tool.EDIT_VERTICES, [ graphic ]);
                    }
                }
```

```
                    elseif (graphic.geometry is Extent)
                    {
                        myEditTool.activate(EditTool.MOVE | Edit-
Tool.SCALE, [ graphic ]);
                    }
                    elseif (graphic.graphicsLayer == myGraphicsLayer)
                    {
                        myEditTool.activate(EditTool.MOVE | Edit-
Tool.EDIT_VERTICES, [ graphic ]);
                    }
                }
                else
                {
                    myEditTool.deactivate();
                      lastActiveEditTypes = "moveRotateScale"; //
make sure move and edit vertices is the 1st mode
                }
            }
        ]]>
    </fx:Script>

    <fx:Declarations>
        <!-- Symbol for all point shapes -->
        <esri:SimpleMarkerSymbol id="sms"
                                 color="0x00FF00"
                                 size="12"
                                 style="square"/>

        <!-- Symbol for all line shapes -->
        <esri:SimpleLineSymbol id="sls"
                               width="3"
                               color="0x00FF00"/>

        <!-- Symbol for all polygon shapes -->
        <esri:SimpleFillSymbol id="sfs"
                               color="0xFFFFFF"
                               style="diagonalcross">
            <esri:outline>
```

```
                    < esri: SimpleLineSymbol width = " 2 " color = "
0x00FF00" />
                </esri:outline>
            </esri:SimpleFillSymbol>

            <esri:DrawTool id="myDrawTool"
                          drawEnd="drawTool_drawEndHandler(event)"
                          fillSymbol="{sfs}"
                          graphicsLayer="{myGraphicsLayer}"
                          lineSymbol="{sls}"
                          map="{myMap}"
                          markerSymbol="{sms}"/>

            <esri:EditTool id="myEditTool" map="{myMap}"/>
        </fx:Declarations>

        <s:controlBarLayout>
            <s:VerticalLayout gap="10"
                              paddingBottom="7"
                              paddingLeft="10"
                                paddingRight="10"
                                paddingTop="7"/>
        </s:controlBarLayout>
        <s:controlBarContent>
            <s:RichText width="100% ">
            //本实例首先实现几何图形的绘制,然后单击几何图形,可对几何图形进行
平移或者顶点的编辑
            </s:RichText>
            <mx:ToggleButtonBar id="tbb"
                                itemClick="tbb_itemClickHandler(event)"
                                labelField="null"
                                selectedIndex="-1"
                                toggleOnClick="true">
                <fx: Object icon = "@ Embed(source ='assets/i_draw_
point.png')"
                          label="MAPPOINT"
                          toolTip="MapPoint"/>
                <fx: Object icon = "@ Embed(source ='assets/i_draw_
```

```
line.png')"
                                label="POLYLINE"
                                toolTip="Polyline"/>
                <fx:Object icon="@Embed(source='assets/i_draw_free-
line.png')"
                                label="FREEHAND_POLYLINE"
                                toolTip="Freehand Polyline"/>
                <fx:Object icon="@Embed(source='assets/i_draw_po-
ly.png')"
                                label="POLYGON"
                                toolTip="Polygon"/>
                <fx:Object icon="@Embed(source='assets/i_draw_freep-
oly.png')"
                                label="FREEHAND_POLYGON"
                                toolTip="Freehand Polygon"/>
                <fx:Object icon="@Embed(source='assets/i_draw_re-
ct.png')"
                                label="EXTENT"
                                toolTip="Extent"/>
                <fx:Object icon="@Embed(source='assets/i_draw_cir-
cle.png')"
                                label="CIRCLE"
                                toolTip="Circle"/>
                <fx:Object icon="@Embed(source='assets/i_draw_el-
lipse.png')"
                                label="ELLIPSE"
                                toolTip="Ellipse"/>
            </mx:ToggleButtonBar>
        </s:controlBarContent>

        <esri:Map id="myMap"
                  level="3"
                  mapMouseDown="myMap_mapMouseDownHandler(event)"
                  wrapAround180="true">
            <esri:ArcGISTiledMapServiceLayer url="http://server.
arcgisonline.com/ArcGIS/rest/services/World_Imagery/MapServer"/>
            <esri:GraphicsLayer id="myGraphicsLayer"/>
        </esri:Map>
```

```
</s:Application>
```

提示：本实例同绘图工具类似，只是多添加了一个编辑工具，同时添加鼠标移动和鼠标抬起事件，在事件中判断是否是目标 Graphic，若是则进行拖动操作，若不是则停止监听鼠标事件，等待下一次鼠标点击或者抬起再次激活。编辑工具示例运行结果如图 14.11 所示。

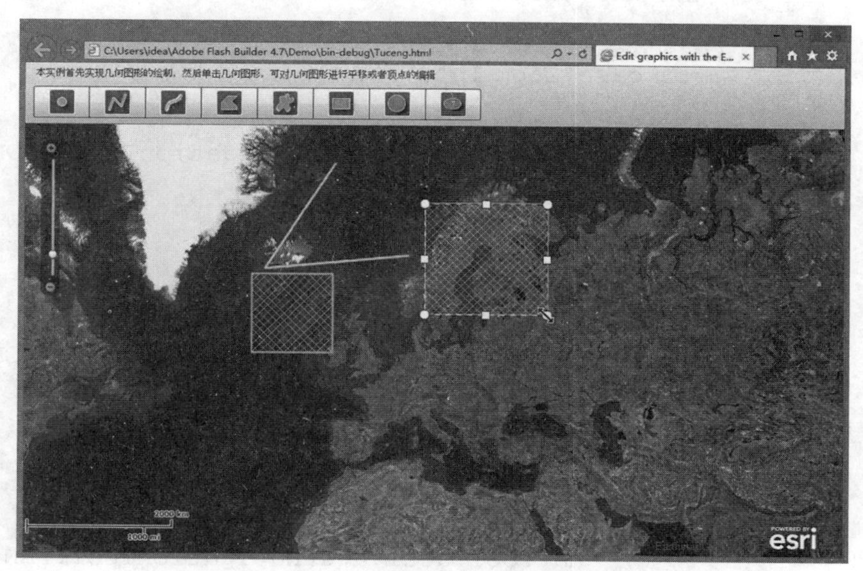

图 14.11　编辑工具示例运行结果图

14.3.3　浏览工具

浏览工具主要实现对地图的放大、缩小、平移、视图记录、全图等功能。

浏览工具示例代码如下：

```
<? xml version="1.0" encoding="utf-8"? >
<s:Application xmlns:fx="http://ns.adobe.com/mxml/2009"
               xmlns:s="library://ns.adobe.com/flex/spark"
               xmlns:mx="library://ns.adobe.com/flex/mx"
               xmlns:esri="http://www.esri.com/2008/ags"
               pageTitle="Using the navigation tools">
    <fx:Script>
    <! [CDATA[
        import com.esri.ags.tools.NavigationTool;

        importmx.logging.LogEventLevel;
```

```
            import spark.events.IndexChangeEvent;

             privatefunction tbb_changeHandler(event:IndexChan-
geEvent):void
            {
              switch(tbb.selectedItem)
              {
                 case"Zoom In":
                 {
                    navTool.activate(NavigationTool.ZOOM_IN);
                    break;
                 }
                 case"Zoom Out":
                 {
                    navTool.activate(NavigationTool.ZOOM_OUT);
                    break;
                 }
                 case"Pan":
                 {
                    navTool.activate(NavigationTool.PAN);
                    break;
                 }
                 default:
                 {
                    navTool.deactivate();
                    break;
                 }
              }
            }
        ]]>
    </fx:Script>

    <fx:Declarations>
        <esri:NavigationTool id="navTool" map="{myMap}"/>

    </fx:Declarations>

    <s:controlBarLayout>
```

```
            <s:VerticalLayout gap="10"
                              paddingBottom="7"
                              paddingLeft="10"
                              paddingRight="10"
                              paddingTop="7"/>
        </s:controlBarLayout>
        <s:controlBarContent>
            <s:RichText width="100%">
                包含对地图的放大、缩小、平移、视图记录、全图等功能
            </s:RichText>
            <s:HGroup width="100%" horizontalAlign="center">
              <s:ButtonBar id="tbb" change="tbb_changeHandler(event)">
                    <s:ArrayList>
                        <fx:String>Zoom In</fx:String>
                        <fx:String>Zoom Out</fx:String>
                        <fx:String>Pan</fx:String>
                    </s:ArrayList>
                </s:ButtonBar>
                <s:HGroup gap="0">
                    <s:Button click="navTool.zoomToPrevExtent()"
                              enabled="{! navTool.isFirstExtent}"
                              label="Previous Extent"/>
                    <s:Button click="navTool.zoomToNextExtent()"
                              enabled="{! navTool.isLastExtent}"
                              label="Next Extent"/>
                </s:HGroup>
                <s:Button click="myMap.zoomToInitialExtent()" label="In-
itial Extent"/>
            </s:HGroup>
        </s:controlBarContent>

        <esri:Map id="myMap">
            <esri:extent>
                <esri:Extent xmin="10404880" ymin="-2003758" xmax="
16960120" ymax="1430405">
                    <esri:SpatialReference wkid="102100"/>
                </esri:Extent>
            </esri:extent>
```

```
        <esri:ArcGISTiledMapServiceLayer
url = " http://server.arcgisonline.com/ArcGIS/rest/services/NatGeo _
World_Map/MapServer"/>
    </esri:Map>

  </s:Application>
```

提示：首先加载页面，添加<esri：NavigationTool id="navTool" map="{myMap}"/>控件，绑定 Map 控件。点击 ButtonBar 中的按钮后，激活 tbb_ changeHandler(event)函数，判断进行哪个操作，激活相应的 NavigationTool 的操作方法。浏览工具示例运行结果如图 14.12 所示。

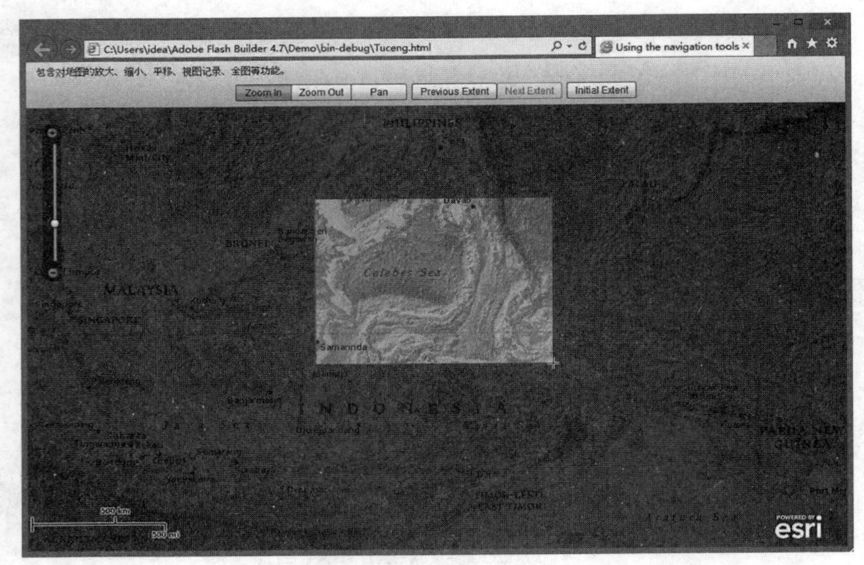

图 14.12　浏览工具示例运行结果图

第15章　查询功能实现

15.1　QueryTask

QueryTask是一个进行空间和属性查询的功能类，它可以在某个地图服务的子图层内进行查询，但QueryTask进行查询的地图服务并不必须加载到Map中进行显示。QueryTask的执行需要两个先决条件：一是需要查询的图层的URL，二是进行查询的过滤条件。QueryTask的方法和事件分别见表15.1和表15.2。

表15.1　**QueryTask的方法**

Method(方法)
QueryTask(url: String = null) Creates a new QueryTask object used to execute a query on the layer or table resource identified by the URL(创建一个新的QueryTask对象，用于在图层或表中执行通过URL定义的查询)
execute(query: Query, responder: IResponder = null): AsyncToken Executes a query against an ArcGIS Server map layer or table(执行一个对ArcGIS Server中图层或表的查询)
executeForCount(query: Query, responder: IResponder = null): AsyncToken Gets the count of features that satisfy the input query(获取满足查询条件的要素数目)
executeForIds(query: Query, responder: IResponder = null): AsyncToken Executes a query against an ArcGIS Server map layer or table(执行一个对ArcGIS Server中图层或表的查询)
executeRelationshipQuery(relationshipQuery: RelationshipQuery, responder: IResponder = null): AsyncToken The executeRelationshipQuery operation is performed on a layer (or table) resource(executeRelationshipQuery查询操作可以在图层或表中进行)

表 15.2 **QueryTask 的事件**

Event(事件)
executeComplete Dispatched when an execute operation successfully completes(当执行操作完成后通过派遣消息执行相应事件)
executeForCountComplete Dispatched when an executeForCount operation successfully completes(当执行计数操作完成后通过派遣消息执行相应事件)
executeForIdsComplete Dispatched when an executeForIds operation successfully completes(当 executeForIds 操作完成后通过派遣消息执行相应事件)
executeRelationshipQueryComplete Dispatched when an executeRelationshipQuery operation successfully completes(当关系查询操作完成后通过派遣消息执行相应事件)
fault Dispatched when a QueryTask fails(当查询任务操作失败后通过派遣消息执行相应事件)

参数 Query 的属性见表 15.3。

表 15.3 **参数 Query 的属性**

Property(属性)
geometry: Geometry The geometry to apply to the spatial filter(可用于空间过滤的几何实体)
groupByFieldsForStatistics: Array This will contain a list of fields to group by during statistical analysis(groupByFieldsForStatistics 通过统计分析形成字段分组列表)
maxAllowableOffset: Number The maximum allowable offset used for generalizing geometries returned by the Query operation(查询操作返回的泛化几何图形的最大偏移量)
objectIds: Array The object IDs of this layer/table to be queried(用于查询的图层或表的对象 id 列表)
orderByFields: Array One or more field names using which the features/records need to be ordered(待排序的要素或记录的一个或多个字段名)
outFields: Array Attribute fields to include in the FeatureSet(包含在要素集中的属性字段)

续表

Property(属性)
outSpatialReference: SpatialReference The well-known ID of the spatial reference for the returned geometry(返回几何体空间参考系的通用 ID)
outStatistics: Array The definitions for one or more field-based statistic to be calculated(用于统计计算的一个或多个字段的定义)
pixelSizeX: Number The size of pixel in X direction used to query visible rasters(用于查询可视栅格图层 X 方向的像素大小)
pixelSizeY: Number The size of pixel in Y direction used to query visible rasters(用于查询可视栅格图层 Y 方向的像素大小)
relationParam: String The spatial relate function that can be applied while performing the query operation(可在执行查询操作时应用的空间相关参数)
returnDistinctValues: Boolean If true then returns distinct values based on the fields specified in outFields(条件符合则返回基于外部指定字段的确定值)
returnGeometry: Boolean If true, each feature in the FeatureSet includes the geometry(如果是真，则要素集中的每个要素都包括几何对象)
returnM: Boolean If true, M values will be included in the geometry if the features have M values(如果是真，则含有 M 值的要素的 M 值将被包含到几何对象中)
returnZ: Boolean If true, Z values will be included in the geometry if the features have Z values(如果是真，则含有 Z 值的要素的 Z 值将被包含到几何对象中)
spatialRelationship: String The spatial relationship to be applied on the input geometry while performing the query(执行查询操作时空间关系可被应用于输入的几何对象中)
text: String Shorthand for a literal search text on the display field, equivalent to: where YourDisplayField like '%SearchText%'(在显示字段中的搜索文本的速记)
timeExtent: TimeExtent The time instant or the time extent to query(查询的时间点或时间范围)
where: String A where clause for the query(查询语句的"where"子句)

QueryTask 功能示例代码如下：

```
<? xml version="1.0" encoding="utf-8"? >
<s:Application xmlns:fx="http://ns.adobe.com/mxml/2009"
               xmlns:s="library://ns.adobe.com/flex/spark"
               xmlns:mx="library://ns.adobe.com/flex/mx"
               xmlns:esri="http://www.esri.com/2008/ags"
               backgroundColor="0xEEEEEE"
               pageTitle="Query, then zoom to results">

    <s:layout>
        <s:VerticalLayout gap="0"/>
    </s:layout>

    <fx:Script>
        <! [CDATA[
            import com.esri.ags.FeatureSet;
            import com.esri.ags.utils.GraphicUtil;

            import mx.collections.ArrayCollection;
            import mx.collections.ArrayList;
            import mx.controls.Alert;
            import mx.rpc.AsyncResponder;

            privatefunction doQuery():void
            {
                //清理结果图层
                myGraphicsLayer.clear();

                queryTask.execute(query, new AsyncResponder(onRe-
sult, onFault));
                function onResult(featureSet:FeatureSet, token:Ob-
ject = null):void
                {

                    if (featureSet.features.length == 0)
                    {
                        Alert.show("No States found. Please try a-
gain.");
```

```
                        }
                        else
                        {
                            var graphicsExtent:Extent = GraphicUtil.
getGraphicsExtent(featureSet.features);
                            if (graphicsExtent)
                            {
                                map.extent = graphicsExtent;
                            }
                        }
                    }
                    function onFault(info:Object, token:Object = null):
void
                    {
                        Alert.show(info.toString());
                    }
                }

                privatefunction formatPopulation(item:Object, dgCol-
umn:GridColumn):String
                {
                    return numberFormatter.format(item[dgColumn.dat-
aField]);
                }
            ]]>
        </fx:Script>

        <fx:Declarations>
            <! -- Symbol for Query Result as Polygon -->
            <esri:SimpleFillSymbol id="sfs" alpha="0.7"/>

            <! -- Layer with US States -->
            <esri:QueryTask id="queryTask" url="http://sampleserv-
er6.arcgisonline.com/arcgis/rest/services/Census/MapServer/3"/>

            <esri:Query id="query"
                        outSpatialReference="{map.spatialReference}"
                        returnGeometry="true"
```

```
                     text="{fText.text}">
            <esri:outFields>
                <fx:String>STATE_NAME</fx:String>
                <fx:String>MED_AGE</fx:String>
                <fx:String>POP2007</fx:String>
            </esri:outFields>
        </esri:Query>
        <s:NumberFormatter id="numberFormatter"/>
    </fx:Declarations>

    <s:controlBarLayout>
        <s:VerticalLayout gap="10"
                          paddingBottom="7"
                          paddingLeft="10"
                          paddingRight="10"
                          paddingTop="7"/>
    </s:controlBarLayout>
    <s:controlBarContent>
        <s:RichText width="100%">
        //根据图层查询关键字,得到结果后返回结果表,并且将地图缩放到返回结
果的可视域
        </s:RichText>
            <s:HGroup width="100%" verticalAlign="baseline">
            <s:Label text="Search for U.S. States:"/>
            <s:TextInput id="fText"
                         enter="doQuery()"
                         text="Ca"/>
            <s:Button click="doQuery()" label="Search"/>
        </s:HGroup>
    </s:controlBarContent>

    <esri:Map id="map">
        <esri:extent>
            <esri:Extent xmin="-14000000" ymin="2800000" xmax="-
7000000" ymax="6400000">
                <esri:SpatialReference wkid="102100"/>
            </esri:Extent>
```

```
            </esri:extent>
            <esri:ArcGISDynamicMapServiceLayer
url = " http://sampleserver6.arcgisonline.com/arcgis/rest/services/
Census/MapServer" visibleLayers="{new ArrayList([3])}"/>
            <esri:GraphicsLayer id="myGraphicsLayer"
                                graphicProvider="{queryTask.
                                executeLastResult.features}"
                                symbol="{sfs}"/>
        </esri:Map>
        <s:DataGrid width="100% " height="40% "
                    dataProvider="{new ArrayCollection(queryTask.
                    executeLastResult.attributes)}">
            <s:columns>
                <s:ArrayList>
                    <s:GridColumn dataField="STATE_NAME" headerText="
State Name"/>
                    <s:GridColumn dataField="MED_AGE" headerText="Me-
dian Age"/>
                    <s:GridColumn dataField="POP2007"
                                  headerText="Population 2007"
                                  labelFunction="formatPopulation"/>
                </s:ArrayList>
            </s:columns>
        </s:DataGrid>
    </s:Application>
```

本程序的关键部分是 Query 控件。具体代码如下：

```
    <esri:Query id="query"
                    outSpatialReference="{map.spatialReference}"   //
参考坐标系,同 Map
                    returnGeometry="true"  //是否返回结果图形,是
                    text="{fText.text}"   //要检索的关键字,同 fText 中
的文字
            <esri:outFields>                  //返回的结果的字段
                <fx:String>STATE_NAME</fx:String>
                <fx:String>MED_AGE</fx:String>
                <fx:String>POP2007</fx:String>
            </esri:outFields>
```

```
</esri:Query>
```

当点击【查询】按钮后激活 doQuery()进行查询，并返回结果，如图 15.1 所示。

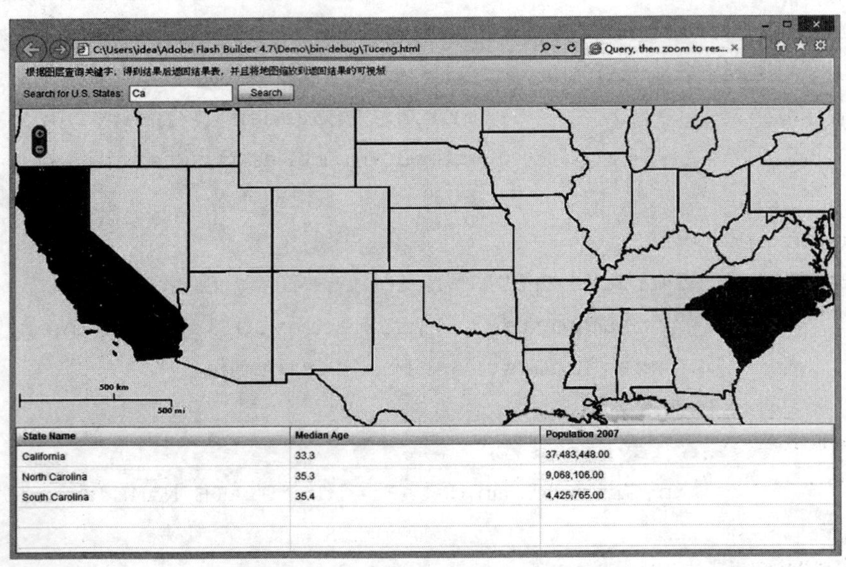

图 15.1 QueryTask 功能示例运行结果图

15.2 FindTask

FindTask 是在某个地图服务中进行属性查询的功能类。FindTask 与 QueryTask 的使用类似，QueryTask 在执行时需要一个 Query 对象作为参数，而 FindTask 则需一个 FindParameters 对象作为参数；FindTask 的 URL 属性指向所查询的地图服务的 REST URL，而 QueryTask 则需要指定子图层的 URL。

1. 属性

①executeLastResult：Array 返回的结果数组。

②gdbVersion：String 使用的 Geodatabase 版本。

2. 方法

①FindTask(url：String = null)：新建一个 FindTask 事件。

②execute(findParameters：FindParameters，responder：IResponder = null)：AsyncToken 根据提供的 ArcGIS REST，发送请求到 ArcGIS REST 地图服务，并进行搜索。

③executeComplete：当返回查询结果后进行的操作。

④Fault：查询失败后进行的操作。

其中，参数 FindParameters 的属性见表 15.4。

表 15.4　**参数 FindParameters 的属性**

属　性
contains：Boolean = true The contains parameter determines whether to look for an exact match of the search text or not(用来表明所包含的参数是否搜索到精确的文本匹配)
dynamicLayerInfos：Array Array of DynamicLayerInfos used to change the layer ordering or to redefine the map(dynamiclayerInfos 返回的列表用来改变图层顺序或者重新定义地图)
layerDefinitions：Array Array of layer definition expressions that allows you to filter the features of individual layers in the results (layerDefinitions 属性允许在结果中过滤单个图层的要素)
layerIds：Array The layers to perform the Find operation on(执行查找操作的图层)
maxAllowableOffset：Number The maximum allowable offset used for generalizing geometries returned by the Find operation(查找操作返回的几何图形的最大允许偏移量)
outSpatialReference：SpatialReference The well-known ID of the spatial reference of the output geometries(输出几何体空间参考系的通用 ID)
returnGeometry：Boolean = false If true, each feature in the FeatureSet includes the geometry(如果是真，则要素集中的每个要素包含几何对象)
returnM：Boolean If true, M values will be included in the geometry if the features have M values(如果是真，则含有 M 值的要素的 M 值将被包含到几何对象中)
returnZ：Boolean If true, Z values will be included in the geometry if the features have Z values(如果是真，则含有 Z 值的要素的 Z 值将被包含到几何对象中)
searchFields：Array A list of names of the fields to search(待搜索字段的名称列表)
searchText：String The search string text that is searched across the layers and the fields as specified in the layers and searchFields parameters(待搜索字符串文本作为特定的图层和搜索字段参数在图层和字段中进行搜索)

FindTask 功能示例代码如下：

```
<?xml version="1.0" encoding="utf-8"?>
<s:Application xmlns:fx="http://ns.adobe.com/mxml/2009"
               xmlns:s="library://ns.adobe.com/flex/spark"
```

```
                xmlns:mx="library://ns.adobe.com/flex/mx"
                xmlns:esri="http://www.esri.com/2008/ags"
                backgroundColor="0xCECECE"
                pageTitle="Find features in Map Layers">

    <s:layout>
    <s:VerticalLayout horizontalAlign="center"/>
  </s:layout>

  <fx:Script>
    <![CDATA[
        import com.esri.ags.Graphic;
        import com.esri.ags.events.FindEvent;
        import com.esri.ags.geometry.Geometry;

        import mx.collections.ArrayCollection;

        privatefunction doFind():void
        {
            findTask.execute(myFindParams);
        }

        privatefunction executeCompleteHandler(event:FindEvent):
void
        {
            myGraphicsLayer.clear();
            var graphic:Graphic;
            resultSummary.text = "Found " + event.findResults.
length + " results.";
            var resultCount:int = event.findResults.length;
            for (var i:Number= 0; i < resultCount; i++)
            {
                graphic = event.findResults[i].feature;
                graphic.toolTip = event.findResults[i].foundField
Name + ": " + event.findResults[i].value;

                switch (graphic.geometry.type)
                {
```

```
                case Geometry.MAPPOINT:
                {
                    graphic.symbol = smsFind;
                    break;
                }
                case Geometry.POLYLINE:
                {
                    graphic.symbol = slsFind;
                    break;
                }
                case Geometry.POLYGON:
                {
                    graphic.symbol = sfsFind;
                    break;
                }
            }
            myGraphicsLayer.add(graphic);
        }
    }
    ]]>
</fx:Script>

<fx:Declarations>
    <!-- Symbol for Find Result as Polyline -->
    <esri:SimpleLineSymbol id="slsFind"
                           width="3"
                           alpha="0.9"
                           color="0xFFFF00"
                           style="solid"/>

    <!-- Symbol for Find Result as Point -->
    <esri:SimpleMarkerSymbol id="smsFind"
                             alpha="0.9"
                             color="0xFFFF00"
                             size="11"
                             style="square">
        <esri:SimpleLineSymbol color="0x000000"/>
    </esri:SimpleMarkerSymbol>
```

```
        <!-- Symbol for Find Result as Polygon -->
        <esri:SimpleFillSymbol id="sfsFind"
                                    alpha="0.7"
                                    color="0xFFFF00"/>

        <!-- Find Task -->
        <esri:FindTask id="findTask"
                            executeComplete="executeCompleteHandler
                            (event)"
url = " http://sampleserver1.arcgisonline.com/ArcGIS/rest/services/
Specialty/ESRI_StatesCitiesRivers_USA/MapServer"/>
        <esri:FindParameters id="myFindParams"
                                contains="true"
                                layerIds="[0,1,2]"
                                outSpatialReference="{map.
                                spatialReference}"
                                returnGeometry="true"
searchFields="[CITY_NAME,NAME,SYSTEM,STATE_ABBR,STATE_NAME]"
                                searchText="{fText.text}"/>
    </fx:Declarations>

    <s:controlBarLayout>
        <s:VerticalLayout gap="10"
                            paddingBottom="7"
                            paddingLeft="10"
                            paddingRight="10"
                            paddingTop="7"/>
    </s:controlBarLayout>
    <s:controlBarContent>
        <s:RichText width="100%">
    //此示例演示如何使用FindTask搜索基于字符串值的地图服务的多层次特征
        </s:RichText>
        <s:HGroup width="100%"
                    horizontalAlign="center"
                    verticalAlign="middle">
            <s:Label text="Search for names of States, Cities, and Riv-
ers:"/>
```

```
            <s:TextInput id="fText"
                         enter="doFind()"
                         maxWidth="400"
                         text="Miss"/>
            <s:Button click="doFind()" label="Find"/>
            <s:Label id="resultSummary"/>
        </s:HGroup>
    </s:controlBarContent>

    <esri:Map id="map">
        <esri:extent>
            <esri:Extent xmin="-125.34967" ymin="25.42261" xmax=
"-66.46295" ymax="49.51035">
                <esri:SpatialReference wkid="4326"/>
            </esri:Extent>
        </esri:extent>
        <esri:ArcGISDynamicMapServiceLayer
    url = " http://sampleserver1.arcgisonline.com/ArcGIS/rest/serv-
ices/Specialty/ESRI_StatesCitiesRivers_USA/MapServer"/>
        <esri:GraphicsLayer id="myGraphicsLayer"/>
    </esri:Map>
    <s:DataGrid width="100%" height="40%"
                dataProvider="{new ArrayCollection(findTask.
                executeLastResult)}">
            <s:columns>
                <s:ArrayList>
                    <s:GridColumn width="70"
                                  dataField="layerId"
                                  headerText="Layer ID"/>
                    <s:GridColumn dataField="layerName" headerText="
Layer Name"/>
                     <s:GridColumn dataField="foundFieldName" header-
Text="Found Field Name"/>
                    <s:GridColumn dataField="value" headerText="Found
Field Value"/>
                </s:ArrayList>
            </s:columns>
        </s:DataGrid>
```

</s:Application>

注释：本程序的主要部分为 FindTask，具体设置如下(详细参数见上面属性说明表)：

<esri:FindTask id="findTask" executeComplete="executeCompleteHandler(event)"//查询结果处理函数

url = " http://sampleserver1.arcgisonline.com/ArcGIS/rest/services/Specialty/ESRI_StatesCitiesRivers_USA/MapServer" //要查询的地图

/>

<esri:FindParameters id="myFindParams" //参数设定

contains="true" //是否完全匹配,是

layerIds="[0,1,2]" //要进行搜索的图层

outSpatialReference="{map.spatialReference}" //参考坐标系

returnGeometry="true" //返回结果图形,是

searchFields="[CITY_NAME,NAME,SYSTEM,STATE_ABBR,STATE_NAME]"检索的字段

searchText="{fText.text}" //搜索的关键字

/>

FindTask 功能实现如图 15.2 所示。

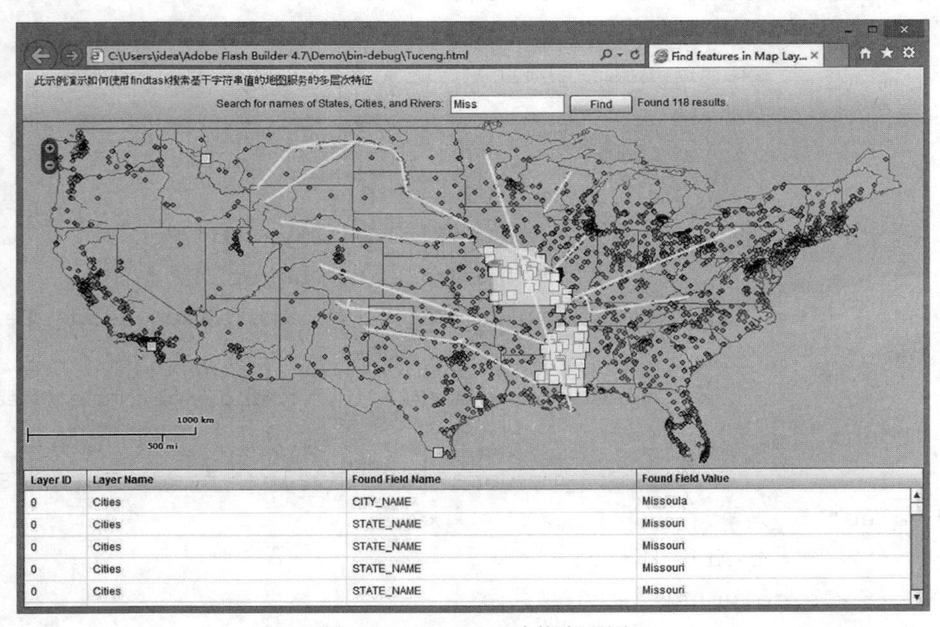

图 15.2 FindTask 功能实现图

15.3 IdentifyTask

IdentifyTask 是一个在地图服务中识别要素的功能类。当用户在客户端使用 Draw 工具绘制了一个几何对象后，此几何对象就可以作为 IdentifyTask 的参数发送到服务器进行识别，满足条件的要素将会被输出到 ArcGIS Flex API 中，这些要素也可以作为 Graphic 被添加到地图上。IdentifyTask 的属性、方法和事件分别见表 15.5、表 15.6 和表 15.7。

表 15.5　**IdentifyTask 属性**

executeLastResult: Array Array of IdentifyResult object returned from last call to the execute function(对最后执行函数请求返回的调用结果进行排序)
gdbVersion: String Geodatabase version to use(地理数据库的在用版本)

表 15.6　**IdentifyTask 方法**

IdentifyTask(url: String = null) Creates a new IdentifyTask object(创建一个新的 IdentifyTask 对象)
execute(identifyParameters: IdentifyParameters, responder: IResponder = null): AsyncToken Sends a request to the ArcGIS REST map service resource to identify features based on the IdentifyParameters specified in the identifyParameters argument(发送请求到 ArcGIS REST 地图资源器以识别在 IdentifyParameters 中指定参数的要素)

表 15.7　**IdentifyTask 事件**

executeComplete Dispatched when execute operation successfully completes(当执行操作完成后通过派遣消息执行相应的事件)
fault Dispatched when an IdentifyTask fails(当执行识别任务失败后通过派遣消息执行相应的事件)

IdentifyTask 功能示例代码如下:

```
<?xml version="1.0" encoding="utf-8"?>
<s:Application xmlns:fx="http://ns.adobe.com/mxml/2009"
               xmlns:s="library://ns.adobe.com/flex/spark"
               xmlns:esri="http://www.esri.com/2008/ags"
               initialize="initializeHandler(event)"
               pageTitle="Identify Features on the Map">
```

```
<fx:Style>
    @ namespace esri "http://www.esri.com/2008/ags";
    esri|ContentNavigator
    {
        headerColor: #FFFFFF;
        headerBackgroundAlpha: 1.0;
        headerBackgroundColor: #353930;
    }

    esri|InfoWindow
    {
        backgroundAlpha: 0.8;
        backgroundColor: #353930;
        borderThickness: 0;
        infoPlacement: top;
        shadowAlpha: 0.5;
        shadowDistance: 10;
        upperRightRadius: 0;
    }
</fx:Style>

<fx:Script>
    <![CDATA[
        import com.esri.ags.Graphic;
        import com.esri.ags.components.ContentNavigator;
        import com.esri.ags.events.MapMouseEvent;
        importcom.esri.ags.geometry.Extent;
        import com.esri.ags.geometry.Geometry;
        importcom.esri.ags.symbols.InfoSymbol;
        import com.esri.ags.tasks.supportClasses.
IdentifyParameters;
        import com.esri.ags.tasks.supportClasses.
IdentifyResult;
        importcom.esri.ags.utils.GraphicUtil;

        import mx.collections.ArrayList;
        import mx.controls.Alert;
        import mx.events.FlexEvent;
```

```
            import mx.rpc.AsyncResponder;
            importmx.utils.GraphicsUtil;

            import spark.events.IndexChangeEvent;

            privatevar contentNavigator:ContentNavigator;

            protectedfunction initializeHandler(event:Flex-
Event):void
            {
                contentNavigator = new ContentNavigator();
                contentNavigator.addEventListener(IndexChangeEvent.
CHANGE, contentNavigator_indexChangeEventHandler, false, 0, true);
                contentNavigator.addEventListener(Event.CLOSE,
contentNavigator_closeEventHandler, false, 0, true);
            }

            privatefunction myMap_mapClickHandler(event:MapMou-
seEvent):void
            {
                myMap.defaultGraphicsLayer.clear();

                var identifyParams:IdentifyParameters = new Identi-
fyParameters();
                identifyParams.layerOption = IdentifyParameters.
LAYER_OPTION_VISIBLE;
                identifyParams.returnGeometry = true;
                identifyParams.tolerance = 3;
                identifyParams.width = myMap.width;
                identifyParams.height = myMap.height;
                identifyParams.geometry = event.mapPoint;
                identifyParams.mapExtent = myMap.extent;
                identifyParams.spatialReference = myMap.
spatialReference;

                clickLocation = event.mapPoint;
                var clickGraphic:Graphic = new Graphic(event.
mapPoint, clickPtSym);
```

```
            myMap.defaultGraphicsLayer.add(clickGraphic);

             identifyTask.execute(identifyParams, new AsyncRe-
sponder(myResultFunction, myFaultFunction));
        }

        privatefunction myResultFunction(results:Array, to-
ken:Object = null):void
        {
            if (results && results.length > 0)
        {
            var list:ArrayList = new ArrayList();
            for (var i:int = 0; i < results.length; i++)
            {
                var result:IdentifyResult = results[i];
                list.addItem(result.feature);
            }

            contentNavigator.dataProvider = list;
            myMap.infoWindowContent = contentNavigator;
            myMap.infoWindow.show(clickLocation);
        }
        myMap.defaultGraphicsLayer.clear();
    }

    privatefunction myFaultFunction(error:Object, token:Ob-
ject = null):void
    {
        Alert.show(String(error), "Identify Error");
    }

    protectedfunction
contentNavigator_indexChangeEventHandler(event:IndexChangeEvent):
void
    {
        var graphic:Graphic =
ContentNavigator(event.currentTarget).selectedItem as Graphic;
   showHighlightCurrentGraphic(graphic);
```

```
        }

        protectedfunction contentNavigator_closeEventHandler(e-
vent:Event):void
        {
            myMap.defaultGraphicsLayer.clear();
            identifyGraphicsLayer.clear();
        }

        protectedfunction showHighlightCurrentGraphic(graphic:
Graphic):void
        {
            identifyGraphicsLayer.clear();
            var currentGraphic:Graphic = graphic;
            switch (currentGraphic.geometry.type)
            {
                case Geometry.MAPPOINT:
                {
                    currentGraphic.symbol = smsIdentify;
                    break;
                }
                case Geometry.POLYLINE:
                {
                    currentGraphic.symbol = slsIdentify;
                    break;
                }
                case Geometry.POLYGON:
                {
                    currentGraphic.symbol = sfsIdentify;
                    break;
                }
            }
            identifyGraphicsLayer.add(currentGraphic);
        }
    ]]>
</fx:Script>

<fx:Declarations>
```

```
        <esri:MapPoint id="clickLocation"/>
        <!-- Symbol for where the user clicked -->
        <esri:SimpleMarkerSymbol id="clickPtSym"
                                 color="0xFF0000"
                                 size="12"
                                 style="cross">
            <esri:SimpleLineSymbol width="2" color="0xFFFFFF"/>
        </esri:SimpleMarkerSymbol>

        <!-- Symbol for Identify Result as Point -->
        <esri:SimpleMarkerSymbol id="smsIdentify"
                                 alpha="0.5"
                                 color="0x00FFFF"
                                 size="20"
                                 style="circle"/>

        <!-- Symbol for Identify Result as Polyline -->
        <esri:SimpleLineSymbol id="slsIdentify"
                               width="10"
                               alpha="0.5"
                               color="0x00FFFF"
                               style="solid"/>

        <!-- Symbol for Identify Result as Polygon -->
        <esri:SimpleFillSymbol id="sfsIdentify"
                               alpha="0.5"
                               color="0x00FFFF"/>

        <!-- Identify Task -->
        <esri:IdentifyTask id="identifyTask"
                           concurrency="last"
                           showBusyCursor="true"
url = " http://sampleserver6.arcgisonline.com/arcgis/rest/services/
USA/MapServer"/>
    </fx:Declarations>

    <s:controlBarContent>
        <s:RichText width="100% ">
```

```
        点击地图上的目标点,程序返回鼠标点中的所有层里面的对象
        </s:RichText>
    </s:controlBarContent>

    <esri:Map id="myMap" mapClick="myMap_mapClickHandler(event)">
        <esri:extent>
            <esri:WebMercatorExtent minlon="-120" minlat="30"
maxlon="-100" maxlat="50"/>
        </esri:extent>
    <esri:ArcGISTiledMapServiceLayerurl="http://server.arcgisonline.
com/ArcGIS/rest/services/Canvas/World_Light_Gray_Base/MapServer"/>
        <esri:ArcGISDynamicMapServiceLayer
url="http://sampleserver6.arcgisonline.com/arcgis/rest/services/USA/
MapServer"/>
        <esri:GraphicsLayer id="identifyGraphicsLayer"/>
        </esri:Map>

    </s:Application>
```

提示：基本原理是，页面加载同时添加一个内容导航，地图加载同时绑定一个点击事件，当地图被点击时激活事件，对地图中的鼠标的点击位置进行检索，将检索的结果返回到地图上高亮显示，同时新建一个 InfoWindow 承载返回的属性信息。功能实现如图 15.3 所示。

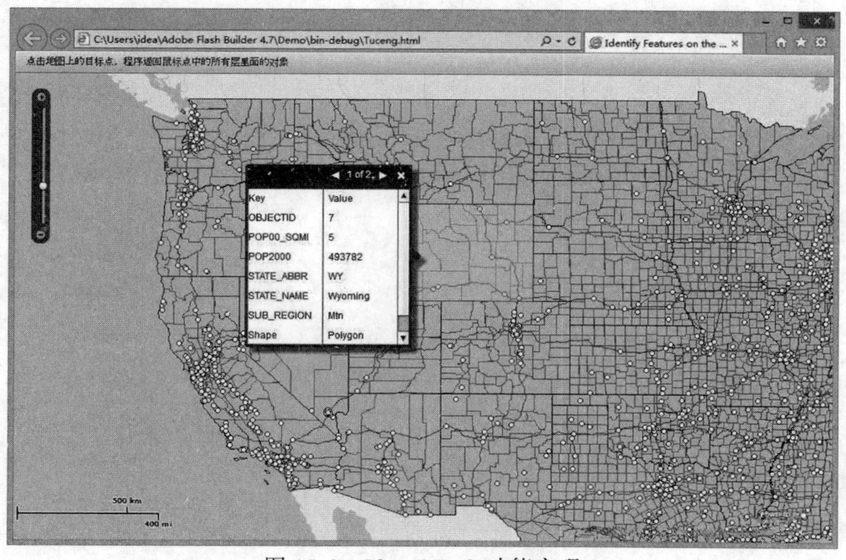

图 15.3　IdentifyTask 功能实现

15.4 InfoWindow

InfoWindow 是 ArcGIS API for Flex 中提供的类似于标注的窗口，可以用来显示用户自定义的信息，可以是文本、图片，也可以是复杂的自定义组件。

使用 InfoWindow，只要设置 Map 的 infoWindow 属性即可，示例代码如下：

```
var canvas:Canvas =new Canvas();
var txtTem:Text =new Text();
txtTem.Text = "aa";
canvas.addChild(txtTem);
myMap.infoWindow.content = canvas;
var mapPnt2:MapPoint =new MapPoint(114.1547298,30.5127677);
myMap.infoWindow.show(mapPnt2);
```

提示：content 设置 infoWindow 的内容，show 方法设置 infoWindow 的显示位置。关于 InfoWindow 的使用已经在 15.3 节中(IdentifyTask)介绍过，示例和代码请参考上例。

第 16 章　地理处理功能实现

16.1　几何服务示例

Geometry Service 提供针对几何层级的服务，如 Project、Simplify、Buffer、Areas And Lengths 等。下面以 Buffer 为例，实现在 ArcGIS API for Flex 中调用 Geometry Service。

首先，使用<esri：GeometryService>标签定义一个 GeometryService 对象：

```
<esri:GeometryService id="myGeometryService"
url="http://sampleserver2.arcgisonline.com/ArcGIS/rest/services/
Geometry/GeometryServer"/>
```

id 唯一标识 Geometry Service，url 指定提供 Geometry Service 的地址。

和 Identify 工具类似，进行 Buffer 操作首先要创建一个需要做 Buffer 的几何，然后定义一个 BufferParameters，执行 Buffer 操作，最后将 Buffer 的结果绘制到 GraphicsLayer 上。BufferParameters 参数的意义：distances 为 Buffer 半径，features 为需要做 Buffer 的要素集合，unit 为单位，BufferSpatialReference 为 Buffer 操作时的空间参照系。

Buffer 调用的方法如下：

```
var bufferParameters : BufferParameters = new BufferParameters();
bufferParameters.features = [point];
bufferParameters.distances = [3000];
bufferParameters.unit = BufferParameters.UNIT_METER;
bufferParameters.bufferSpatialReference = newSpatialReference (
602113);
myGeometryService.addEventListener
(GeometryServiceEvent.BUFFER_COMPLETE, bufferCompleteHandler);
myGeometryService.buffer( bufferParameters );
```

注意：关键的部分是 BufferParameters. features 参数设置的 features 一定要有空间参考系，不然 Buffer 就不成功。

完整的示例代码如下：

```
<? xml version="1.0" encoding="utf-8"? >
<s:Application xmlns:fx="http://ns.adobe.com/mxml/2009"
               xmlns:s="library://ns.adobe.com/flex/spark"
               xmlns:esri="http://www.esri.com/2008/ags"
```

```
                initialize="initializeHandler(event)"
                pageTitle=" Perform geodesic buffering using the
GeometryService">

        <fx:Style>
        @namespace s "library://ns.adobe.com/flex/spark";
        s|Panel
        {
            backgroundColor: #93AED7;
            chromeColor: #6784BE;
            color: #FFFFFF;
        }
        s|Button
        {
            chromeColor: #000000;
            color: #EDEDE6;
        }
    </fx:Style>

    <fx:Script>
        <![CDATA[
            import com.esri.ags.Graphic;
            import com.esri.ags.events.DrawEvent;
            import com.esri.ags.geometry.Polygon;
            import com.esri.ags.tasks.supportClasses.BufferParameters;

            import mx.controls.Alert;
            import mx.events.FlexEvent;
            import mx.rpc.AsyncResponder;
            import mx.rpc.Fault;

            privatevar distance:Number = 750;
            privatevar unit:Number = GeometryService.UNIT_KILOME-
TER;

            privatevar euclideanBuffer:BufferParameters = new
BufferParameters();
            privatevar geodesicBuffer:BufferParameters = new Buff-
```

```
erParameters();

                    protectedfunction initializeHandler(event:Flex-
Event):void
                {
                    drawTool.activate(DrawTool.FREEHAND_POLYLINE);
                    drawGraphicsLayer.clear();
                    userResultsGraphicsLayer.clear();
                    legend.layers = [graphicsLayer, euclideanGraphics-
Layer, geodesicGraphicsLayer];
                }

                    protectedfunction esriService_faultHandler(fault:
Fault, token:Object = null):void
                {
                    Alert.show("Error: " + fault.faultString, "Error
code: "+ fault.faultCode);
                }

                protectedfunction drawTool_drawEndHandler(event:Draw-
Event):void
                {
                    var userInputPolyline:Polyline = event.graphic.
geometry as Polyline;
                    drawTool.deactivate();

                    euclideanBuffer.geometries=[userInputPolyline];
                    euclideanBuffer.distances = [ distance ];
                    euclideanBuffer.unit = unit;
                    euclideanBuffer.bufferSpatialReference=map.
spatialReference;
                    euclideanBuffer.outSpatialReference=map.
spatialReference;
                    euclideanBuffer.unionResults = true;

                    geodesicBuffer.geometries = [ userInputPolyline ];
                    geodesicBuffer.distances = [ distance ];
                    geodesicBuffer.unit = unit;
```

```
                geodesicBuffer.outSpatialReference = map.
spatialReference;
                geodesicBuffer.geodesic = true;
                geodesicBuffer.unionResults = true;

                geometryService.buffer(euclideanBuffer, new
AsyncResponder(euclideanBufferCompleteHandler, esriService_fault-
Handler, { userBufferType:'euclidean'}));
                geometryService.buffer(geodesicBuffer, new
AsyncResponder(geodesicBufferCompleteHandler, esriService_fault-
Handler, { userBufferType:'geodesic'}));
            }

            privatefunction euclideanBufferCompleteHandler(item:
Object, token:Object = null):void
            {
                //Note: As of version 2.0, GeometryService returns
geometries (instead of graphics)
                var euclideanBufferResultsArray:Array = item as Ar-
ray; //array of geometries
                if (token && token.userBufferType)
                {
                    for (var j:int = 0; j < euclideanBufferResultsAr-
ray.length; j++)
                    {
                        var euclideanUserGraphic:Graphic = new
Graphic(euclideanBufferResultsArray[j] as Polygon, euclideanSym-
bol);
                        userResultsGraphicsLayer.add(euclidean-
UserGraphic);
                    }

                }
                else
                {
                    for (var i:int = 0; i < euclideanBufferResultsAr-
ray.length; i++)
                    {
```

```
                        euclideanGraphicsLayer.add(new
Graphic(euclideanBufferResultsArray[i] as Polygon, null, null));
                    }
                }
            }

                privatefunction geodesicBufferCompleteHandler
(item:Object, token:Object = null):void
                {
                    var geodesicBufferResultsArray:Array = item as
Array; //array of geometries
                    if (token && token.userBufferType)
                    {
                        for (var j:int = 0; j < geodesicBufferResult-
sArray.length; j++)
                        {
                            var geodesicUserGraphic:Graphic = new
Graphic(geodesicBufferResultsArray[j] as Polygon, geodesicSymbol);
                            userResultsGraphicsLayer.add(geodesi-
cUserGraphic);
                        }
                    }
                    else
                    {
                        for (var i:int = 0; i < geodesicBufferResult-
sArray.length; i++)
                        {
                            geodesicGraphicsLayer.add(new
Graphic(geodesicBufferResultsArray[i] as Polygon, null, null));
                        }
                    }
                }

        ]]>
    </fx:Script>

    <fx:Declarations>
        <esri:GeometryService id="geometryService"
```

```
                                showBusyCursor="true"
url="http://sampleserver6.arcgisonline.com/arcgis/rest/services/U-
tilities/Geometry/GeometryServer"/>
        <esri:DrawTool id="drawTool"
                        drawEnd="drawTool_drawEndHandler(event)"
                        graphicsLayer="{drawGraphicsLayer}"
                        lineSymbol="{lineSymbol}"
                        map="{map}"/>
        <esri:SpatialReference id="sr" wkid="102100"/>
        <esri:Extent id="initialExtent"
                      xmin = "-22437625" ymin = "-12705101" xmax = "
25581950" ymax="15472645"
                      spatialReference="{sr}"/>

        <esri:Polyline  spatialReference="{sr}"/>

        <!-- Symbol for line to be buffered -->
        <esri:SimpleLineSymbol id="lineSymbol"
                               width="4"
                               color="0xC7B878"/>
        <!-- Symbol for regular buffer operation -->
        <esri:SimpleFillSymbol id="euclideanSymbol" style="null">
            <esri:SimpleLineSymbol width="3" color="0xDB2B2B"/>
        </esri:SimpleFillSymbol>
        <!-- Symbol for geodesic buffer operation -->
        <esri:SimpleFillSymbol id="geodesicSymbol" style="null">
            <esri:SimpleLineSymbol width="3" color="0x287D09"/>
        </esri:SimpleFillSymbol>
    </fx:Declarations>

    <s:controlBarContent>
        <s:RichText width="100%">
            //点击鼠标左键开始画线,左键抬起后自动进入缓冲区分析
        </s:RichText>
    </s:controlBarContent>

    <esri:Map id="map"
            extent="{initialExtent}"
```

```
            wrapAround180="true">
        <esri:ArcGISTiledMapServiceLayer
url=" http://services.arcgisonline.com/arcgis/rest/services/Ocean/
World_Ocean_Base/MapServer"/>
        <esri:ArcGISTiledMapServiceLayer
url=" http://services.arcgisonline.com/arcgis/rest/services/Ocean/
World_Ocean_Reference/MapServer"/>
        <esri:GraphicsLayer id="geodesicGraphicsLayer" name="测
地线缓冲"
                           symbol="{geodesicSymbol}"/>
        <esri:GraphicsLayer id="euclideanGraphicsLayer" name="欧
式缓冲"
                           symbol="{euclideanSymbol}"/>
        <esri:GraphicsLayer id="graphicsLayer" name="画图层
(lines being buffered)"
                           symbol="{lineSymbol}"/>
        <esri:GraphicsLayer id="userResultsGraphicsLayer"/>
        <esri:GraphicsLayer id="drawGraphicsLayer"/>
    </esri:Map>
    <s:Panel width="300"
            right="20" top="20"
            title="Geodesic Buffer using the Geometry Service">
        <s:layout>
            <s:VerticalLayout gap="10"
                              paddingLeft="5"
                              paddingRight="5"
                              paddingTop="5"/>
        </s:layout>

        <s:Line width="100%">
            <s:stroke>
                <s:SolidColorStroke color="0xFFFFFF" weight="1"/>
            </s:stroke>
        </s:Line>
        <esri:Legend id="legend" map="{map}"/>
    </s:Panel>
</s:Application>
```

示例实现结果如图16.1所示。

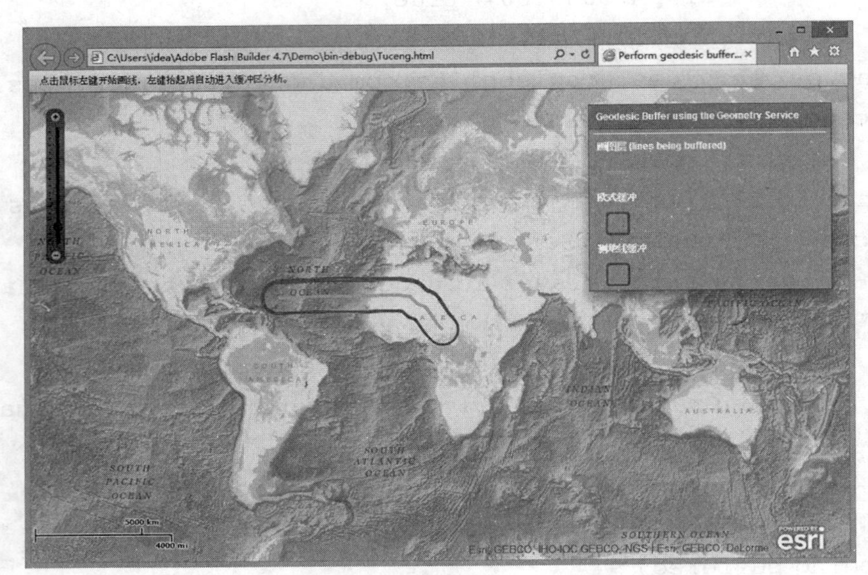

图 16.1　几何服务调用实现图

注意：在安装完成 ArcGIS Server 后，本机自带几何服务，如图 16.2 所示。右击【服务属性】，弹出【服务编辑器】，如图 16.3 所示，即在功能选项中的 REST URL 填写调用的几何服务地址。

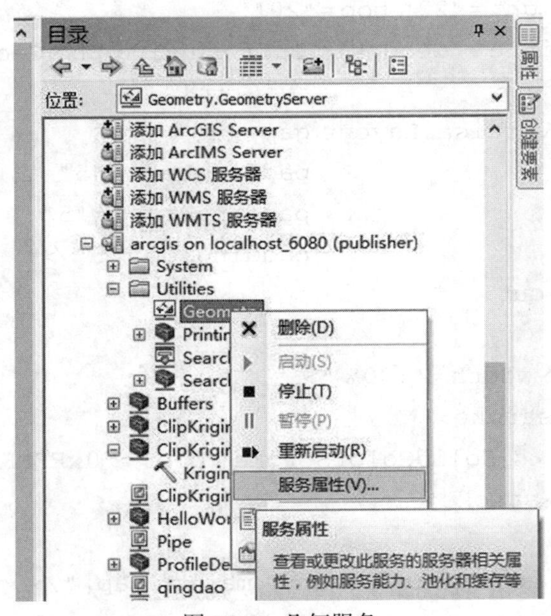

图 16.2　几何服务

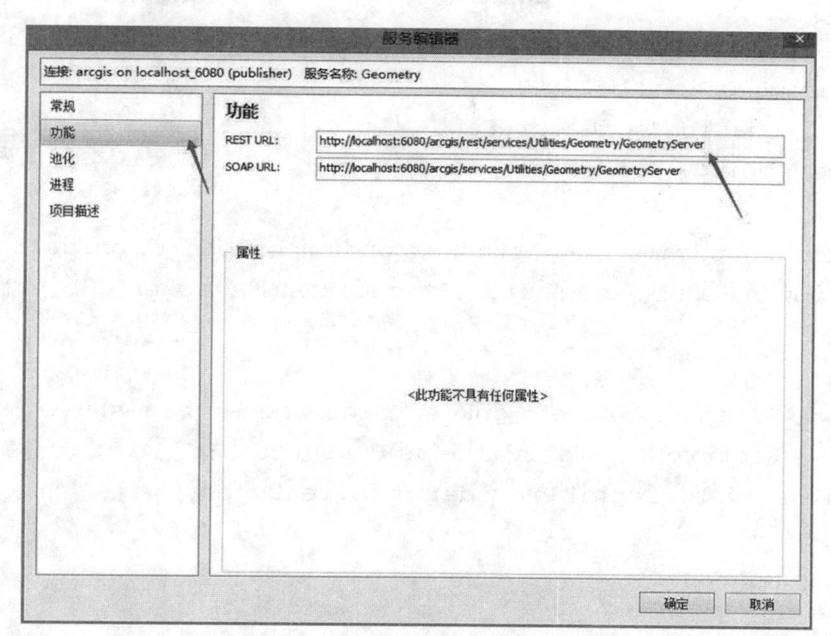

图 16.3　服务编辑器

16.2　GP 服务调用

在 ArcGIS API for Flex 中可以使用 GP 服务来进行地理处理。在 ArcGISOnline 上的 GP 服务有 CreateDriveTimePolygons 和 Viewshed 等，下面以 CreateDriveTimePolygons 服务为例来实现在 ArcGIS API for Flex 中调用 GP 服务。

①首先，使用<esri：Geoprocessor >标签定义一个 GP 服务，url 指向提供 GP 服务的地址。

```
<esri:Geoprocessor id="gp"
url = " http://sampleserver1.arcgisonline.com/ArcGIS/rest/services/
Network/ESRI_DriveTime_US/GPServer/CreateDriveTimePolygons" />
```

②定义 GP 对象后，再定义一个 GP 的参数对象来传递参数。

```
var params:Object = {
"Input_Location" : featureSet,
"Drive_Times" : driveTimes
};
```

其中，参数 Input_Location 和 Drive_Times 与发布的 GP 服务相关，不同的 GP 服务有不同的参数。

③定义好 GP 对象和 GP 参数对象之后，就可实现在地图上单击，然后计算 DriveTimes，并把得到的结果绘制在地图上的功能。

其中，GP 调用的代码如下：

```
var featureSet:FeatureSet = new FeatureSet([graphic]);
var params:Object = {
"Input_Location" : featureSet,
"Drive_Times" : driveTimes
};
gp.execute(params, new AsyncResponder( onResult, onFault ));
function onResult(gpResult : ExecuteResult, token : Object = null)
: void
{
var pv : ParameterValue = gpResult.parameterValues[0];
var fs : FeatureSet = pv.value as FeatureSet;
graphicsLayer.graphicProvider = fs.features;
}
function onFault( info : Object, token : Object = null ) : void
{
Alert.show( info.toString() );
}
```

定义好 GP 之后，直接调用 GP 的 execute 的方法即可，成功响应 onResult，失败响应 onFault。

GP 服务调用完整示例代码如下：

```
<?xml version="1.0" encoding="utf-8"?>
<s:Application xmlns:fx="http://ns.adobe.com/mxml/2009"
               xmlns:s="library://ns.adobe.com/flex/spark"
               xmlns:mx="library://ns.adobe.com/flex/mx"
xmlns:supportClasses="com.esri.ags.skins.supportClasses.*" minWidth="
955" minHeight="600" xmlns:esri="http://www.esri.com/2008/ags">
    <s:layout>
        <supportClasses:AttachmentLayout/>
    </s:layout>
    <fx:Script>
        <![CDATA[
            import com.esri.ags.FeatureSet;
            import com.esri.ags.Graphic;
            import com.esri.ags.geometry.MapPoint;
            import com.esri.ags.symbols.Symbol;
            import com.esri.ags.tasks.supportClasses.ExecuteResult;
            import com.esri.ags.tasks.supportClasses.ParameterValue;
```

```
        import mx.controls.Alert;
        import mx.rpc.AsyncResponder;
        privatevar driveTimes:String = "1 2 3";
        privatefunction computeServiceArea(event : MouseEvent):
void
        {
            graphicsLayer.clear();
            var mapPoint : MapPoint = myMap.toMapFromStage
(event.stageX, event.stageY);

            var graphic : Graphic = new Graphic(mapPoint,
sms_circleAlphaSizeOutline);

            graphicsLayer.add(graphic);

            var featureSet:FeatureSet = new FeatureSet([graphic]);

            var params:Object = {
                "Input_Location" : featureSet,
                "Drive_Times" : driveTimes
            };

            gp.execute(params, new AsyncResponder(onResult,
onFault ));

            function onResult(gpResult :ExecuteResult,token:
Object = null) : void
            {
                var pv : ParameterValue = gpResult.results[0];
                var fs : FeatureSet = pv.value as FeatureSet;
                graphicsLayer.graphicProvider = fs.features;
            }
            function onFault(info:Object, token:Object =null):void
            {
                Alert.show( info.toString() );
            }
        }
```

```
            privatefunction fillFunc(g: Graphic) : Symbol
            {
                var toBreak : Number = g.attributes.ToBreak;
                if ( toBreak == 1 )
                {
                    return rFill;
                }
                if ( toBreak == 2 )
                {
                    return gFill;
                }
                return bFill;
            }
        ]]>

    </fx:Script>
    <fx:Declarations>
        <esri:SimpleMarkerSymbol id="sms_circleAlphaSizeOutline"
color="0xFF0000"  alpha="0.5" size="15" style="circle"/>
        <esri:SimpleFillSymbol id="rFill" alpha="0.5" color="
0xFF0000"/>
        <esri:SimpleFillSymbol id="gFill" alpha="0.5" color="
0x00FF00"/>
        <esri:SimpleFillSymbol id="bFill" alpha="0.5" color="
0x0000FF"/>
        <esri:UniqueValueRenderer id="uniqueValueRenderer"
field="ToBreak">
            <esri:UniqueValueInfo symbol="{rFill}" value="1"/>
            <esri:UniqueValueInfo symbol="{gFill}" value="2"/>
            <esri:UniqueValueInfo symbol="{bFill}" value="3"/>
        </esri:UniqueValueRenderer>
        <esri:Geoprocessor id="gp"
url = " http://sampleserver1.arcgisonline.com/ArcGIS/rest/services/
Network/ESRI_DriveTime_US/GPServer/CreateDriveTimePolygons"  />
    </fx:Declarations>
    <esri:Map id="myMap" width="500" height="500" click="comp-
uteServiceArea(event)"  openHandCursorVisible="false">
        <esri:extent>
```

```
            <esri:Extent xmin="-95.41" ymin="38.86" xmax="-95.1"
  ymax="39.06"/>
        </esri:extent>
        <esri:ArcGISTiledMapServiceLayer
url = " http://server.arcgisonline.com/ArcGIS/rest/services/ESRI_
StreetMap_World_2D/MapServer"/>
        <esri:GraphicsLayer id="graphicsLayer" renderer="
{uniqueValueRenderer}"/>
    </esri:Map>
  </s:Application>
```

GP 服务调用示例实现结果如图 16.4 所示。

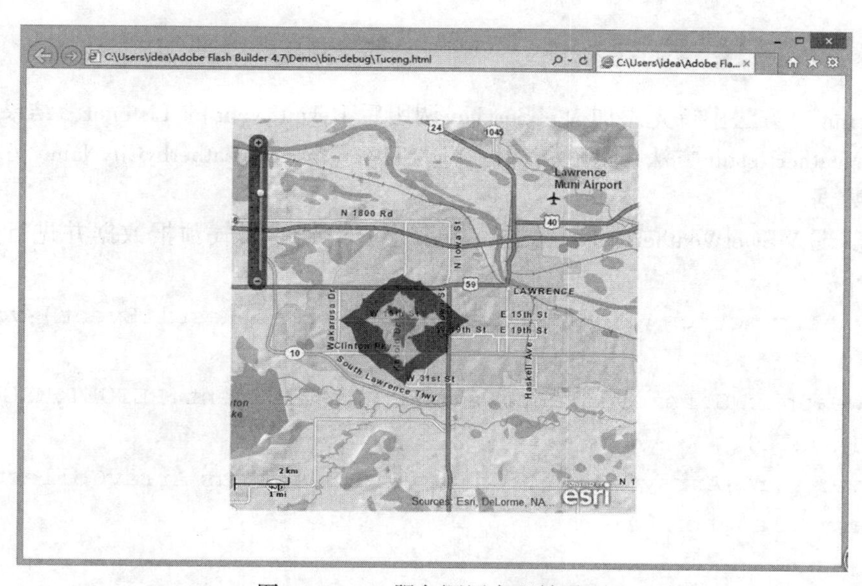

图 16.4　GP 服务调用实现结果图

16.3　Web Service 调用

Flex 本身对 WebServices 有着良好的支持，可以调用互联网上的各种 WebServices 来结合 ESRI 的 Map 做出自己想要的东西。

下面以在地图上显示天气预报为例，介绍 Flex 中调用 WebServices 的方法：

①首先，找到提供天气预报的 WebServices。

(http://www.webxml.com.cn/WebServices/WeatherWebService.asmx?)

②然后，使用<mx: WebService>标签定义一个 WebService 对象。

```
<mx:WebService id="weatherWS"
```

```
wsdl = "http://www.webxml.com.cn/WebServices/WeatherWebService.asmx?
WSDL"
showBusyCursor = "true" />
```

id 唯一标识 WebService，wsdl 指向提供 WebService 的地址。

③Application 创建完成后，调用这个 WebService，得到结果后直接显示到 Map 上。在 <s：Application>标签的 creationComplete 事件绑定 Init( )方法。

creationComplete = "Init( )"

使用 ActionScript 脚本实现 Init( )方法：

```
private function Init():void
{
weatherWS.addEventListener(ResultEvent.RESULT, WSGetWeatherRe-
sult);
weatherWS.getWeatherbyCityName("武汉");
}
```

④在 Init( )方法中首先添加 WebService 调用后 ResultEvent 的 Listener，结果返回后响应 WSGetWeatherResult 方法。然后，调用 WebService 的 getWeatherbyCityName 方法获取天气预报的数据。

下面实现 WSGetWeatherResult 方法，用来读取得到的天气预报数据并进行处理。具体代码如下：

```
private function WSGetWeatherResult(event:ResultEvent):void
{
    weatherWS.removeEventListener(ResultEvent.RESULT,WSGet-
WeatherResult);
    var arrC:ArrayCollection =event.result as ArrayCollection;
if(arrC.length >0)
{
        var str:String = arrC.getItemAt(0).toString();
        var str2:String = arrC.getItemAt(1).toString();
        var vbox:VBox =new VBox();
        var vbox2:VBox =new VBox();
        var hbox:HBox =new HBox();
        var canvas:Canvas =new Canvas();
        var path:String ="assets\\weather\\";
        var img1:Image =new Image;
        var index1:int=newint(arrC.getItemAt(8).toLocaleString());
        img1.load(picArray[index1]);
        hbox.addChild(img1);
        var img2:Image =new Image;
```

```
        var index2:int =newint(arrC.getItemAt(9).toLocaleString
());
        img2.load(picArray[index2]);
        hbox.addChild(img2);
        var txtTem:Text =new Text();
        txtTem.text = arrC.getItemAt(5).toString();
        var txtWea:Text =new Text();
        txtWea.text = arrC.getItemAt(6).toString();
        var txtWind:Text =new Text();
        txtWind.text = arrC.getItemAt(7).toString();
        vbox.addChild(txtTem);
        vbox.addChild(txtWea);
        vbox.addChild(txtWind);
        vbox2.addChild(hbox);
        vbox2.addChild(vbox);
        canvas.addChild(vbox2);
        myMap.infoWindow.content = canvas;
        var mapPnt2:MapPoint =new MapPoint(114.31,30.52);
        myMap.infoWindow.show(mapPnt2);
    }
    }
```

在上面函数中对得到的天气预报数据进行解析，把对应的天气图片、气温等信息分类整理，使用 infoWindow 显示出来即可。

第三编 ArcGIS API for JavaScript

第17章　相关技术

17.1　JavaScript 简介

JavaScript 是一种基于对象(Object)和事件驱动(Event Driven)并具有相对安全性能的脚本语言。它是通过嵌入或调入到标准的 HTML 语言中实现的。它具有以下几个基本特点:

①解释性脚本语言。JavaScript 是一种脚本语言，它采用小程序段的方式实现编程。它的基本结构形式与 C、C++等类似，但它不像这些语言，需要先编译，而是在程序运行过程中被逐行地解释。它与 HTML 标识结合在一起，方便用户的使用操作。

②基于对象。JavaScript 是一种基于对象的语言，或者可以看作一种面向对象的开发语言。这就意味着它能运用自己已经创建的对象实现操作。

③简单性。JavaScript 的简单性主要体现在，首先它是一种基于 Java 基本语句和控制流之上的简单而紧凑的设计，对有 Java 基础的学习者非常有利，或对日后学习 Java 语言大有帮助。其次它的变量类型是采用弱类型，并未使用严格的数据类型。

④安全性。JavaScript 是一种安全性语言，它不允许访问本地资源，并不能将数据存入到服务器，不允许对网络文档进行修改和删除，用户只能通过浏览器实现信息浏览或动态交互，从而有效防止数据的丢失。

⑤动态性。JavaScript 是动态的，它可以直接对用户或客户的输入做出响应。它对用户的反映响应是按事件驱动方式进行的。事件发生后，可能会引起相应的事件响应。

⑥跨平台性。JavaScript 依赖于浏览器本身，与操作环境无关，只要能运行浏览器并支持 JavaScript 的计算机，浏览器就可正确执行。

17.2　Dojo 简介

Dojo 是一个强大的面向对象的 JavaScript 框架，主要由 4 个包组成：Dojo、Dijit、DojoX、Util。其中，Dojo 是 Dojo 框架的核心，它提供 Ajax、events、packaging、CSS-based querying、animations、JSON 等相关操作 API。Dijit 是一系列小部件，基于 Dojo 核心库。DojoX 包括一些新颖的代码和控件：DateGrid、charts、离线应用、跨浏览器矢量绘图等。Util 是支持剩余功能的工具包，例如，构建、测试、文档编码等功能。Dojo 的特点包括以下几个方面：

①Dojo 是一个纯 JavaScript 库，后台只要提供相应的接口就能够将数据以 Json 的格式

输出给前台。

②Dojo 自身定义了完整的函数库，屏蔽了浏览器的差异。

③Dojo 自身定义了界面组件库，其组件代码采用了面向对象的思想，便于继承及扩展。

④当对前端界面联动需求较为复杂的时候，基于 Dojo 的页面组件将是首选，因为其可以将界面中某一个具有共性的区域抽象出来，封装这一区域的界面行为以及数据，可以用搭积木的方式完成复杂页面的开发。

17.3 REST 简介

REST(Representational State Transfer)表述性状态转移，是 Roy Fielding 博士于 2000 年在其博士论文中提出来的一种软件架构风格。REST 本身并不涉及任何新的技术，它基于 HTTP 协议，比起 SOAP 和 XML-RPC，它更加简洁、高效。REST 最突出的特点就是用 URI 来描述互联网上所有的资源，Roy Fielding 博士对其进行了抽象，并认为：设计良好的网络应用表现为一系列的虚拟"网页"，或者说这些虚拟网页就是资源状态的表现(Representational)；用户选择这些链接导致下一个虚拟的"网页"传输到用户端展现给用户，而这正代表了资源状态的转发(State Transfer)。

REST 主要有以下的特点：

①资源通过 URI 来指定和操作；

②对资源的操作包括获取、创建、修改和删除资源，这些操作正好对应 HTTP 协议提供的 GET、POST、PUT 和 DELETE 方法；

③连接是无状态性的；

④能够利用 Cache 机制来提高性能；

⑤网络上的所有事物都被抽象为资源(resource)；

⑥对资源的各种操作不会改变资源标识。

17.4 JSON 简介

使用 REST API 进行信息传输的时候，有必要了解其数据传输格式，这种格式称为 JSON(Javascript Object Notation)。JSON 是一种轻量级的数据交换格式，易于人阅读和编写。JSON 能够描述 4 种简单的类型(字符串、数字、布尔值及 null)和两种结构化类型(对象及数组)。下面是一个 JSON 对象的例子：

```
var zhang = {
"name": "张筱",
"school": "山东科技大学",
"age": 22,
"married": false,
"friends": [
```

```
{ "name": "A", "age": 20 },
{ "name": "B", "age": 22 }
]}
```

17.5 ArcGIS API for JavaScript

17.5.1 ArcGIS API for JavaScript 简介

ArcGIS API forJavaScript 是由美国 ESRI 公司推出的基于 Dojo 框架和 REST 风格实现的一套编程接口(目前最新版本为3.13，Dojo1.10.4)。通过 ArcGIS API for JavaScript 可以对 ArcGIS for Server 进行访问，并且将 ArcGIS for Server 提供的地图资源和其他资源(ArcGIS Online)嵌入到 Web 应用中。

17.5.2 ArcGIS API for JavaScript 的特点

ArcGIS API forJavaScript 主要特点如下：

①一切基于服务；

②简单易学的语言基础；

③多种多样的开发方式；

④丰富的网络资源；

⑤基于功能强大的 Dojo JavaScript 工具包；

⑥开发和部署都是完全免费的。

17.6 ArcGIS for Server 服务

服务，简单来说就是 ArcGIS for Server 发布的 GIS 资源，不同的资源可以被发布为不同的服务，不同的服务具有不同的功能，详细信息见表17.1、表17.2。在使用 ArcGIS API for JavaScript 的时候，其实就是在使用这些 REST API，使用这些服务对外的功能，了解每种服务的具体功能，在开发的时候就可以根据需要做到游刃有余。

表17.1 **ArcGIS for Server 资源所对应的服务**

GIS 资源	在 ArcGIS for Server 中的功能
地图文档	制图、网络分析、网络覆盖服务(WCS)发布、网络要素服务(WFS)发布、网络地图切片服务(WMTS)发布、移动数据发布、KML 发布、地理数据库数据提取和复制要素访问发布、Schematics 发布
地址定位	地理编码
地理数据库	地理数据库查询、提取及复制；WCS 发布；WFS 发布

续表

GIS 资源	在 ArcGIS for Server 中的功能
地理处理模型或工具	地理处理、网络处理服务(WPS)发布
ArcGlobe 文档	3D 制图
栅格数据、镶嵌数据集，或者引用栅格数据集或镶嵌数据集的图层文件	影响发布、WCS 或 WMS 发布
GIS 内容所在的文件夹和地理数据库	创建组织的 GIS 内容的可搜索索引

表 17.2 **ArcGIS for Server 的服务功能**

功能服务	用　途	暴露此功能的服务
要素访问	用于访问地图中的矢量要素	地图服务
地理编码	用于访问地址定位器，发布地理编码服务时总会启用此功能	地理编码服务
地理数据	用于访问地理数据库的内容以进行数据查询、提取和复制，发布地理数据服务时总是会启用此功能	地理数据服务
地理处理	用于访问地理处理模型，发布地理处理服务时总是会启用此功能	地理处理服务
Globe	用于访问 Globe 文档的内容，发布 Globe 服务时总会启用此功能	Globe 服务
影像	用于访问栅格数据集或镶嵌数据集的内容，包括像素值、属性、元数据和波段。发布影像服务时总会启用此功能	影像服务
JPIP	在使用 JPEG 2000 文件或 NITF 文件(使用 JPEG2000 压缩类型)并通 ITTVIS 配置了 JPIP 服务时提供 JPIP 数据流功能	影像服务
KML	使用地图文档创建 keyhole 标记语言(KML)要素	地图服务
制图	用于访问地图的内容，如图层及其属性，发布地图服务时总是会启用此功能	地图服务
移动数据访问	可以将数据从地图中提取到移动设备	地图服务
网络分析	使用 ArcGIS Network Analyst 扩展模块求解交通网分析问题	地图服务
Schematics	允许查看、生成、更新和编辑逻辑示意图	地图服务
WCS	创建符合开发地理空间联盟(OGC)网络覆盖服务(WCS)规范的服务	地图服务、影像服务、地理数据服务

续表

功能服务	用　途	暴露此功能的服务
WFS	创建符合 OGC 网络要素服务(WFS)规范的服务	地图服务、地理数据服务
WMS	创建符合 OGC 网络要素服务(WMS)规范的服务	地图服务、影像服务
WMTS *	创建符合 OGC 的 Web 地图切片服务(WMTS)规范的服务	地图服务、影像服务
WPS	创建符合 OGC 的网络处理服务(WPS)规范的服务	地理处理服务

发布完成一个地图服务时，进入到 ArcGIS for Server 的管理页面，可以看到非常详细的信息，图 17.1 是作者发布的一个叫做 JsMap 的 2D 地图动态服务，在功能选项卡中可以看到该服务可以支持的功能以及每种功能支持的操作。

图 17.1　地图服务支持的功能

第18章 开发基础

18.1 基本概念

18.1.1 Map

Map(地图)是承载图层的容器，一个图层只有被添加到Map中，才能被显示出来。

18.1.2 Layer

Layer(图层)是承载服务的载体(GraphicsLayer除外)，ArcGIS for Server将GIS资源作为服务发布出来，要将这些资源以图层的形式加载到地图上，才能在浏览器端看到这些服务。不同的服务对应不同的图层类型，表18.1列出了不同服务和ArcGIS API for JavaScript中图层的对应关系。

表18.1 **图层和服务的对应关系**

图层	服务
ArcGISDynamicMapServiceLayer	动态地图服务
ArcGISTiledMapServiceLayer	缓存地图服务
ArcGISImageServiceLayer	影像地图服务
GraphicsLayer	不对应ArcGIS for Server发布的服务
FeatureLayer	要素服务或者地图服务
WMSLayer	调用OGC(Open Geospatial Consortium)矢量地图服务
WMTSLayer	OGC(Open Geospatial Consortium)地图切片服务
KMLLayer	Keyhole Markup Language描述和保存地理信息文件
VETiledLayer	微软的Bing地图服务
GeoRssLayer	支持GeoRss服务

图层加入到地图后的顺序如图18.1所示。

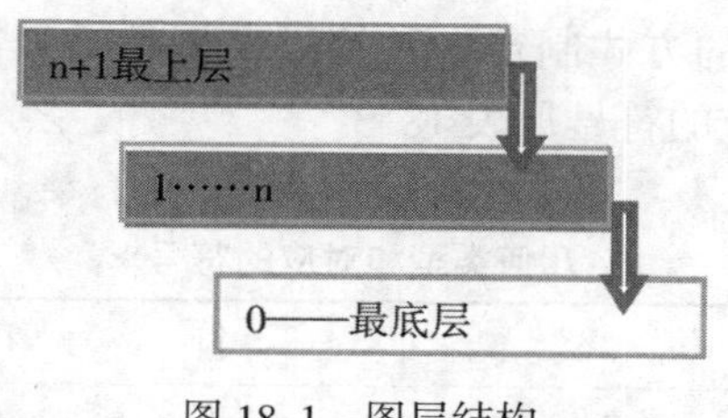

图 18.1 图层结构

18.1.3 Geometry

Geometry(几何对象)用于表示对象的显示型式，在 ArcGIS API for JavaScript 中 Geometry 大体上可以分为下面几类：点、多点、线、矩形、多边和 ScreenPoint，见表 18.2。其中，ScreenPoint 对象是最新版本增加的，是以像素的方式表示的点，而点、多点、线、矩形、多边形都是继承 Geometry 的，其关系如图 18.2 所示。

表 18.2 **Geometry 类**

几　何	说　明　名
Geometry	抽象类，定义几何体的图形
MapPoint	点对象
MultiPoint	多点对象
Polyline	多义线
Envelop	矩形
Polygon	多边形
ScreenPoint	用屏幕 x，y 表示的点，相对屏幕左上角

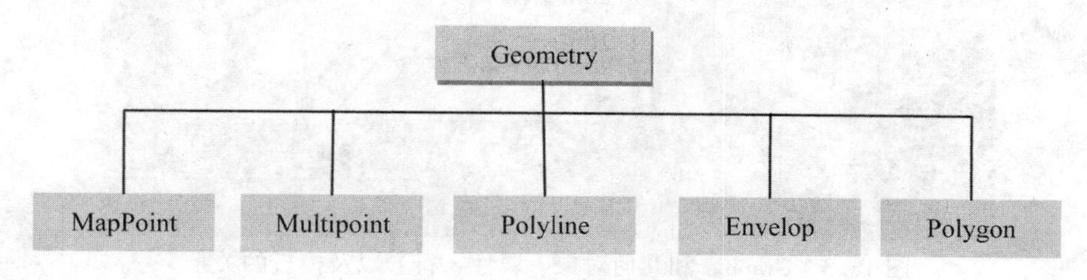

图 18.2 Geometry 对象结构

18.1.4 Symbol

Symbol 定义了如何在 GraphicLayer 上显示点、线、面和文本，符号定义了几何对象的外观，包括图形的颜色、边框线宽度、透明度等。ArcGIS API for JavaScript 包含很多符号类，每个类都允许使用唯一的方式制定一种符号，每种符号都特定于一种类型(点、线、面和文本)。几何类型和对应的符号见表 18.3。

表 18.3　　几何类型和对应的符号

类　型	符　号
点	SimpleMarkerSymbol，PictureMarkerSymbol
线	SimpleLineSymbol，CartographicLineSymbol
面	SimpleFillSymbol，PictureFillSymbol
文本	TextSymbol，Font

18.1.5　Graphic

Geometry 定义了对象的形状，Symbol 定义了图形是如何显示的，Graphic 可以包含一些属性信息，并且在 JavaScript 中还可以使用 infoTemplate 定义如何对属性信息进行显示，最终的 Graphic 则是被添加到 GraphicsLayer 中，对于 Graphic 的描述可以用一个数学表达式来表示：

Graphic = Geometry+Attribute+Symbol+infoTemplate

Graphic 和几何对象、属性、符号以及模板的关系如图 18.3 所示(底图来自 ArcGIS online 案例)。

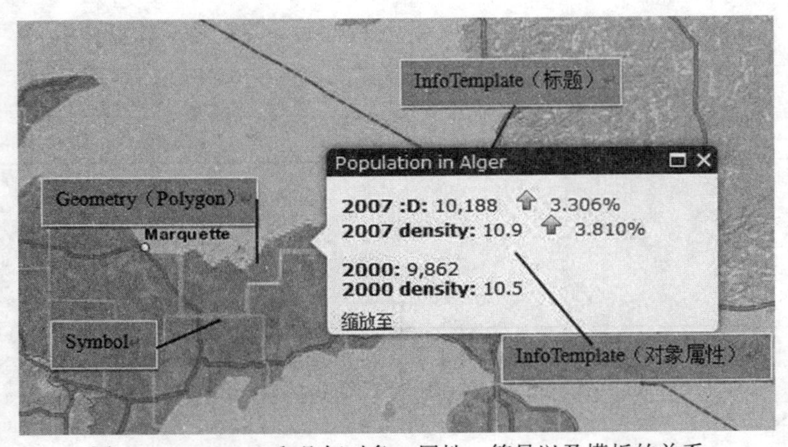

图 18.3　Graphic 和几何对象、属性、符号以及模板的关系

18.2　常用控件

ArcGIS API forJavaScript 提供了很多快速开发的控件，这些控件除了工具条之外，其

余都位于 esri. dijit 中，而工具条位于 esri. toolbars 中，现对常用控件进行介绍。

18.2.1　鹰眼图

OverviewMap 控件用于在其关联的主地图内较清楚地查看当前鸟瞰图的范围，其属性见表 18.4。

表 18.4　鹰眼图主要属性

属　性	说　明
Hide	隐藏鹰眼图
Show	显示鹰眼图
Startup	当构造函数创建成功后，使用该方法后就可以进行用户交互
Destroy	当应用程序不再需要比例尺控件的时候，摧毁该对象

其构造方法 esri. dijit. OverviewMap(params, srcNodeRef)，在创建一个鹰眼图时需要传入关联的地图对象和一个用于呈现鹰眼图控件的 HTML 元素，另外包括的其他参数见表 18.5。

表 18.5　鹰眼图构造函数的部分参数

参　数	说　明
attachTo	指定鹰眼图附加到地图的位置
baseLayer	指定鹰眼图空间地图的底图
expandFactor	设置鹰眼地图控件和矩形之间的比例
opacity	指定鹰眼图控件上矩形的透明度

在 Dijit 一系列生命周期中，一个重要方法就是启动方法 startup。它会在 DOM 节点被创建并添加到网页之后执行，同时也会等待当前控件中所包含的子控件被创建并正确启动之后才执行。

鹰眼图示例代码如下：

```
function OverviewMap(){
    var over =
{
map: Map,
attachTo:"top-left",
color:"black",
expandFactor:2,
opacity:0.5,
```

```
    baseLayer:map,
    };
        var MapViewer =new esri.dijit.OverviewMap(over, dojo.byId
("OverViewDiv"));
        MapViewer.startup();
    }
```

运行结果如图 18.4 所示。

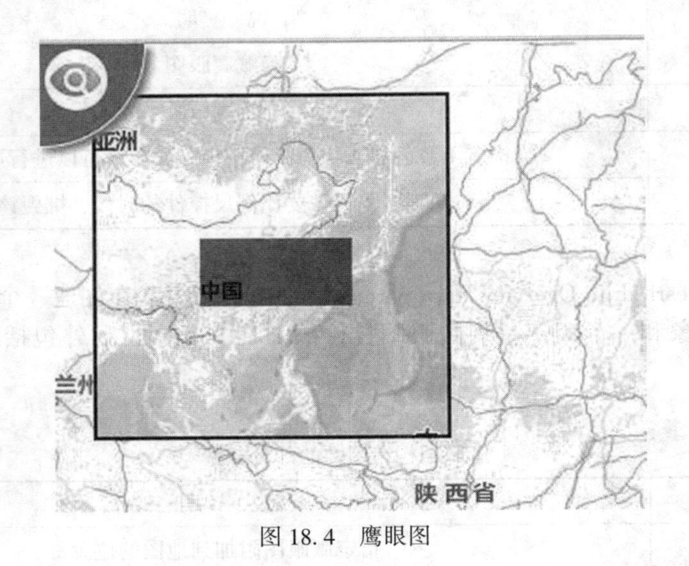

图 18.4　鹰眼图

18.2.2　InfoWindow

InfoWindow 控件是一个带有小尾巴的窗口，小尾巴指向一个位置或感兴趣的要素，其本质上就是一个 HTML 弹出框，InfoWindow 包括 Graphic 的属性信息。

18.2.3　编辑控件

编辑控件(editor)是一个高度定制的对象，该对象提供了要素编辑的功能，编辑控件往往不是孤立存在的而是和编辑模板、附件编辑控件一起搭配用来完成一个编辑任务。使用编辑控件，可以对要素进行删除、分割、更新以及为要素添加和删除附件等。在使用编辑控件的时候，一定要给编辑控件设置一个几何服务。

编辑工具条提供了一个针对要素服务的可编辑图层的即拿即用控件，同时该控件还结合 TemplatePicker（编辑模板选择器）、AttachmentEditor、AttributeInspector 3 个控件以及几何服务对要素的图形和属性进行编辑。编辑控件的构造方法 esri.dijit.Editor(params, srcNodeRef)，在创建一个编辑工具条时需要传入关联的地图对象、几何服务和一个用于呈现编辑工具条的 HTML 元素，另外还包括很多可选参数，表 18.6 中为常用的可选参数。

表 18.6　编辑控件构造方法的可选参数

参　数	说　明
enableUndoRedo	Undo 和 Redo 是否可用
maxOperations	当 Undo 和 Redo 可用时，指定可操作的最大次数
toolbarVisible	指定一个图层子集用于在图例中显示
layerInfos	FeatureLayer 的定义信息
templatePicker	是否指定编辑模板选择器
toolbarVisible	编辑工具条是否可见
undoManager	指定一个重做管理器实例
createOptions	当 toolbar 可见的时候，可以通过设置 createOperations、polylineDrawTools 和 polygonDrawTools
map	为 editor 指定关联地图

18.2.4　图例

Legend 控件用于动态显示全部或者部分图层的标签和符号信息，图例控件支持下面 4 种图层：ArcGISDynamicMapServiceLayer、ArcGISTiledMapServiceLayer、FeatureLayer 和 KMLLayer。

图例的构造方法 esri. dijit. Legend(params, srcNodeRef)，在创建一个图例时需要传入关联的地图对象和一个用于呈现图例控件的 HTML 元素，另外还包括很多可选参数，表 18.7 中为常用的可选参数。

Legend 控件的方法说明见表 18.8。

表 18.7　图例构造方法的可选参数

参　数	说　明
autoUpdate	当地图的比例尺发生变化或者图层发生变化的时候，图例控件是否自动更新
respectCurrentMapScale	图例控件是否自动更新
layerInfos	指定一个图层子集用于在图例中显示
arrangement	指定图例在 HTML 元素中的对齐方式

表 18.8　Legend 控件的方法

方　法	说　明
Refresh	当在构造函数中用了 layerInfos，用这个方法刷新图例以替换构造函数中的图层

图例控件示例代码如下：

```
function Maplegend() {
var legendPar = {
        map: Map,
        arrangment: esri.dijit.Legend.ALIGN_RIGHT,
        autoUpdate: true
    };
var legendDijit = new esri.dijit.Legend(
    legendPar,
"legendDiv");
    legendDijit.startup();
  }
```

代码运行结果如图 18.5 所示。

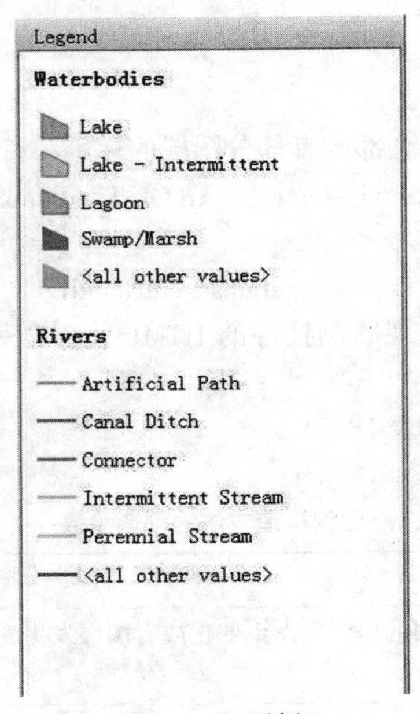

图 18.5　图例

18.3　环境部署和 API 准备

开发 JavaScript 的程序，有许多工具可以选择，如 webstorm、notepad++、Aptana、Visual Studio 等。本教材采用集成开发环境——Visual Studio 2010。开发环境安装完成后，需要引入 ArcGIS API for JavaScript 开发包，API 获取可以从 ESRI 官网下载。对于 ArcGIS API for JavaScript，不仅仅提供了 API，还提供了 SDK(SDK 里面含有 API 的帮助以及例

子)。将下载后的 API 压缩包解压，可以看到其结构如图 18.6 所示。

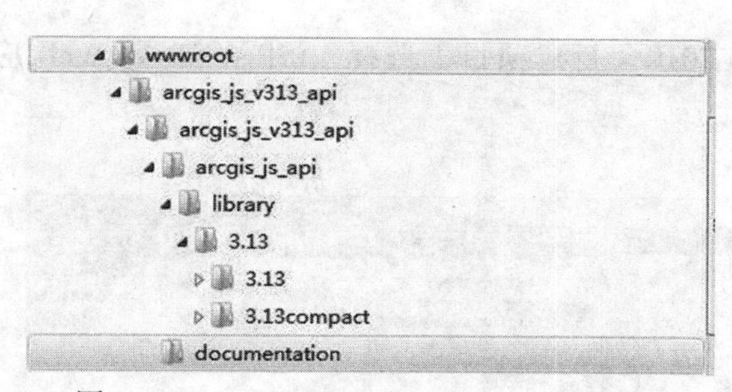

图 18.6　ArcGIS API for JavaScript 文件结构

在这里可以看到 3.13 和 3.13compact 两个目录，其中紧凑版本(3.13compact)去掉了 dojo dijit 的依存关系，并最大限度地减少了不必要的 ArcGISJavaScript API 类，紧凑版本是为了在网络慢和有网络延迟的环境中使用。

可以引用在线的 JavaScript API。如果只能连接内网，无法连接互联网或者网速较慢的情况下，建议使用本地部署的 JavaScript。

在 arcgis_js_v313_api \ arcgis_js_v313_api 目录下的 install.html 中具体介绍了部署的步骤。

将修改后的文件连同解压目录内的所有文件拷贝到 Web 服务器根目录下，以 IIS 为例，拷贝为 wwwroot 目录下的 arcgis_js_api，最终的目录结构如图 18.7 所示。

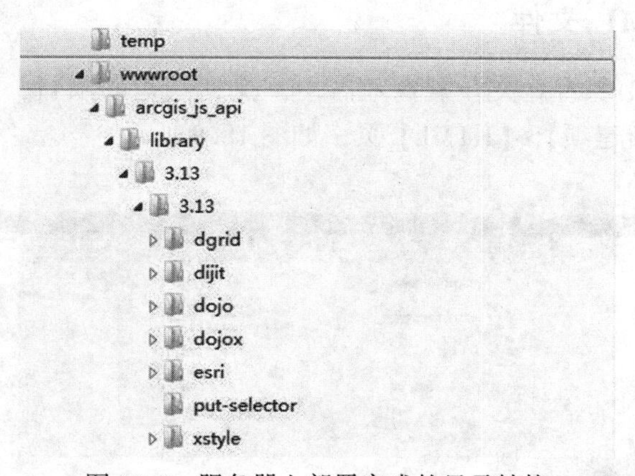

图 18.7　服务器上部署完成的目录结构

18.4　构建第一个应用

下面给出构建第一个应用程序——“helloWorld”的详细步骤和过程。

18.4.1 建立项目

启动 Visual Studio 2010，新建项目，选择“ASP.NET 空 Web 应用程序”，给项目命名，如图 18.8 所示。

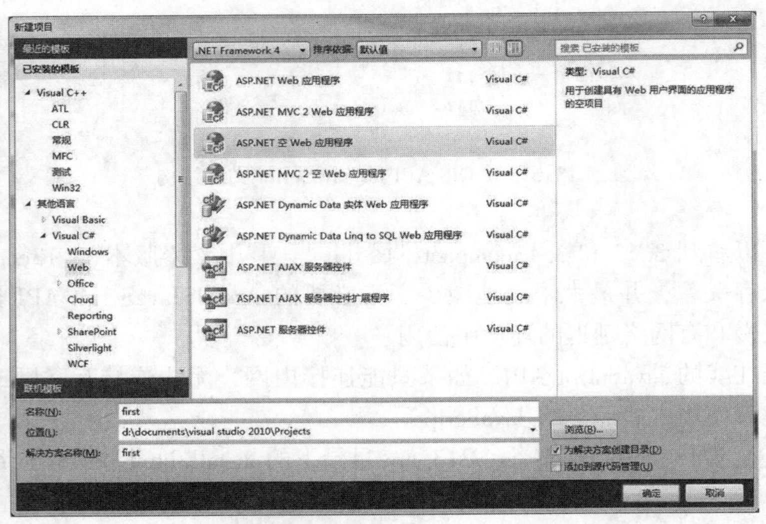

图 18.8 新建 ASP.NET 空 Web 应用程序

18.4.2 添加 HTML 文件

在 VisualStudio 2010 的解决方案管理器中，找到新建的项目，在项目上点击鼠标右键，选择【添加】≫【新建项】≫【HTML】页，如图 18.9 所示。

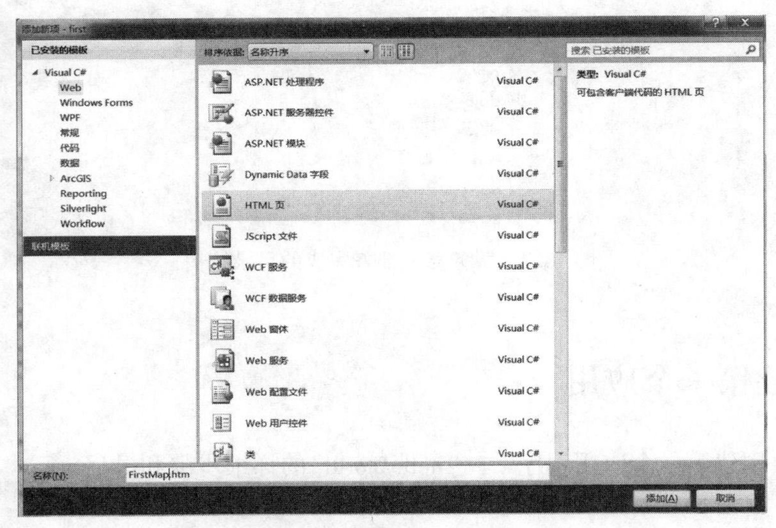

图 18.9 添加 HTML 页面

18.4.3　引入 ArcGIS API for JavaScript 的智能提示文件

在项目中，创建 dojo 文件夹，引入 ArcGIS API for JavaScript 的智能提示文件(下载地址：https：//developers.arcgis.com/javascript/jsapi/api_codeassist.html)，如图 18.10 所示。

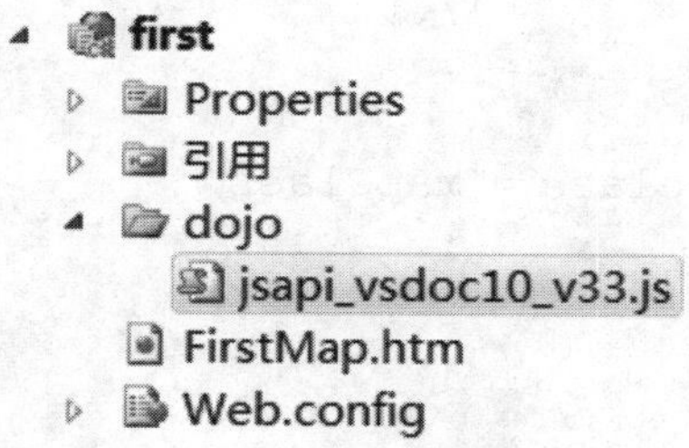

first
Properties
引用
dojo
jsapi_vsdoc10_v33.js
FirstMap.htm
Web.config

图 18.10　引入智能提示文件

如果使用 Aptana3，直接将智能提示文件拷贝到工程里面就可以了。

18.4.4　编写代码

打开 FirstMap.htm 页面，输入以下代码：

```
<html><head><meta http-equiv="Content-Type" content="text/html; charset=utf-8"/>
    <title>第一个应用</title>
    <link rel="stylesheet" type="text/css"
href = " http://localhost/arcgis _ js _ api/library/3.13/3.13/dijit/themes/tundra/tundra.css"/>
    <link rel="stylesheet"
type="text/css"href="http://localhost/arcgis_js_api/library/3.13/3.13//esri/css/esri.css"/>
    <script type="text/Javascript"
src="http://localhost/arcgis_js_api/library/3.13/3.13/init.js"></script>
    <script  type="text/Javascript"></script>
    <style type="text/css">
    .MapClass { width:900px; height:600px; border:1px solid #000;}
    </style>
    <script type="text/Javascript">
    dojo.require("esri.map");
    dojo.addOnLoad(function (){
        var MyMap =new esri.Map("MyMapDiv");
        var MyTiledMapServiceLayer =new
```

```
esri.layers.ArcGISTiledMapServiceLayer("http://www.arcgisonline.
cn/ArcGIS/rest/services/ChinaOnlineCommunity/MapServer");
    MyMap.addLayer(MyTiledMapServiceLayer)
})
</script>
</head>
<body class="tundra">
<div id="MyMapDiv"class="MapClass">
</div>
</body>
</html>
```

18.4.5　代码解释

(1)dojo.require("esri.map")

ArcGIS API for Javascript 是基于 dojo 的，dojo 有一个特点就是模块化，因为所用的地图对象在 esri.map 当中，所以需要通过 dojo.require("esri.map")引入。其实 dojo.require 就是加载了 JavaScript 文件。

(2)dojo.addOnLoad

当页面内容加载完成后就会触发 addOnLoad 事件，执行加载地图等函数操作。

18.4.6　运行结果

运行 Visual Studio 后，可以看到图 18.11 所示的效果。

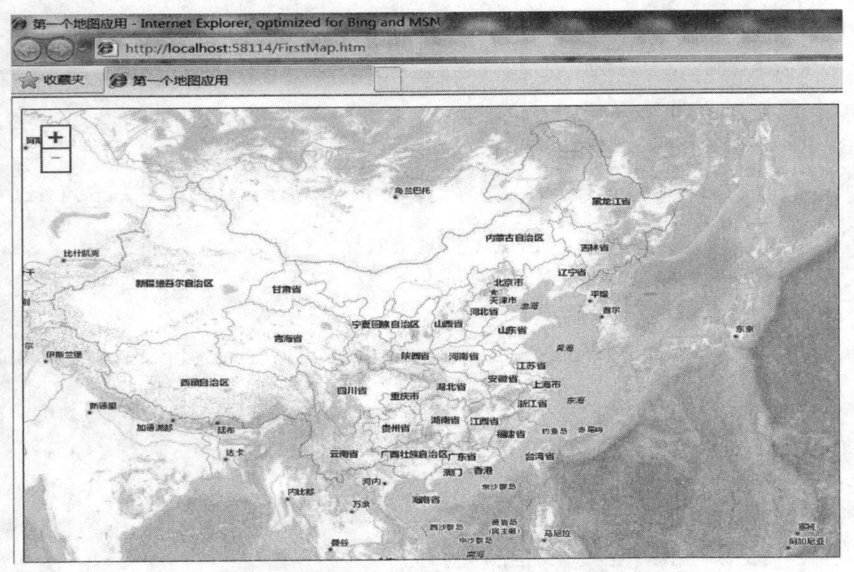

图 18.11　程序运行效果图

第 19 章　服务访问

19.1　基本函数

ArcGIS API forJavaScript 是基于 dojo 框架的，在开发的过程中会使用 dojo 或者 dijit 的一些函数，下面介绍一些常用的函数。

19.1.1　dojo. require

dojo. require 是 dojo 包的核心函数，加载除了 dojo. js 以外的其他功能包。格式如下：

```
// 加载 esri/map. js
dojo. require( " esri. map" ) ;
```

19.1.2　dojo. addOnLoad

页面加载完毕后调用的函数，用法如下：

```
dojo.addOnLoad (function(){
var MyMap =new esri.Map("MyMapDiv");
var MyTiledMapServiceLayer =new esri.layers.ArcGISTiledMapService-
Layer ( " http://www.arcgisonline.cn/ArcGIS/rest/services/ChinaOnli-
neCommunity/MapServer"); MyMap.addLayer(MyTiledMapServiceLayer);
});
```

一般使用一个自定义函数：

```
var init =function(){
var MyMap =new esri.Map("MyMapDiv");
var MyTiledMapServiceLayer =new esri.layers.ArcGISTiledMapService-
Layer ("http://www.arcgisonline.cn/ArcGIS/rest/services/
ChinaOnlineCommunity/MapServer");
MyMap.addLayer(MyTiledMapServiceLayer)
}
dojo.addOnLoad(init);
```

dojo. byId 的作用和 document. getElementsById 相同，但是简化了很多。用法如下：

```
dojo.addOnLoad(function(){
var mymap  = dojo.byId("map");
```

```
});
```

19.1.3　dojo.byId

dojo.byId 是针对 Dom 节点元素的，Dijit 是针对 dojo 的小部件，每个小部件都会有唯一的 ID，dijit.byId 可以通过 ID 返回小部件对象。

19.1.4　dojo.create

dojo.create()用来创建一个 DOM 对象，并设置一些列操作，原型为：

dojo.create(tag, attrs, refNode, pos);

参数 tag 可以是字符串或 DOM 节点，如果是字符串，函数会将其视作节点的标签名，以此来新建节点。建立节点时，会以 refNode 作为父节点。如果 refNode 为 null 或并未指定，则默认以 dojo.doc 作为父节点。

参数 attrs 是一个 JavaScript 对象，其中包含了用以赋予节点的一组属性信息。该参数会在节点创建成功后被原封不动地传给 dojo.attr。attrs 参数可以 null，也可以不指定，亦即"不设置任何属性"。如果用户想指定函数余下的传入参数，则应该为其显示的指定 null 值。

参数 refNode 作为创建节点的父节点对象，该参数为 DOM 节点对象或节点的 ID。此参数可以省略，即表示"不立即安置该节点"。

参数 pos 为可选，表示安置创建的节点到给定的位置上。取值可以是数字，或如下字符串之一："before"，"after"，"replace"，"only"，"first"，或"last"。如果省略，则默认取"last"。

```
dojo.create("div",{id:"mapbtm"})
dojo.query
```

返回 DOM 节点的列表，以 css 选择器来实现。用法如下：

```
dojo.addOnLoad(function(){
    dojo.query(".blueButton").forEach(function(node, index, rr){
      });
      });
```

19.1.5　dojo.connect

dojo.connect 用于为指定的元素添加事件，如当地图发生 onload 事件的时候，调用 mapload 函数，代码如下：

```
dojo.connect(map,'onload',mapload);
var mapload=function{
map.centerAt(esri.geometry.Point(116, 34));
}
dojo.forEach
```

dojo.forEach 遍历数组里的每一个数值相当于 JavaScript 中的

```
    for(var i in geometries){
alert(geometries [i]);
}
```

用 dojo.forEach 则可以这样写：

```
dojo.forEach(geometries,
function (element, index) {
var graphic = new esri.Graphic(element, PolygonSymbol);
Map.graphics.add(graphic); });
dojo.hasClass, dojo.addClass
```

dojo.hasClass 用于判断给定的 DOM 节点是否有指定的 CSS class
dojo.addClass 用于为给定的 DOM 节点增加指定的 CSS class

19.2　动态地图服务加载

动态地图服务由 ArcGISDynamicMapServiceLayer 承载，提供了对由 ArcGIS Server REST API 暴露的动态地图服务资源的访问，动态地图服务实时生成地图图片。

19.2.1　动态 2D 地图服务属性和方法

动态 2D 地图服务的主要属性见表 19.1。

表 19.1　　**动态 2D 地图服务主要属性**

属　性	说　　明
capabilities	获取地图的能力，如 Map，Query 或者 Data
disableClientCaching	是否启用客户端缓存
Dpi	设置输出图片的 dpi
imageTransparency	动态图片的背景透明
DynamicLayerInfos	获取一组 DynamicLayerInfos，DynamicLayerInfos 用来改变服务中图层的顺序
ImageFormat	获取 ArcGISDynamicServiceLayer 生成的图片格式
LayerDefinitions	为服务中的每个图层设置过滤信息
LayerDrawingOptions	设置 LayerDrawingOptions 用来覆盖 Layer 的绘制
layerInfos	获取服务中的图层以及它们默认的可见性
maxImageHeight	导出图片的最大高度
maxImageWidgth	导出图片的最大宽度
maxRecordCount	返回查询的最大记录数

续表

属　性	说　明
maxScale	2D 动态服务的最大比例尺
minScale	2D 动态服务的最小比例尺
visibleAtMapScale	是否在当前地图比例尺中可见
visibleLayers	获取可见图层

动态 2D 地图服务的主要方法见表 19.2。

表 19.2　**动态 2D 地图服务主要方法**

方　法	说　明
createDynamicLayerInfosFromLayerInfos	根据当前的一组 LayerInfo 创建一组 DynamicLayerInfos based
exportMapImage	导出地图
setDPI	设置导出地图的 DPI，默认是 96
setDisableClientCaching	设置客户端是否缓存
setDynamicLayerInfos	设置一组 DynamicLayerInfos 用来改变服务的图层顺序
setImageFormat	设置图片的格式
setLayerDefinitions	设置图层的过滤条件，用于过滤图层中的要素
setLayerDrawingOptions	设置 LayerDrawingOptions 用来覆盖 Layer 的绘制
setScaleRange	设置比例尺范围
setVisibleLayers	设置图层的可见性

19.2.2　动态 2D 地图服务加载实例

该实例用于说明如何加载一个动态 2D 地图服务，代码如下：

```
dojo.require("esri/layers/ArcGISDynamicMapServiceLayer");
var DynamicLayer =new
ArcGISDynamicMapServiceLayer ( " http://localhost: 6080/arcgis/rest/
services/YSLX/MapServer");
map.addLayer(DynamicLayer);
```

运行结果如图 19.1 所示。

图 19.1　动态 2D 地图服务效果

19.3　切片服务加载

ArcGISTiledMapServiceLayer 切片地图服务图层(即缓存 2D 地图服务)，提供对由 ArcGIS Server REST API 所暴露的缓存地图服务资源的访问。缓存服务访问缓存文件夹中预先创建好的切片图片，而不是动态生成图片。

切片服务与动态服务加载方式类似，只是用 ArcGISTiledMapServiceLayer 替换了 ArcGISDynamicMapServiceLayer。

19.4　要素服务加载

FeatureLayer(要素图层)是在 ArcGIS 10.0 的时候增加的，是一种特殊的 GraphicsLayer，它继承了 GraphicsLayer，用来对服务图层中的要素服务进行显示，同时还提供了支持表达式过滤，要素的关联查询以及在线编辑等功能。

在 ArcGIS API for JavaScript 中提供了针对要素服务的图层 FeatureLayer。FeatureLayer 有很多属性和方法，用于对要素服务实现查询、渲染、编辑等操作。通过设置 FeatureLayer 的 setDefinitionExpression 属性，还可以实现对数据的过滤。

19.5　影像服务加载

影像服务的数据源可以是栅格数据集(来自磁盘上的地理数据库或文件)、镶嵌数据集或者引用栅格数据集或镶嵌数据集的图层文件。对定义了动态处理的栅格数据集或栅格图层(如符号系统或栅格函数)进行共享是影像服务的核心功能，它不需要任何扩展模块。在共享镶嵌数据集或包含镶嵌函数的栅格图层时，则需要 ArcGIS Image 扩展模块。例如，如果有一个包含镶嵌数据集的地图文档，则需要 ArcGIS Image 扩展模块。

19.5.1　ArcGIS 影像服务功能

ArcGIS 影像服务发布的影像服务，使用动态镶嵌和在线处理技术，客户端不仅可以快速访问影像，还能对其元数据进行查询，处理影像数据。下面是详细功能：

①快速显示(动态服务、缓存服务)；

②数据输出，包括像素值、原始数据和处理后的数据；

③动态镶嵌和影像目录；

④动态影像处理；

⑤影像量测(2D、3D)；

⑥影像服务编辑(增加、删除、更新)；

⑦影像服务高速缓存；

⑧结合时间滑块，还可以实现展示不同时期影像的变化。

除此之外，ArcGIS 的影像服务不仅能完成 TB~PB 级的海量影像数据的访问，百万级别的数据查询和检索，还能实现对企业级影像管理和应用，满足在线业务的影像更新、统计、下载、共享和业务应用。影像切片服务为用户提供了即拿即用的服务，能快速地显示影像，适合公众用户作为底图使用。影像动态服务则提供了专业的分析和处理能力。

19.5.2　ArcGISImageServiceLayer

ArcGIS API for Javascipt 提供的 ArcGISImageServiceLayer 对应的是 ArcGIS for Server 发布的影像服务，ArcGISImageServiceLayer 允许使用由 ArcGIS Server REST API 暴露的影像服务资源。ArcGISImageServiceLayer 的主要属性见表 19.3，其主要方法见表 19.4。

表 19.3　**ArcGISImageServiceLayer 的主要属性**

属　性	说　明
bandCount	获取波段个数
bandIds	获取波段 ID
CompressionQuality	设置压缩比，只对 JPG 格式有效
Interpolation	获取内插方式
Format	输出图片的格式
MosaicRule	指定镶嵌规则
RenderingRule	指定渲染规则
maxImageHeight	最大图片高度
maxImageWidth	最大图片宽度
maxRecordCount	查询的最大记录
pixelSizeX	X 方向像素的大小

续表

属　　性	说　　明
pixelSizeY	Y 方向像素的大小
pixelType	服务的像素类型
timeInfo	如果支持时态 GIS，那么可以获取时间信息

表 19.4　　**ArcGISImageServiceLayer 的主要方法**

方　　法	说　　明
setBandIds	设置影像服务 R，G，B 对应的波段
exportMapImage	导出图片
setCompressionQuality	设置压缩比，只对 JPG 格式有效
setDisableClientCaching	设置客户端是否缓存
setInterpolation	设置内插格式
setImageFormat	设置图片的格式
setMosaicRule	设置镶嵌规则
setRenderingRule	设置渲染规则

19.5.3　OGC 标准服务

OGC(Open Geospatial Consortium)开放地理信息联盟，是一个非营利的志愿的国际标准化组织。OGC 的服务标准分别有 WCS、WFC、WMS、WMTS 和 WPS，ArcGIS 支持全部 OGC 的服务，其服务对应关系见表 19.5。

表 19.5　　**OGC 服务和 ArcGIS for Server 服务的对应关系**

	WCS	WFS	WMS	WMTS	WPS
MapService	✓	✓	✓	✓	
Geodata service	✓	✓			
Image service	✓		✓	✓	
Geopressing service					✓

发布服务的时候在 Service Editor 中可以对 OGC 服务进行选择，如图 19.2 所示。OGC 服务和 ArcGIS API for JavaScript 提供的图层对应关系见表 19.6。

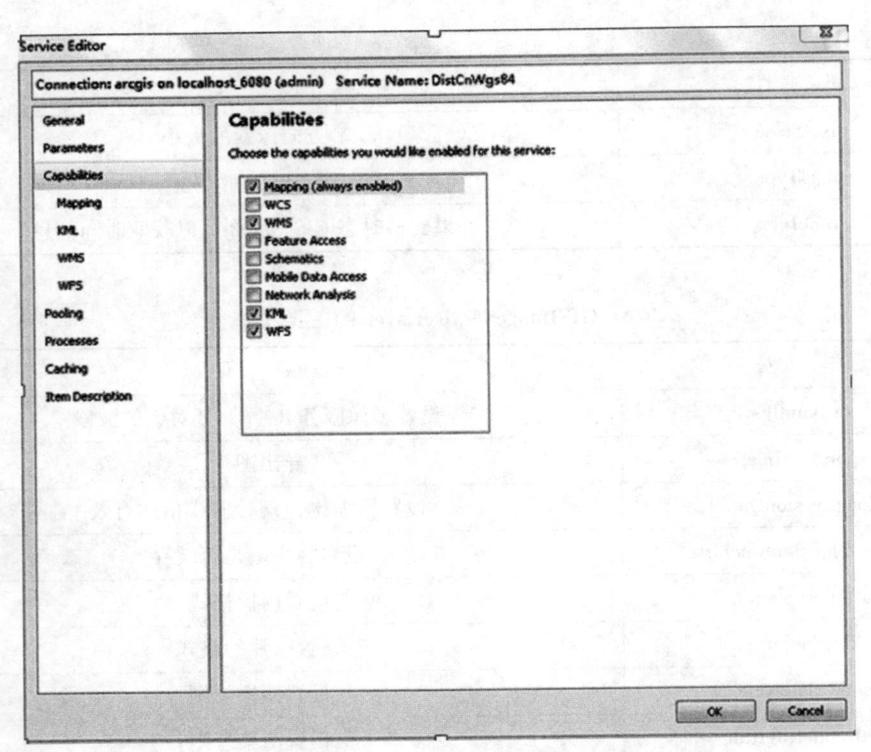

图 19.2　Service Editor 查看 OGC 服务

表 19.6　　**OGC 服务和 ArcGIS API for JavaScript 提供的图层对应关系**

服　务	图　层
WMS	WMSLayer(使用 setVisibility 方法决定显示哪个图层)
WMTS	WMTSLayer

第20章 地图操作

开发程序首先要解决地图的显示问题，以及如何显示地图相关的控件，如图例等。本章对地图的显示进行介绍。

20.1 地图

Map(地图)的对象是 ArcGIS API for JavaScript 的核心对象，其他控件或多或少的都将 Map 对象作为其参数，它主要用于呈现地图服务、影像服务等。一个地图对象需要通过一个 DIV 元素显示，通常地图的宽度和高度通过 DIV 控制。Map 对象不仅用来承载地图服务和 GraphicsLayer，同时还可以监听用户在地图上的各种操作事件，并做出响应。Map 对象提供了非常丰富的事件，通过这些事件可以实现地图与用户的交互。

20.1.1 Map 的属性

Map 对象的主要属性见表 20.1。

表 20.1 **Map 对象的主要属性**

属 性	说 明
extent	地图的显示范围
autoResize	当浏览器窗口或地图容器大小变化时，地图范围是否自动调整
attribution	地图属性
fadeOnZoom	当地图缩放时，是否启用淡入淡出的效果
infoWindow	在地图上显示消息框
isClickRecenter	按住 Shift 建单击地图时，是否将单击点设为地图中心
isDoubleClickZoom	双击地图时是否缩放地图
isPan	是否可以平移
spatialReference	获取地图的空间参考信息
isKeyboardNavigation	是否利用键盘控制缩放
isRubberBandZoom	是否启用橡皮筋缩放模式
isScrollWheelZoom	是否利用滚轮缩放

续表

属　性	说　明
isShiftDoubleClickZoom	按住 Shift 键双击地图时是否缩放地图
geoGraphicExtent	地理坐标范围
layerIds	获取所有地图图层 ID
loaded	地图是否加载完成
graphics	获取地图的 GraphicsLayer
position	地图左上角坐标
root	容纳图层、消息框等的容器的 dom 节点
showAttribution	是否允许显示地图属性
snappingManager	捕捉管理器
isZoomSlider	设置或获取地图的缩放滑动条
navigationMode	设置或获取地图的导航模式
timeExtent	获取或设置地图的时间范围

20.1.2 Map 的方法

Map 对象的构造方法 esri. Map(divId, options)，用来创建一个 Map 对象时，需传入一个 div 元素作为其容器，此外这个构造方法还包括一系列可选的参数用来描述地图的相关行为，表 20.2 为其常用的可选参数。

Map 对象的常用方法见表 20.3。

表 20.2　　**Map 对象构造方法的相关参数**

可选参数	说　明
extent	设置地图的初始显示范围
logo	是否显示 esri 的 logo
wrapAround180	是否连续移动地图，即通过日期变更线，对地图进行横向旋转 360°
lods	设置地图初始比例级别
maxScale	地图最大比例尺
sliderStyle	是指 slider 的样式，large 或 small

表 20.3　　**Map 对象的常用方法**

方　法	说　明
toScreen/toMap	地图或屏幕坐标系之间转换

续表

方　法	说　明
setScale	设置地图显示比例尺
setZoom	缩放到指定级别
setLevel	缩放到指定层级
setExtend	指定地图显示范围
disablePan	禁止地图上的平移事件
removeAllLayers	清除地图中所有图层
addLayer	向地图中添加一个图层
getBasemap	获得地图的基础底图
getLayersVisibleAtScaleRange	获取某一比例尺下的可见图层
getScale	获取当前比例尺
hideZoomSlider	隐藏缩放时的鼠标的箭头
panRight panUp	向右平移 向上平移
removeLayer	移除指定图层
reorderLayer	改变指定图层的顺序
reposition	复位地图，该方法在地图的 div 被复位时用到
setTimeExtent	设置地图的时间范围
setTimeSlider	设置和地图关联的时间滑块
setZoom	设置放大级别
showPanArrows	设置平移是鼠标的显示样式
showZoomSlider	显示缩放工具条

20.1.3　Map 的事件

Map 对象的主要事件见表 20.4。

表 20.4　**Map 对象的主要事件**

事　件	说　明
onExtentChange	地图范围改变事件
onBasemapChange	地图底图改变事件
onLoad	地图加载完成时发生

续表

事　件	说　明
onClick	鼠标单击事件
onLayerAdd	添加一个图层时发生
onLayersAddResult	addLayers 方法完成时
onLayersRemoved	当所有图层移除后
onMouseDown	当鼠标在地图上单击时
onMouseMove	当鼠标在地图上移动时
onMouseOut	当鼠标移出地图时
onMouseOver	当鼠标在地图中移动时

20.2 导航

Navigation(导航)用于控制地图导航操作，支持平移、缩放和视图回退和前进等操作。

20.2.1 Navigation 的方法

Navigation 对象的构造方法为：esri.toolbars.Navigation(map)，用来创建一个 Navigation 对象时，需传入一个 map 对象作为参数，无可选参数。

Navigation 对象的常用方法见表 20.5。

表 20.5 **Navigation 对象的常用方法**

方　法	说　明
active	激活导航
deactive	取消导航
isFirsExtent	是否为初始范围
isLastExtent	是否为最后一个范围
zoomToFullExtent	缩放到全图
zoomToNextExtent	缩放到前一视图
zoomToPreExtent	缩放到后一视图

20.2.2 Navigation 的事件

Navigation 对象的事件见表 20.6。

表 20.6　**Navigation 对象的事件**

事　件	说　明
onExtentHistoryChange	历史范围发生变化时

20.2.3　导航实例

实例具体代码如下：

```
var navToolbar;
    navToolbar =new navigation(map);
    navToolbar.activate(navigation.PAN);
//放大
    on(dom.byId("in"),"click",function(){
        navToolbar.activate(navigation.ZOOM_IN);
});
//缩小
    on(dom.byId("out"),"click",function(){
        navToolbar.activate(navigation.ZOOM_OUT);
});
//平移
    on(dom.byId("pan"),"click",function(){
        navToolbar.activate(navigation.PAN);
});
//全图
    on(dom.byId("full"),"click",function(){
        map.setExtent(initialExtent);
//map.extent = initialExtent;
//map.setextent();
//navToolbar.zoomToFullExtent();
});
//前一视图
    on(dom.byId("p"),"click",function(){
        navToolbar.zoomToPrevExtent();
});
//后一视图
    on(dom.byId("n"),"click",function(){
navToolbar.zoomToNextExtent();});
```

运行结果如图 20.1 所示。

图 20.1　导航实例运行效果

20.3　绘图

在地图上进行绘图操作，主要是借助于 Toolbar 上的 Draw(绘图)工具，绘图工具支持几何对象的创建，绘图工具还支持点、线、面的绘制。使用绘图工具的时候常常伴随鼠标的操作，对于不同的几何对象，鼠标的操作也有所不同。

①点：单击添加一个点。

②多点：单击添加点，双击添加最后一个点的多点。

③多线和多边形：单击添加顶点，双击添加最后一个顶点。

④徒手折线和徒手画的多边形，从按下鼠标开始绘制，到释放鼠标结束绘制。

20.3.1　绘图的属性

绘图工具的主要属性见表 20.7。

表 20.7　**绘图工具的主要属性**

属　性	说　明
RespectDrawingVetexOrder	是否设置绘制的图形拓扑顺序
MarkerSymbol	获取或设置点或多点的符号
FillSymbol	获取或设置面或 extent 的符号
LineSymbol	获取或设置线符号

20.3.2　绘图的方法

绘图工具的构造方法 esri. toolbars. Draw(map，srcNodeRef)，在创建绘图对象时，需要传递地图对象以及一些可选参数，表 20.8 中为主要可用参数。绘图工具的常用方法见

表 20.9。

表 20.8　　绘图工具构造方法的参数

参　　数	说　　明
DrawTime	在使用徒手工具，多长时间可以添加下一个点
showTooltips	是否显示提示信息
Tolerant	徒手绘图工具设置添加下一个点的容差
tooltipOffset	提示信息的偏差位置

表 20.9　　绘图工具的常用方法

方　　法	说　　明
Active	激活绘制图形的类型，点、线、面等
deactive	取消激活的绘制工具
FinishDrawing	绘制结束并导致 onDrawEnd 事件发生
setFillSymbol	设置面的符号
setLineSymbol	设置线符号

20.3.3　绘图的事件

绘图工具的主要事件见表 20.10。

表 20.10　　绘图工具的主要事件

事　　件	说　　明
onDrawEnd	当图形绘制结束时发生

20.4　图形图层

GraphicsLayer 是一种客户端图层，并不对应到服务器端的某个地图服务，用于在客户端展现各种数据，如绘制的图形、查询返回的结果等。GraphicsLayer 在客户端数据表达方面有非常重要的作用，它可以根据各种请求动态地在客户端绘制一些符号化的几何对象——Graphic。

在使用 GraphicsLayer 时，可以新建一个图层对象，也可以使用地图默认的 GraphicsLayer，默认对象通过 Map. graphics 获取。GraphicsLayer 经常和 Draw 工具搭配使用，GraphicsLayer 用来将 Draw 工具绘制的图形进行显示和符号化。在 GraphicsLayer 图层上，还可以响应一些事件，例如，鼠标单击、双击、鼠标移动等，单击事件在查看某一个具体的 graphic 的时候很有帮助。

20.4.1 GraphicsLayer 的属性和方法

GraphicsLayer 的主要属性见表 20.11，其主要方法见表 20.12。

表 20.11 GraphicsLayer 的主要属性

方　法	说　明
graphics	获取所有 graphic
renderer	设置图层渲染器

表 20.12 GraphicsLayer 的主要方法

方　法	说　明
add	添加 graphic
clear	清除所有 graphic
disableMouseEvents	禁止响应鼠标事件
enableMouseEvents	启用鼠标响应事件
remove	删除特定 graphic
setInfortemplate	设置 Infortemplate
setRender	设置图层渲染器

20.4.2 GraphicsLayer 实例

该实例将查询结果绘制到默认的 graphics 图层中，代码如下：

```
queryTask.execute(query,function(result){
var heilightS =new
   SimpleLineSymbol(SimpleLineSymbol.STYLE_DASH,
new Color([0,0,255]),3);
   map.graphics.clear();
   array = result.features;
for(i =0; i < array.length; i++){
      array[i].setSymbol(heilightS);
      map.graphics.add(array[i]);}
});
```

运行结果如图 20.2 所示。

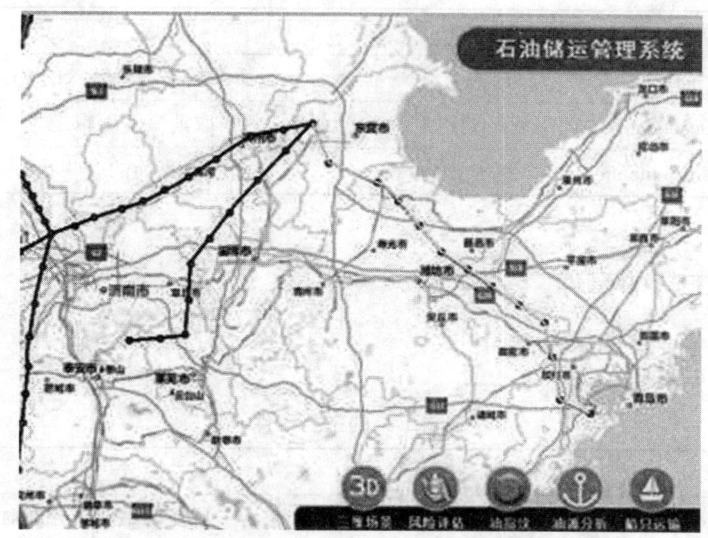

图 20.2　在 GraphicsLayer 上绘制图形

20.5　图形编辑

对地图上显示的图形元素(Graphic)的几何对象进行编辑要借助于编辑工具条(Edit Toolbar)，编辑工具条不是一个可视化的用户小部件，而是一个用来帮助编辑的类，它提供了移动图形或编辑现有几何对象的功能。如果要添加几何对象，就要使用绘图工具条(Draw Toolbar)。使用编辑工具条的时候常常伴随着鼠标的操作，使用鼠标可以移动要素，对现有的几何对象添加点、删除点以及对几何对象进行旋转和缩放操作。

20.5.1　编辑工具的方法

编辑工具的构造方法 esri. toobars. Edit(params，srcNodeRef)，其可选参数见表 20.13。编辑工具的常用方法见表 20.14。

表 20.13　　**编辑工具构造方法的可选参数**

参　数	说　明
AllowAddVertices	是否设置绘制的图形拓扑正确
AllowDeleteVertices	获取或设置点或多点的符号
VertexSymbol	获取或设置面或 extent 的符号

表 20.14　　　　编辑工具的常用方法

方　法	说　明
activate	激活工具并编辑转入的 Graphic
deactivate	取消激活编辑工具
getCurrentState	获取当前的一些状态

20.5.2　编辑工具的事件

编辑工具的主要事件见表 20.15。

表 20.15　　　　编辑工具的主要事件

事　件	说　明
onActivate	激活工具并禁用地图导航时触发
onDeactivate	取消激活编辑工具
onGraphicClick	点击地图中的 graphic 要素时触发
onGraphicMove	graphic 移动时触发
onRotateStop	旋转停止时
onScaleStop	缩放停止时
onVertexAdd	添加顶点时
onVertexClick	点击顶点时
onVertexDelete	删除顶点时

第21章 任　务

在ArcGIS API for JavaScript中，可以看到各种各样的"任务"。不同的任务都会有一个执行对象，还有一个对象的参数设置对象，执行对象在整个执行过程中可能返回一些状态信息，在执行结束后返回任务执行信息。任务成功，则得到结果；任务失败，则通过检查任务失败的错误提示信息进行检查。本章主要介绍两大常用任务：查询检索和网络分析。

21.1 查询检索

查询检索是重要的任务之一，在每一个应用程序中几乎都有查询的功能，如属性查询、空间查询、矢量数据查询、影像数据查询等。

21.1.1 QueryTask

QueryTask是一个进行空间和属性查询的功能类，QueryTask以Query为执行参数，空间查询和属性查询的设置都是在Query对象上进行，QueryTask的过滤条件除了属性过滤，还支持空间过滤。QueryTask进行查询的地图服务并不必须加载到Map中进行显示，在QueryTask执行成功后，可以从其返回结果中获取查询到的空间数据并绘制到GraphicsLayer中。

(1)属性查询

在属性查询的时候，只需要设置Query对象的Where过滤语句，输出字段参数、是否返回几何对象等。具体示例代码如下：

```
_doSearch:function(){
var txt;
if(dom.byId("AttrName").value =="编号"){
        txt ="FID"+"="+dom.byId("AttrValue").value;
}
else{
        txt = dom.byId("AttrName").value +"='"+dom.byId("AttrVal-
ue").value +"'";
}
var queryTask =new
QueryTask("http://localhost:6080/arcgis/rest/services/YSLX/MapServ-
er/3");
```

```
//查询参数
var query =new Query();
//需要返回 Geometry
    query.returnGeometry =true;
//需要返回的字段
    query.outFields =["FID","地区","所属管线","公称通径","制作材料","工作温度","试验压力","工作压力"];
//查询条件
    query.where = txt;
//信息模板
var infoTemplate =new InfoTemplate();
//设置 Title
    infoTemplate.setTitle("阀门信息");
//设置 Content
    infoTemplate.setContent("<div ><label>编号:</label>${FID}<br /><label>所属地区:</label>${地区}<br />"
+"<label>所在管道:</label>${所属管线}<br />"+"<label>公称通径:</label>${公称通径}cm<br />"
+"<label>试验压力:</label>${试验压力}Mpa<br />"+"<label>制作材料:</label>${制作材料}<br />"+
"<label>工作压力:</label>${工作压力}<br />"+"<label>工作温度:</label>${工作温度}<br />"+"</div>");
//设置 infoWindow 的尺寸
    map.infoWindow.resize(245,125);
//进行查询,完成后调用 showResults 方法
    console.log("查询前:", infoTemplate);
    queryTask.execute(query, showResults);
function showResults(results){
//清除上一次的高亮显示
        map.graphics.clear();
//查询结果样式
var symbol =new SimpleMarkerSymbol({"color":[0,255,0],"size":12,"type":"esriSMS","style":"esriSMSCircle"});
//遍历查询结果
var info ="<label >编号</label><label>所属地区</label><br>";
for(var i =0; i < results.features.length; i++){
var graphic = results.features[i];
```

```
//设置查询到的graphic的显示样式
            graphic.setSymbol(symbol);
//设置graphic的信息模板
            graphic.setInfoTemplate(infoTemplate);
            info +="<label> "+graphic.attributes.FID +"</label>
<label >   "+graphic.attributes.地区+"</label><br>";
//把查询到的结果添加到map.graphics中进行显示
            map.graphics.add(graphic);
}
        info +="";
        dom.byId("famen").innerHTML ="";
        dom.byId("famen").innerHTML = info;
}
```

(2)空间查询

在空间查询的时候，必须设置Query对象的geometry属性和spatialRelationship属性，这里设置的是空间包含关系。具体示例代码如下：

```
var queryTask =new
QueryTask("http://localhost:6080/arcgis/rest/services/YSLX/MapServ-
er/3");
//查询参数
var query =new Query();
//需要返回Geometry
query.returnGeometry =true;
//需要返回的字段
query.outFields =["FID","地区","所属管线","公称通径","制作材料","工
作温度","试验压力","工作压力"];
//查询条件
//信息模板
var infoTemplate =new InfoTemplate();
//设置Title
infoTemplate.setTitle("阀门信息");
//设置Content
infoTemplate.setContent("<div ><label>编号:</label>${FID} <br />
<label>所属地区:</label>${地区}<br />"
+"<label>所在管道:</label>${所属管线}<br />"+"<label>公称通径:
</label>${公称通径}cm<br />"
+"<label>试验压力:</label>${试验压力}Mpa<br />"+"<label>制作材料:
</label>${制作材料}<br />"+
```

```
"<label>工作压力:</label>${工作压力}<br />"+"<label>工作温度:</la-
bel>${工作温度}<br />"+"</div>");

//设置 infoWindow 的尺寸
map.infoWindow.resize(245,125);
//进行查询,完成后调用 showResults 方法
tb =new Draw(map);
tb.on("draw-end", addGraphic);
tb.activate(esri.toolbars.Draw.FREEHAND_POLYGON);
function addGraphic(geometry){
    console.log("geometry");
    symbol =new SimpleFillSymbol(SimpleFillSymbol.STYLE_SOLID,
new SimpleLineSymbol(SimpleLineSymbol.STYLE_DASHDOT,new Color([255,
0,0]),2),new Color([255,255,0,0.5]));
var handgraphic =new Graphic(geometry, symbol);
    query.geometry = handgraphic.geometry;
    queryTask.execute(query,function(results){
//清除上一次的高亮显示
        map.graphics.clear();
//查询结果样式
var symbol =new
SimpleMarkerSymbol({"color":[0,255,0],"size":12,"type":"esriSMS","
style":"esriSMSCircle"});
//遍历查询结果
var info ="<label >编号</label><label>所属地区</label><br>";
for(var i =0; i < results.features.length; i++){
var graphic = results.features[i];
//设置查询到的 graphic 的显示样式
            graphic.setSymbol(symbol);
//设置 graphic 的信息模板
            graphic.setInfoTemplate(infoTemplate);
            info +="<label> "+ graphic.attributes.FID +"</label><
label >   "+ graphic.attributes.地区+"</label><br>";
//把查询到的结果添加到 map.graphics 中进行显示
            map.graphics.add(graphic);
}
//将信息显示在信息框
        dom.byId("famen").innerHTML ="";
```

```
        dom.byId("famen").innerHTML = info;
        tb.deactivate();
    });
```

实例运行结果如图 21.1 所示。

图 21.1　空间查询功能效果图

21.1.2　FindTask

FindTask 是在某个地图服务中进行属性查询的功能类，FindTask 以 FindParameters 对象作为参数，能查询同一个地图服务的一个或者多个图层，并且可以在多个字段中进行查询，FindTask 仅仅用于属性信息的查询，在 FindTask 执行结束后，可以从其返回结果中获取查询的对象来自哪个图层和哪个字段。

21.1.3　IdentifyTask

IdentifyTask 跟桌面软件中的 Identify 类似，是在某个地图服务中进行空间查询的类。IdentifyTask 以 IdentifyParameters 对象作为参数，能查询同一个地图服务的一个或者多个图层，IdentifyTask 仅仅用于空间信息查询。通常使用 Toolbar 的 Draw 工具绘制一个几何对象，然后将这个几何对象作为 IdentifyTask 的参数发送到服务器进行识别，执行结束后可以通过结果获取查询对象所属的图层。

以下实例用来说明如何使用 IdentifyTask 进行查询，即通过获取鼠标点击位置的几何对象，对指定图层查询输入对象处几何对象的属性。实例具体代码如下：

```
    //鼠标点击查询
    var Task =new
IdentifyTask("http://localhost:6080/arcgis/rest/services/YSLX/MapS-
erver");
    var identifyParams =new IdentifyParameters();
```

```
        identifyParams.tolerance =3;
        identifyParams.returnGeometry =true;
        identifyParams.layerOption = IdentifyParameters.LAYER_OP-
TION_ALL;
        identifyParams.width = map.width;
        identifyParams.height = map.height;
        identifyParams.mapExtent = map.extent;
        map.graphics.clear();
        identifyParams.geometry = event.mapPoint;
    //根据图层显示查询结果显示窗
    var infoTemplate =new InfoTemplate("查询信息");
        console.log(infoTemplate);
    //设置 identifyParams.layerIds = [0]和查询图层
    switch(LayerName){
    //陆地阀门
    case'layerlg':
            identifyParams.layerIds =[3];
            infoTemplate.content ="阀门 ID:${FID}<br/><br/>所在管线号:
${gid}<br/><br/>所在地区:${地区}";
    break;
    //海上阀门
    case'layerhdgx':
            identifyParams.layerIds =[6];
            infoTemplate.content ="阀门 ID:${FID}<br/><br/>所在管线号:
${gid}<br/><br/>所在地区:${地区}";
    break;
    //炼油厂图层
    case'layeryc':
    {
                identifyParams.layerIds =[0];
                infoTemplate.content ="阀门 ID:${FID}<br/><br/>所在管
线号:${gid}<br/><br/>所在地区:${地区}";
    break;
    }
    default:return;}
      Task.execute(identifyParams,function(idResults){
            console.log(idResults)
    if(idResults.length >0){
```

```
            console.log("idResults.features.length:"+ idResults.
length);
    for(var i =0, il = idResults.length; i < il; i++){
    var Psymbol =new SimpleMarkerSymbol({
    "color":[255,255,0],
    "size":12,
    "type":"esriSMS",
    "style":"esriSMSCircle"
    });
    //初始化信息模板
    var point =new esri.geometry.Point(idResults[i].feature.geometry,
new esri.SpatialReference({ wkid:3857}));
    //
                    infoTemplate.content +="<br /><br />X 坐标:"+ point.
x +"<br /><br />Y 坐标:"+ point.y;
                    graghics = idResults[i].feature;
                    graghics.setSymbol(Psymbol);
                    graghics.setInfoTemplate(infoTemplate);
                    map.graphics.add(graghics);
                    setTimeout(function(){
                        map.graphics.clear(graghics);
    },4000);
                    console.log(graghics)
                    console.log("map.graphic.add(graghics)");
                    map.infoWindow.setTitle(graghics.getTitle());
                    map.infoWindow.setContent(graghics.getContent());
    //var screenPnt = map.toScreen(location);
                    map.infoWindow.resize(350,400);
                    map.infoWindow.show(event.screenPoint);

    }
    }
    else{
    returnfalse;
    }
    });
```

实例运行结果如图21.2所示。

图 21.2　IdentifyTask 查询功能运行效果图

提示：文中对接口的属性、方法、事件的介绍均由 ESRI ArcGIS API for JavaScript 官方网站总结而来。

21.2　网络分析

21.2.1　网络分析类别

ArcGIS 提供了两种网络分析，即基于 Geometric Network 的有向网络或者设施网络和基于 Network Dataset 的无向网络，在这里网络分析指的是后者。在最短路径分析中，ArcGIS 还提供了设置障碍点、线、面等丰富的功能。ArcGIS API for JavaScript 支持网络分析中的最短路径分析，服务区分析和临近设置分析，分别为这 3 种分析提供了不同的执行对象和参数对象。它们的区别详见表 21.1。

表 21.1　　　　网络分析类别

执行对象	参数对象	返回对象	使用范围
最短路径 RouteTask	RouteParameters	RouteSolveResult	计算最短路径或最少时间
服务区分析 ServiceAreaTask	ServiceAreaParameters	ServiceAreaSolveResult	计算某设施在指定时间内能服务的范围，如加油站一小时内能到达的范围
临近设施分析 ClosestFacilityTask	ClosestFacilityParameters	ClosestFacilitySolveResult	计算距离事件点最近的设施并给出行驶路线

21.2.2　ESRI 开发竞赛获奖案例

“石油储运管理系统”(研发者：张淑珍、谭丽、张志超；指导教师：柳林)是 2014 年度“ESRI 杯”中国大学生 GIS 软件开发竞赛 Web 开发组获奖作品，下面以此作品为案例来演示如何使用网络分析中的最短路径分析等功能。“石油储运管理系统”案例更多内容请查阅随书所赠光盘。

1. 数据准备

要使用网络分析功能，必须在制作地图文档时将网络分析的数据加载到地图中，同时还要有相应的网络分析图层(最短路径网络图层，临近设施网络图层和服务区分析网络图层)。这里以最短路径为例，在地图文档中可以看到除了网络数据集之外的最短路径网络分析图层，如图 21.3 所示。

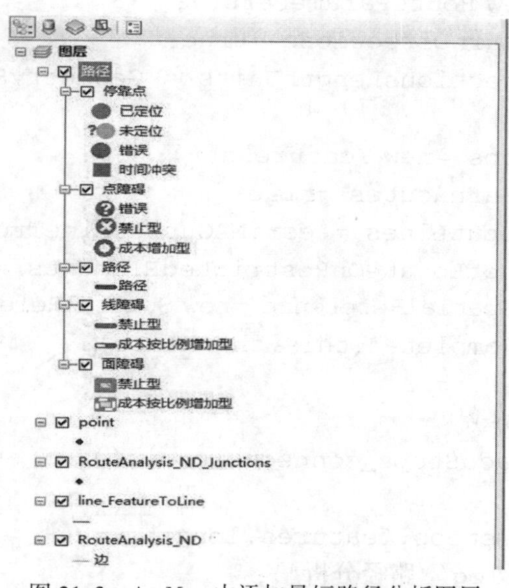

图 21.3　ArcMap 中添加最短路径分析图层

2. 发布服务

在发布地图服务的时候，给服务配置网络分析功能，如图 21.4 所示。

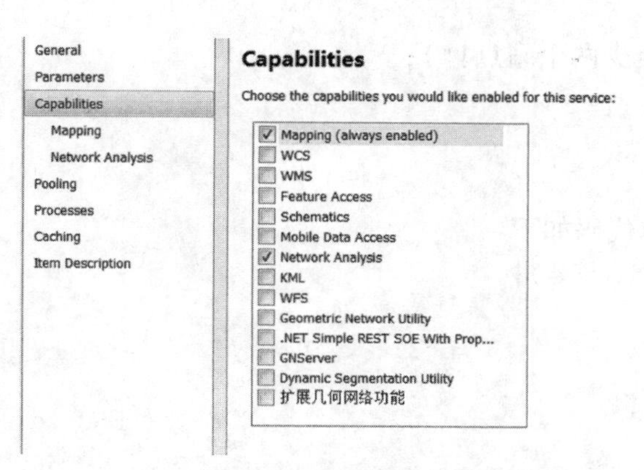

图 21.4　配置网络分析服务

3. 核心代码

初始化执行对象 RouteTask 和参数对象 RouteParameters。具体代码如下：

```
Task =new
RouteTask( "http://localhost:6080/arcgis/rest/services/Route/NAServer/%E8%B7%AF%E5%BE%84");
//查询功能
//identifyTask = new
IdentifyTask("http://localhost:6080/arcgis/rest/services/Route/MapServer");
routeParams =new RouteParameters();
routeParams.returnDirections =true;
routeParams.directionsLengthUnits = GeometryService.UNIT_KILOMETER;
routeParams.stops =new FeatureSet();
routeParams.returnRoutes =true;
routeParams.outputLines ="esriNAOutputLineTrueShape";
routeParams.doNotLocateOnRestrictedElements =true;
routeParams.outSpatialReference =new SpatialReference({ wkid:3857});
on(Task,"SolveComplete",this.showRoute);
```

执行最短路径分析

```
_Solve:function(){
    mapOnClick_addStops_connect.remove();

if(routeParams.stops.features.length >=2){
        console.log("路径分析");
        console.log(routeParams);
        Task.solve(routeParams);

}
else{
        alert("至少两个站点!");
returnfalse;
}
}
```

获取执行结果的具体代码如下：

```
showRoute:function(e){
    routes =[];
var dataset ={
identifier:"id",
items:[]
};
```

```
    var data =[];
    var rows =60;
        directions = e.routeResults[0].directions;
    //每段路径
    var routeSymbol =new SimpleLineSymbol().setColor(new Color([0,0,
255,0.5])).setWidth(4);

    //Zoom to results.
        map.setExtent(directions.mergedGeometry.getExtent(),true);
    //Add route to the map.
    var routeGraphic =new Graphic(directions.mergedGeometry, route-
Symbol);
        map.graphics.add(routeGraphic);

        routes.push(routeGraphic);

    //显示路径信息
    var directionsInfo = e.routeResults[0].directions.features;

    for(i =0; i < directionsInfo.length; i++){
            data[i]={
    one: directionsInfo[i].attributes.text,
    two: Math.round(parseFloat(directionsInfo[i].attributes.length,{
places:2})*100)/100,
    three:"千米"
    }
    }
    }
```

4. 运行结果

“石油储运管理系统”案例中，最短路径的运行结果如图 21.5 所示。

图 21.5　网络分析功能效果图

第四编 ArcGIS for Android 移动开发

ESRI 一直致力于将 ArcGIS 打造成一个平台，为不同用户提供从内容到功能再到应用的全方位的服务。ArcGIS 目前支持 Android、iOS、Windows Phone 等移动平台，提供各个平台上集成特定功能的原生 SDK[24]，其中面向 Android 终端的移动 GIS 开发套件为 ArcGIS Runtime SDK for Android。鉴于目前市面上的开发书籍以及网络上的参考文档，大部分编写时间早于 2013 年，随着 Android SDK 以及 ArcGIS SDK 版本的更新，已渐渐不能适应开发者当前需求，本文以官方提供的最新 API 为基础，结合程序案例，介绍 Android 环境下的移动 GIS 开发。

第 22 章　移动开发基础

22. 1　ArcGIS 移动开发基础

22. 1. 1　ArcGIS 移动开发 SDK

ArcGIS Runtime SDKs for Smartphones and Tablets 是 ESRI 为开发者提供的移动应用开发包，目前支持 iOS、Android、Windows Phone 三大主流移动操作系统。注册 ESRI 全球账号，可以免费下载各个版本的开发包以及其他相关资料。

1. iOS SDK

ArcGIS Runtime SDK for iOS 是 ESRI 公司专门为 iOS 平台提供的移动开发包，支持使用 Objective-C 和 Swift 语言开发地图应用程序，开发完成的应用能够部署在各种 iPhone 手机、iPad 平板和 iPod Touch 设备上。其提供的主要功能包括：

①离线数据，支持直接加载本地的 tpk 切片包和 Runtime Geodatabase 等数据。

②地图浏览，支持缩放、平移等基本操作和捏夹、双击等手势操作。

③地图查询，支持对地图进行查询，包括关键字搜索、identity、find、query 查询、多条件联合查询、范围查询等。

④地图定位，使用智能设备自带的 GPS 定位模块，可以实现快速地图定位。

⑤数据展示，使用多种渲染方式显示加载的数据，并使用 popups、callout 等方式显示数据的属性信息。

⑥外业数据采集，使用 SDK 开发 APP，或者使用即拿即用的 APP 实现数据的快速采集、实时同步、本地编辑，包括属性信息的录入和现场坐标数据的采集，还可以采集现场的照片、视频等。

⑦数据编辑，可在移动端进行细粒度的业务数据编辑，包括修改属性信息、移动要素的位置、改变要素的形状、添加/删除要素等操作。

⑧数据同步，在移动端编辑的数据，可采用在线实时更新的方式同步到远程数据库中。

⑨地理编码，支持在线和离线的地理编码和反地理编码功能。

⑩路径规划，可实现在线的或者离线的、单点或多点的路径规划。

在连接 ArcGIS for Server 数据或者服务时，ArcGIS Runtime SDK for iOS 要求安装 ArcGIS for Server 10.0 及以上版本，如果需要离线使用，则需要安装 ArcGIS for Server 10. 2. 2 及以上版本。

2. Android SDK

ArcGIS Runtime SDK for Android 是 ESRI 为 Android 平台提供的移动开发包，Android SDK 支持使用 Java 语言开发应用程序，便于开发者开发功能丰富的地图应用，开发好的应用能够部署在 Android 智能手机、平板电脑和其他智能终端上，支持众多国内外品牌。相比于 ArcGIS Runtime SDK for iOS，Android 版本提供了两个额外功能：

①视域分析，支持移动端离线的视域分析。

②视线分析，支持移动端离线的视线分析。

ArcGIS Runtime SDK for Android 通过 ArcGIS Server REST 服务获取数据和服务资源。ESRI 发布了 GeoServices REST Specification，这一标准规定了 ArcGIS REST Service 各种接口的访问参数及返回数据的结构，ArcGIS Runtime SDK for Android 正是基于这一标准封装的。ArcGIS 基于 REST 接口的 API，包括 ArcGIS Runtime SDK for Android/IOS/Windows Phone、ArcGIS API for Flex/ Silverlight/ JavaScript 以及 ArcGIS Runtime SDK for Java/. NET，都是基于这一标准进行封装的。尽管不同平台、不同语言的开发包有其自己的特性，但其对应服务端的编程模型是一致的。

3. Windows Phone SDK

ArcGIS Runtime SDK for WindowsPhone 是 ESRI 为 WindowsPhone 平台提供的移动开发包。Windows Phone SDK 支持使用 C#语言开发应用程序，允许开发者在 Windows Phone 平台上集成 ArcGISServer 的服务以及微软的 Bing 地图服务。目前，ArcGIS Runtime SDK for WindowsPhone 的最新版版本为 3. 0，较之前版本增加了对 10. 1 新特性的支持，如动态图层，服务端打印等。ArcGIS Runtime SDK for Windows Phone 功能与 ArcGIS Runtime SDK for Android 以及 ArcGIS Runtime SDK for iOS 功能基本一致。

22. 1. 2 Android 系统

Android 是一种基于 Linux 的自由及开放源代码的操作系统，主要应用于移动设备，如智能手机和平板电脑，由 Google 公司和开放手机联盟领导及开发，于 2007 年 11 月 5 日发布内部测试版，2008 年 9 月发布第一版正式手机操作系统。获取 Android 相关知识的网络渠道众多，由于篇幅限制，下面着重介绍初级开发中涉及的一些概念或要点。

1. Android 版本

Android 自发布以来版本更迭迅速，API 的频繁升级给实际开发带来了版本控制的问题。近年来 Android 发布的主要版本如下：

Android 2. 2——Froyo；

Android 2. 3——Gingerbread；

Android 4. 0——Ice Cream Sandwich；

Android 4. 2——Jelly Bean；

Android 4. 4——KitKat；

Android 5. 0/5. 1——Lollipop；

需要注意的是，最早版本的 ArcGIS SDK 对 Android 版本的要求是 2. 2 及以上，目前大多数的参考文档仍停留在 2012 年前，实际上最新发布的 SDK 对 Android 版本的要求已经提高到了 4.0 及以上。

2. Android 文件结构

.apk 是安卓应用的后缀，是 AndroidPackage 的缩写，即 Android 安装包(apk)。.apk 是类似 Symbian Sis 或 Sisx 的文件格式。通过将 apk 文件直接传到 Android 模拟器或 Android 手机中执行即可安装。apk 文件和 sis 一样，把 AndroidSDK 编译的工程打包成一个安装程序文件，格式为 .apk。apk 文件其实是 zip 格式，但后缀名被修改为 .apk，通过 UnZip 解压后，可以看到 Dex 文件，Dex 是 Dalvik VM executes 的全称，即 Android Dalvik 执行程序，并非 Java ME 的字节码而是 Dalvik 字节码。

一个 apk 文件结构为：

①META-INF \ （注：Jar 文件中常可以看到）；

②res \ （注：存放资源文件的目录）；

③AndroidManifest. xml（注：程序全局配置文件）；

④classes. dex（注：Dalvik 字节码）；

⑤resources. arsc（注：编译后的二进制资源文件）。

3. Android 应用组件

Android 开发 4 大组件分别是：

①活动(Activity)：用于表现功能。

②服务(Service)：后台运行服务，不提供界面呈现。

③广播接收器(Broadcast Receiver)：用于接收广播。

④内容提供商(Content Provider)：支持在多个应用中存储和读取数据，相当于数据库。

活动(Activity)是所有程序的根本，所有程序的流程都运行在 Activity 之中，Activity 可以算是开发者遇到的最频繁，也是 Android 当中最基本的模块之一。Activity 概念和网页的概念类似，在 Android 的程序当中，Activity 一般代表手机屏幕的一屏，如果把手机比作一个浏览器，那么 Activity 就相当于一个网页。在 Activity 当中可以添加 Button、Check box 等控件。

服务(Service)是 Android 系统中的一种组件，它跟 Activity 的级别差不多，但是不能自己运行，只能后台运行，并可以和其他组件进行交互。Service 是没有界面的长生命周期的代码，是一种程序，它可以运行很长时间，但是没有用户界面。

4. Android 开发环境

在 Windows 下，长期以来最流行的 Android 开发平台一直是 Eclipse。除此之外，今年 Google 公司发布了 Android Studio 的正式版，是专门面向 Android 开发的平台，具有诸多方便于 Android 开发的特性，但由于目前存在的不稳定性，本书不做重点推荐。

22.2　ArcGIS SDK for Android 开发环境

ArcGIS SDK for Android 的开发环境其实就等于“Android 开发环境”+“ArcGIS 开发控件”，Android 开发环境是基于 Eclipse 的，因此，开发环境搭建的思路是在 Eclipse 的基础上完成 Android 开发工具的扩展。

22.2.1 基础环境要求

Windows 环境下的移动 GIS 开发所需的系统环境包括：

①操作系统：Windows 8 旗舰版 64 位(要求 Windows XP 以上)。

②PC 硬件：CPU Core2 2.2G、内存 6G、硬盘 500G SSD、显卡 GT240(如果使用 Android 虚拟机进行开发，建议 CPU 使用 intel i 系列及以上)。

③移动设备：支持 OpenGL ES 2.0、版本 4.0 以上的 Android 虚拟机供开发使用，建议通过真机调试。

④网络条件：无线 Wifi(真机调试)+网络宽带。

⑤ArcGIS for Server：版本 10.1(要求 Server 9.3.1 及以上)。

22.2.2 JDK 的版本要求及安装配置

环境的配置首先需要安装 Java SE Development Kit(JDK6 及以上)，要注意的是 JDK 分 32 位和 64 位，需要跟操作系统对应，安装完成还需手动进行环境变量修改，具体过程如下：

①安装包获取：可以进入官方网站：http://java.sun.com/javase/downloads，或者使用本书附带的离线版本。

②安装注意事项：安装过程不再赘述，注意安装路径不要有中文，路径、名称不要随意更改。

③系统环境变量修改：在“我的电脑”右键点击【属性】，进入系统管理页面，如图 22.1 所示。

图 22.1 系统管理页面

点击【系统高级设置】≫【高级】≫【环境变量】，在【系统变量】下进行设置，如图 22.2 所示。

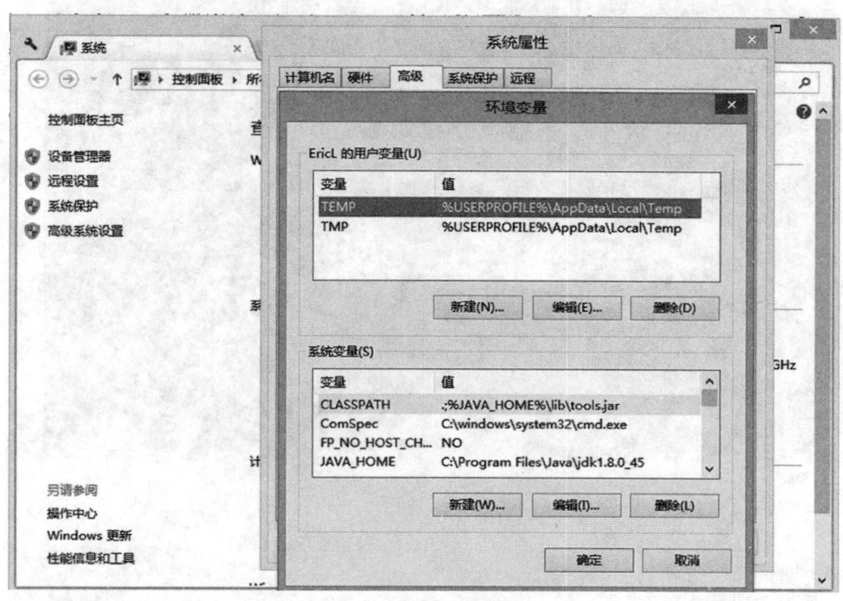

图 22.2　系统变量

点击【新建系统变量】，变量名为 JAVA_HOME，变量值是 java 的安装路径 jdk 路径。

添加完 JAVA_HOME 系统变量后，需要添加两个包文件。在系统变量中，选中 classpath，点击【编辑】，在 classpath 后边添加%JAVA_HOME%\lib\dt.jar;%JAVA_HOME%\lib\tools.jar；添加 path，如果没有 path，则需要新建 pathpath=%JAVA_HOME%\bin。

④Java 环境检验：如果 JDK 安装成功，在命令行窗口下，键入 java -version 命令可以看到安装的 JDK 版本信息；键入 java 命令可以看到此命令的帮助信息；键入 javac 命令可以看到此命令的帮助信息，如图 22.3 所示。

22.2.3　Eclipse 及 Android SDK 的版本要求及安装配置

Eclipse 版本要求 Eclipse 3.6.2(Helios)以上，推荐 3.7Indigo 或者 4.2 Juno。(本书版本为 3.7Indigo)下载地址 http://www.eclipse.org/downloads/packages/eclipse-ide-java-ee-developers/indigosr2。对于 Android SDK 的安装，建议直接使用本书附带的离线文件，将文件解压放到安装目录下。

将 Eclipse 和 Android SDK 联系起来，在 Eclipse 下配置好 Android SDK 的路径。所以，在 Eclipse 中安装 Android 的 Eclipse 插件——Android Development Tool(ADT)，安装可以通过 Android 在线升级地址(https://dl-ssl.google.com/android/eclipse/)进行，通过 Eclipse 中的【Help】≫【Install New Software...】，全部选择进行安装即可，如图 22.4 所示。

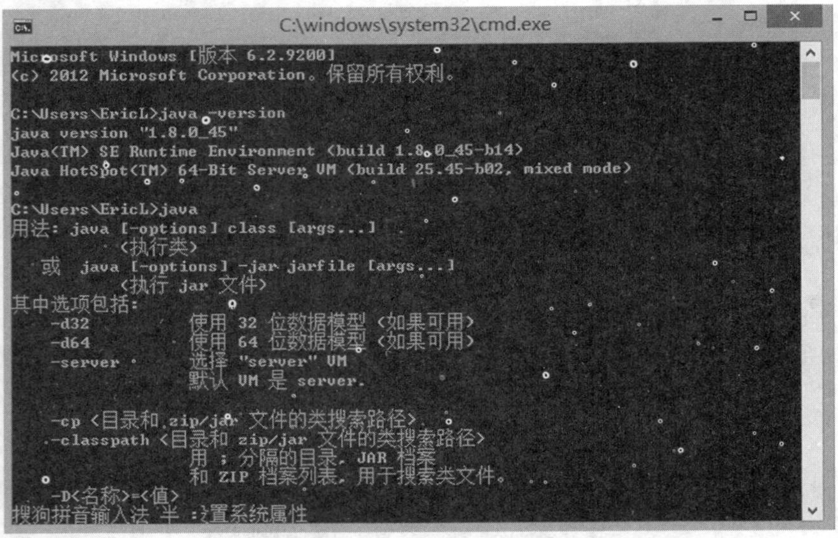

```
C:\windows\system32\cmd.exe
Microsoft Windows [版本 6.2.9200]
(c) 2012 Microsoft Corporation。保留所有权利。

C:\Users\EricL>java -version
java version "1.8.0_45"
Java(TM) SE Runtime Environment (build 1.8.0_45-b14)
Java HotSpot(TM) 64-Bit Server VM (build 25.45-b02, mixed mode)

C:\Users\EricL>java
用法: java [-options] class [args...]
           (执行类)
   或  java [-options] -jar jarfile [args...]
           (执行 jar 文件)
其中选项包括:
    -d32          使用 32 位数据模型 (如果可用)
    -d64          使用 64 位数据模型 (如果可用)
    -server       选择 "server" VM
                  默认 VM 是 server.

    -cp <目录和 zip/jar 文件的类搜索路径>
    -classpath <目录和 zip/jar 文件的类搜索路径>
                  用 ; 分隔的目录, JAR 档案
                  和 ZIP 档案列表, 用于搜索类文件。
    -D<名称>=<值>
搜狗拼音输入法 半 :设置系统属性
```

图 22.3 环境检验

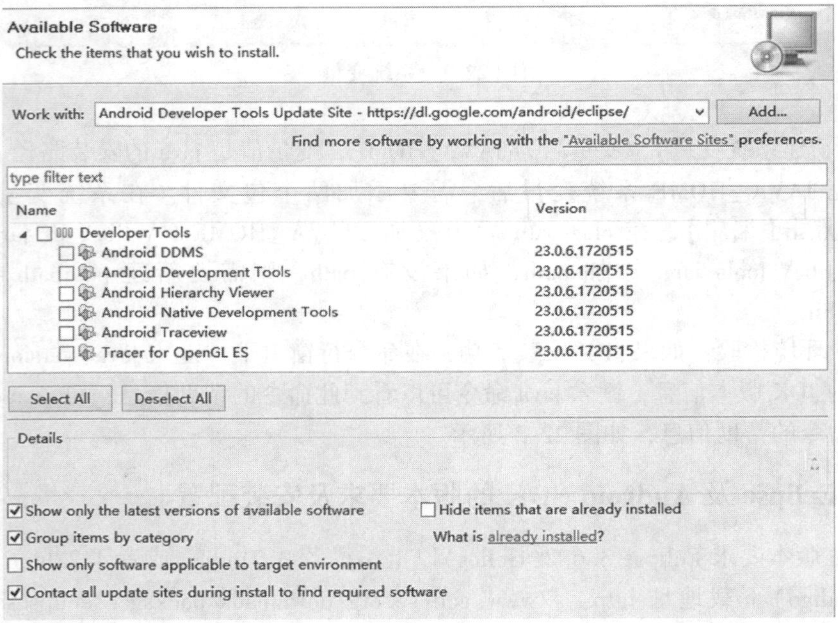

Available Software

Check the items that you wish to install.

Work with: Android Developer Tools Update Site - https://dl.google.com/android/eclipse/ Add...

Find more software by working with the "Available Software Sites" preferences.

type filter text

Name	Version
Developer Tools	
Android DDMS	23.0.6.1720515
Android Development Tools	23.0.6.1720515
Android Hierarchy Viewer	23.0.6.1720515
Android Native Development Tools	23.0.6.1720515
Android Traceview	23.0.6.1720515
Tracer for OpenGL ES	23.0.6.1720515

Select All Deselect All

Details

- [x] Show only the latest versions of available software
- [ ] Hide items that are already installed
- [x] Group items by category
- What is already installed?
- [ ] Show only software applicable to target environment
- [x] Contact all update sites during install to find required software

图 22.4 ADT 安装

安装完成后，通过【Window】≫【References】菜单可以看到 Eclipse 中已经存在 Android 的选项，需要指定 Android SDK 的路径，如图 22.5 所示。

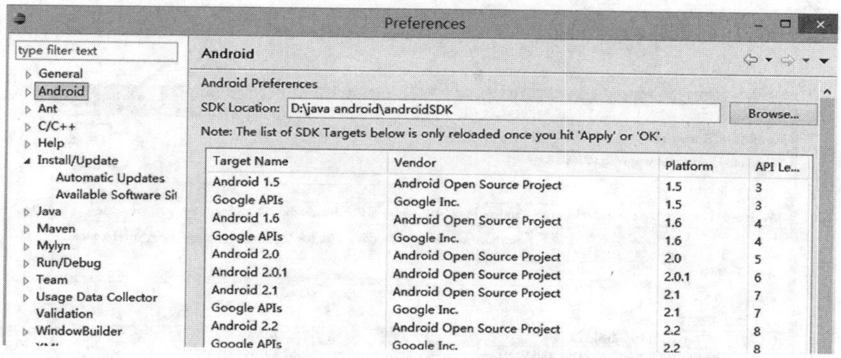

Preferences

type filter text

- General
- Android
- Ant
- C/C++
- Help
- Install/Update
  - Automatic Updates
  - Available Software Sit
- Java
- Maven
- Mylyn
- Run/Debug
- Team
- Usage Data Collector
- Validation
- WindowBuilder

Android

Android Preferences

SDK Location: D:\java android\androidSDK　Browse...

Note: The list of SDK Targets below is only reloaded once you hit 'Apply' or 'OK'.

Target Name	Vendor	Platform	API Le...
Android 1.5	Android Open Source Project	1.5	3
Google APIs	Google Inc.	1.5	3
Android 1.6	Android Open Source Project	1.6	4
Google APIs	Google Inc.	1.6	4
Android 2.0	Android Open Source Project	2.0	5
Android 2.0.1	Android Open Source Project	2.0.1	6
Android 2.1	Android Open Source Project	2.1	7
Google APIs	Google Inc.	2.1	7
Android 2.2	Android Open Source Project	2.2	8
Google APIs	Google Inc.	2.2	8

图 22.5　Android SDK 路径指定

在 Android SDK Manager 中可以查看已安装的 SDK 版本，并对 SDK 进行管理，如图 22.6 所示。至此，Android 开发环境已经搭建完毕。

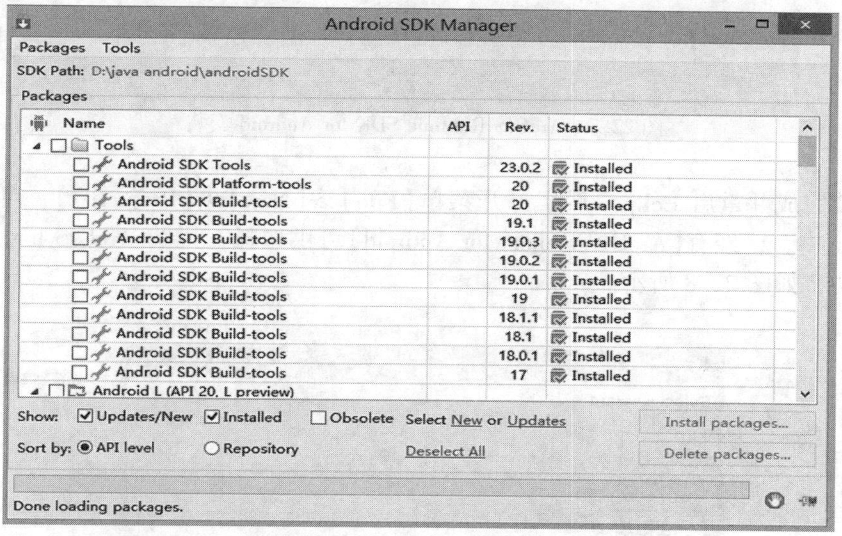

Android SDK Manager

Packages　Tools

SDK Path: D:\java android\androidSDK

Packages

Name	API	Rev.	Status
Tools			
Android SDK Tools		23.0.2	Installed
Android SDK Platform-tools		20	Installed
Android SDK Build-tools		20	Installed
Android SDK Build-tools		19.1	Installed
Android SDK Build-tools		19.0.3	Installed
Android SDK Build-tools		19.0.2	Installed
Android SDK Build-tools		19.0.1	Installed
Android SDK Build-tools		19	Installed
Android SDK Build-tools		18.1.1	Installed
Android SDK Build-tools		18.1	Installed
Android SDK Build-tools		18.0.1	Installed
Android SDK Build-tools		17	Installed
Android L (API 20, L preview)			

Show: Updates/New　Installed　Obsolete　Select New or Updates　Install packages...

Sort by: API level　Repository　Deselect All　Delete packages...

Done loading packages.

图 22.6　Android SDK 查看和管理

22.2.4　ArcGIS Runtime SDK for Android 的版本要求及安装配置

ArcGIS Developer 网站提供 ArcGISRuntime SDK for Android 的在线安装地址：http://resources.arcgis.com/en/communities/runtime-android/，但是该地址下的 SDK 版本较老，在这里推荐使用本书附带的新版 SDK，采用离线安装包的形式进行安装。

①准备 SDK 包，本书提供版本为 arcgis-android-sdk-v10.2.3。需进行解压，离线安装 jar 文件的路径为 ./tools/eclipse-plugin/arcgis-android-eclipse-plugin.jar。

②打开菜单【Help】≫【Install New Software】，点击【Add】按钮，在弹出窗口中点击【Archive】按钮，在本地安装已经下载的 ArcGIS Android 插件压缩包，完成安装即可，如

图 22.7 所示。

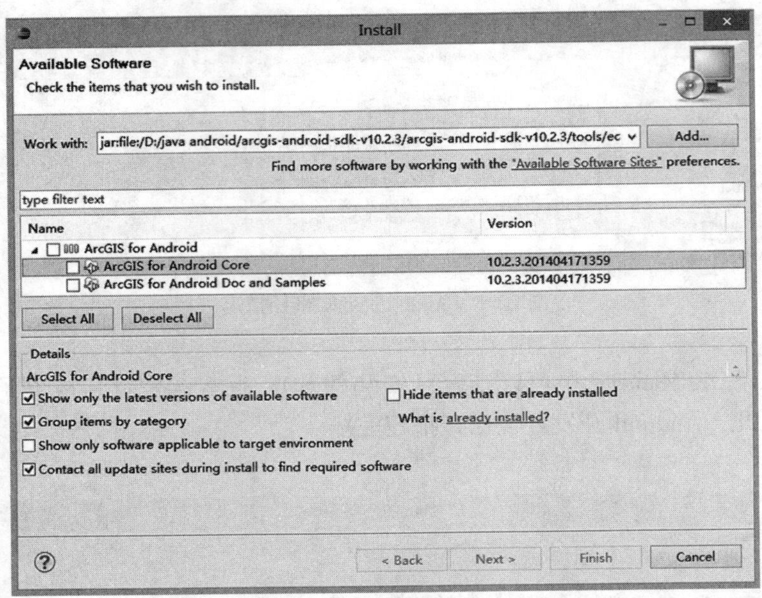

图 22.7　ArcGIS Runtime SDK for Android 安装

③安装完成并重启 Eclipse 后，打开菜单【File】≫【New】≫【Project…】，可以看到在 New Project 中，已经有【ArcGIS Project for Android】可供选择，说明 ArcGIS for Android 已经安装成功，如图 22.8 所示。

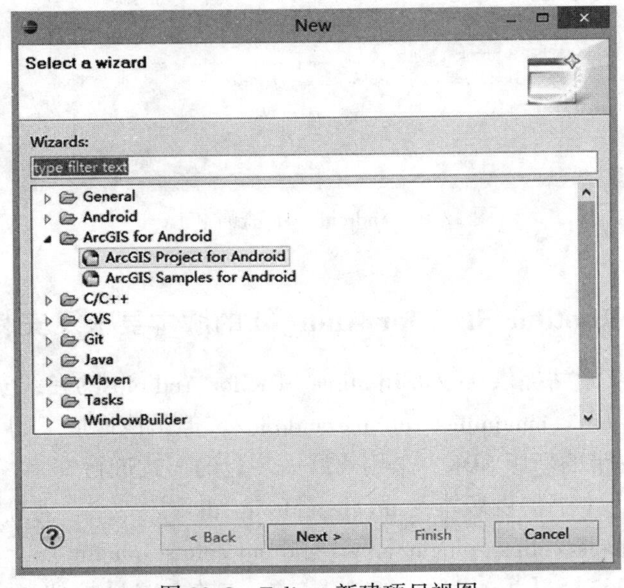

图 22.8　Eclipse 新建项目视图

22.2.5　Android 模拟器配置

①在 Eclipse 中选择【Window】菜单下的【AVD Manager】。

②在弹出的【Android Virtual Device Manager】对话框中选择【New】。

③在打开的【Create new Android Virtual Device】对话框中，配置 AVD(Android 模拟器)的名称、Hardware 等属性；选择 Hardware 右侧的【New】按钮，添加“GPU emulation”，如图 22.9 所示。

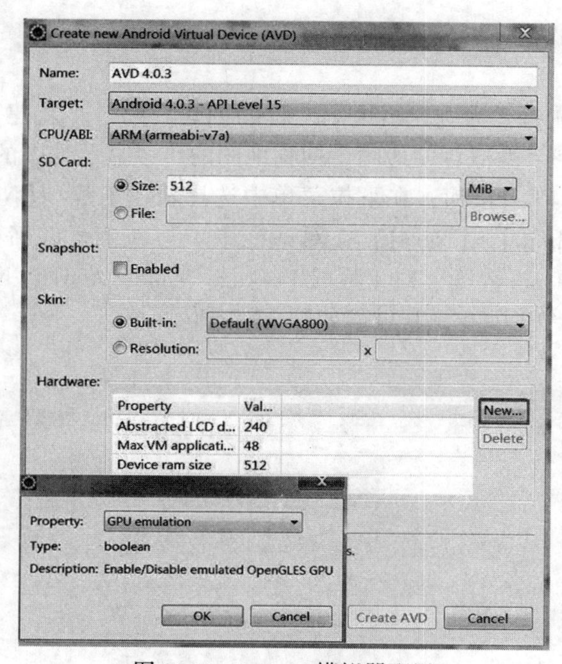

图 22.9　Android 模拟器配置

④将【属性】的值改为“yes”，点击【Create AVD】，创建完成，如图 22.10 所示。

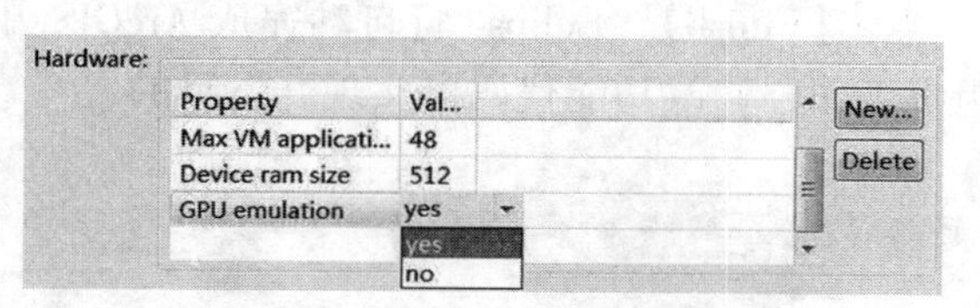

图 22.10　Android 模拟器创建完成

第23章　创建地图工程

23.1　新建 HelloWorldMap

本节以“HelloWorldMap”为例介绍地图工程的创建。第一步新建一个工程，并命名为HelloWorldMap，完成一个类似百度地图、搜狗地图的可以显示地图的手机APP。

①点击 Eclipse 的【File】按钮，在下拉菜单中选择新建工程项【New】≫【Other】，找到【ArcGIS Project for Android】≫【Next】，在对话框中填写新工程的名称及存储目录，点击【Next】，在 Application Info 界面，对工程的 Package Name、Activity Name 进行修改，并选择进行编译的 Android API 版本，其过程如图 23.1 所示。

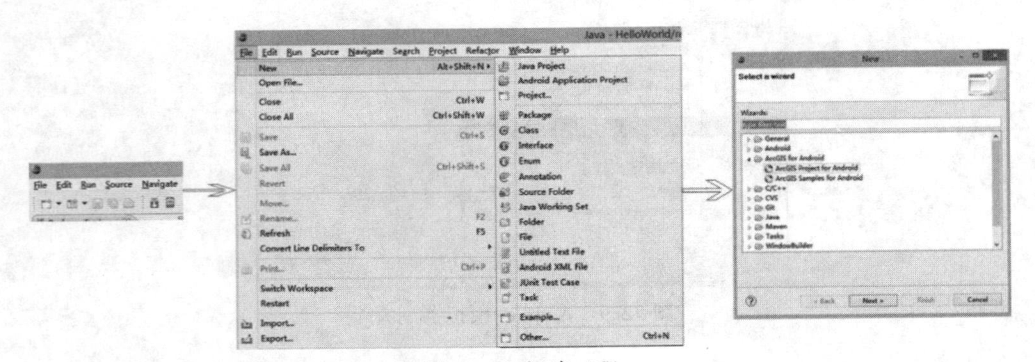

图 23.1　新建工程

②如图 23.2 所示，点击【Finish】，Eclipse 将自动完成 ArcGIS 开发包的依赖引用、程序基本代码等工程的组织工作，修改功能代码可实现开发目的。

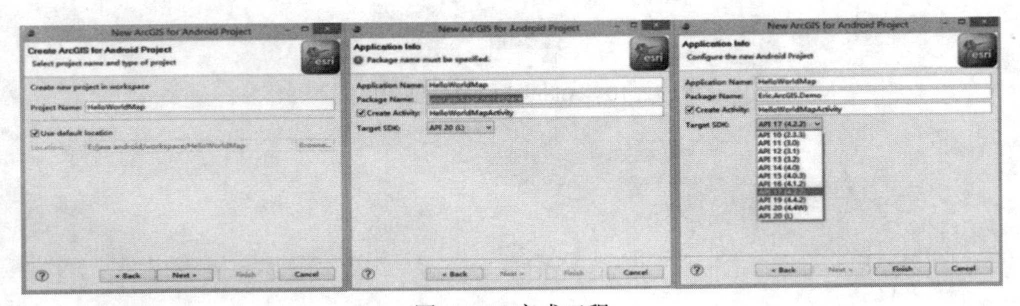

图 23.2　完成工程

23. 2　地图工程组织结构

本节通过“HelloWorldMap”示例工程对 Eclipse 开发环境进行介绍，并对工程结构进行分析。

23. 2. 1　Eclipse IDE

如图 23. 3 所示，Eclipse IDE 包括以下几个部分：

①菜单栏：与普通 Windows 程序一样，菜单栏集成程序整体菜单，包括 File、Edit、Source、Navigate、Search、Project、Refactor、Window、Help 等。

②工具条(toolbar)：继承了开发中较常使用的一些功能的快捷按钮，包括新建、保存、SDK manager、AVD manager、运行调试工具、撤销恢复更改等。

③代码区：提供代码文件编辑修改的主要区域，代码文件以标签的形式在此区域进行组织。

④Package Explorer：Eclipse 下打开的项目在此区域以树形结构展示。

⑤信息显示区：Eclipse 提供的信息显示工具在激活状态下在此区域按标签存放，包括 Problem 标签下 Bug 的显示、Console 标签下控制台信息、Logcat 标签下程序运行信息等。

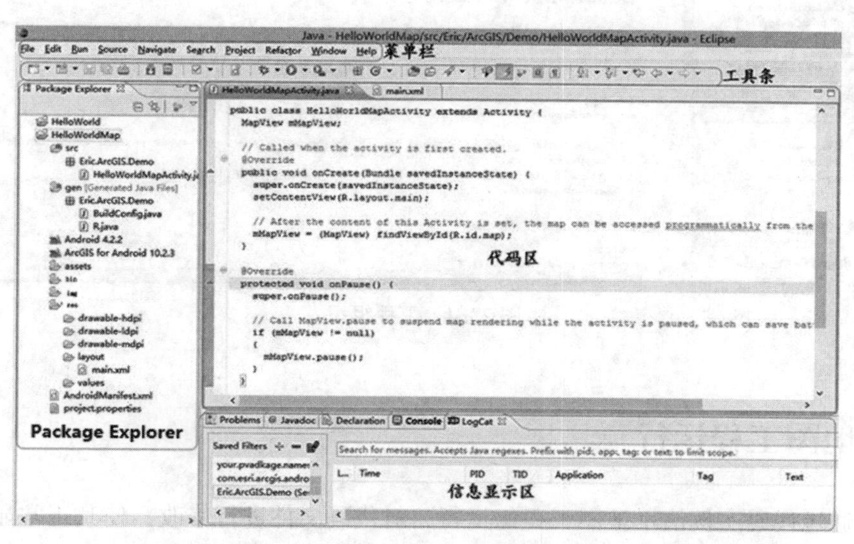

图 23. 3　Eclipse IDE 环境

23. 2. 2　工程组织介绍

如图 23. 4 所示，ArcGIS 工程的组织结构与普通的 Android 工程基本相同，包括以下组成部分：

①src：src 目录下存放着程序的主要代码文件。Activity 文件默认以 Activity 结尾，不同功能模块的文件可以用不同的 Package 进行分类存放。

②gen：此目录下的文件是自动生成的，一般不进行人为修改。其中的 R 文件是系统 Android Framework 负责管理的，随着代码的修改自动变更。

③库文件：程序运行所需要的库文件，在 Android 4.2 的库文件以外，还增加了 ArcGIS for Android 10.2.3 的库。

④libs：libs 目录下存放的是一些项目所需的动态链接库。对于 ArcGIS 项目，它默认存放了两个 GIS 所需的动态链接库，当然也可以添加所需的其他动态链接库。

⑤res：存放资源文件的目录。存放了所需的大部分的资源，默认目录下有三类资源：drawable 目录主要存放一些图片；layout 目录主要存放一些布局文件；values 目录主要存放一些项目中所需的参数值文件。除此之外还有其他分类，如 anim 和 xml 目录。

⑥AndroidManifest：xml 文件是项目的一个系统配置文件，它包含了 activity（行为）、view（视图）、service（服务）等的信息，以及运行 Android 应用程序需要的用户权限列表，同时也描述了 Android 应用的项目结构。

图 23.4 工程组织

23.3 地图工程运行

本节通过代码实例演示地图工程的运行，对代码文件进行修改，使其正常显示地图。

①打开 HelloWorldMapActivity.java、main.xml 两个文件。在 HelloWorldMapActivity.java 文件中，向空白的 Activity 完善如下代码：

```
MapViewmMapView;
publicvoid onCreate(Bundle savedInstanceState) {
super.onCreate(savedInstanceState);
    setContentView(R.layout.main);
```

```
mMapView = (MapView) findViewById(R.id.map);
  }

@ Override
protectedvoid onPause() {
super.onPause();
if (mMapView ! = null)
    {
mMapView.pause();
    }
  }

@ Override
protectedvoid onResume() {
super.onResume();
if (mMapView ! = null)
    {
mMapView.unpause();
    }
  }
```

②在 main. xml 文件中，【LinearLayout】标签内添加一个 MapView。代码如下：

```
<com.esri.android.map.MapView
android:id="@ +id/map"
android:layout_width="fill_parent"
android:layout_height="fill_parent"
mapoptions.MapType="Topo"
mapoptions.center="34.056215, -117.195668"
mapoptions.ZoomLevel="16" />
```

③修改后保存，按照错误提示器修改代码中的错误，检查无误后连接手机设备(或是运行一个 AVD 实例)，点击“运行”按钮，作为 Android 程序运行，如图 23. 5 所示。

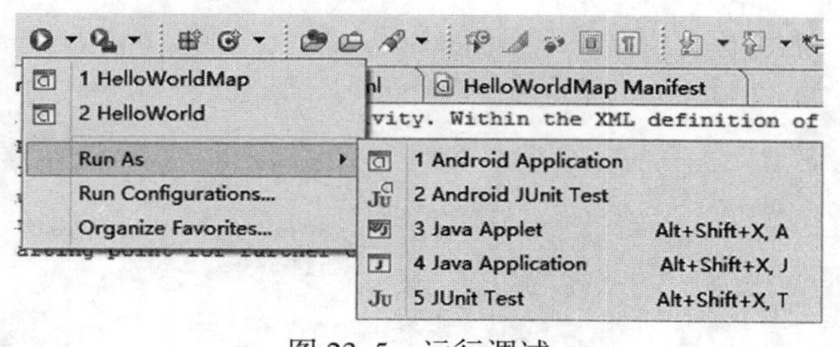

图 23. 5　运行调试

④在弹出的可运行设备列表下选择目标，在 Internet 接入的条件下，地图可以在手机上正常显示，如图 23.6 所示。

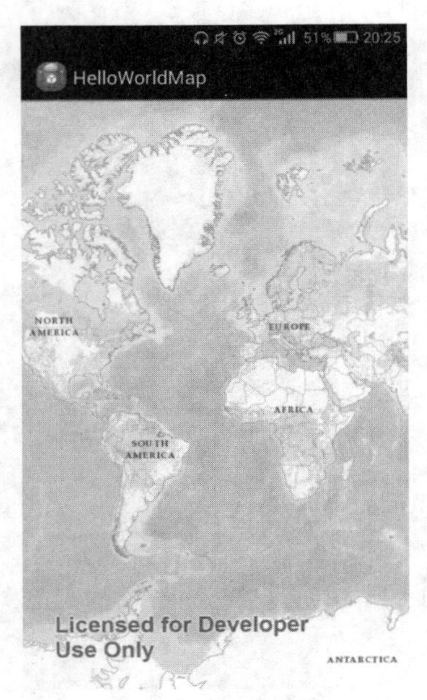

图 23.6　运行截图

第 24 章　地图浏览功能

本章主要介绍基本地图浏览功能的开发，Android 的开发也将在本章的示例中进行说明。在 24.1 节中以 MapView 作为例子，将详细讲解控件的开发使用，为提高效率，之后的功能开发就只进行重要内容的讲解，不再详细叙述基础性的内容。

24.1　MapView 控件

MapView 是 ArcGIS Runtime SDK for Android 的核心组件，通过 MapView 可以呈现地图服务的数据，并且在 MapView 中定义丰富的属性、方法和事件。MapView 是 Android 中 ViewGroup 的子类，也是 ArcGIS Runtime SDK for Android 中的地图容器，与 ArcGIS API 中的 Map、MapControl 类的作用是一样的。由于 MapView 直接继承自 Android 的 ViewGroup，因此 MapView 类继承了 ViewGroup 的所有方法和属性，其操作方式同 ViewGroup 极其相似。为使 MapView 的功能正确实现，需要至少加载一个 layer 图层。

MapView 扩展自 Android 中的 ViewGroup，继承了 ViewGroup. OnHierarchy ChangeListener 接口。继承结构如图 24.1 所示。

```
java.lang.Object
  ↳android.view.View
      ↳android.view.ViewGroup
          ↳com.esri.android.map.MapView
```

图 24.1　MapView 继承结构

MapView 的强大不仅因为它是呈现地图数据的容器，而且它能提供丰富的功能：

①MapView 具有呈现地图的能力，MapView 可以添加一个或多个图层。图层分很多种，如切片图层、动态图层、本地图层等，图层只有添加到 MapView 容器中才能进行显示。

②通过 MapView 可以设置地图的显示范围、是否允许被旋转、地图背景、地图的最大/最小分辨率及指定当前显示的分辨率/比例尺。

③MapView 提供了丰富的手势监听接口，通过这些监听器，可以监听各种手势动作，如点击、双击、移动或长按等操作。

用户通过 MapView 可以操作设备的触摸屏，默认 MapView 可以响应用户的各类手势的操作。MapView 默认支持以下的触摸手势：

①单指双击或者双指屏幕打开，会在 MapView 中实现地图的放大；

②双指单击或者双指屏幕捏合，会在 MapView 中实现地图的缩小；

③单指拖动，会在 MapView 中实现地图的视图拖拽。

24.1.1　MapView 控件提供的方法

MapView 提供的方法主要包括：地图缩放、旋转和坐标转换。

①MapView 提供了多种地图缩放的方式，示例代码如下：

```
mMapView.zoomin();
mMapView.zoomout();
mMapView.zoomToResolution(centerPt, res);
mMapView.zoomToScale(centerPt, scale);
```

提示：前两种主要功能是逐级缩放，调用一次 zoomin()方法地图将放大一级，调用一次 zoomout()方法地图将缩小一级。后两种缩放是按照不同的分辨率或比例尺进行的，调用 zoomToResolution(centerPt, res)方法进行缩放时需要传入两个参数，第一个参数 centerPt 是要按照哪个中心点进行缩放，因此需要传入一个 Point 对象，第二个参数是要缩放到的分辨率；zoomToScale(centerPt, scale)和 zoomToResolution- -(centerPt, res)很类似，第一个参数是相同的，第二个参数不再是分辨率，而是传入地图的比例尺。

分辨率或比例尺值与发布的服务密切相关，发布的地图服务如果生成了缓存切片，则在服务目录中可以看到不同级别对应的比例尺和分辨率，并通过服务提供的参数进行参数传递。例如，一个公共 World_Street_Map 地图服务的参数格式如图 24.2 所示。

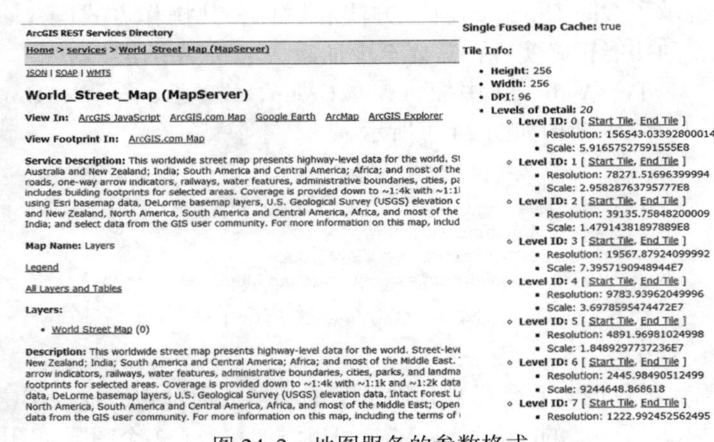

图 24.2　地图服务的参数格式

②MapView 还可以设置地图的显示范围、比例尺、分辨率、旋转角度和地图背景色，示例代码如下：

```
mMapView = (MapView)findViewById(R.id.map);
Envelope env = new Envelope(12957628.58241, 4864247.2803126,
12958114.4225065, 4864490.20036087); //范围
mMapView. setExtent(env) //设置地图显示范围
mMapView. setScale(295828763); //当前显示的比例尺
mMapView. setResolution(9783.93962049996); //设置当前显示的分辨率
//上面三种方法都可以改变地图的显示范围,在代码中是不会同时使用的
```

mMapView.setMapBackground (0xffffffff, Color.TRANSPARENT, 0, 0);//设置地图背景

mMapView.setAllowRotationByPinch(true); //是否允许使用 Pinch 方式旋转地图

mMapView.setRotationAngle(15.0);//初始化时将地图旋转 15 度,参数为正时按逆时针方向旋转

24.1.2　MapView 控件的添加、绑定

在工程中加入 MapView 控件有两种方式，一种如前面章节所述通过布局文件(main.xml)添加，此方法可以方便布局调整，并可以直接利用 mapoptions 进行初始属性的设置，例如以下代码：

```
<com.esri.android.map.MapView
android:id="@+id/map"
android:layout_width="fill_parent"
android:layout_height="fill_parent"
mapoptions.MapType="Topo"
mapoptions.center="36.010821,120.161335"
mapoptions.ZoomLevel="16"/>
```

以上代码中，设置了 MapView 的 id 是 map，在手机屏幕中的布局为 fill_parent，地图类型为 Topo，显示中心进行了修改(此经纬度是青岛市所在地区)，如图 24.3 所示，则程序将自动定位到青岛市所在地区，以展示利用布局文件添加控件的优势。因此，通常采用这种方式添加控件。

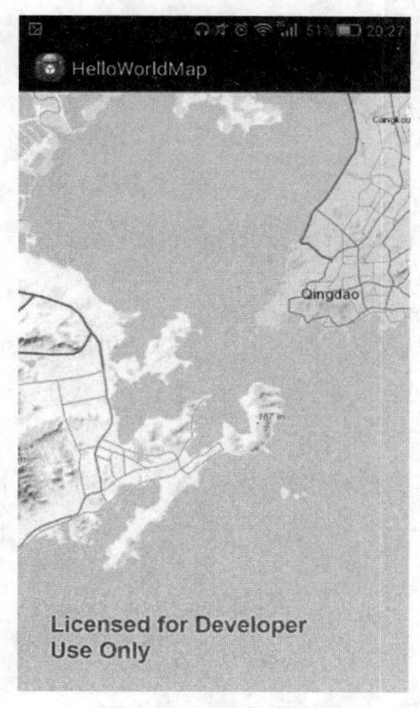

图 24.3　运行截图

另外一种方法是采用硬编码的方式，直接在 Java 文件中加载 MapView，载入地图服务，示例代码如下：

```
MapViewmap = new MapView(this);
map.setLayoutParams(new LayoutParams(LayoutParams.FILL_PARENT,
LayoutParams.FILL_PARENT));
tileLayer =new
ArcGISTiledMapServiceLayer( " http://services.arcgisonline.com/Arc-
GIS/rest/services/World_Street_Map/MapServer");
map.addLayer(tileLayer);
setContentView(map);
```

与桌面开发相同，MapView 控件添加入工程后，必须经过 Java 文件声明，才能绑定使用，代码如下：

```
MapView mMapView;
mMapView = (MapView) findViewById(R.id.map);
```

map 对应了布局文件 xml 中的 id 标签，变量 mMapView 就获取到了 MapView 控件，然后利用 ArcGIS SDK 封装好的各种方法(function)，以及绑定各种监听器进行需要的操作(.setOnClickListener)。

24.1.3　MapView 监听

移动端手势的监听是一个重要的环节，地图的手势操作由 MapView 来管理，主要有以下几种监听：

①地图单击监听：OnSingleTapListener；

②平移监听：OnPanListener；

③长按监听：OnLongPressListener；

④缩放监听：OnZoomListener；

⑤状态监听：OnStatusChangedListener；

⑥pinch 监听：OnPinchListener。

利用这些监听用户就可以与地图进行多种交互。例如，通过屏幕单击的监听，实现该点空间坐标和屏幕坐标的转换，示例代码如下：

```
mMapView.setOnSingleTapListener(new OnSingleTapListener() {
publicvoid onSingleTap(float x, float y) {
//TODO Auto-generated method stub
   Point pt =mMapView.toMapPoint(x,y);//屏幕坐标转换成空间坐标
Point screenPoint =mMapView. toScreenPoint(pt);//转换成屏幕坐标对
象
        }
});
```

24.2　Layer 地图图层

Layer 图层是空间数据的载体，通过它可将各种类型的地图数据进行加载显示。在 Android 移动 GIS 的开发中，图层需要被加到 MapView 对象中进行显示使用。ArcGIS for Android SDK 提供了多种图层，不同种类的 Layer 具有不同的作用和使用方式。Layer 类及继承类的关系如图 24.4 所示，以 Layer 为基类的继承类都可以被添加到 MapView 中使用。

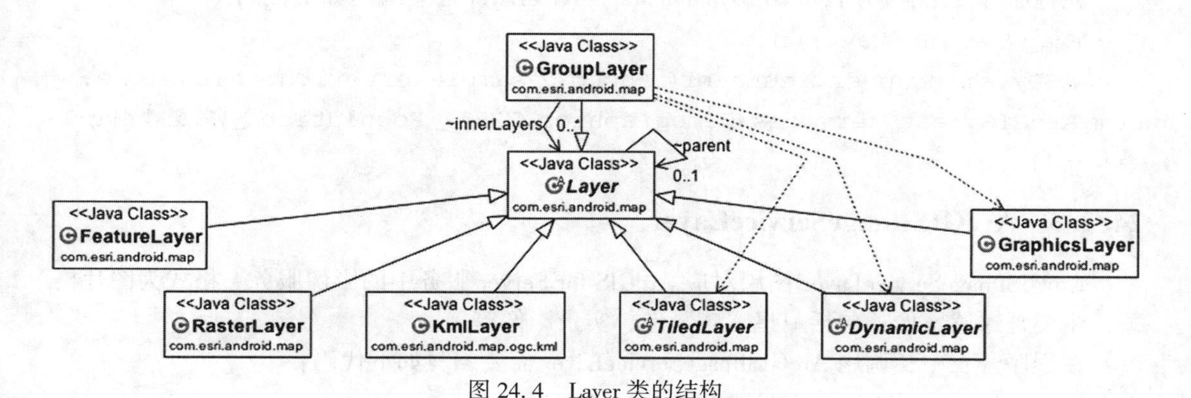

图 24.4　Layer 类的结构

24.2.1　ArcGISTiledMapServiceLayer

在 ArcGIS for Server 中可以发布多种地图服务，移动端需要有不同的图层来对应这些服务。ArcGISTiledMapServiceLayer 图层对应 ArcGIS for Server 服务中的切片服务，由于切片提前生成，所以 ArcGISTiledMapServiceLayer 不能对图层中的数据进行更改，除非更新服务缓存。在 ArcGIS for Android 中，也不允许对此类型的数据查询，通常做底图使用。同时，由于加载的是缓存切片，这个类型的图层是 ArcGIS for Android 中相应请求最快的图层之一，它采用多个线程，通常是每个图片使用一个线程来处理请求和绘制图片，并且异步处理。

在程序开发中，调用 ArcGISTiledMapServiceLayer 需要编写如下代码：

```
MapView mMapView =new MapView(this);
mMapView.addLayer(new
ArcGISTiledMapServiceLayer( "http://services.arcgisonline.com/ArcGIS/rest/services/World_Topo_Map/MapServer"));
```

24.2.2　ArcGISDynamicMapServiceLayer

ArcGISDynamicMapServiceLayer 图层对应 ArcGIS for Server 服务中的动态服务，动态地图服务的数据是按照移动设备范围读取的，用法与 ArcGISTiledMapServiceLayer 图层相同。

ArcGISDynamicMapServiceLayer 图层通常包含多个图层，这些图层由 Server 根据请求来渲染并以 raster image 的形式返回，且返回的 raster 并不包含 feature 的额外信息。

ArcGISDynamicMapServiceLayer 的功能特点：更新方便，并能及时呈现给用户；图层的可视性和空间参考可以改变，可以给图层添加单独的 layer definition，每个 image 可获得 ArcGIS Server 动态投影，地图的空间参考由第一个加载的图层决定。其性能特点：渲染时间取决于请求数据的数量和复杂度，相比 tiled map server 较慢。因此，它适合数据经常发生改变或者需要针对不同的用户呈现不同的数据，且不需要要素信息(如 attributes, geometry, symbol 等)的情况。

在程序开发中，调用 ArcGISDynamicMapServiceLayer 需要编写如下代码：

```
mMapView.addLayer(new
ArcGISDynamicMapServiceLayer( "http://sampleserver1.arcgisonline.
com/ArcGIS/rest/services/Demographics/ESRI_Population_World/MapS-
erver"));
```

24.2.3　ArcGISImageServiceLayer

ArcGISImageServiceLayer 图层对应 ArcGIS for Server 服务中的影像服务。格式为图片格式，主要是遥感影像，包括卫星遥感影像、实景影像等。

在程序开发中，调用 ArcGISImageServiceLayer 需要编写如下代码：

```
mMapView.addLayer(newArcGISImageServiceLayer(
"http://myserver/arcgis/rest/services/MyImage/ImageServer",null));
```

24.2.4　ArcGISFeatureLayer

ArcGISFeatureLayer 图层对应 ArcGIS for Server 服务中的 Feature Service，该图层与其他图层类型相比具有更为丰富的功能，它继承自 GraphicsLayer，因此也能够执行该图层的所有操作。它包含了丰富的要素信息，其中的每个要素都能被查询，可以通过空间查询或者 SQL 语句过滤；每个要素都单独渲染，从 ArcGIS Server Feature Service 或者 map service(此种 feature layer 不能编辑)中请求要素，并返回 JSON 格式的数据。

该图层可以是空间图层，也可以是非空间的表，虽然需要一定的响应时间，如 ArcGIS Server 处理请求的时间、请求返回的时间、渲染速度等，但能够实现各种丰富的功能。但只有 Feature Service 才具备在线数据编辑功能，因此，如果想要对某个数据进行在线编辑或同步，需要将其发布成 Feature Service，并在移动端新建一个 ArcGISFeatureLayer 图层以加载该服务。

该图层有一些设定，这些设定会影响图层的性能，包括请求时间、返回大小、处理和响应时间等(设置需要在图层 initialized 完成之后进行)。Out fields：可以通过 outFields() 方法来限制返回的属性字段的数量，可以减少网络传输和响应处理的时间。Max Allowable offset：使用 Max Allowable offset()方法可以限制要素返回的顶点数量，以地图单位设置的数值将用来在 server 返回要素之前简化要素的顶点。正在编辑要素时不能使用此方法，否则可能会导致顶点丢失和 topo 关系的破坏。

ArcGISFeatureLayer 在调用时可以设置三种模式(mode)，不同模式返回数据的方式和执行的效率不同：

①Snapshot mode：快照模式能从 server 上快速取回要素，减少响应时间，但如果图层包含大量、复杂的要素时，可能会导致设备停止响应，因为有大量的要素需要绘制，也可能会导致 ArcGIS server 达到它返回要素数量的上限。

②On demand mode：On demand mode 需要在客户端和服务器之间传递更多的请求，适合展示变化的数据，该模式适合包含复杂要素的大数据量的 dataset，因为它只请求和返回需要的要素，但是为图层设置比例尺限制仍然是有必要的。

③Selection mode：该模式下，开始时并不向服务器请求任何要素，只有当选择集出现时，要素才被加载，该模式适合于不需要将所有要素展示在客户端，而只是强调一些特殊要素的情况。例如，一个 road 图层，使用了 feature layer 不支持的自定义的符号，但又需要对该图层进行 Web edit，此时可以使用 dynamic map service layer 和 feature layer in selection mode 将需要编辑的 road 选出来，以简单的符号绘制并存储以实现编辑，当编辑完成后，清除选择并刷新 dynamic map service 来查看更新。

调用快照模式下 ArcGISFeatureLayer 图层的示例代码如下：

```
String url
="https://servicesbeta.esri.com/ArcGIS/rest/services/SanJuan/Trail
Conditions/FeatureS/erver/0";
mMapView.addLayer(new ArcGISFeatureLayer(url,MODE.SNAPSHOT));//
按照快照方式
```

24.3　导航与触屏操作

在介绍了地图操作最重要的 MapView 与图层相关知识后，本节将介绍 Android 程序下地图的导航与触屏操作，将通过几个示例程序帮助读者进行手机地图操作的开发。

如果要实现触屏操作后地图响应，需要绑定触屏事件监听器。前文中 MapView 已经封装了几个简单的事件监听，但对于移动 GIS 开发中复杂的地图操作来说是不够的。ArcGIS API for Android 提供了如下几个常用的事件监听器：MapOnTouchListener、OnLongPressListener、OnPanListener、OnPinchListener、OnzoomListener。

24.3.1　监听器的使用

1. MapOnTouchListener

MapOnTouchListener 是 MapView 最为重要的监听器之一，它实现了 OnTouchListener 和 MapGestureDetector. OnGestureListener 接口，对于地图的所有操作 MapOnTouchListener 都可以进行响应，使用方便。在使用前只需扩展这个类并重写该类中的方法即可，用法如下：

```
class MyTouchListener extendsMapOnTouchListener {
public MyTouchListener(Context context, MapView view) {
super(context, view);
        }
```

```
publicvoid setType(String geometryType) {
this.type = geometryType;
    }
public String getType() {
returnthis.type;
    }
publicboolean onSingleTap(MotionEvent e) {
returntrue;
    }
publicboolean onDragPointerMove(MotionEvent from, MotionEvent to){
returnsuper.onDragPointerMove(from,to);
    }
@ Override
publicboolean onDragPointerUp(MotionEvent from, MotionEvent to){
returnsuper.onDragPointerUp(from, to);
    }
  }
```

通过完成代码，即可实现监听不同的手势操作，并实现对于不同的手势操作执行不同的方法，通过这些方法可以添加所需的操作功能。例如，可以在 onSingleTap( )方法中完成点的获取、窗体的创建及弹出操作相关代码，即可实现地图上点击时弹出一个窗体的功能定制。

2. OnLongPressListener

OnLongPressListener 接口可以实现监听 MapView 中的地图长按事件。用法如下：

```
//为地图添加一个长按监听器
    mMapView.setOnLongClickListener(new View.OnLongClickListener(){
    //长按后自动执行的方法
    publicboolean onLongClick(View v) {
    //TODO Auto-generated method stub
    returnfalse;
    }
});
```

3. OnPanListener

OnPanListener 接口可以实现监听 MapView 中的地图平移事件。用法如下：

```
//为地图添加一个平移监听器
mMapView.setOnPanListener(new OnPanListener() {
publicvoid prePointerUp(float fromx, float fromy, float tox, float
toy) { }
publicvoid prePointerMove ( float fromx, float fromy, float tox,
```

```
float toy) {      }
    publicvoid postPointerUp ( float  fromx,  float  fromy,  float  tox,
float toy) { }
    publicvoid postPointerMove ( float  fromx,  float  fromy,  float  tox,
float toy) {     }
    });
```

4. OnPinchListener

OnPinchListener 接口可以实现监听 MapView 中的两指捏合拉伸事件。用法如下：

```
//为地图添加夹/捏监听器
mMapView.setOnPinchListener(new OnPinchListener() {
publicvoid prePointersUp(float x1, float y1, float x2, float y2,
double factor) {
}
publicvoid prePointersMove(float x1, float y1, float x2, float y2,
double factor) {
}
publicvoid prePointersDown(float x1, float y1, float x2, float y2,
double factor) {
}
publicvoid postPointersUp(float x1, float y1, float x2, float y2,
double factor) {
}
publicvoid postPointersMove(float x1, float y1, float x2, float y2,
double factor) {
}
publicvoid postPointersDown(float x1, float y1, float x2, float y2,
double factor) {
}
});
```

5. OnZoomListener

OnZoomListener 接口可以实现监听 MapView 中的两指捏合拉伸事件。用法如下：

```
mMapView.setOnZoomListener(new OnZoomListener() {
//缩放之前自动调用的方法
publicvoid preAction(float pivotX, float pivotY, double factor) {
}
//缩放之后自动调用的方法
publicvoid postAction(float pivotX, float pivotY, double factor) {
}
});
```

24.3.2 AddaLayer 示例程序

新建一个工程，命名为 AddaLayer，准备工作参考 24.2 节中的内容。AddaLayerActivity.java 文件中，在系统自动创建的 AddaLayerActivity 下完成如下代码：

```
publicclass AddaLayerActivity extends Activity {
    privatestaticfinal String TILED_WORLD_STREETS_URL = "http://services.arcgisonline.com/ArcGIS/rest/services/World_Street_Map/MapServer";
    privatestaticfinal String DYNAMIC_USA_HIGHWAY_URL = "http://sampleserver1.arcgisonline.com/ArcGIS/rest/services/Specialty/ESRI_StateCityHighway_USA/MapServer";
     MapView mMapView ;
     ArcGISDynamicMapServiceLayermStreetsLayer = null;
    @ Override
    publicvoid onCreate(Bundle savedInstanceState) {
    super.onCreate(savedInstanceState);
       setContentView(R.layout.main);
    mMapView = (MapView) findViewById(R.id.map);
    //向 MapView 容器中添加基础 map
     mMapView.addLayer ( new ArcGISTiledMapServiceLayer ( TILED_WORLD_STREETS_URL));
    //创建第二个 map,在触屏操作触发时加入 MapView
    mStreetsLayer = new ArcGISDynamicMapServiceLayer(DYNAMIC_USA_HIGHWAY_URL);
    //向 MapView 添加动态图层
    mMapView.addLayer(mStreetsLayer);
    //向 MapView 添加监听器
    mMapView.setOnLongPressListener(new OnLongPressListener() {
    privatestaticfinallongserialVersionUID = 1L;
    publicboolean onLongPress(float x, float y) {
    //检查 map 是否已经加载
    if (mMapView.isLoaded()) {
    if (mStreetsLayer ! = null) {
       mStreetsLayer.setVisible(! mStreetsLayer.isVisible());
               }
           }
    returntrue;
         }
```

```
        });
      }
    @ Override
    protectedvoid onPause() {
    super.onPause();
    mMapView.pause();
      }
    @ Override
    protectedvoid onResume() {
    super.onResume();
    mMapView.unpause();
      }
}
```

代码各部分的意义和功能见代码注释，这里就不再详细讲解。以上示例完成了程序主要功能的编写和触屏监听事件的添加。然后需要进行布局文件 xml 的代码编写。打开 main. xml，完成如下的代码：

```
<? xmlversion="1.0"encoding="utf-8"? >
< LinearLayoutxmlns: android = " http://schemas.android.com/apk/
res/android"
android:layout_width="fill_parent"
android:layout_height="fill_parent"
android:orientation="vertical">

<TextView
android:id="@ +id/textView1"
android:layout_width="wrap_content"
android:layout_height="wrap_content"
android:text="长按添加或移除图层"
android:textAppearance="? android:attr/textAppearanceMedium"/>
<! -- MapView layout and initial extent -->
<com.esri.android.map.MapView
android:id="@ +id/map"
android:layout_width="fill_parent"
android:layout_height="fill_parent"
initExtent="-19332033.11, -3516.27, -1720941.80, 11737211.28">
</com.esri.android.map.MapView>

</LinearLayout>
```

如果代码编写正确，在 Graphical Layout 视图下就会出现手机布局的预读，用户可以以可视化的形式修改和添加屏幕控件。在 MapView 之外添加一个 TextView，以便显示一些辅助性的说明文字。作为底图的 MapView 中，依然在 xml 属性中对 MapView 进行必要的设置。整体布局放置在一个 LinearLayout(线性布局)中，如图 24.5 所示。

图 24.5　Graphical Layout 视图

运行程序，通过真机运行来对程序进行必要的修改，程序运行结果如图 24.6 所示。

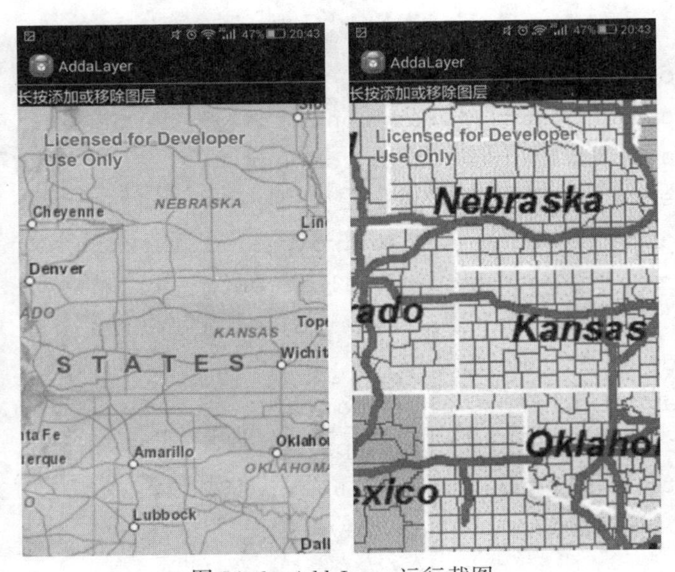

图 24.6　AddaLayer 运行截图

24.4　空间要素绘制

通过 MapView，用户可以接受地图数据并进行显示，与此同时，获取的各种地理数据也可以通过要素图层绘制的方式加载到地图中，更新绘制的要素图层就可以被赋予有关的地理意义。ArcGIS API 提供的要素图层是 GraphicsLayer，本节将详细介绍通过 GraphicsLayer 实现空间要素可视化的方法。加载到图层中的空间要素主要是：点、线、面和文字类型等，下面主要介绍 API 中的常用要素。

24.4.1　Graphic

Graphic 是承载空间几何要素的载体，Graphic 对象可以添加到 GraphicsLayer 图层中进行展示。Graphic 主要由 4 部分组成：Geomtry、Symbol、Attributes 和 InfoTemplate，通过这 4 个部分可以获取 Graphic 对象的 Geometry、符号、属性信息和模板信息，但查看 Graphic 的 API 文档时会发现，通过这个对象可以获取相应的属性信息，却没有相应的方法来修改这些属性。这是因为 Graphic 对象一经构造就不可修改，构造成功后会为每个 Graphic 生成唯一的 UID 以便调用。修改 Graphic 的属性，通常使用 GraphicsLayer 的 updateGraphic ()方法间接地修改 Graphic 对象的属性，用法如下：

```
publicbooleanonSingleTap(MotionEvent e) {
if (type.length() > 1 &&type.equalsIgnoreCase("POINT")) {
//创建 Graphic 对象,添加几何结构,样式
Graphic graphic =new Graphic(mapView.toMapPoint(new Point(e.
getX(), e.getY())),
new SimpleMarkerSymbol(Color.RED,25,STYLE.CIRCLE));
graphicsLayer.addGraphic(graphic);//添加到图层中
returntrue;
}
returnfalse;
}
```

1. Point

Point 是针对空间要素的点对象，Point 既可代表二维的点也可以是三维点对象，可以通过它自身的方法获取 x 或 y 坐标。用法如下：

```
Point point =new Point();//创建点对象
Point.setX(114);//设置 x 坐标
Point.setY(32);//设置 y 坐标
Graphic gp =new Graphic(point, new SimpleMarkerSymbol(Color.RED,
25,STYLE.CIRCLE));
graphicsLayer.addGraphic(gp);//添加到图层中显示
```

2. MultiPoint

MultiPoint 表示多点对象，MultiPoint 通常存储一系列的基础点，这些点按照一定的顺序存储，并且每个点都可以获取它的索引位置。可以通过每个点的索引位置对 MultiPoint 对象进行增、删或改操作，用法如下：

```
Point point1 =new Point(114,32);//创建点对象
Point point2 =new Point(112,28);//创建点对象
MultiPoint multipoint =new MultiPoint();
multipoint.add(point1);//添加点
multipoint.add(point2);//添加点
multipoint.removePoint(1);//移除第二点
```

3. MultiPath

MultiPath 是 Polygons 和 Polylines 的基类，MultiPath 与 MultiPoint 很类似，只是 MultiPoint 存储点数据集，而 MultiPath 存储一条或多条轨迹线。MultiPath 提供了丰富的操作接口，可以通过这些接口来操作 MultiPath 对象里的任何轨迹上的点。用法如下：

```
Point startPoint =new Point(114,28);
MultiPath path =newMultiPath();
path.startPath(startPoint);//设置路径的初始位置
path.lineto(new Point(113,32));//给路径添加点
```

4. Envelope

Envelope 代表一个矩形要素，可以通过 Envelope 对象获取矩形窗口的中心点、矩形的上下 4 个点、宽和高等。用法如下：

```
Envelope env =new Envelope(112,28,113,32);//创建矩形对象
map.setExtent(env);//设置地图显示范围
Point point= env.getCenter();//获取矩形框的中心点
```

5. Polygon

Polygon 是 MultiPath 子类，表示多边形或多多边形，Polygon 里的所有 path 都是闭合的环。Polygon 对象也是 ArcGIS for Android 开发中常用的对象，如标绘多边形或做空间查询时，Polygon 对象中至少存在 3 个点并且不能同时在一条直线上。用法如下：

```
Polygon poly =new Polygon();//创建多边形对象
poly.startPath(new Point(0,0));//添加初始点
poly.lineto(new Point(10,0));
poly.lineto(new Point(10,10));
poly.lineto(new Point(0,0));//多边形是闭合的,因此最后还要添加初始点的
位置
```

24.4.2　Symbol

1. PictureMarkerSymbol

PictureMarkerSymbol 是对于点或多点要素的 Graphic 对象进行样式设置的类，

PictureMarkerSymbol 主要通过图片的 URL 或 Drawable 等方式来设置图片符号，对于 PictureMarkerSymbol 还可以设置图片符号的旋转角度和位置偏移，用法如下：

```
//创建图片样式符合
PictureMarkerSymbol pic =new
PictureMarkerSymbol(getResources().getDrawable(R.drawable.icon));
Point pt =new Point(113,32);//创建一个点对象
Graphic gp =new Graphic(pt,pic); //设置样式
graphicsLayer.addGraphic(gp); //添加到图层中
```

2. SimpleMarkerSymbol

SimpleMarkerSymbol 也是针对点状要素的 Graphic 对象进行样式设置的类，SimpleMarkerSymbol 与 PictureMarkerSymbol 类很相似，但一个渲染成矢量点，另一个通过图片来替换该点。使用 SimpleMarkerSymbol 可以设置点的样式，如点的大小、颜色和类型等，用法如下：

```
Point point =new Point();//创建点对象
Point.setX(114);//设置 x 坐标
Point.setY(32);//设置 y 坐标
//设置点样式的颜色、大小和点类型
SimpleMarkerSymbol sms = new SimpleMarkerSymbol ( Color.RED, 25,
STYLE.CIRCLE)
Graphic gp =new Graphic(point,sms);
graphicsLayer.addGraphic(gp);//添加到图层中显示
```

3. SimpleLineSymbol

SimpleLineSymbol 是针对线状要素的 Graphic 对象进行样式设置的类，通过它可以设置线状要素的样式，包括线型、线颜色、线宽和线的透明度等，用法如下：

```
Polyline poly =new Polyline ();//创建多边形对象
poly.startPath(new Point(0,0));//添加初始点
poly.lineto(new Point(10,0));
poly.lineto(new Point(10,10));
SimpleLineSymbol sls =new SimpleLineSymbol(Color.RED,25, Simple-
LineSymbol.SOLID);//线样式对象,包括颜色、线宽和线型等参数
sfs.setAlpha(50);//设置透明度
Graphic gp =new Graphic(poly, sls);
graphicsLayer.addGraphic(gp);//添加到图层中
```

4. SimpleFillSymbol

SimpleFillSymbol 是对于面状要素的 Graphic 对象进行样式设置的类，通过它设置面状要素的填充颜色和透明度，也可以为面状要素添加边界的样式设置。用法如下：

```
Polygon poly =new Polygon();//创建多边形对象
poly.startPath(new Point(0,0));//添加初始点
```

```
poly.lineto(new Point(10,0));
poly.lineto(new Point(10,10));
poly.lineto(new Point(0,0));//多边形是闭合的,因此最后还要添加初始点的位置
SimpleFillSymbol sfs =new SimpleFillSymbol(Color.RED);//面样式对象
sfs.setAlpha(50);//设置透明度
Graphic gp =new Graphic(poly,sfs);
graphicsLayer.addGraphic(gp);//添加到图层中
```

5. TextSymbol

TextSymbol 也是针对点状要素的 Graphic 对象进行样式设置的类，它将点的位置替换成文字进行标绘显示，通过 TextSymbol 可以设置文字的大小、颜色、内容和排列方式。标绘的文字排列方式有两种：横向排列和纵向排列，默认横向居中显示。用法如下：

```
Point point =new Point();//创建点对象
Point.setX(114);//设置 x 坐标
Point.setY(32);//设置 y 坐标
//设置点样式的颜色、大小和文本内容
TextSymbol ts =new TextSymbol (12,"点样式",Color.RED);
Graphic gp =new Graphic(point,ts);
graphicsLayer.addGraphic(gp);//添加到图层中显示
```

24.4.3 GraphicElements 示例程序

新建一个工程，并命名为 GraphicElements，这个示例工程的目的是实现在一个要素图层上进行要素绘制的功能。新加入两个 Button 控件，一个是“选择”，选择绘制要素的种类，然后在地图上进行要素绘制工作；另一个是“清空”，清空之前绘制的要素，以便进行新的绘制。示例只给出关键代码，其余相关代码请开发者自行补全。

新建好工程之后，打开 GraphicElementsActivity.java，首先定义需要用到的控件和变量，以及后面用到的地图链接，示例代码如下：

```
MapView mapView = null;
ArcGISTiledMapServiceLayer tiledMapServiceLayer = null;
GraphicsLayer graphicsLayer = null;
MyTouchListener myListener = null;

/*
 * 接下来是 Android UI 元素
 */
Button geometryButton = null;
Button clearButton = null;
```

```
TextView label = null;
//分别通过控件 ID 获取到布局中的控件,相关代码略去。为两个 Button 设置监听:
clearButton.setOnClickListener(new View.OnClickListener() {
            publicvoid onClick(View v) {
            graphicsLayer.removeAll();
            clearButton.setEnabled(false);
            }
        });
```

为 MapView 控件添加图层:

```
mapView.addLayer(tiledMapServiceLayer);
mapView.addLayer(graphicsLayer);
```

分别设置触屏操作事件监听:

在 class MyTouchListener extends MapOnTouchListener;下设置监听:

```
publicboolean onSingleTap(MotionEvent e)
publicboolean onDragPointerMove(MotionEvent from, MotionEvent to)
publicboolean onDragPointerUp(MotionEvent from, MotionEvent to)
```

补完相关代码，完成 Java 主代码文件的编写。打开 main. xml 进行布局文件的代码编写。添加一个 RelativeLayout 容器，在 RelativeLayout 容器内嵌套一个 LinearLayout，LinearLayout 内部放置两个 Button 控件和一个 TextView 控件。LinearLayout 内相关代码如下:

```
<LinearLayout
android:orientation="horizontal"
android:gravity="left"
android:layout_width="wrap_content"
android:layout_height="wrap_content"
>
<Buttonandroid:id="@ +id/geometrybutton"
       android:text="选择"
       android:layout_width="wrap_content"
       android:layout_height="wrap_content" />
   <TextViewandroid:id="@ +id/label"
       android:layout_width="wrap_content"
       android:layout_height="wrap_content"
       android:maxLength="25"
       android:singleLine="false"
       android:editable="false"
       android:gravity="center_horizontal" />
   <Buttonandroid:id="@ +id/clearbutton"
       android:text="清空"
```

```
        android:layout_width="wrap_content"
        android:layout_height="wrap_content" />
</LinearLayout>
```

进行真机运行调试，运行结果如图 24.7 和图 24.8 所示。

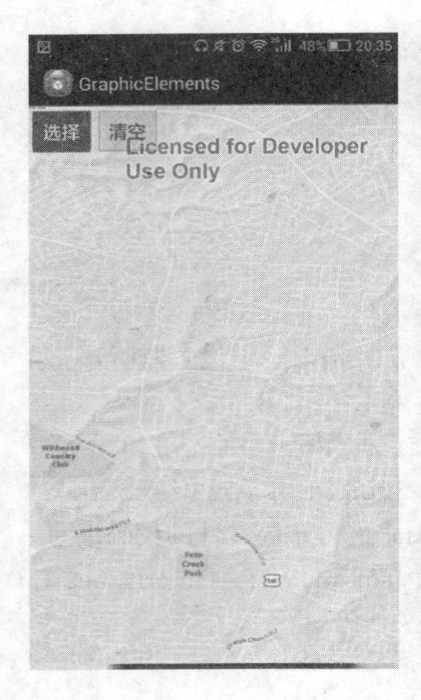

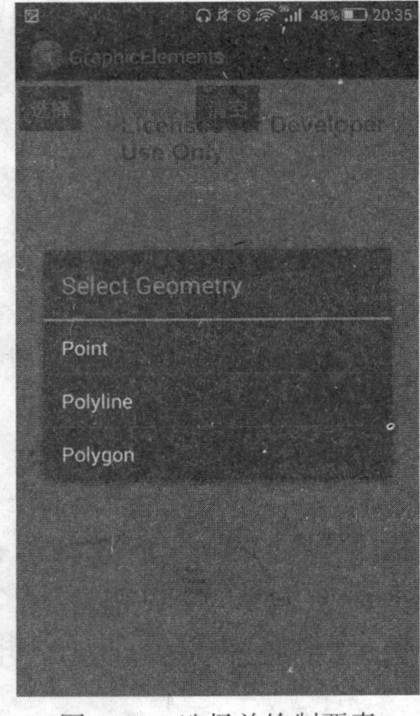

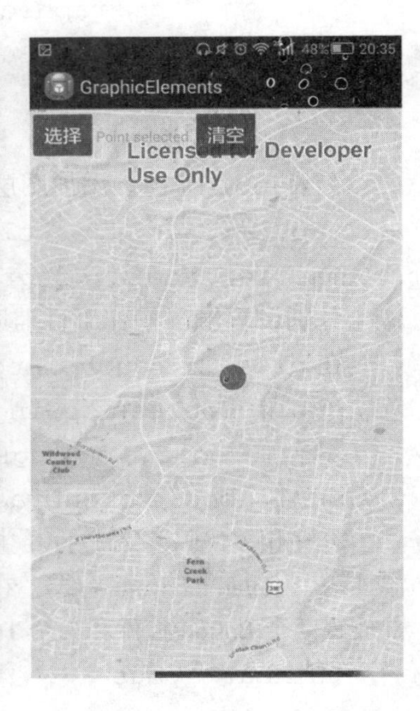

图 24.7　选择并绘制要素

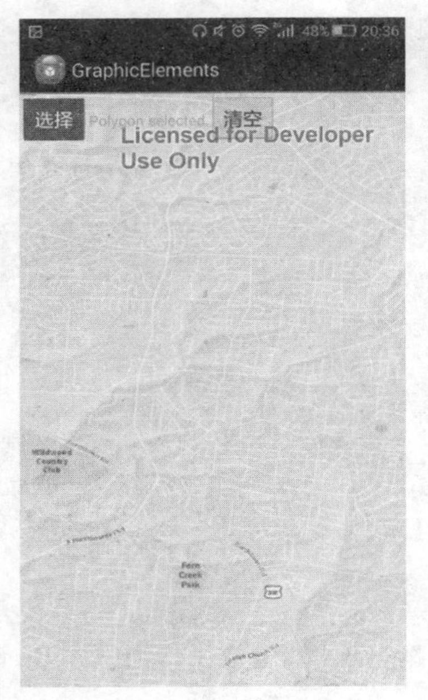

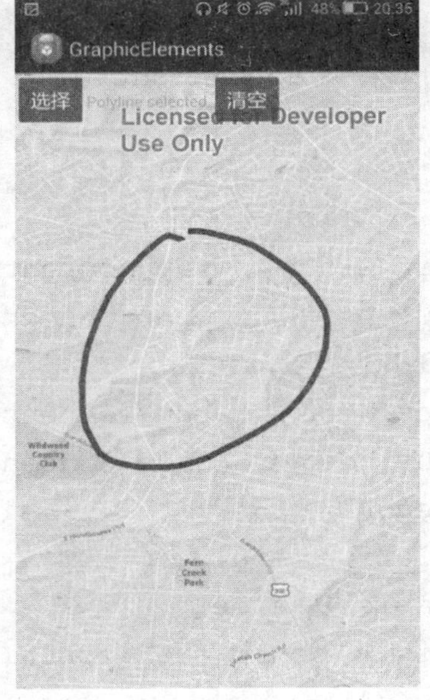

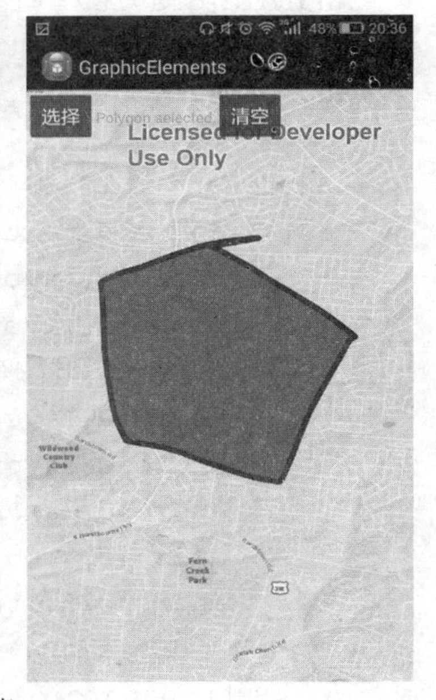

图 24.8　切换绘制要素最后清除要素

第 25 章　查询和检索功能

查询和检索是 GIS 开发中不可或缺的部分，本节内容将介绍 API Task 下的 Identify 和 Query 任务。IdentifyTask 用来识别图层中的要素，QueryTask 用来查询图层中的要素。

25.1　要素识别

25.1.1　IdentifyTask

IdentifyTask 在 ArcGIS API 中是一个识别任务类，它的作用是当通过手指点击地图时获取地图上的要素信息。但是在进行识别操作前必须首先为 IdentifyTask 任务设置一组参数信息，IdentifyTask 可以接受的参数必须是 IdentifyParameters 类型的对象，在参数 IdentifyParameters 对象中可以设置相应的识别条件。

IdentifyTask 作用于服务中的多个图层的识别，返回的结果是一个 IdentifyResult[ ]数组，存储所需要素信息。执行识别任务需要以下几个步骤：

①创建识别任务所需的参数对象 IdentifyParameters。

②为参数对象设置识别条件。

③定义 MyIdentifyTask 类并继承 AsyncTask。

④在 MyIdentifyTask 的 oInBackground( )方法中执 IdentifyTask 的 execute( )。

IdentifyTask 任务在要素识别参数设置中存在三种模式：

①ALL_LAYERS：该模式表示在识别时检索服务上的所有图层的要素。

②VISIBLE_LAYERS：该模式表示在识别时只检索服务上的可见图层的要素。

③TOP_MOST_LAYER：该模式表示在识别时只检索服务上最顶层的要素。

参数设置方法代码如下：

```
params=newIdentifyParameters();//识别任务所需参数对象
params.setTolerance(20);//设置容差
params.setDPI(98);//设置地图的 DPI
params.setLayers(newint[]{4});//设置要识别的图层数组
params.setLayerMode(IdentifyParameters.ALL_LAYERS);//设置识别模式
```

IdentifyResult[ ]数组示例代码如下：

```
protectedIdentifyResult[]doInBackground(IdentifyParameters...params){
```

```
IdentifyResult[]mResult=null;
if(params! =null&&params.length>0){
IdentifyParametersmParams=params[0];
try{
mResult=mIdentifyTask.execute(mParams);//执行识别任务
}catch(Exceptione){
//TODOAuto-generatedcatchblock
e.printStackTrace();
}
}
returnmResult;
}
```

25.1.2 Identify 示例程序

通过 Identify 示例程序介绍其用法。设计一个简单的工程 Identify，进入程序之后会有一个带有城市信息的 Map 图层，当点击地图上某城市的地理位置时，程序将对该点进行属性查询，弹窗显示城市信息。主程序代码参考前面 IdentifyTask 介绍，只需把参数设置部分按照示例程序需求进行修改即可。布局 xml 文件代码也不需赘述，程序运行成功后，结果如图 25.1 所示。

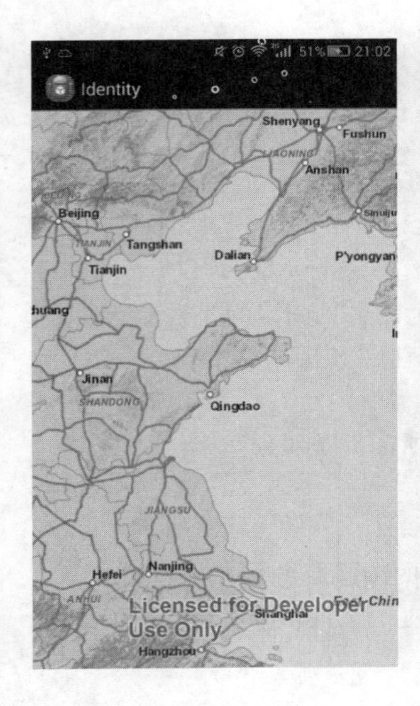

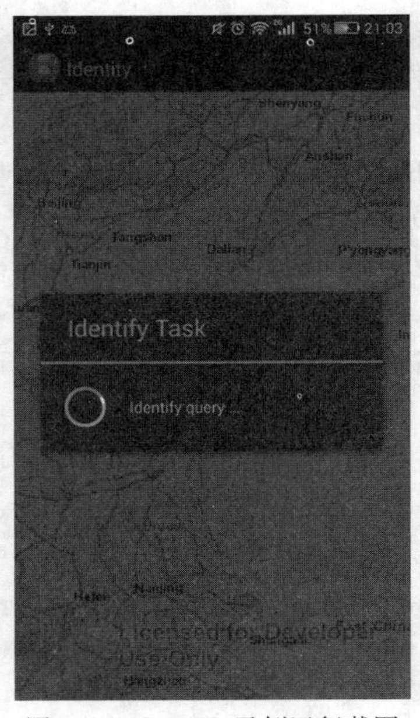

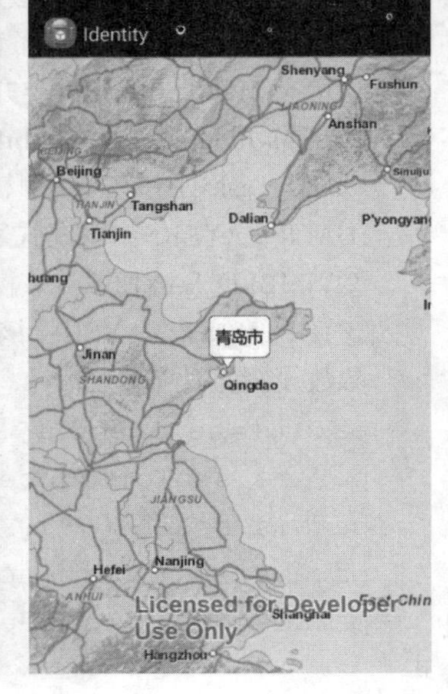

图 25.1 Identify 示例运行截图

25.2　要素查询

25.2.1　QueryTask

QueryTask 可以对图层进行属性查询、空间查询以及属性与空间联合查询。QueryTask 只针对服务中的一个图层进行查询。在执行查询前需要构建 Query 参数对象，该参数主要包含查询的设置条件。

QueryTask 的使用步骤如下：

①创建 Query 参数对象。

②为参数对象设定查询条件。

③通过 AsyncTask 的子类来执行查询任务。

示例代码如下：

```
Protected FeatureSet doInBackground(String...queryParams){
if(queryParams==null‖queryParams.length<=1)
returnnull;

Stringurl=queryParams[0];
Queryquery=newQuery();//创建查询参数对象
StringwhereClause=queryParams[1];
SpatialReferencesr=SpatialReference.create(102100);
query.setGeometry(newEnvelope(-20147112.9593773,557305.257274575,
-6569564.7196889,11753184.6153385));//设置空间查询条件
query.setOutSpatialReference(sr);//设置输出坐标系
query.setReturnGeometry(true);//指定是否返回几何对象
query.setWhere(whereClause);//设置属性查询条件
QueryTaskqTask=newQueryTask(url);
FeatureSetfs=null;
try{
fs=qTask.execute(query);//执行查询任务
}catch(Exceptione){
//TODOAuto-generatedcatchblock
e.printStackTrace();
return fs;
}
Return fs;
}
}
```

25.2.2　Query 示例程序

设计一个简单的工程 Query，进入程序之后会有一个 Button，点击 Button 时 MapView 中加载的地图会自动跳转到青岛市所在的位置，并弹窗显示。主程序代码参考前面 QueryTask 的介绍，只需把查询条件按照示例程序需求进行修改即可。布局 xml 文件代码也不需赘述，程序运行成功后，结果如图 25.2 所示。

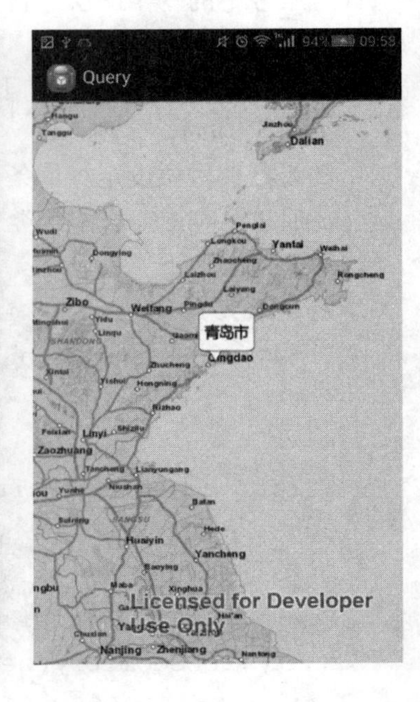

图 25.2　Query 示例运行截图

第 26 章　移动开发应用案例

案例名称：基于 ArcGIS for Android SDK 的大众性软件《神图》。该案例获得 2013 年度“ESRI 杯”中国大学生 GIS 软件开发竞赛奖项。《神图》软件的更多内容参见随书所赠光盘。

功能列表：《神图》软件功能包括以下几项：

①地图浏览。软件展示了青岛市区(包括黄岛区的部分区域)的地图，里面含有大量兴趣点和功能点，包含绿地、道路、餐饮、旅馆、旅游景点、车站等各类基本要素。

②地图操作功能。在图层上实现地图漫游、地图放大、地图缩小与地图刷新。

③HOT 分布。主要是为神瓶服务的，它在地图上显示大量的已经被放置过神瓶的地点，让用户在放置的时候适当地选择位置，使自己的瓶子更容易遇到他/她。

④搜索查询。主要是查询餐饮、购物、KTV 等功能性场所。

⑤最短路径查询：主要为现实版的淘宝、搜索查询等功能服务，实现最短路线查询。

⑥聊天功能。软件对于神瓶相遇的人可以进行聊天。

26.1　地图浏览与操作

《神图》软件展示了青岛市区(包括黄岛区的部分区域)的地图，包含大量兴趣点、功能点，如绿地、道路、餐饮、旅馆、旅游景点、车站等各类基本要素。地图浏览界面同时显示地图操作控件，包括地图漫游、地图放大、地图缩小与地图刷新。案例代码如下：

```
map.addLayer(new
ArcGISDynamicMapServiceLayer ( " http://192.168.69.56: 6080/arcgis/
rest/services/huangdao/MapServer"));
Envelope initextext = new Envelope(120.1200, 36.051, 120.147,
35.951);
map.setExtent(initextext);
mGraphicsLayer =new GraphicsLayer();
map.addLayer(mGraphicsLayer);
lGraphicsLayer=new GraphicsLayer();
map.addLayer(lGraphicsLayer);
```

运行结果如图 26.1 所示。

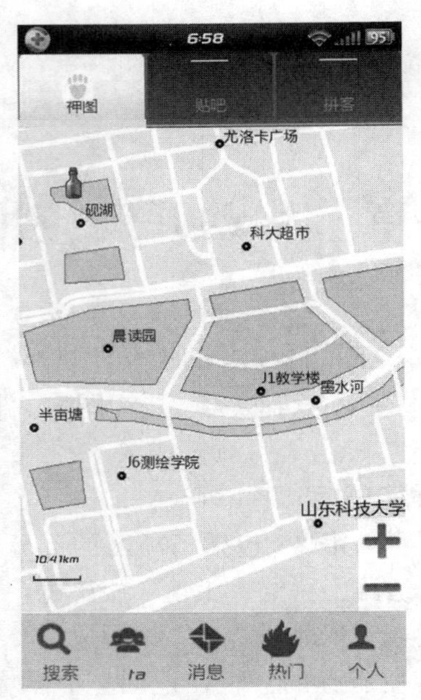

图 26.1 地图浏览与地图操作

26.2 HOT 分布

《神图》软件的 HOT 分布模块是为神瓶服务的，它在地图上显示大量的已经被放置过神瓶的地点，让用户在放置的时候适当地选择位置，使自己的瓶子更容易相遇到他或她。案例代码如下：

```
Int k=multipoint.size()/13;
if(k==0){
System.out.println("没有信息点!");

}
intm=0;
Doublex1=Double.parseDouble(multipoint.get(11));
Doubley1=Double.parseDouble(multipoint.get(12));
doublemin=Math.sqrt((choosex-x1)*(choosex-x1)+(choosey-y1)*
(choosey-y1));
doublethan;
System.out.println(k);
for(inti=0;i<k;i++){

Doublex=Double.parseDouble(multipoint.get(i*13+11));
```

```
Doubley=Double.parseDouble(multipoint.get(i*13+12));
System.out.println(x);
System.out.println(y);
than=Math.sqrt((choosex-x)*(choosex-x)+(choosey-y)*(choosey-y));
if(than<min){
min=than;
m=i;
}

}
```

运行结果如图 26.2 所示。

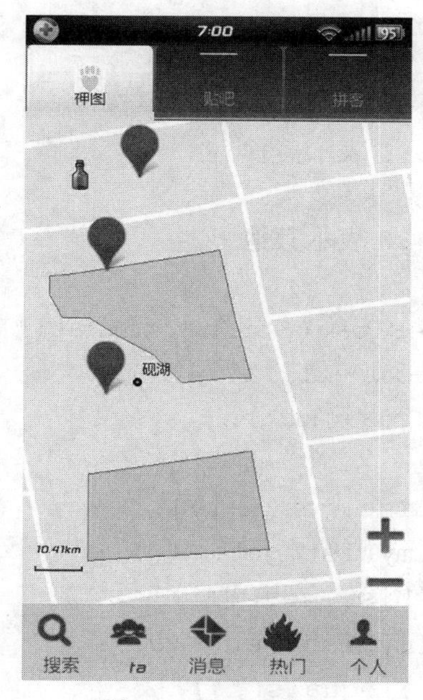

图 26.2　HOT 分布

26.3　搜索查询

《神图》软件的搜索查询模块主要是查询餐饮、购物、KTV 等功能性场所。以搜索餐馆为例，其实现代码如下：

```
Handler handler=newHandler(){
publicvoidhandleMessage(Messagem){
progressDialog.dismiss();
ArrayList<String>myList=(ArrayList<String>)m.getData().
```

```
getStringArrayList("pinlist");
    ArrayList<Map<String,Object>>data = newArrayList<Map<String,Ob-
ject>>();

    //第一个值为餐馆数目
    if(! myList.get(0).equals("")){
    for(inti = 1;i<myList.size();i++){
    System.out.println("myList.get(i).toString()"+i+myList.get(i).
toString()+""+myList.get(i+1).toString());
    Map<String,Object>item = newHashMap<String,Object>();
    doublet1 = Double.parseDouble(myList.get(i).toString());
    doublet2 = Double.parseDouble(myList.get(++i).toString());
    points.add(t1);
    points.add(t2);
    if(type_temp.equals("eatery")){
    points.add(3.0);//3.0 表示美食
    }else{
    points.add(4.0);//4.0 表示其他
    }
    floatt3 = Float.parseFloat(String.format("% .2f",Math.abs((t1-x)
*96.403)+Math.abs((t1-x)*111.17747)));
    item.put("distance",t3+"km");

    item.put("name",myList.get(++i));
    item.put("viewNum",myList.get(++i));
    item.put("price",myList.get(++i)+"元");
    item.put("special",myList.get(++i));
    if(! myList.get(++i).equals("")){

    byteb[] = android.util.Base64.decode(myList.get(i),Base64.DEFAULT);
    if(b.length! =0){
    Bitmapimg;
    img = BitmapFactory.decodeByteArray(b,0,b.length);
    item.put("iimg",img);
    }
    }else{

    Bitmapbmp = BitmapFactory.decodeResource(getResources(),R.drawable.
noimg);
```

```
item.put("iimg",bmp);
}

item.put("id",myList.get(++i));
data.add(item);
}

SimpleAdaptersimpleAdapter = newSimpleAdapter(search.this,data,
R.layout.item_eatery,newString[]{"distance","name","viewNum","
price"," special "," iimg "," id "}, newint [ ] { R.id.distance,
R.id.name, R.id.viewNum, R.id.price, R.id.special, R.id.img,
R.id.id});
lv_eatery.setAdapter(simpleAdapter);
//以下的方法可加载服务器的用户头像到 ListView
simpleAdapter.setViewBinder(newViewBinder(){
publicbooleansetViewValue(
Viewview,
Objectdata,
StringtextRepresentation){
//判断是否为要处理的对象
if(viewinstanceofImageView&&datainstanceofBitmap){
ImageViewiv=(ImageView)view;
iv.setImageBitmap((Bitmap)data);
returntrue;
}else
returnfalse;
}
});

lv_eatery.setOnItemClickListener(newOnItemClickListener(){

//@Override
publicvoidonItemClick(AdapterView<?>parent, Viewarg1, intposi-
tion,longid)
{
//System.out.println("position,id是"+position+id);
ListViewlistView=(ListView)parent;
HashMap < String, String > map = (HashMap < String, String >) list-
View.getItemAtPosition(position);
```

```
    Stringseatery_id=map.get("id");
    inteatery_id=Integer.parseInt(seatery_id);
    Intentintent=newIntent(search.this,shop_detail.class);
    intent.putExtra("eatery_id",eatery_id);
    search.this.startActivity(intent);
    }
    });
    }else{
    Toast.makeText(search.this,"亲,没有您需要的内容",Toast.LENGTH_
SHORT).show();
    }


    }
    };
```

运行结果如图 26.3 所示。

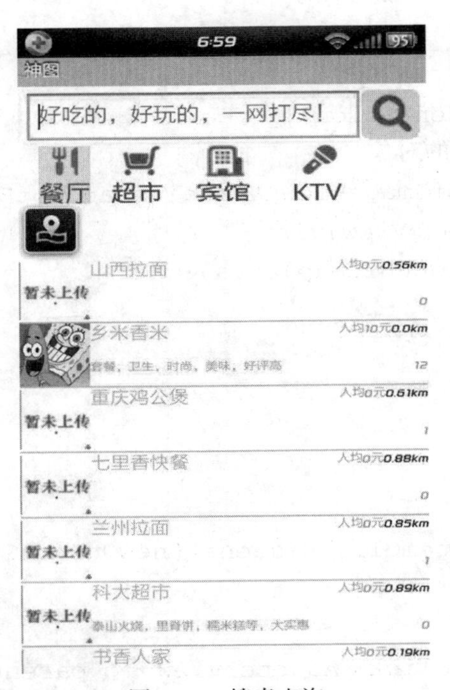

图 26.3　搜索查询

26.4　最短路径查询

《神图》软件的最短路径查询模块，主要为淘宝、搜索等功能服务，实现最短路线查询。案例代码如下：

```
if(stopPoints.size()<2){
Toast.makeText(getApplicationContext(), "至少选择两个端点", 0).show();
return;
}

Toast.makeText(getApplicationContext(),"开始查找最短路径",1).show();
newThread(){//网络分析不能在主线程中进行
publicvoidrun(){
//准备参数
RoutingParametersrp=newRoutingParameters();
NAFeaturesAsFeaturenaferture=newNAFeaturesAsFeature();
//设置查询停靠点,至少要两个
for(Pointp:stopPoints){
StopGraphicsg=newStopGraphic(p);
naferture.addFeature(sg);
Log.e("=============================","=="+p.getX()+
p.getY());}
System.out.println(stopPoints.size()+"stopPoints.size()");
rp.setStops(naferture);
//设置查询输入的坐标系跟底图一样
//naferture.setSpatialReference(map.getSpatialReference());
rp.setOutSpatialReference(mapSR);
RoutingTaskrt = newRoutingTask ( " http:// 192.168.69.34/ArcGIS/
rest/services/Routeservice/NAServer/Routelayer",null);
System.out.println("123");
try{
//执行操作
RoutingResultrr=rt.solve(rp);
System.out.println("1234");
runOnUiThread(newMyRun(rr));
System.out.println("runOnUiThread(newMyRun(rr))");
}catch(Exceptione){
e.printStackTrace();
Looper.prepare();
System.out.println(e.getMessage());
//text.setText(e.getMessage());
Toast.makeText(getApplicationContext(),e.getMessage(),1).show();
```

```
Looper.loop();
}
}
}.start();
```

运行结果如图 26.4 所示。

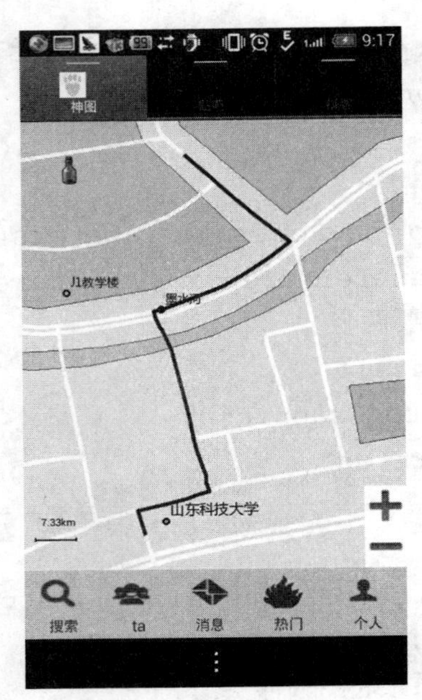

图 26.4 最短路径查询

26.5 聊天功能

《神图》软件的聊天模块是对于神瓶相遇的人可以进行聊天，用"你的 TA"功能找到的人也可以打招呼、聊天等。案例代码如下：

```
/**
 * 发送消息
 * /
privatevoidsend(){
StringcontString=mEditTextContent.getText().toString();
if(contString.length()>0){
ChatMsgEntityentity=newChatMsgEntity();
entity.setName(me);
entity.setDate(MyDate.getDateEN());
entity.setMessage(contString);
entity.setImg(0);
```

```
entity.setMsgType(false);

messageDB.saveMsg(user.getName(),entity);

mDataArrays.add(entity);
mAdapter.notifyDataSetChanged();//通知 ListView,数据已发生改变
mEditTextContent.setText("");//清空编辑框数据
mListView.setSelection(mListView.getCount()-1);//发送一条消息时,
ListView 显示选择最后一项
RecentChatEntityentity2=newRecentChatEntity(
user.getName(),user.getImg(),0,MyDate.getDate(),
contString);

application.getmRecentAdapter().remove(entity2);//先移除该对象,
目的是添加到首部
application.getmRecentList().addFirst(entity2);//再添加到首部
application.getmRecentAdapter().notifyDataSetChanged();

//创建一个线程
HttpThreadthread=newHttpThread(handler_send);

//构造请求参数
HashMap<String,Object>params=newHashMap<String,Object>();

params.put("user",me);
params.put("to_user",user.getName());
params.put("msg",contString);

Stringurl=WebInfo.url;
StringnameSpace="http://wld2010.com/";
StringmethodName="faxin";

//开始新线程进行 WebService 请求

thread.doStart(url,nameSpace,methodName,params);

}
}
```

```
publicvoidgetMessage()
{
//创建一个线程
HttpThreadthread=newHttpThread(handler);

//构造请求参数
HashMap<String,Object>params=newHashMap<String,Object>();
SharedPreferencessp=this.getSharedPreferences("userInfo",Con-
text.MODE_WORLD_READABLE);
Stringme=sp.getString("USER_NAME","none");

params.put("user",me);

Stringurl=WebInfo.url;
StringnameSpace="http://wld2010.com/";
StringmethodName="msgGet";

//开始新线程进行 WebService 请求

thread.doStart(url,nameSpace,methodName,params);
}
```

运行结果如图 26.5 所示。

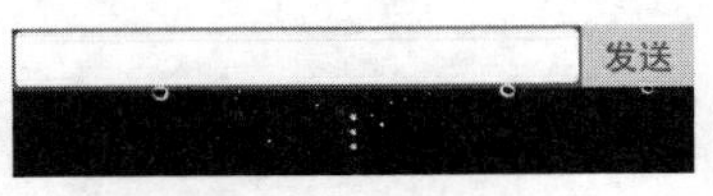

图 26.5　聊天功能

26.6　特色功能

《神图》软件的特色功能包括邂逅、个人信息展示、游记分享等，运行结果分别如图 26.6、图 26.7、图 26.8 所示。

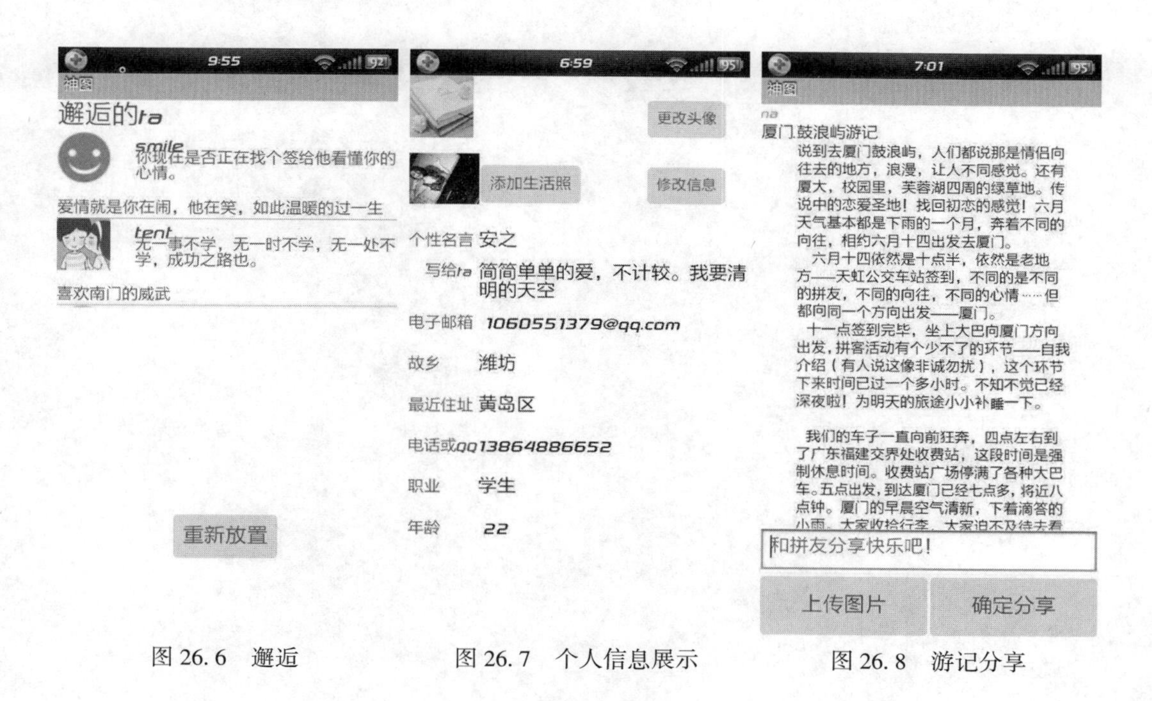

图 26.6　邂逅　　　图 26.7　个人信息展示　　　图 26.8　游记分享

第五编　MapGIS IGServer for Flex 开发

第 27 章　MapGIS for Flex 初级开发

27.1　创建第一个应用

27.1.1　配置开发环境

MapGIS IGServer 作为新一代的网络 GIS 开发平台，可提供丰富多样的二次开发方式，包括客户端的 Flex 开发。MapGIS IGServer 全面支持 Flex 的二次开发方式，即在客户端采用功能强大、优势突出的 Flex 实现。依托全新的 MapGIS IGServer 平台的 GIS 服务，采用 Flex 作为客户端，以用户为中心，为企业级 WebGIS 应用提供强大可靠的支持，为用户提供个性化网络 GIS 服务和丰富惊炫的视觉体验。基于 Flex 的 MapGIS IGServer 开发环境，包括操作系统、MapGIS IGServer 平台、数据库、浏览器、Flex 集成开发环境。

下面以 .NET 版本的 MapGIS IGServer 平台为例，主要介绍在 Windows 系统下的开发环境与相关工具：

①操作系统：Windows 系列，包括 Windows XP、Windows 2003、Windows 2008、Windows 7 旗舰版等系列版本。

②.NET 环境：Microsoft Visual Studio .NET 2005(.NET Framework 2.0 等)。

③Web 服务器：Microsoft Internet Information Server 5.0(IIS5.0)或更高版本。

④数据库：Microsoft SQL Server 2000【SP3】/2005、Oracle 等。

⑤WebGIS 开发平台：MapGIS IGServer 平台。

⑥浏览器：IE6 以上系列版本、Firefox 等主流浏览器。

⑦开发工具：Microsoft Visual Studio 2005。

⑧集成开发工具：Flex Builder 3、Adobe Flash Builder 4 及以上版本。

1. 配置 MapGIS IGServer 开发环境

MapGIS IGServer 环境配置，需要先安装 MapGIS IGServer 平台，并根据开发需求进行 GIS 服务器与数据库的配置。总体环境配置如下：

①安装平台：安装 MapGIS IGServer 平台，构建 GIS 服务器环境。

②服务器配置：安装平台后，在服务管理器(MapGIS Server Manager)中已默认配置本机器 GIS 服务器(包括 DCServer 服务器与 IGServer 服务器等各项配置)。根据实际开发环境进行配置。

③数据组织与发布：根据开发中采用的地图加载方式(地图类型)，在 MapGIS 10 平台中准备数据，并在 MapGIS Server Manager 中发布相应的地图服务。

④二次开发前，必须确保 GIS 服务器的系统服务中数据存储服务以及 MapGIS Server Manager 中的 DCServer 服务、IGServer 服务都处于启动状态，如图 27.1、图 27.2 所示。

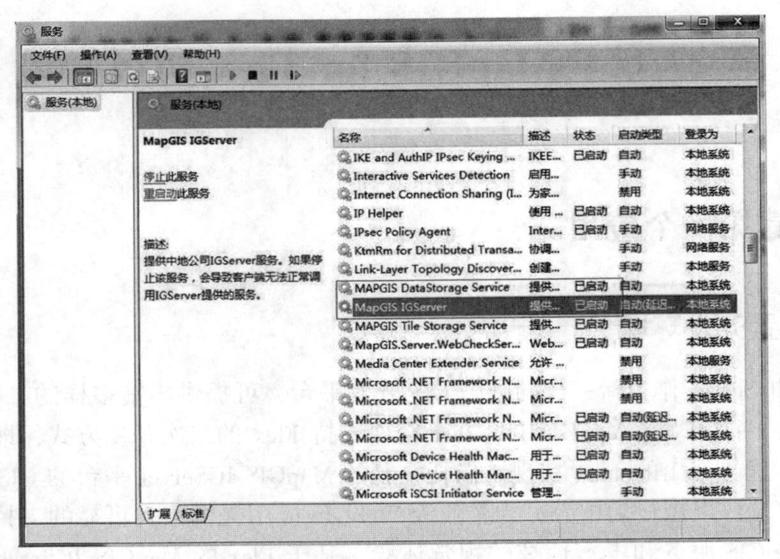

图 27.1　启动数据存储服务与 IGS 服务

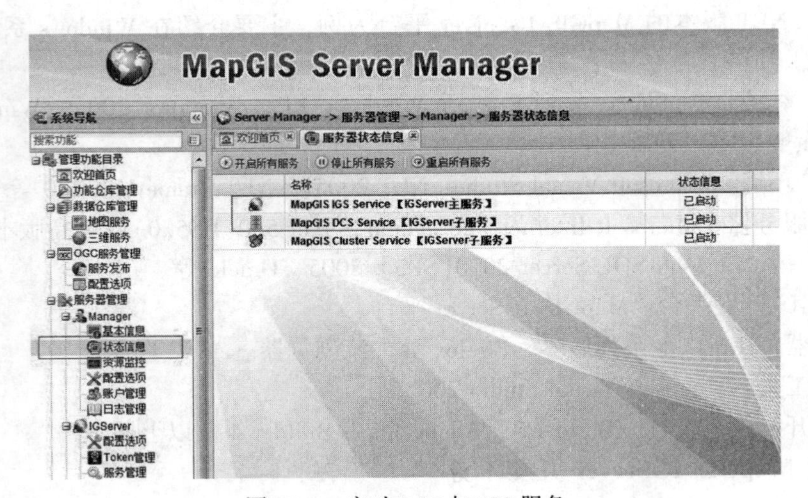

图 27.2　启动 DCS 与 IGS 服务

2. 配置 Flex 开发环境

基于 Flex 的二次开发环境，需要安装配置 Flex 集成开发环境和两个 FlashPlayer 插件。

①安装 Flex 集成开发环境——Flash Builder4。

②安装 FlashPalyer 插件，针对不同的浏览器内核选择相应的安装文件进行安装。针对 IE 内核的 flash 播放插件，安装 Adobe Flash Player Active 10.0.22.87.exe；针对非 IE 内核的 flash 播放插件，安装 Adobe Flash Player Plugin 10.0.22.87.exe。

③安装 FlashPlayer 的 debug 程序，用于调试 AS 脚本程序。安装 flashplayer_10_ax_debug 程序。

27.1.2　Flex SDK 简介

MapGIS IGServer 平台提供的 Flex SDK 包括两个部分：一部分为基础开发库，不允许用户修改；另一部分为开源库，允许用户在此开发库上进行修改或扩展开发。目前，分别提供 2.0、4.0、4.5 版本的 SDK，其中 4.5 版本为新版本(做了很多优化和更新)，推荐使用，基础库与开源库的详细说明分别见表 27.1 和表 27.2。

表 27.1　基础库

SDK(开发包)	说明	功能说明
Zdims2.0.swc	Adobe Flex Builder 3.0 对应的基础开发库	包括地图显示、基本操作、绘制等基础功能
Zdims4.0.swc	Adobe Flex Builder 4.0 对应的基础开发库	
Zdims4.5.swc	Adobe Flex Builder 4.5 对应的基础开发库	

表 27.2　开源库

SDK(开发包)	说明	功能说明
地图查询	ConditionInput、MapDocDataViewer	条件查询相关控件
地图编辑	AnnotationStyle、Editor、PointStyle、LineStyle、PolygonStyle	编辑功能相关控件
统计分析	Chart	统计图窗口分析控件
空间分析	BufferAnalyse、ClipAnalyse、NetAnalyse、OverLayAnalyse、TopAnalyse	空间分析相关控件
投影转换	Project	投影转换控件
公交换乘	BusAnalyse	公交换乘窗口控件
GPS 导航	GPS	GPS 定位控件
OGC 功能	OGCCatalog、OGCConditionInput、OGCDataViewer、OGCEditor、OGCToolBar	OGC 服务相关控件
基本操作	Measure、Scale、IMSCatalog、Magnifier、NavigationBar	基本操作相关控件
图层类控件	LayerConditionInPut、LayerDataViewer、LayerDisplaySet、LayerEditor、LayerInfo	图层显示、编辑、查询结果显示等图层类控件

27.1.3　创建案例应用

在 Flex 集成开发环境(Adobe Flash Builder 4)中创建 MapGIS IGServer Flex 的新工程，

具体步骤如下：

①从开始菜单中打开 Adobe Flash Builder 4，进入 Adobe Flash Builder 4 主界面，选择【新建】菜单，创建一个 Flex Project 工程，如图 27.3 所示。

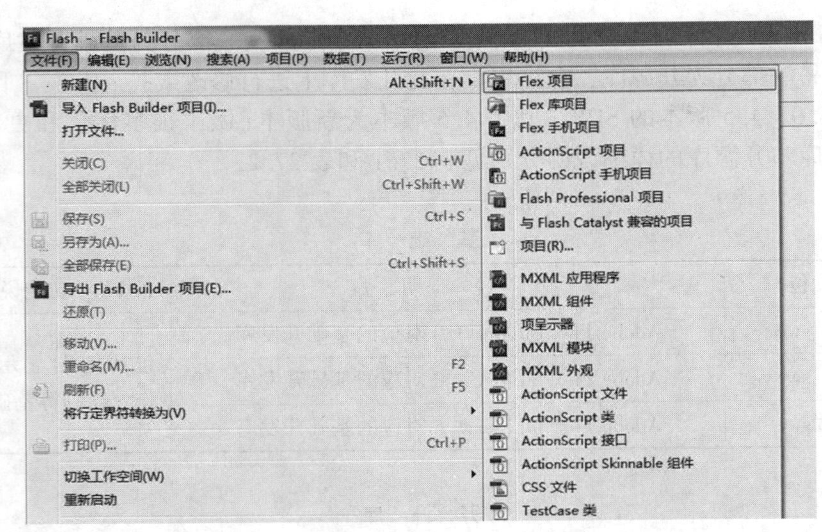

图 27.3　Adobe Flash Builder 4 中新建工程

②在弹出的窗口中输入工程名，点击【浏览】，选择工程存放的路径，如图 27.4 所示。

图 27.4　创建 Flex 项目设置

设置好之后，点击【下一步】，弹出配置输出的对话框，如图 27.5 所示。

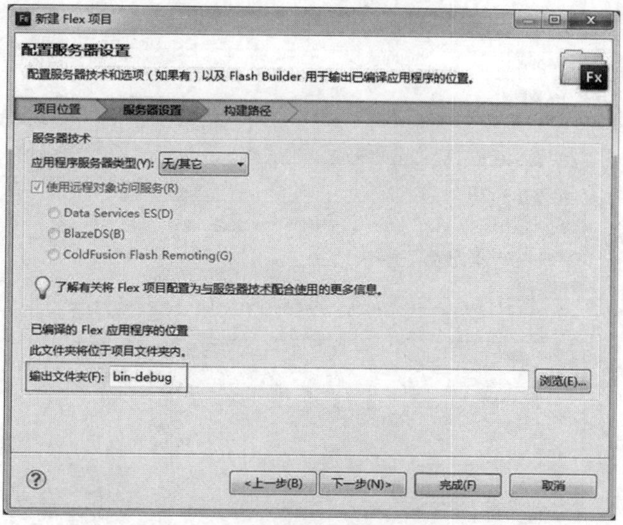

图 27.5　设置配置输出路径

③使用默认的输出路径(bin-debug)，点击【下一步】进入到下一个页面，如图 27.6 所示。

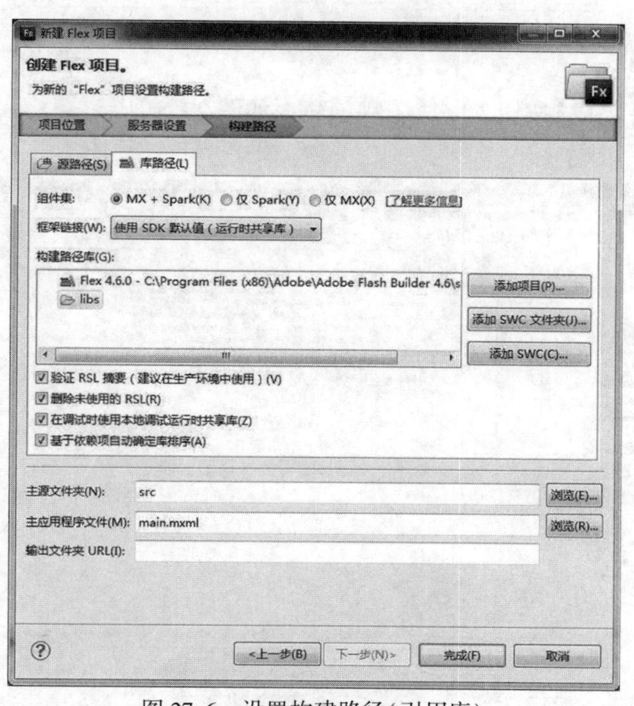

图 27.6　设置构建路径(引用库)

④点击右边的【添加 SWC】按钮，弹出添加 SWC 的对话框，单击【浏览】按钮，将 Flex 二次开发 SDK 的基础库(如 zdims4. 5. swc)添加进来，如图 27. 7 所示。

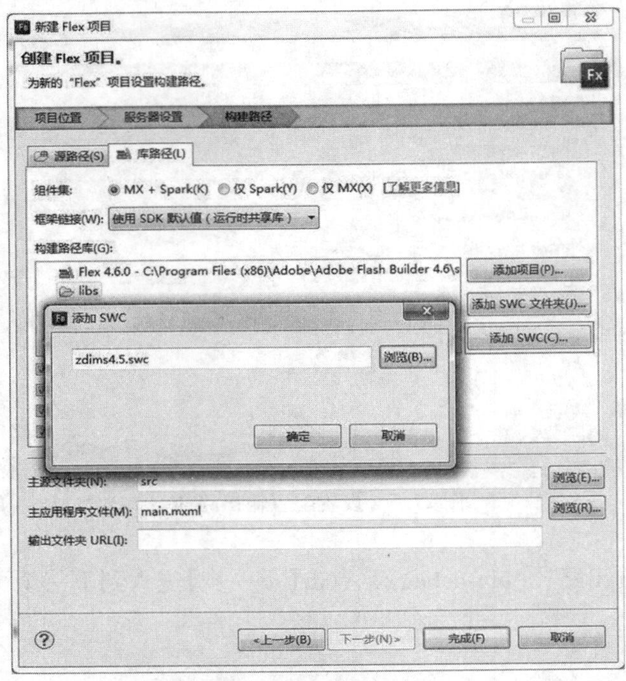

图 27. 7　引用 SWC 文件内容

⑤最后，点击【完成】按钮，完成工程创建，如图 27. 8 所示。

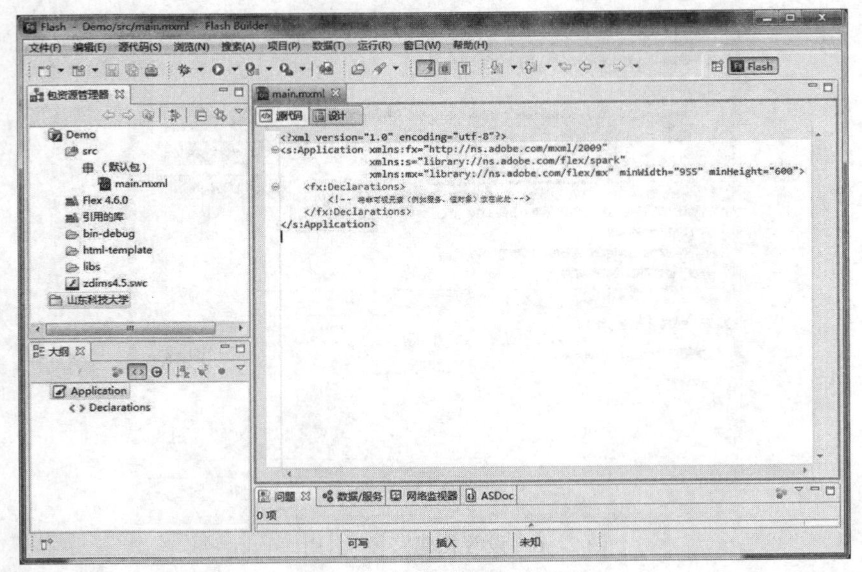

图 27. 8　新建工程的主界面

27.1.4　地图加载运行

单击页面的【设计】按钮，切换到设计模式，这时左边组件(Components)窗口会在自定义(Custom)目录下添加 GIS 功能控件，如 IMSMAP 等，切换到左侧【组件】窗口，在【自定义】目录中找到 IMSMap 控件，直接将其拖到项目的设计页面里，如图 27.9 所示。

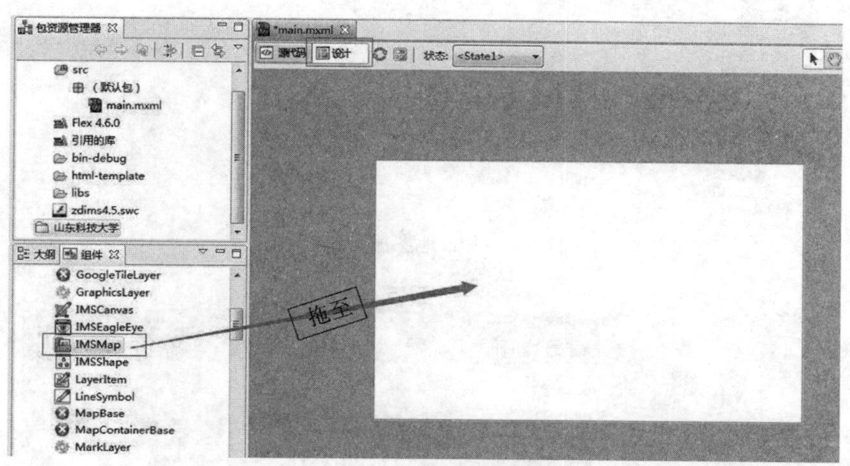

图 27.9　新建工程的主界面

然后切换到右侧的 Flex 属性视窗，根据整体布局需求调整该控件的大小、位置等属性。此控件的大小为地图域的大小，可根据需要自行设置(此处设置为布满网页)，如图 27.10 所示。

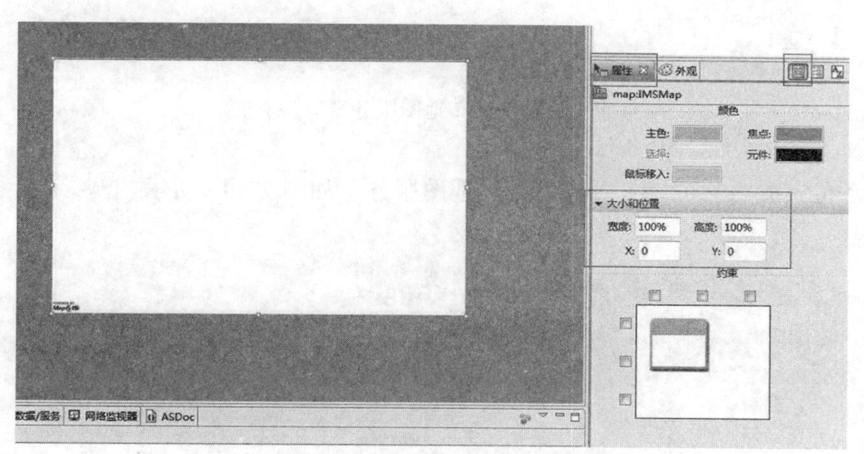

图 27.10　设置 IMSMap 控件的属性

登录 MapGIS Server Manager，在左侧导航条中点击【数据仓库管理】下的【地图服务】，在右侧面板中点击【发布地图文档】，如图 27.11 和图 27.12 所示。

图 27.11 发布地图服务窗口(1)

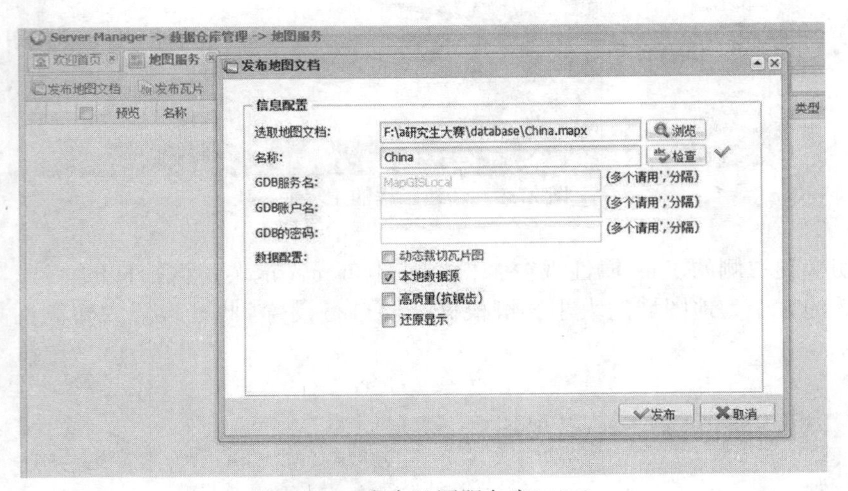

图 27.12 发布地图服务窗口(2)

服务发布完成后，可以查看已发布的地图服务，如图 27.13 所示。

图 27.13 查看已发布的地图服务

切换到左侧【组件】窗口，在【自定义】目录中找到 VectorMapDoc 控件，直接将其拖到项目设计页面的 IMSMap 控件里，如图 27.14 所示。

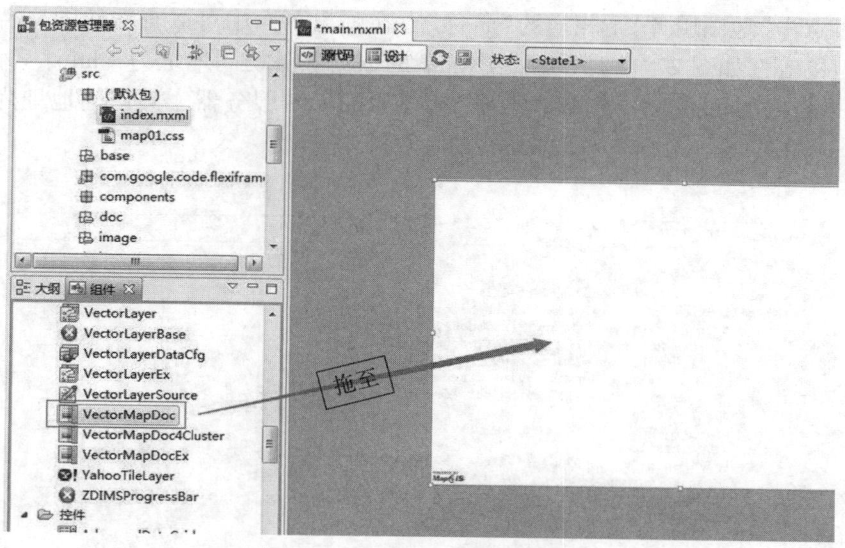

图 27.14　添加矢量地图文档控件

然后切换到右侧的 Flex 属性视窗，根据整体布局需求调整该控件的大小、位置等属性。可根据需要自行设置(此处设置为布满地图域)，如图 27.15 所示。

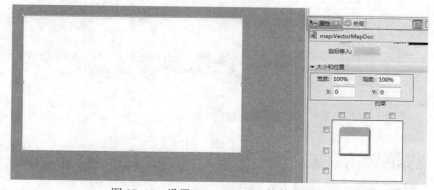

图 27.15　设置 VectorMapDoc 控件的属性

地图容器添加地图需要注意以下两点：

①如果地图容器包含瓦片地图，一定要给地图容器的终止显示级数 levelNum 和显示范围(xMinMap，yMinMap，xMaxMap，yMaxMap)赋值，该情况下显示范围必须是瓦片地图的裁图范围，否则不能正常显示和使用瓦片地图。

②如果地图容器里只包含矢量地图，则不需要设置地图容器的终止显示级数 levelNum，但显示范围(xMinMap，yMinMap，xMaxMap，yMaxMap)属性一定要设置，该情

况下的显示范围属性可以设置为任何自己想要初始化时显示的地图范围。

切换至【源代码】视图，如图 27.16 所示。由于地图容器里只包含矢量地图，所以应该按第二条进行设置。红色框内的代码是需要添加的内容，如图 27.17 所示。地图显示范围可以通过点击【地图服务】下相应的“名称”，在弹出页面中可以查看地图文档的详细信息，包含范围等，如图 27.18 所示。地图加载的数据服务访问地址为：http：// IP：端口号/igs/rest/ims/relayhandler。例如，本地 .NET 版的 GIS 服务器，其访问地址为 Http：//127. 0. 0. 1：6163/igs/rest/ims/relayhandler。

```
*main.mxml
源代码 设计
1  <?xml version="1.0" encoding="utf-8"?>
2  <s:Application xmlns:fx="http://ns.adobe.com/mxml/2009"
3                 xmlns:s="library://ns.adobe.com/flex/spark"
4                 xmlns:mx="library://ns.adobe.com/flex/mx"
5                 xmlns:map="zdims.map.*"
6                 xmlns:othermap="zdims.othermap.*"
7                 minWidth="955" minHeight="600">
8      <fx:Declarations>
9          <!-- 将非可视元素（例如服务、值对象）放在此处 -->
10     </fx:Declarations>
11     <map:IMSMap x="0" y="0" width="100%" height="100%">
12         <map:VectorMapDoc  x="0" y="0"  width="100%" height="100%">
13         </map:VectorMapDoc>
14     </map:IMSMap>
15 </s:Application>
```

图 27.16　切换至源代码视图

```
*main.mxml
源代码 设计
1  <?xml version="1.0" encoding="utf-8"?>
2  <s:Application xmlns:fx="http://ns.adobe.com/mxml/2009"
3                 xmlns:s="library://ns.adobe.com/flex/spark"
4                 xmlns:mx="library://ns.adobe.com/flex/mx"
5                 xmlns:map="zdims.map.*"
6                 xmlns:othermap="zdims.othermap.*"
7                 minWidth="955" minHeight="600">
8      <fx:Declarations>
9          <!-- 将非可视元素（例如服务、值对象）放在此处 -->
10     </fx:Declarations>
11     <map:IMSMap x="0" y="0" width="100%" height="100%"
12                 xMinMap="6933046.115037738" yMinMap="1075039.825021435" xMaxMap="16345196.216909475" yMaxMap="8134152.401425237">
13         <map:VectorMapDoc  x="0" y="0" width="100%" height="100%"
14                            mapDocName="China" serverAddress="http://127.0.0.1:6163/igs/rest/ims/relayhandler">
15         </map:VectorMapDoc>
16     </map:IMSMap>
17 </s:Application>
```

图 27.17　补充代码

图 27.18　查看地图范围

保存工程，保证没有报错后，单击 ，选择 1 main 点击运行。如图 27.19、图 27.20 所示。

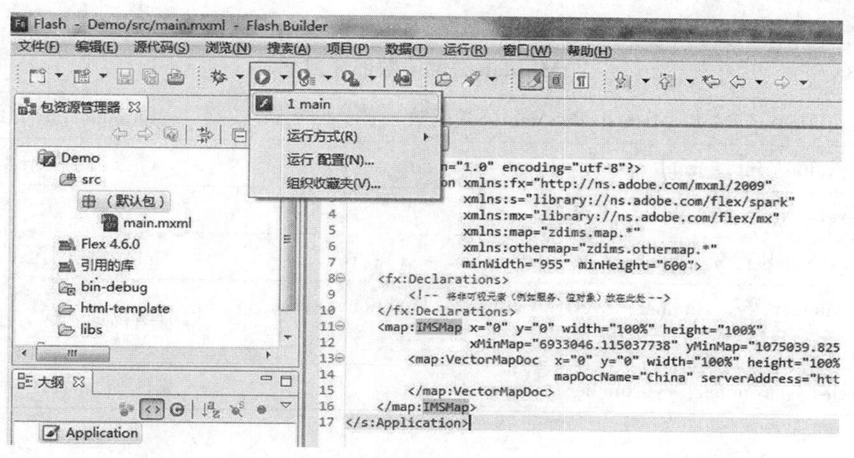

图 27.19　点击运行

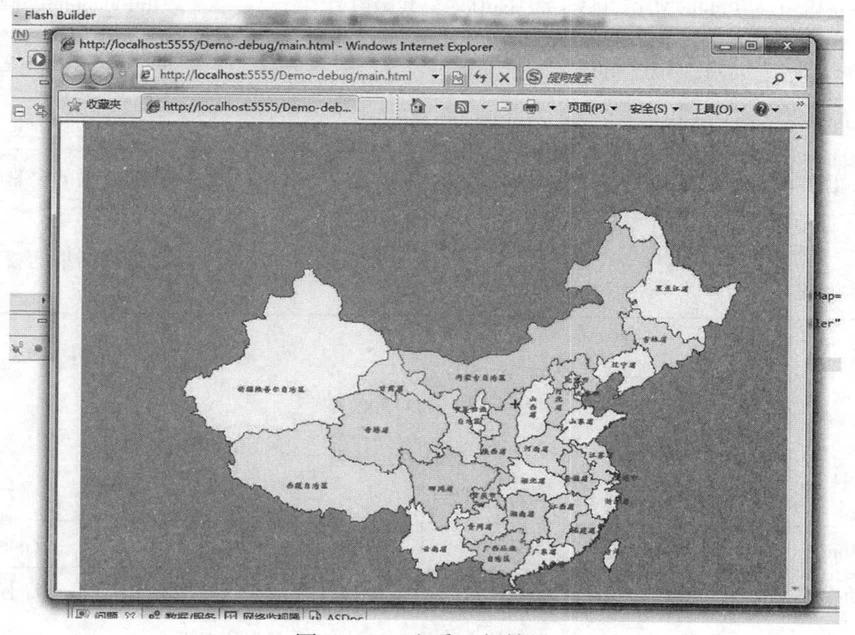

图 27.20　查看运行结果

27.2　地图事件

地图基本操作(放大、缩小、移动、复位、更新)，地图控制(地图显示级别、地图居中、地图移至某一位置等)，获取地图文档图层列表等信息均与地图功能函数事件绑定，直接调用函数实现。相关函数详见 MapGIS《Flex 二次开发库接口文档》zdims. map. IMSMap、zdims. map. VectorMapDoc 等相关类。此处仅列出部分函数供参考，见表 27.3、表 27.4。地图基本操作 IMSOperType 常量列举如图 27.21 所示。

表 27.3　**IMSMap 公共函数部分列举**

IMSMap Public Method(IMSMap 公共函数)	Introduction(介绍)
setCurOper(value: int): void	设置当前操作类型
setLevel(level: Number): void	设置显示级数
setCenter(x: Number, y: Number, level: Number = -1): void	移动地图至某处
panTo(x: Number, y: Number, oveOverCallback: Function = null): void	将地图移动到屏幕中心位置
screenToLogic(x: Number, y: Number): Point	窗口坐标转逻辑坐标

表 27.4　**VectorMapDoc 公共函数部分列举**

VectorMapDoc Public Method(VectorMapDoc 公共函数)	Introduction(介绍)
getLayerInfo(index: int): CLayerInfo	获取图层信息
addFeature(featureInfo: CMapFeatureInfo, callBack: Function): void	添加要素
onSelectAndGetFeature(e: Event): CMapSelectAndGetFeature	获取查询结果的要素集合结果
selectAndGetFeature(selectParam: CMapSelectParam, callback: Function): void	获取查询结果的要素集合

IMSOperType
(Constant)

↓

Drag	None	Refresh	Restore	ZoomIn	ZoomOut
移动操作	无操作	刷新操作	复位操作	放大操作	缩小操作

图 27.21　地图基本操作 IMSOperType 常量列举

提示：窗口坐标转逻辑坐标 可以利用此方法进行当前地图窗口内地物的查询操作。

地图基本操作前台界面交互设计示例代码如下：

```
<mx:Canvas horizontalCenter="0" y="10" width="269" height="30"
backgroundAlpha="0.35"
        backgroundColor="#CCCCCC" borderStyle="solid" cornerRa-
```

```
dius="16">
            <mx:LinkButton x="21" y="4" label="复位" click="MapBase-
Function('Restore')"/>
            <mx:LinkButton x="67" y="4" label="放大" click="MapBase-
Function('ZoomIn')"/>
            <mx:LinkButton x="116" y="4" label="缩小"click="MapBase-
Function('ZoomOut')"/>
            <mx:LinkButton x="164" y="4" label="平移" click="MapBase-
Function('Drag')"/>
            <mx:LinkButton x="214" y="4" label="刷新" click="MapBase-
Function('Refresh')"/>
    </mx:Canvas>
```

功能事件示例代码如下：

```
    <fx:Script>
        <![CDATA[
            import zdims.util.IMSOperType;
            publicfunction MapBaseFunction(fun:String):void
            {
                switch(fun)
                {
                    case"ZoomIn":containerMap.setCurOper(IMSOper-
Type.ZoomIn);break;
                    case"ZoomOut":containerMap.setCurOper(IMSOper
Type.ZoomOut);break;
                    case"Restore":containerMap.setCurOper(IMSOper
Type.Restore);break;
                    case"Drag":containerMap.setCurOper(IMSOperType.
Drag);break;
                    case"Refresh":containerMap.setCurOper(IMSOper
Type.Refresh);break;
                    default:containerMap.setCurOper(IMSOperType.
None);
                }
            }
        ]]>
    </fx:Script>
```

运行效果(点击功能名称可以实现操作)如图 27.22 所示。

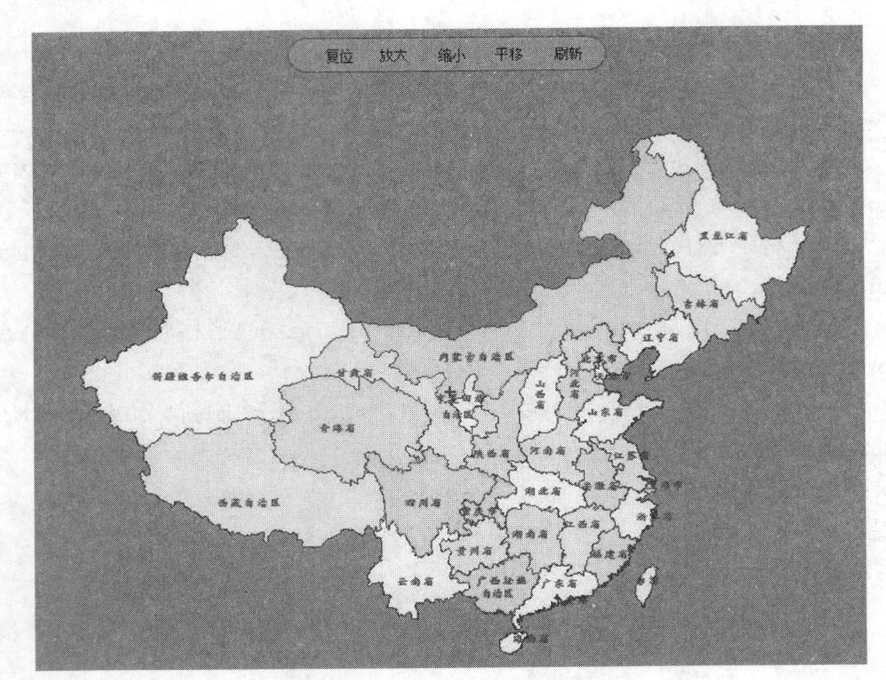

图 27.22 地图基本操作

地图控制中移动地图至某一位置操作前台界面交互设计示例代码如下：

```
<mx:Canvas verticalCenter="0" x="10" width="211" height="34"
backgroundAlpha="0.35"
          backgroundColor="#CCCCCC" borderStyle="solid" cornerRa-
dius="16">
     <s:Label x="17" y="4" width="79" height="23" text="地图移至"
textAlign="center"
             verticalAlign="middle"/>
     <s:ComboBox id="MapMoveTo" x="100" y="4" width="77"
change="MapMoveToTarget(event)"
                dataProvider="{MapPlacePosition}" labelField="name"/>
</mx:Canvas>
```

功能事件示例代码如下：

```
<fx:Script>
     <![CDATA[
          import mx.collections.ArrayCollection;
          import spark.events.IndexChangeEvent;
          import zdims.util.IMSOperType;
          [Bindable]
          private var MapPlacePosition:ArrayCollection=new Ar-
```

```
rayCollection(
                    [{name:"北京",x:"12958216.53112934",y:"4867876.
858935801"},
                 {name:"济南",x:"13038934.032998485",y:"4395801.
772246553"},
                 {name:"青岛",x:"13397882.317825677",y:"4312271.
387736512"}]
        );
        protected function MapMoveToTarget(event:IndexChange
Event):void
        {
            this.containerMap.panTo(Number(MapMoveTo.
selectedItem.x),
                         Number(MapMoveTo.selectedItem.y));
        }
      ]]>
      </fx:Script>
```

运行效果(选择“青岛”名称可以实现操作)如图 27.23 所示。

图 27.23　移动地图至某一位置

27.3　图形绘制添加

图形绘制执行流程：首先添加一个绘图层的控件(GraphicsLayer)，Flex4 里绘图都是在绘图层上完成的，然后通过绘制命令函数，用来设置画图的类型，同时将所画图形的信息传递给一个回调函数。回调函数 drawingOverCallback( )里面传递了 3 个参数，该函数可以设置点的信息，以及将所画的图形显示出来。

图形添加：只需先创建一个图形对象并对点数据进行赋值存储，将点设置给图形对象，然后将其作为子节点添加到 GraphicsLayer 中即可。

相关类详见 MapGIS《Flex 二次开发库接口文档》zdims. drawing 包中的相关类。此处仅列出部分类组织供参考，如图 27. 24 所示。

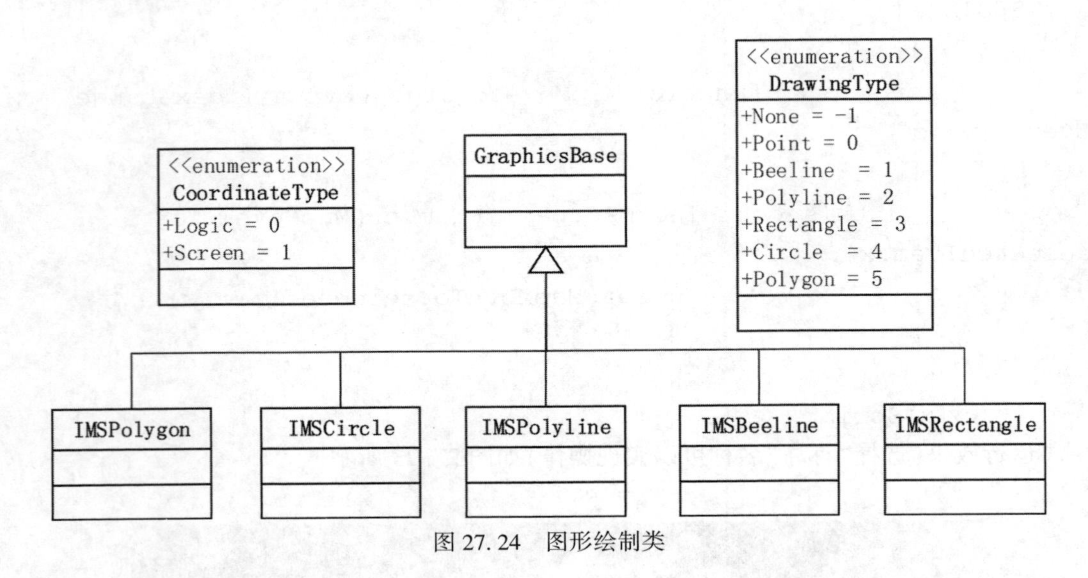

图 27. 24　图形绘制类

提示：可以将绘制的图形点坐标序列、类型存入数据库中，也可将已存的图形数据添加到图层上。

图形鼠标绘制(以多边形为例)、图形添加(以矩形为例)、清楚绘图操作前台交互界面设计示例代码如下：

```
    <map:IMSMap id = " containerMap "  x = " 0 "  y = " 0 "  width = " 100% "
height = "100% "
                xMinMap = "6933046.115037738" yMinMap = "1075039.
825021435"
                xMaxMap = "16345196.216909475" yMaxMap = "8134152.
401425237">
        <map:VectorMapDoc id = "vecMap"   x = "0" y = "0" width = "100% "
height = "100% "
            mapDocName = "China"
    serverAddress = " http:// 127.0.0.1: 6163 /igs /rest /ims /relayhan-
dler">
        </map:VectorMapDoc>
        <drawing:GraphicsLayer id = "graphic">
```

```
        </drawing:GraphicsLayer>
    </map:IMSMap>
    <mx:Canvas verticalCenter="0" right="10" width="245" height="
30" backgroundAlpha="0.35"
              backgroundColor="#CCCCCC" borderStyle="solid" corner
Radius="16">
            <mx:LinkButton x="21" y="4" label="绘制多变形" click="
drawPolygon(event)"/>
            <mx:LinkButton x="97" y="5" label="添加矩形" click="ad-
dRect(event)"/>
            <mx:LinkButton x="169" y="5" label="清除" click="graphic-
Clear()"/>
    </mx:Canvas>
```

功能事件示例代码如下：

```
    <fx:Script>
        <![CDATA[
              import zdims.drawing.CoordinateType;
              import zdims.drawing.DrawingType;
              import zdims.drawing.IMSPolygon;
              import zdims.drawing.IMSRectangle;
              import zdims.interfaces.IGraphics;
              public function drawPolygon(event:MouseEvent):void{
                  graphic.drawingType=DrawingType.Polygon;
                  graphic.drawingOverCallback=DrawOverCallback;
              }
              public functionDrawOverCallback(gLayer:GraphicsLayer,
graphics:IGraphics,
              polygonArr:Vector.<Point>):void{
                  var polygon:IMSPolygon=new IMSPolygon(Coordinate-
Type.Logic);
                  gLayer.addGraphics(polygon);
                  polygon.draw(polygonArr);
                  polygon.enableEdit=true;
              }
              public function addRect(event:MouseEvent):void
              {
                  var rect:IMSRectangle=new IMSRectangle(Coordinate
Type.Logic);
```

```
                graphic.addGraphics(rect);
                rect.startPoint=new Point(13038934.032998485,
4395801.772246553);
                rect.endPoint=new Point(13397882.317825677,
4312271.387736512);
                rect.enableEdit=true;
            }
            public function graphicClear():void
            {
                graphic.removeAllElements();
            }
        ]]>
    </fx:Script>
```

运行操作效果如图 27.25 所示。

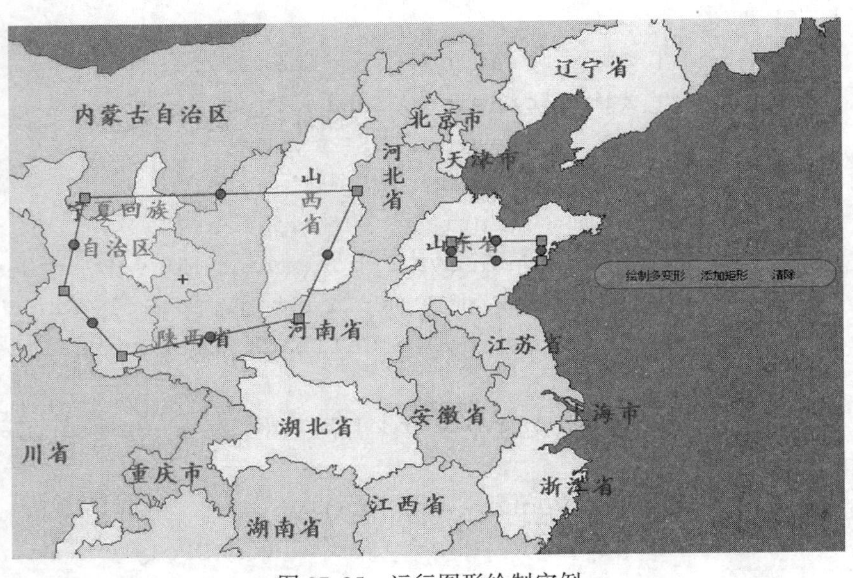

图 27.25　运行图形绘制实例

27.4　地图标注

地图标注是将空间位置信息点与地图关联，通过图标、窗口等形式把点相关的信息展现到地图上。地图标注也是 WebGIS 中的比较重要的功能之一，在大众应用中较为常见。地图标注能丰富 GIS 应用，可以为用户提供更多个性化的地图服务，如标注兴趣点等。

常用的地图标注分为：图片标注、添加标注点、统计图标注、添加热区等。本节将通过实现添加标注点，并可点击标注点图标定位到视图中心，同时会显示当前选中的标注点名称，使读者理解功能的实现过程。

前台交互界面设计示例代码如下：

```
<map: IMSMap id = " containerMap " x = " 0 " y = " 0 " width = " 100% " height = "100% "
                    xMinMap = "6933046.115037738" yMinMap =
                    "1075039.825021435"
    xMaxMap = "16345196.216909475" yMaxMap = "8134152.401425237">
        <map:VectorMapDoc id = "vecMap"  x = "0" y = "0" width = "100% height = "100% "mapDocName = "China" serverAddress = "http://127.0.0.1:6163/igs/rest/ims/relayhandler">
        </map:VectorMapDoc>
        <mark:MarkLayer id = "markLayer"></mark:MarkLayer>
    </map:IMSMap>
    <mx:Canvas horizontalCenter = "0" y = "50" width = "237" height = "30" backgroundAlpha = "0.35"
        backgroundColor = "#CCCCCC" borderStyle = "solid" cornerRadius = "16">
        <mx:LinkButton x = "11" y = "4" label = "兴趣点" click = "drawMarks()"/>
        <mx:LinkButton x = "62" y = "4" label = "清除" click = "markLayerClear()"/>
        <s:Label x = "109" y = "9" text = "当前选中:"/>
        <s:Label id = "POIName" x = "170" y = "8"width = "46"color = "#167CAC"fontWeight = "bold"/>
    </mx:Canvas>
```

运行操作效果如图 27.26 所示。

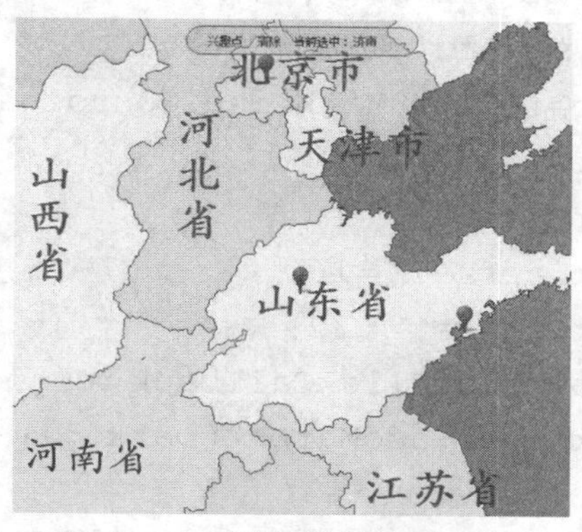

图 27.26　添加地图兴趣点标注

功能事件示例代码如下：

```
<fx:Script>
        <![CDATA[
import mx.collections.ArrayCollection;
            import mx.controls.Image;
            import zdims.drawing.CoordinateType;
            import zdims.mark.IMSMark;
             private var MapPlacePosition:ArrayCollection=new Ar-
rayCollection(
                            [{name:"北京",x:"12958216.53112934",y:"
4867876.858935801"},
                            {name:"济南",x:"13038934.032998485",y:"
4395801.772246553"},
                            {name:"青岛",x:"13397882.317825677",y:"
4312271.387736512"}]
            );
  private function drawMarks():void{
                for(var i:int=0;i<MapPlacePosition.length;i++){
                    addMark(new Number(MapPlacePosition[i].x),
  new Number(MapPlacePosition[i]——y),i);
                }
                markLayer.enableMarkHiden=false;
            }
            [Embed("img/widget/point.png")]
            public static var icon:Class;
            private function addMark(x:Number,y:Number,j:int):void{
                var mark:IMSMark;
                var img:Image;
                img=new Image();
                img.source=icon;
                mark=new IMSMark(img,CoordinateType.Logic);
                mark.x=x;
                mark.y=y;
                mark.offsetX=16;
                mark.offsetY=32;
                mark.mouseClickCallback=changeBaseContent;
                markLayer.addMarkAt(mark,j);
            }
              private function changeBaseContent(mark:IMSMark,e-
```

```
vent:MouseEvent):void{
                    containerMap.panTo(mark.x,mark.y);
                    POIName.text=MapPlacePosition[markLayer.indexOf
(mark)]——name;
                }
                public function markLayerClear():void{
                    this.markLayer.removeAllElements();
                }
    ]]>
    </fx:Script>
```

27.5 空间查询

空间查询是指基于给定属性或空间约束条件从地理数据库中查找指定地理对象及其属性的过程。空间查询属于数据库的范畴，其属性约束条件一般用带比较运算符的逻辑表达式来描述；空间约束条件用带空间谓词的逻辑表达式来描述，空间谓词由地理对象间空间关系演变而来，如包含、相交、分离、重叠等。因此，空间查询是作用在库体上的函数返回用户请求的内容，属于咨询式分析。查询是 GIS 用户使用最频繁的功能，用户提出的很大一部分问题都可以通过查询的方式解决，查询的方法和查询的范围在很大程度上决定了 GIS 的应用程度和应用水平。查询是 GIS 的一个非常重要的功能，定位空间对象、提取对象信息，是地理信息系统进行高层次空间分析的基础。空间查询实现流程如图 27.27 所示。

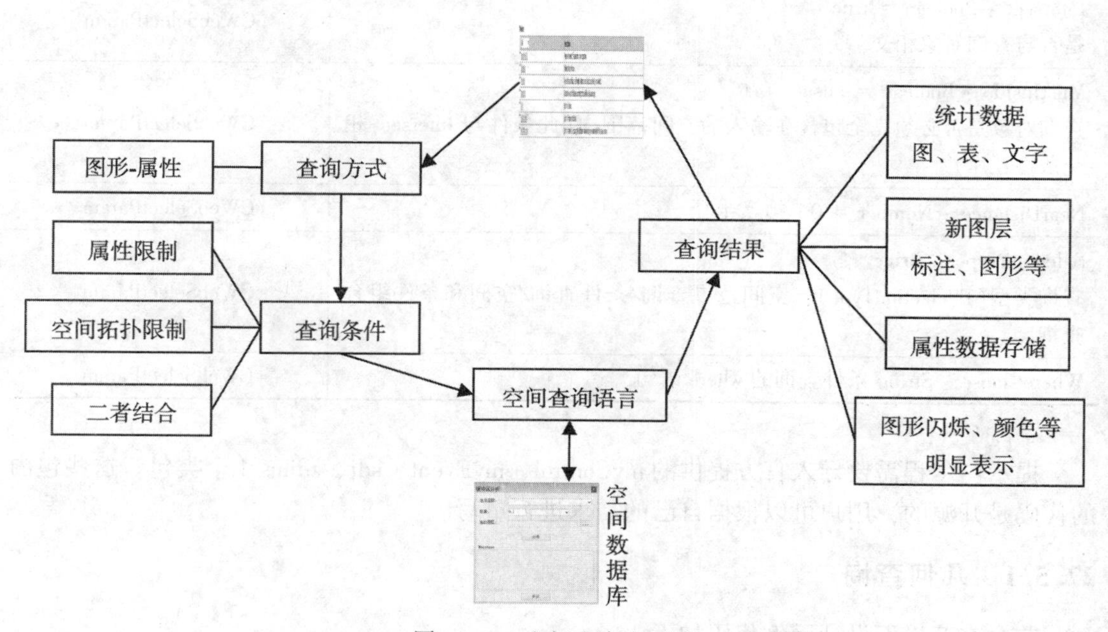

图 27.27 空间查询流程图

空间查询的方式：

①给出图形信息：如鼠标点取、拉框等方式，检索其相应属性或检索其空间拓扑关系。

②给出属性特征条件：如选择或输入属性值，检索对应的空间实体或查询属性。

单纯查询是指单纯地查询属性，或者只查询几何空间拓扑关系。复合查询是指将空间数据与属性数据联合查询。

查询地图文档中的数据，调用 VectorMapDoc 的 select(selectParam：CMapSelectParam，callback：Function)：void 方法可以实现，其中参数 selectParam ：CMapSelectParam 的对象属性 SelectParam ：CWebSelectParam 为 CWebSelectParam 对象。CWebSelectParam 类构造见表 27.6。

Mapgis7. WebService. BasLib. CWebSelectParam 构造函数为：CWebSelectParam()。

表 27.6　**CWebSelectParam()函数属性**

Public Properties(属性)	Define By(定义)
CompareRectOnly ：Boolean = false 是否仅仅比较查询对象的外包络矩形框	CWebSelectParam
Geometry ：Object 查询范围，接受的类型有 Dot_2D，AnyLine，Circle，Rect，Polygon	CWebSelectParam
GeomType ：String 所输入空间范围的几何类型(EWebGeomType)	CWebSelectParam
Intersect ：Boolean = true 是否与查询对象相交	CWebSelectParam
MustInside ：Boolean = false 查询对象是否必须完全包含在输入的空间范围内(此条件与 Intersect 相交条件互斥)	CWebSelectParam
NearDistance ：Number = 0　搜索半径	CWebSelectParam
SelectionType ：String 查询类型(ESelectionType)：空间范围查询/条件查询/空间和条件组合查询	CWebSelectParam
WhereClause ：String 条件查询的 where 语句	CWebSelectParam

提示：工程需要导入官方提供的 myControl、myEvent、sdj、zdims 4 个类包。这些包内的代码是开源的，用户可以根据自己的需求进行重写。

27.5.1　几何查询

前台交互界面设计示例代码如下：

```
<map: IMSMap id = " containerMap " x = " 0 " y = " 0 " width = " 100% "
```

```
height="100%"
                xMinMap="6933046.115037738" yMinMap="1075039.
825021435"
    xMaxMap="16345196.216909475" yMaxMap="8134152.401425237">
            <map:VectorMapDoc id="vecMap"  x="0" y="0" width="100%"
height="100%"
        mapDocName="China" serverAddress="http://127.0.0.1:6163/
igs/rest/ims/relayhandler">
            </map:VectorMapDoc>
            <drawing:GraphicsLayer id="graphic"> </drawing:Graphic-
sLayer>
            <control:MapDocDataViewer x="170" y="200" id="mapDocDat-
aViewer"
                                imsmap="{containerMap}">
            </control:MapDocDataViewer>
    </map:IMSMap>
    <mx:Canvas horizontalCenter="0" y="10" width="300" height="30"
backgroundAlpha="0.35"
                backgroundColor="#CCCCCC" borderStyle="solid" cor-
nerRadius="16">
            <mx:LinkButton x="21" y="4" label="点选" click="draw-
Graph('Point')"/>
            <mx:LinkButton x="67" y="4" label="线选" click="draw-
Graph('Polyline')"/>
            <mx:LinkButton x="116" y="4" label="矩形选" click="draw-
Graph('Rectangle')"/>
            <mx:LinkButton x="169" y="4" label="圆选" click="draw-
Graph('Circle')"/>
            <mx:LinkButton x="214" y="4" label="多边形选" click="
drawGraph('Polygon')"/>
    </mx:Canvas>
```

功能事件示例代码实现如下：

```
<fx:Script>
        <![CDATA[
            import Mapgis7.WebService.BasLib.AnyLine;
            import Mapgis7.WebService.BasLib.Arc;
            import Mapgis7.WebService.BasLib.CMapSelectParam;
            import Mapgis7.WebService.BasLib.COperResult;
```

```
            import Mapgis7.WebService.BasLib.CWebSelectParam;
            import Mapgis7.WebService.BasLib.Circle;
            import Mapgis7.WebService.BasLib.Dot_2D;
            import Mapgis7.WebService.BasLib.ESelectionType;
            import Mapgis7.WebService.BasLib.EWebGeomType;
            import Mapgis7.WebService.BasLib.EnumLayerStatus;
            import Mapgis7.WebService.BasLib.IWebGeometry;
            import Mapgis7.WebService.BasLib.Polygon;
            import Mapgis7.WebService.BasLib.Rect;
            import flashx.textLayout.events.UpdateCompleteEvent;
            import zdims.drawing.CoordinateType;
            import zdims.drawing.DrawingType;
            import zdims.drawing.IMSBeeline;
            import zdims.drawing.IMSCircle;
            import zdims.drawing.IMSPolygon;
            import zdims.drawing.IMSRectangle;
            import zdims.interfaces.IGraphics;
            import zdims.interfaces.control.INavigationBar;
            public function drawGraph(fun:String):void
            {
                switch(fun)
                {
                    case"Point":graphic.drawingType=DrawingType.
Point;break;
                    case"Polyline":graphic.drawingType=DrawingType.
Polyline;break;
                    case"Rectangle":graphic.drawingType=DrawingType.
Rectangle;break;
                    case"Circle":graphic.drawingType=DrawingType.
Circle;break;
                    case"Polygon":graphic.drawingType=DrawingType.
Polygon;break;
                    default:graphic.drawingType=DrawingType.None;
                }
                graphic.drawingOverCallback=DrawingOverCallback;
            }
             public function DrawingOverCallback(gLayer:Graphics-
Layer,graphics:IGraphics,pntArr:Vector.<Point>):void
```

```
            {
                gLayer.removeAllElements();
                //设置图层为可查询状态
    this.vecMap.getMapLayerInfo(3).LayerStatus = EnumLayerStatus.
Selectable;
                this.vecMap.updateAllLayerInfo();
                this.containerMap.activeMapDoc = this.vecMap;
                var mapsel:CMapSelectParam = new CMapSelectParam();
                var websel:CWebSelectParam = new CWebSelectParam();
                websel.CompareRectOnly = this.vecMap.compareRectOnly;
                websel.MustInside = this.vecMap.mustInside;
                switch(gLayer.drawingType)
                {
                    case DrawingType.Point:
                            var circleDraw:IMSCircle = new IMSCir-
cle(CoordinateType.Logic);
                            circleDraw.radius = 0.02;
                            gLayer.addGraphics(circleDraw);
                            circleDraw.draw(pntArr);
                            var dot:Dot_2D = new Dot_2D();
                            dot.x = pntArr[0].x;
                            dot.y = pntArr[0].y;
                            websel.Geometry = dot;
                            websel.GeomType = (dot as IWebGeometry).
GetGeomType();
                            break;
                    case DrawingType.Polyline:
                            var bline:IMSBeeline = new
IMSBeeline(CoordinateType.Logic);
                            gLayer.addGraphics(bline);
                            bline.draw(pntArr);
                            var line:AnyLine = new AnyLine();
                            var arc:Arc = new Arc();
                            for(var i:int = 0;i<pntArr.length;i++)
                            {
                                arc.Dots[i] = new Dot_2D(pntArr
```

```
[i].x,pntArr[i].y);
                                                    line.Arcs[i]=arc;
                                                }
                                                websel.Geometry=line;
                                                websel.GeomType=(line as IWebGeometry)
.GetGeomType();
                                                break;
                                    case DrawingType.Circle:
                                            varcircle_:IMSCircle=newIMSCircle
(CoordinateType.Logic);
                                                gLayer.addGraphics(circle_);
                                                circle_.draw(pntArr);
                                                var circle:Circle=new Circle();
                                                circle.Center=new Dot_2D();
                                                circle.Center.x=pntArr[0].x;
                                                circle.Center.y=pntArr[0].y;
                                          circle.Radius=Math.sqrt(Math.pow(pntArr
[0].x-pntArr[1].x,2)+Math.pow(pntArr[0].y-pntArr[1].y,2));
                                                websel.Geometry=circle;
                                                websel.GeomType=(circleas IWebGeom-
etry).GetGeomType();
                                                break;
                                    case DrawingType.Rectangle:
        var polygon:IMSRectangle=new IMSRectangle(CoordinateType.Logic);
                                                gLayer.addGraphics(polygon);
                                                polygon.draw(pntArr);
                                                var rec:Rect=new Rect();
                                                    rec.xmax=Math.max(pntArr[0].x,
pntArr[1].x);
                                                    rec.xmin=Math.min(pntArr[0].x,
pntArr[1].x);
                                                    rec.ymax=Math.max(pntArr[0].y,
pntArr[1].y);
                                                    rec.ymin=Math.min(pntArr[0].y,
pntArr[1].y);
                                                websel.Geometry=rec;
```

```
                              websel.GeomType=(rec as IWebGeometry).
GetGeomType();
                              break;
                      case DrawingType.Polygon:
                              var polygon_:IMSPolygon=new
IMSPolygon(CoordinateType.Logic);
                              gLayer.addGraphics(polygon_);
                              polygon_.draw(pntArr);
                              var pol:Polygon=new Polygon();
                              for(var j:int=0;j<pntArr.length;j++)
                              {
                                  pol.Dots[j]=new Dot_2D(pntArr
[j].x,pntArr[j].y);
                              }
                              pol.Dots[pol.Dots.length-1]=pol.
Dots[0];
                              websel.Geometry=pol;
                              websel.GeomType=(pol as IWebGeometry).
GetGeomType();
                              break;
                      default:break;
                  }
                  websel.SelectionType=ESelectionType.SpatialRange;
                  mapsel.SelectParam=websel;
                  if(websel.Geometry! =null){
                      if(websel.GeomType=="AnyLine"){
                          websel.NearDistance=this.vecMap.nearDistanse;
                      }
                      this.vecMap.select(mapsel,this.mapDocDataViewer.
selectCallback);
                  }
              }
          ]]>
    </fx:Script>
```

运行操作效果如图 27.28、图 27.29、图 27.30、图 27.31、图 27.32 所示。

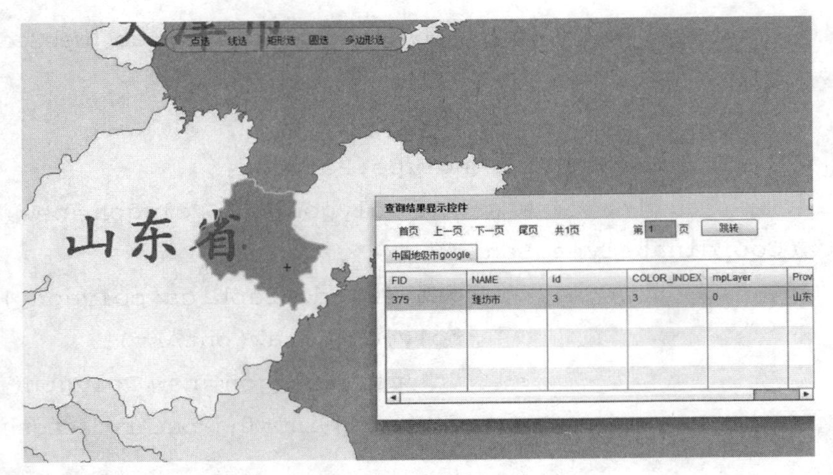

图 27.28　点空间查询

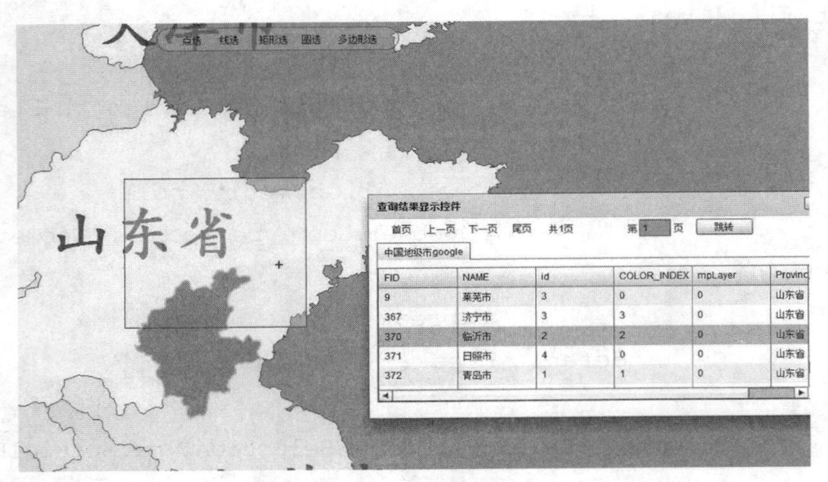

图 27.29　矩形空间查询

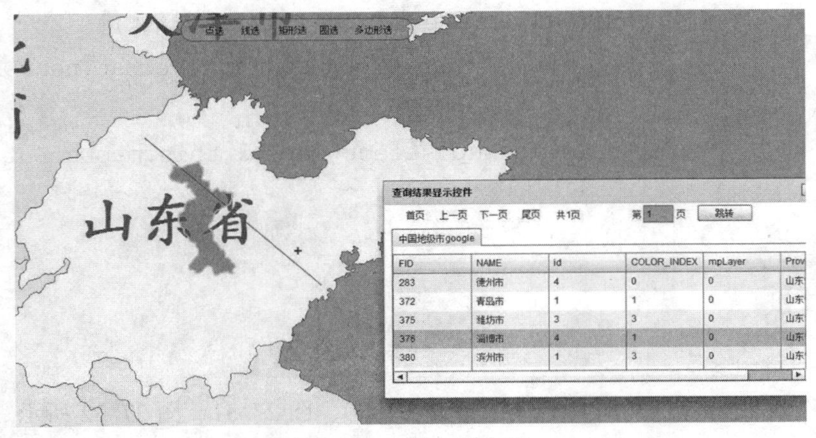

图 27.30　线空间查询

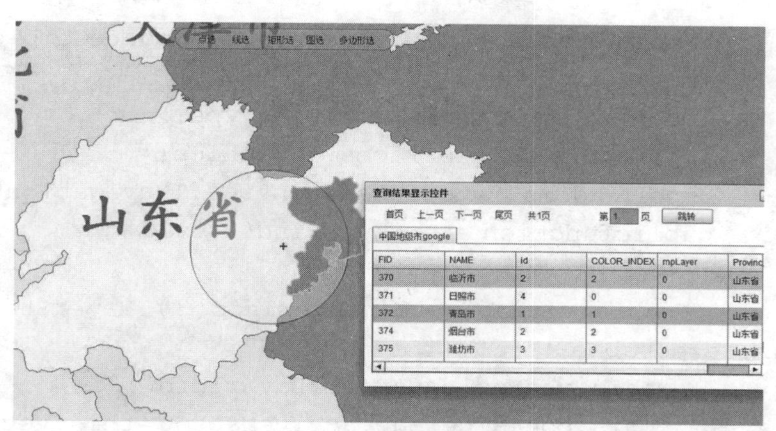

图 27.31　圆空间查询

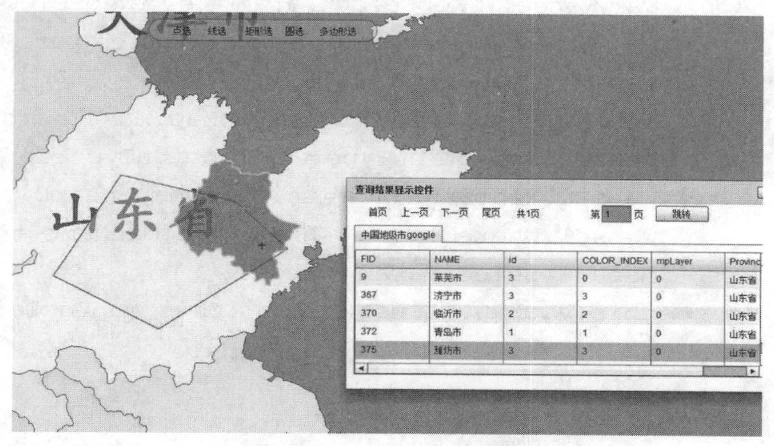

图 27.32　多边形空间查询

27.5.2　属性查询

通过 CWebSelectParam 设置查询参数，参数里设置条件的函数为 WhereClause 进行设置查询的条件，类型设置为 Condition 条件查询。

功能实现示例代码如下：

```
<fx:Script>
      <![CDATA[
          import Mapgis7.WebService.BasLib.CMapSelectParam;
          import Mapgis7.WebService.BasLib.COperResult;
          import Mapgis7.WebService.BasLib.CWebSelectParam;
          import Mapgis7.WebService.BasLib.ESelectionType;
          import Mapgis7.WebService.BasLib.EWebGeomType;
          import Mapgis7.WebService.BasLib.EnumLayerStatus;
          import Mapgis7.WebService.BasLib.IWebGeometry;
```

```
            import flashx.textLayout.events.UpdateCompleteEvent;
            import mx.controls.Alert;
            import zdims.drawing.DrawingType;
            import zdims.interfaces.IGraphics;
            import zdims.interfaces.control.INavigationBar;
            publicfunction attributeFind():void
            {
                this.vecMap.getMapLayerInfo(3).LayerStatus = Enum-
LayerStatus.Selectable;
                this.vecMap.updateAllLayerInfo();
                this.containerMap.activeMapDoc=this.vecMap;
                var mapsel:CMapSelectParam=new CMapSelectParam();
                var websel:CWebSelectParam=new CWebSelectParam();
                websel.CompareRectOnly=this.vecMap.compareRectOnly;
                websel.Geometry=null;
                websel.MustInside=this.vecMap.mustInside;
                websel.SelectionType=ESelectionType.Condition;
                websel.NearDistance=this.vecMap.nearDistanse;
                websel.WhereClause="NAME='"+placeName.text+"'";
                mapsel.SelectParam=websel;
                this.vecMap.select(mapsel,this.mapDocDataViewer.
selectCallback);
            }
        ]]>
    </fx:Script>
```

交互操作界面如图 27.33 所示。

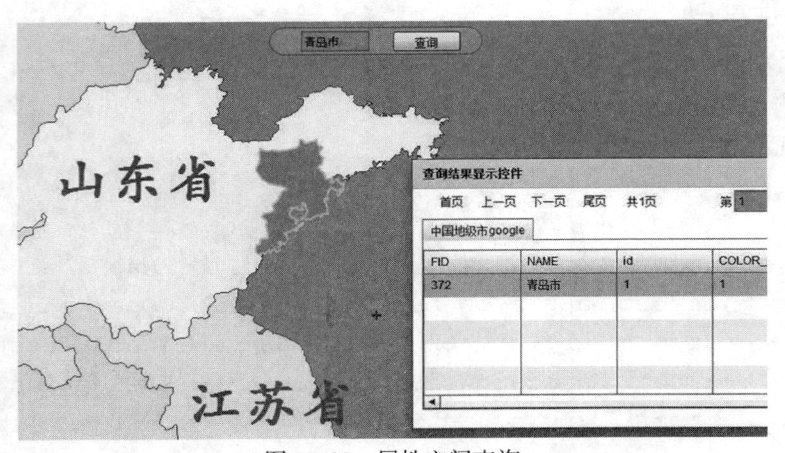

图 27.33　属性空间查询

27.5.3 复合查询

先实现几何查询，要变为几何条件查询，就需要将查询的类型从“SpatialRange”改为“Both”，另外在设置 CWebSelectParam 查询参数的时候，添加一个查询的条件，在 WhereClause 里设置条件，其他的都与几何查询相同。代码与几何查询代码几乎相同，只需将 A 代码段删除，添加 B 代码段即可。WhereClause 属性值可自行改变。

A 代码：websel.SelectionType = ESelectionType.SpatialRange

B 代码：websel.SelectionType = ESelectionType.Both

websel.WhereClause = "**ProvinceName** = '山东省'";

交互操作界面如图 27.34 所示。

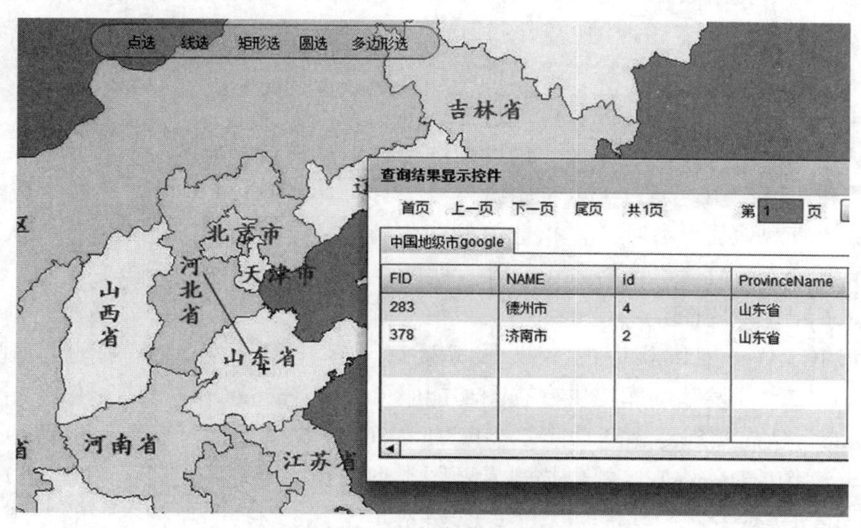

图 27.34 复合查询

第 28 章　MapGIS for Flex 中级开发

28.1　专题图

专题地图是指突出而尽可能完善、详尽地表示制图区域内的一种或几种自然或社会经济(人文)要素的地图。专题地图的制图领域宽广，凡具有空间属性的信息数据都可用其来表示。其内容、形式多种多样，专题地图和普通地图相比，又有自己的特征，包含专题要素的空间分布特征、时态特征、可示特征。

在信息化程度越来越高的今天，借助地图专题信息进行相应的分析预测越来越被各行各业所接受和广泛应用。通过 Flex 提供的专题图接口，用户可以在客户端生成各类专题图，以便更加直观地显示各项关键指标值的变化情况、分布情况、地域特征，从而为行业应用提供重要参考。Flex 可支持各类统计专题图、点密度专题图、分段专题图、等级符号专题图、统一配置专题图、四色专题图、单值专题图、随机专题图等。

MapGIS IGServer 提供相应的 Web 服务接口与客户端开发 API，用于实现专题图。zdims4.5.swc 二次开发基础库提供专题图功能服务接口 ThemeOper 和专题信息结构对象接口 ThemesInfo，从而实现专题图功能。接口 ThemeOper 提供添加、获取、更新和删除方法的对象接口。接口 ThemesInfo 提供描述专题图类型和相关专题图信息的对象接口，与接口 ThemeOper 共同实现专题图功能。无论实现 zdims4.5.swc 支持的哪种类型的专题图，都需要调用 ThemeOper 及其 4 种方法，以及基于接口 ThemesInfo 建立的专题信息对象。通过构建专题信息对象，针对指定图层，可以构造多个专题图，且专题图类型可以不同；针对一个专题图可以构造多个专题信息，而专题信息的意义和表现都可以由开发者根据实际需求设计和实现。因此，专题图功能实现的关键步骤如下：

①调用专题图功能服务接口 ThemeOper(即专题图操作类)；

②调用专题图添加、删除、获取、更新这几种操作专题图的方法；

③调用接口 ThemesInfo 中提供的对象，构建专题图信息对象，作为传递给第一步中调用的方法的对应参数的值。

Flex 提供的专题图 API 对应 MapGIS 10 平台提供的专题图属性设置项，用户可以在 MapGIS 10 平台中打开原地图文档，通过在地图文档上点击右键，在右键菜单中选择【专题图】，根据提示创建对应的专题图，创建完之后，会在右侧自动打开专题图属性窗口，可以在此窗口中对专题图各项属性进行设置，当调整到一个满意的显示结果之后，根据这里的设置项信息，对应在 Flex 代码里设置相关的显示信息就可以了，如四色专题图所用的颜色数目、颜色条等信息。专题图的创建、种类选择、参数设置等功能实现如图 28.1、

图 28.2、图 28.3 所示。

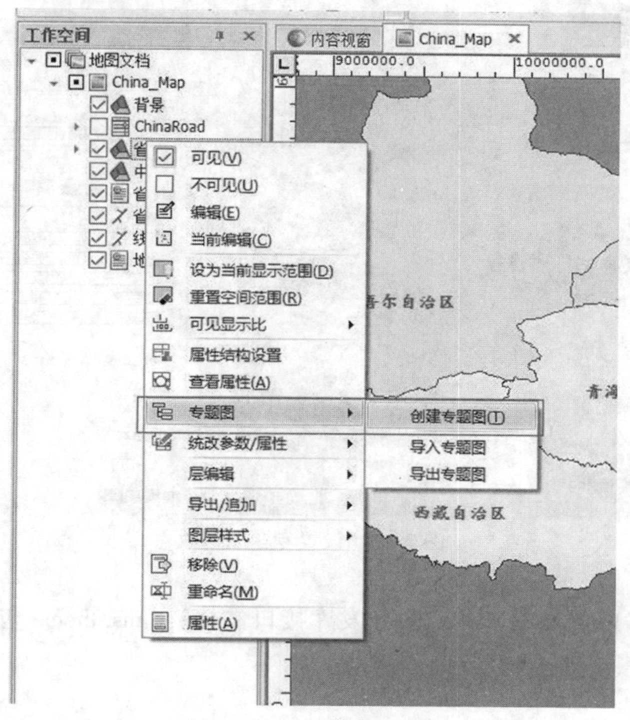

图 28.1　创建专题图

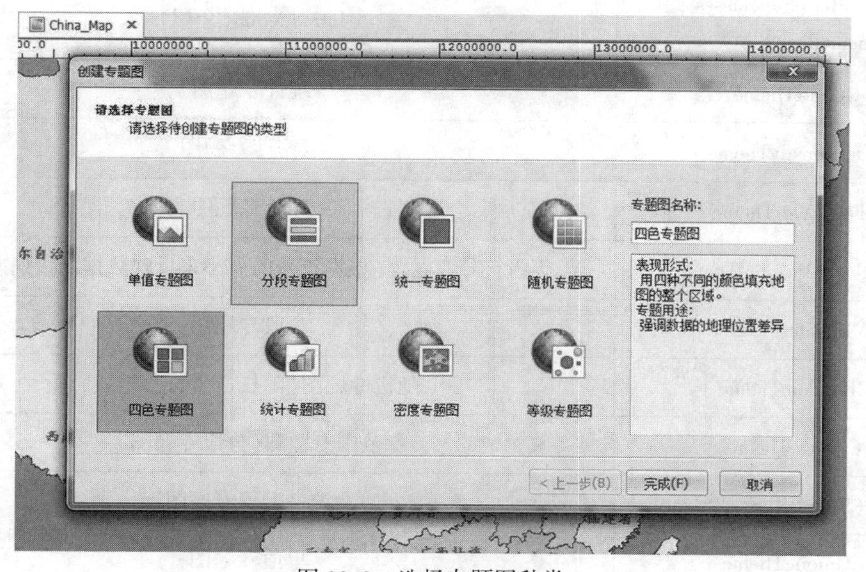

图 28.2　选择专题图种类

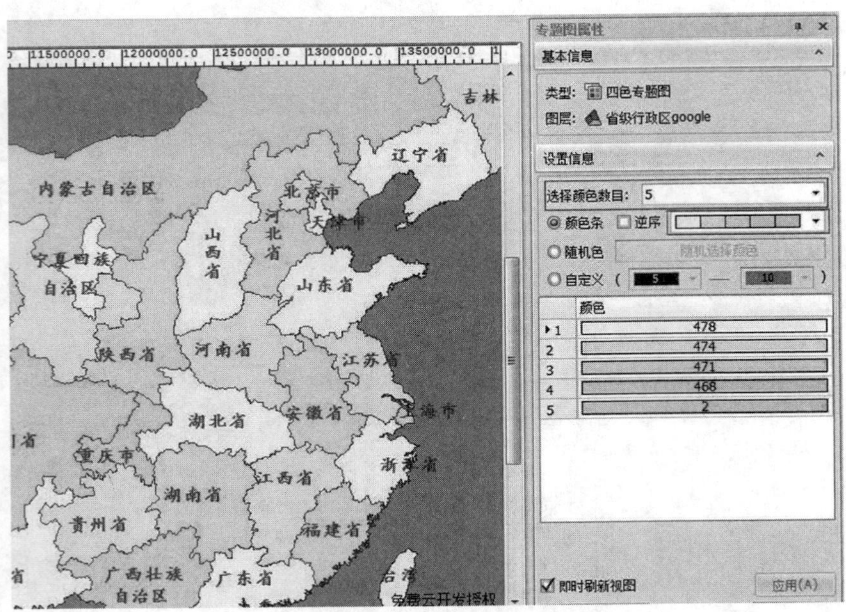

图 28.3　设置专题图参数

相关函数详见 MapGIS《Flex 二次开发库接口文档》zdims. theme 包内相关类。此处仅列出部分类供参考，见表 28.1、表 28.2。

表 28.1　**专题图基类子类**

ThemeBaseSubclasses (ThemeBaseSub 类)	Introduction(介绍)
CChartTheme	统计专题图
CDotDensityTheme	点密度专题图
CFourColorTheme	四色专题图
CGraduatedSymbolTheme	等级符号专题图(根据符号的大小来反映数据的级别差异)
CMultiClassTheme	分段专题图
CRandomTheme	随机专题图(随机产生符号样式)
CRangeTheme	范围专题图(分段专题图)
CSimpleTheme	统一配置专题图
CUniqueTheme	单值专题图

表 28.2　**ThemeOper Public Class 的方法**

ThemeOper Public Class(ThemeOper 类方法)
ThemeOper(ip: String, port: String, clientUID4VectorMapDoc: String = null) 构造方法
addThemesInfo(mapDocName: String, idxArr: String, themesInfoArr: Array, onSuccess: Function): void 添加专题图信息 mapDocName 地图文档名称 idxArr 专题图索引数组(层次从地图开始，索引从 0 开始，例如:"0/0, 1/1, 2/2")themesInfoArr: 添加的数据(ThemesInfo[ ]) onSuccess(themesInfoArr)获取成功回调方法
getThemesInfo(mapDocName: String, idxArr: String, onSuccess: Function): void 获取专题图信息 mapDocName: 地图文档名称 idxArr: 专题图索引数组(层次从地图开始，索引从 0 开始，例如:"0/0, 1/1, 2/2") onSuccess(themesInfoArr) : 获取成功回调方法
removeThemesInfo(mapDocName: String, idxArr: String, onSuccess: Function): void 删除专题图信息 mapDocName : 地图文档名称 idxArr: 专题图索引数组(层次从地图开始，索引从 0 开始，例如:"0/0, 1/1, 2/2") onSuccess(themesInfoArr) : 获取成功回调方法
updateThemesInfo (mapDocName: String, idxArr: String, themesInfoArr: Array, onSuccess: Function): void 更新专题图信息 mapDocName : 地图文档名称 idxArr: 专题图索引数组(层次从地图开始，索引从 0 开始，例如:"0/0, 1/1, 2/2") themesInfoArr: 更新的数据(ThemesInfo[ ]) onSuccess(themesInfoArr) : 获取成功回调方法

下面对统计专题图和分段专题图进行实现介绍，其他类型专题图实现思路大同小异。

28.1.1　统计专题图

统计专题图提供多种统计类型，如直方图、折线图、饼状图等，主要用于分析统计多个数值变量，即地理要素属性字段。前台交互设计界面示例代码如下：

```
<map:IMSMap id="containerMap" x="0" y="0" width="100%"
height="100%"
xMinMap="6933046.115037738" yMinMap="1075039.825021435"
xMaxMap="16345196.216909475" yMaxMap="8134152.401425237">
<map:VectorMapDoc id="vecMap"  x="0" y="0" width="100%"
height="100%"
mapDocName="China" serverAddress="http://127.0.0.1:6163/igs/
rest/ims/relayhandler">
        </map:VectorMapDoc>
</map:IMSMap>
<!--选项控制设置-->
<mx:Canvas id="isotree" x="350" width="241" y="50"
          backgroundColor="#020000" backgroundAlpha="0.67"
```

```
        dropShadowVisible = "true" showEffect = "{show}" hideEffect = "
{hide}"
height = "56" borderColor = "#FDFAFA" borderStyle = "solid">
            <mx:Button x = "19" y = "10" width = "90" height = "34" label = "
生成专题图"
chromeColor = "#EEEEF7" click = "addtheme()" fontFamily = "微软雅黑" />
            <mx:Button x = "140" y = "10" width = "85" height = "34" label =
"删除专题图"
          chromeColor = "#EEEEF7" click = "deleteSect()" fontFamily = "微
软雅黑" />
    </mx:Canvas>
```

功能实现示例代码如下:

```
    <fx:Declarations>
            <!-- 将非可视元素(例如服务、值对象)放在此处 -->
            <s:Resize id = "mouseon" widthTo = "40" heightTo = "50" dura-
tion = "500" />
            <s:Resize id = "mouseout" widthTo = "30" heightTo = "30" dura-
tion = "500" />
            <mx:Glow id = "glow" blurXFrom = "0" blurXTo = "30" blurYFrom =
"0" blurYTo = "30"
                    duration = "1000" />
            <mx:WipeRight id = "show" duration = "1000" />
            <mx:WipeLeft id = "hide" duration = "1000" />
        </fx:Declarations>
        <fx:Script>
            <![CDATA[
                import mx.collections.ArrayCollection;
                import mx.controls.Alert;
                import mx.events.FlexEvent;
                import spark.events.IndexChangeEvent;
                import zdims.othermap.GoogleLayerType;
                import zdims.theme.CAnnInfo;
                import zdims.theme.CChartLabelFormat;
                import zdims.theme.CChartTheme;
                import zdims.theme.CChartThemeInfo;
                import zdims.theme.CChartThemeRepresentInfo;
                import zdims.theme.CChartType;
                import zdims.theme.CPntInfo;
```

```
            import zdims.theme.CRegInfo;
            import zdims.theme.ThemeOper;
            import zdims.theme.ThemesInfo;
            private var oper:ThemeOper;
            protected function toollogo_clickHandler(event:
MouseEvent):void
            {
                isotree.visible=! isotree.visible;
            }
            private function addtheme():void
            {
                oper=new ThemeOper("127.0.0.1","6163",vecMap);
                var themeinfoArr:Array=new Array();
                themeinfoArr[0]=new ThemesInfo();
                themeinfoArr[0].LayerName="省级行政区 google";
                themeinfoArr[0].ThemeArr=new Array();
                themeinfoArr[0].ThemeArr[0]=new CChartTheme();
                themeinfoArr[0].ThemeArr[0].Name="全国市级人口柱状图";
                themeinfoArr[0].ThemeArr[0].ChartType=1;
    themeinfoArr[0].ThemeArr[0].ChartThemeInfoArr=new Array();
      themeinfoArr[0].ThemeArr[0].ChartThemeInfoArr[0]=new CChar-
tThemeInfo();
                themeinfoArr[0].ThemeArr[0].ChartThemeInfoArr[0]
.Expression="人口数";
                themeinfoArr[0].ThemeArr[0].ChartThemeInfoArr[0]
.RegInfo=new CRegInfo();
                themeinfoArr[0].ThemeArr[0].ChartThemeInfoArr[0]
.RegInfo.FillClr=11;
            themeinfoArr [ 0 ] .ThemeArr [ 0 ] .ChartThemeInfoArr [ 0 ]
.RegInfo.OutPenW=1;
      themeinfoArr[0].ThemeArr[0].RepresentInfo=
   new CChartThemeRepresentInfo();
                themeinfoArr[0].ThemeArr[0].RepresentInfo.
IsDrawLabel=true;
                themeinfoArr[0].ThemeArr[0].RepresentInfo.
DigitLabel=2;
                themeinfoArr[0].ThemeArr[0].RepresentInfo.
FormatLabel=
```

```
  CChartLabelFormat.Value;
      themeinfoArr[0].ThemeArr[0].RepresentInfo.AnnInfoLabel=new
CAnnInfo();
                 themeinfoArr[0].ThemeArr[0].RepresentInfo.
AnnInfoLabel.Ovprnt=true;
                 themeinfoArr[0].ThemeArr[0].RepresentInfo.
MaxLength=30;
                 themeinfoArr[0].ThemeArr[0].RepresentInfo.
ThickPersent=5;
                 themeinfoArr[0].ThemeArr[0].RepresentInfo.Width=2;
                 oper.addThemesInfo("China","2",themeinfoArr,onTheme);
             }
             private function onTheme(e):void
             {
                 if(e) { vecMap.refresh();}
                 else{ Alert.show("操作失败！"); }
             }
             public function deleteSect():void
             {vecMap.clear();}
         ]]>
     </fx:Script>
```

实现结果如图 28.4 所示。

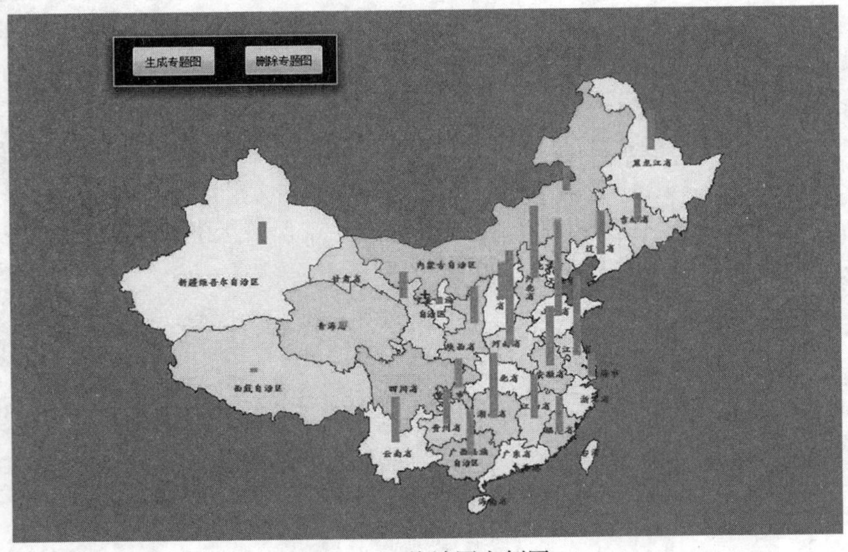

图 28.4 统计图实例图

28.1.2　分段专题图

分段专题图根据每个要素属性值所在的分段范围赋予相对应的显示风格，一般用于分析统计多个数值变量。前台交互设计界面示例代码如下：

```
<map: IMSMap id = " containerMap " x = " 0 " y = " 0 " width = " 100% "
height = "100% "
        xMinMap = "6933046.115037738" yMinMap = "1075039.825021435"
  xMaxMap = "16345196.216909475" yMaxMap = "8134152.401425237">
      <map:VectorMapDoc id = "vecMap"  x = "0" y = "0" width = "100% "
height = "100% "
    mapDocName = " China " serverAddress = " http:// 127.0.0.1: 6163/
igs/rest/ims/relayhandler">
      </map:VectorMapDoc>
    </map:IMSMap>
    <! --选项控制设置-->
  <mx:Canvas id = "isotree" x = "350" width = "241" y = "50"
              backgroundColor = "#020000" backgroundAlpha = "0.67"
              dropShadowVisible = "true"
              showEffect = "{show}" hideEffect = "{hide}" height = "56"
              borderColor = "#FDFAFA" borderStyle = "solid">
    <mx:Button x = "19" y = "10" width = "90" height = "34" label = "生成
专题图" chromeColor = "#EEEEF7" click = "addtheme( )" fontFamily = "微软雅
黑"/>
  <mx:Button x = "140" y = "10" width = "85" height = "34" label = "删除专题
图" chromeColor = "#EEEEF7" click = "deleteSect( )" fontFamily = "微软雅
黑"/>
  </mx:Canvas>
```

功能实现示例代码如下：

```
<fx:Script>
    <! [CDATA[
        import mx.controls.Alert;
        import mx.events.FlexEvent;
        import zdims.othermap.GoogleLayerType;
        import zdims.theme.CDotDensityTheme;
        import zdims.theme.CMultiClassTheme;
        import zdims.theme.CPntInfo;
        import zdims.theme.CRegInfo;
        import zdims.theme.CThemeInfo;
```

```
            import zdims.theme.ExpInfo;
            import zdims.theme.ItemValue;
            import zdims.theme.ThemeOper;
            import zdims.theme.ThemesInfo;
            protected function toollogo_clickHandler(event:
MouseEvent):void
            {
                isotree.visible=! isotree.visible;
            }
    private var oper:ThemeOper;
            private function addtheme():void
            {
    oper=new ThemeOper("127.0.0.1","6163",vecMap);
    oper.removeThemesInfo("China","2",onTheme);
    var themeinfoArr:Array=new Array();
    themeinfoArr[0]=new ThemesInfo();
    themeinfoArr[0].LayerName="省级行政区 google";
    themeinfoArr[0].ThemeArr=new Array();
    themeinfoArr[0].ThemeArr[0]=new CMultiClassTheme();
    themeinfoArr[0].ThemeArr[0].Name="分段专题图";
    themeinfoArr[0].ThemeArr[0].Visible=true;
    themeinfoArr[0].ThemeArr[0].ExpInfoArr=new Array();
    themeinfoArr[0].ThemeArr[0].ExpInfoArr[0]=new ExpInfo();
    themeinfoArr[0].ThemeArr[0].ExpInfoArr[0].Expression="人口数";
    themeinfoArr[0].ThemeArr[0].ExpInfoArr[0].ItemValueArr=new Ar-
ray();
    themeinfoArr[0].ThemeArr[0].ExpInfoArr[0].ItemValueArr[0]=new
ItemValue();
    themeinfoArr[0].ThemeArr[0].ExpInfoArr[0].ItemValueArr[0].Star-
tValue=0;
    themeinfoArr[0].ThemeArr[0].ExpInfoArr[0].ItemValueArr[0].End-
Value=1000000;
    themeinfoArr[0].ThemeArr[0].ExpInfoArr[0].ItemValueArr[0].Clas-
sItemType=2;
    themeinfoArr[0].ThemeArr[0].ExpInfoArr[0].ItemValueArr[1]=new
ItemValue();
    themeinfoArr[0].ThemeArr[0].ExpInfoArr[0].ItemValueArr[1].Star-
tValue=1000000;
```

```
themeinfoArr[0].ThemeArr[0].ExpInfoArr[0].ItemValueArr[1].EndValue=5000000;
themeinfoArr[0].ThemeArr[0].ExpInfoArr[0].ItemValueArr[1].ClassItemType=2;
themeinfoArr[0].ThemeArr[0].ExpInfoArr[0].ItemValueArr[2]=new ItemValue();
themeinfoArr[0].ThemeArr[0].ExpInfoArr[0].ItemValueArr[2].StartValue=5000000;
themeinfoArr[0].ThemeArr[0].ExpInfoArr[0].ItemValueArr[2].EndValue=10000000;
themeinfoArr[0].ThemeArr[0].ExpInfoArr[0].ItemValueArr[2].ClassItemType=2;
themeinfoArr[0].ThemeArr[0].ExpInfoArr[0].ItemValueArr[3]=new ItemValue();
themeinfoArr[0].ThemeArr[0].ExpInfoArr[0].ItemValueArr[3].StartValue=10000000;
themeinfoArr[0].ThemeArr[0].ExpInfoArr[0].ItemValueArr[3].EndValue=40000000;
themeinfoArr[0].ThemeArr[0].ExpInfoArr[0].ItemValueArr[3].ClassItemType=2;
themeinfoArr[0].ThemeArr[0].ExpInfoArr[0].ItemValueArr[4]=new ItemValue();
themeinfoArr[0].ThemeArr[0].ExpInfoArr[0].ItemValueArr[4].StartValue=40000000;
themeinfoArr[0].ThemeArr[0].ExpInfoArr[0].ItemValueArr[4].EndValue=80000000;
themeinfoArr[0].ThemeArr[0].ExpInfoArr[0].ItemValueArr[4].ClassItemType=2;
themeinfoArr[0].ThemeArr[0].ExpInfoArr[0].ItemValueArr[5]=new ItemValue();
themeinfoArr[0].ThemeArr[0].ExpInfoArr[0].ItemValueArr[5].StartValue=80000000;
themeinfoArr[0].ThemeArr[0].ExpInfoArr[0].ItemValueArr[5].EndValue=120000000;
themeinfoArr[0].ThemeArr[0].ExpInfoArr[0].ItemValueArr[5].ClassItemType=2;
themeinfoArr[0].ThemeArr[0].GeoInfoType="Reg";
themeinfoArr[0].ThemeArr[0].MultiClassThemeInfoArr=new Array();
```

```
    themeinfoArr[0].ThemeArr[0].MultiClassThemeInfoArr[0] = new
CThemeInfo();
    themeinfoArr[0].ThemeArr[0].MultiClassThemeInfoArr[0].Caption="
0~100万";
    themeinfoArr[0].ThemeArr[0].MultiClassThemeInfoArr[0].IsVisi-
ble=true;
    themeinfoArr[0].ThemeArr[0].MultiClassThemeInfoArr[0].RegInfo=
new CRegInfo();
    themeinfoArr[0].ThemeArr[0].MultiClassThemeInfoArr[0].
RegInfo.FillClr=15;
    themeinfoArr[0].ThemeArr[0].MultiClassThemeInfoArr[1] = new
CThemeInfo();
    themeinfoArr[0].ThemeArr[0].MultiClassThemeInfoArr[1].Caption="
100万~500万";
    themeinfoArr[0].ThemeArr[0].MultiClassThemeInfoArr[1].IsVisi-
ble=true;
    themeinfoArr[0].ThemeArr[0].MultiClassThemeInfoArr[1].RegInfo=
new CRegInfo();
    themeinfoArr[0].ThemeArr[0].MultiClassThemeInfoArr[1].
RegInfo.FillClr=14;
    themeinfoArr[0].ThemeArr[0].MultiClassThemeInfoArr[2] = new
CThemeInfo();
    themeinfoArr[0].ThemeArr[0].MultiClassThemeInfoArr[2].Caption="
500万~1000万";
    themeinfoArr[0].ThemeArr[0].MultiClassThemeInfoArr[2].IsVisi-
ble=true;
    themeinfoArr[0].ThemeArr[0].MultiClassThemeInfoArr[2].RegInfo=
newCRegInfo();
    themeinfoArr[0].ThemeArr[0].MultiClassThemeInfoArr[2].
RegInfo.FillClr=13;
    themeinfoArr[0].ThemeArr[0].MultiClassThemeInfoArr[3] = new
CThemeInfo();
    themeinfoArr[0].ThemeArr[0].MultiClassThemeInfoArr[3].Caption="
1000万~4000万";
    themeinfoArr[0].ThemeArr[0].MultiClassThemeInfoArr[3].IsVisi-
ble=true;
    themeinfoArr[0].ThemeArr[0].MultiClassThemeInfoArr[3].RegInfo=
new CRegInfo();
```

```
    themeinfoArr [ 0 ]. ThemeArr [ 0 ]. MultiClassThemeInfoArr [ 3 ].
RegInfo.FillClr=12;
    themeinfoArr[0].ThemeArr[0].MultiClassThemeInfoArr[4] = new
CThemeInfo();
    themeinfoArr[0].ThemeArr[0].MultiClassThemeInfoArr[4].Caption="
4000万~8000万"
    themeinfoArr[0].ThemeArr[0].MultiClassThemeInfoArr[4].IsVisi-
ble=true;
    themeinfoArr[0].ThemeArr[0].MultiClassThemeInfoArr[4].RegInfo=
new CRegInfo();
    themeinfoArr [ 0 ]. ThemeArr [ 0 ]. MultiClassThemeInfoArr [ 4 ].
RegInfo.FillClr=11;
    themeinfoArr[0].ThemeArr[0].MultiClassThemeInfoArr[5] = new
CThemeInfo();
    themeinfoArr[0].ThemeArr[0].MultiClassThemeInfoArr[5].Caption="
8000万~12000万";
    themeinfoArr[0].ThemeArr[0].MultiClassThemeInfoArr[5].IsVisi-
ble=true;
    themeinfoArr[0].ThemeArr[0].MultiClassThemeInfoArr[5].RegInfo=
newCRegInfo();
    themeinfoArr [ 0 ]. ThemeArr [ 0 ]. MultiClassThemeInfoArr [ 5 ].
RegInfo.FillClr=10;
    themeinfoArr[0].ThemeArr[0].DefaultInfo=new CThemeInfo();
    themeinfoArr[0].ThemeArr[0].DefaultInfo.RegInfo=new CRegInfo();
    themeinfoArr[0].ThemeArr[0].DefaultInfo.RegInfo.FillClr=1;
    oper.addThemesInfo("China","2",themeinfoArr,onTheme);
            }
    private function onTheme(e):void
            {
                if(e) { vecMap.refresh(); }
                else { Alert.show("操作失败! "); }
            }
            public function deleteSect():void
            {  vecMap.clear();  vecMap.refresh(); }
        ]]>
    </fx:Script>
```

示例实现结果如图 28.5 所示。

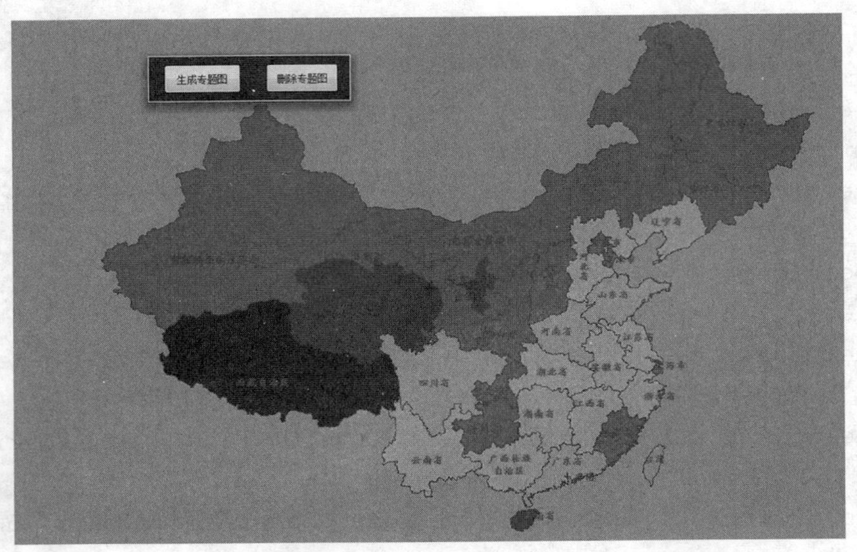

图 28.5　分段专题图实例图

28.2　空间分析

GIS 与一般电子地图最重要的区别之一，就是提供强大的查询统计、空间分析功能，而这些特性让其在各行业领域应用中发挥着重要作用，为生产生活提供了更多的便利与服务。

从 GIS 应用角度看，空间分析大致可以归纳为两大类：①基于点、线、面基本地理要素的空间分析，通过空间信息查询与量测、缓冲区分析、叠加分析、网络分析、地理统计分析等空间分析方法挖掘出新的信息；②地理问题模拟，解决应用领域对空间数据处理与输出的特殊要求，地理实体和空间关系通过专业模型得到简化和抽象，而系统则通过模型进行深入分析操作。

MapGIS 二次开发提供了大量的集成控件，所以实现空间分析功能，只需要调用相应控件，配置相应的参数即可。且空间分析控件是开源的，用户可以根据自己的需求进行重写。

28.2.1　缓冲区分析

要实现图层查询功能，先拖入 LayerDataViewer，再拖入一个缓冲区分析 BufferAnalyse 控件，将 LayerDataViewer 和 BufferAnalyse 进行关联，查询完成后，再进行缓冲区分析。前台交互设计界面示例代码如下：

```
<map:IMSMap id="mymap" x="0" y="0" width="100%" height="100%"
xMinMap="6933046.115037738" yMinMap="1075039.825021435"
xMaxMap="16345196.216909475" yMaxMap="8134152.401425237">
        <map:VectorMapDoc id="mapDoc"  x="0" y="0" width="100%"
```

```
height="100%"
   mapDocName="China" serverAddress="http://127.0.0.1:6163/igs/
rest/ims/relayhandler">
         </map:VectorMapDoc>
      <s:Button x="10" y="87" label="第1个图层要素缓冲区分析"
   click="buffer_onclick(event)"/>
         <drawing:GraphicsLayer x="42" y="188" id="graphicsLayer">
         </drawing:GraphicsLayer>
      <layer:LayerDataViewer x="332" y="10" id="layerData"
imsmap="{mymap}">
         </layer:LayerDataViewer>
         <control:BufferAnalyse x="0" y="192" id="buffer"
          layerDataViewer="{layerData}" imsmap="{mymap}">
         </control:BufferAnalyse>
   </map:IMSMap>
```

功能实现示例代码如下：

```
<fx:Script>
            <![CDATA[
               import Mapgis7.WebService.BasLib.CAttDataRow;
               import Mapgis7.WebService.BasLib.CAttDataSet;
               import Mapgis7.WebService.BasLib.CAttDataTable;
               import Mapgis7.WebService.BasLib.CGetObjByID;
               import Mapgis7.WebService.BasLib.CMapSelectAndGetAtt;
               import Mapgis7.WebService.BasLib.CMapSelectParam;
               import Mapgis7.WebService.BasLib.COperResult;
               import Mapgis7.WebService.BasLib.CWebSelectParam;
               import Mapgis7.WebService.BasLib.Dot_2D;
               import Mapgis7.WebService.BasLib.ESelectionType;
               import Mapgis7.WebService.BasLib.EWebGeomType;
               import Mapgis7.WebService.BasLib.EnumLayerStatus;
               import Mapgis7.WebService.BasLib.IWebGeometry;
               import Mapgis7.WebService.BasLib.Polygon;
               import mx.controls.Alert;
               import zdims.drawing.DrawingType;
               import zdims.interfaces.IGraphics;
               import zdims.interfaces.control.INavigationBar;
               public function buffer_onclick(event:MouseEvent):
void
```

```
                {
            this.mapDoc.getMapLayerInfo(2).LayerStatus=Enum-
LayerStatus.Editable;
            this.mapDoc.updateAllLayerInfo();
            this.graphicsLayer.drawingType=DrawingType.Point;
            this.graphicsLayer.drawingOverCallback=Drawing-
OverCallback;
                }
                public functionDrawingOverCallback
                (gLayer:GraphicsLayer,graphics:IGraphics,pntArr:
Vector.<Point>):void
                {
                    var websel:CWebSelectParam=new CWebSelect-
Param();
                    websel.SelectionType=ESelectionType.
SpatialRange;
                    websel.GeomType=EWebGeomType.Point;
                    var pnt:Dot_2D=new Dot_2D();
                    pnt.x=pntArr[0].x;
                    pnt.y=pntArr[0].y;
                    websel.Geometry=pnt;
                    var mapsel:CMapSelectParam=new CMapSelect-
Param();
                    mapsel.SelectParam=websel;
                    this.mapDoc.select(mapsel,callBack);
                }
                public function callBack(e:Event):void{
                    this.mymap.activeMapDoc=this.mapDoc;
                    this.mymap.activeMapDoc.activeLayerIndex=1;
                    var result:CMapSelectAndGetAtt=this.mapDoc.
onSelect(e);
                    if(result==null || result.Count[0][1]==0)
                    {
                        return;
                    }
                    else
                    {
                        var targetObj:CGetObjByID=new CGetObj-
```

```
ByID();
            targetObj.FeatureID = (((result.AttDS[0] as CAttDataSet)
.attTables[1] asCAttDataTable).Rows[0] as CAttDataRow).FID;
                                targetObj.LayerIndex = 1;
                                this.buffer.setTargetFeature(targetObj);
                                this.buffer.visible = true;
                            }
                        }
                    ]]>
                </fx:Script>
```

28.2.2　网络分析

首先显示地图文档，然后添加一个 NetAnalyse 网络分析控件进行路径分析，在地图上设置相应的路径点和障碍点即可进行分析。网络分格 UI 界面设计如图 28.6 所示。

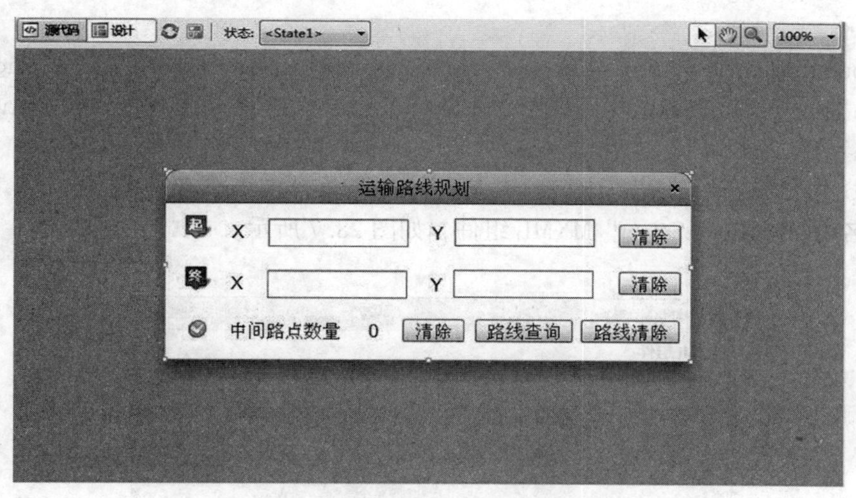

图 28.6　网络分析 UI 界面设计

Web 端交互设计示例代码如下：

```
<map:IMSMap id="mapContainer" width="100%" height="100%" leve-
lNum="10" initShowLevel="0" enableProgressBar="false"  enableAni-
mate="false"
restoreCenterX = " 11574400.571054548 "  restoreCenterY = " 4363637.
0707441345">
        <othermap:GoogleTileLayer  x = "0" y = "0" width = "100%"
height="100%"
          id = "googleMap" startLevel = "4" viewBeginLevel = "0" view-
```

```
EndLevel="22"
        googleLayerType="{GoogleLayerType.BasicMap}" creationComplete=
        "googleMapInit()">
        </othermap:GoogleTileLayer>
            <map:VectorMapDoc  x="0" y="0" display="false" id="vectorMap"
        mapDocName="China"enableFillImg="true" enableZoomAnimation="true"
        serverAddress="http://127.0.0.1:6163/igs/rest/ims/relayhandler">
        </map:VectorMapDoc>
        <drawing:GraphicsLayer x="0" y="0"  id="graphicsLayer"/>
        <mark:MarkLayer x="0" y="0" id="mk"/>
        <mx:Image id="routeSelectBtn" x="195" y="3" width="32" height="32"
        buttonMode="true"source="@ Embed('img/widget/car1.png')"
        toolTip="查询最优路线" click="routeSelectBtn_clickHandler(event)"/>
    </map:IMSMap>
```

新建名为"BestRouteSelect" MXML组件，如图28.7所示。

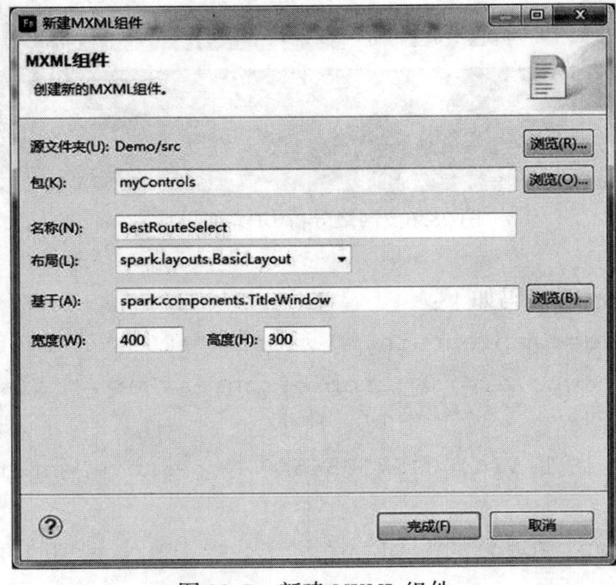

图28.7　新建MXML组件

组件设计示例代码如下：

```
<? xml version="1.0" encoding="utf-8"? >
<s:TitleWindow xmlns:fx="http://ns.adobe.com/mxml/2009"
xmlns:s="library://ns.adobe.com/flex/spark"
xmlns:mx="library://ns.adobe.com/flex/mx"
width="478" height="178" cornerRadius="13" textAlign="center"
fontSize="17" title="运输路线规划"
creationComplete="init()">
    <s:Image x="18" y="11" source="@ Embed('img/widget/qidi-
an.gif')" click="startPoint()"/>
    <s:Image x="18" y="59" source="@ Embed('img/widget/zhongdi-
an.gif')" click="endPoint()"/>
    <s:Image x="18" y="102" source="@ Embed('img/widget/v1.gif
')" click="middlePoint()"/>
    <s:TextInput id="aXId" x="93" y="14" width="127"/>
    <s:TextInput id="aYId" x="262" y="13" width="127"/>
    <s:TextInput id="bXId" x="92" y="62" width="127"/>
    <s:TextInput id="bYId" x="261" y="61" width="127"/>
    <s:Label x="55" y="62" width="19" height="27" text="X" ver-
ticalAlign="middle"/>
    <s:Label id="midCount" x="174" y="106" width="29" height="
27" text="0" verticalAlign="middle"/>
    <s:Label x="46" y="105" width="123" height="27" text="中间路
点数量" verticalAlign="middle"/>
    <s:Label x="236" y="63" width="19" height="27" text="Y" ver-
ticalAlign="middle"/>
    <s:Label x="56" y="14" width="19" height="27" text="X" ver-
ticalAlign="middle"/>
    <s:Label x="237" y="15" width="19" height="27" text="Y" ver-
ticalAlign="middle"/>
    <s:Button x="281" y="107" label="路线查询" click="submit()"/>
  <s:Button x="411" y="18" width="56" label="清除" click="start-
Clear(event)"/>
    <s:Button x="411" y="62" width="56" label="清除" click="end-
Clear(event)"/>
    <s:Button x="215" y="107" width="56" label="清除" click="
middleClear(event)"/>
    <s:Button x="377" y="107" width="89" label="路线清除"
```

```
click="clearRoad()"/>
    </s:TitleWindow>
```

发布地图文档中网络数据集如图 28.8 所示。

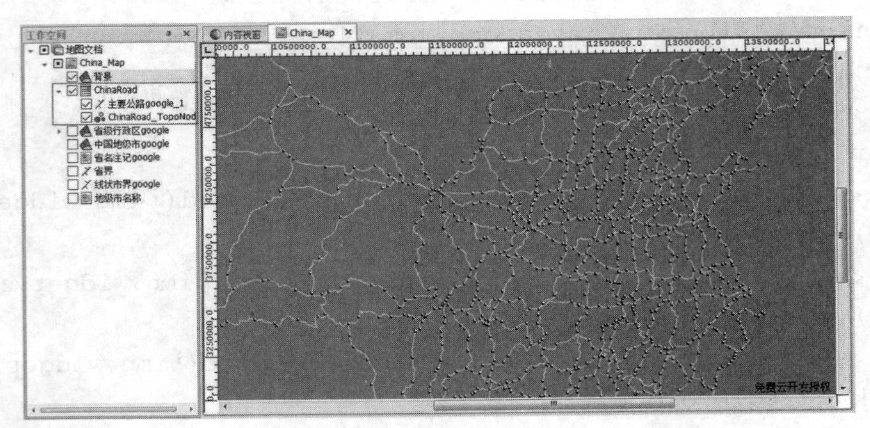

图 28.8 发布网络数据集

功能实现示例代码如下：

```
protected function googleMapInit():void
{
    googleLayerWin= new GoogleLayerWin();
    googleLayerWin.id ="googleSetWin";
    googleLayerWin.y = 200;
    googleLayerWin.x = 450;
    googleLayerWin.imsmap =this.mapContainer;
    googleLayerWin.googletilelayer =this.googleMap;
    this.addElement(googleLayerWin);
    googleLayerWin.visible=false;
}
protected function routeSelectBtn _ clickHandler ( event: Mou-
seEvent):void
{
    if(bestRouteSelect==null){
        bestRouteSelect=new BestRouteSelect();
        bestRouteSelect.imsmap=this.mapContainer;
 bestRouteSelect.imMapDoc=this.vectorMap;
 bestRouteSelect.addEventListener ( CloseEvent.CLOSE, removeBe-
stRouteWindow);
        PopUpManager.addPopUp(bestRouteSelect,this,false);
```

```
        PopUpManager.centerPopUp(bestRouteSelect);
    }elseif(bestRouteSelect.visible==false){
        bestRouteSelect.visible=true;
    }else{
        bestRouteSelect.visible=false;
    }
}
<fx:Script>
        <![CDATA[
            import Mapgis7.WebService.BasLib.CGdbInfo;
            import Mapgis7.WebService.BasLib.CNetAnalyse;
            import Mapgis7.WebService.BasLib.CNetEdge;
            import Mapgis7.WebService.BasLib.CNetInputDot;
            import Mapgis7.WebService.BasLib.CNetNode;
            import Mapgis7.WebService.BasLib.CNetPath;
            import Mapgis7.WebService.BasLib.CPathAnalyzeResult;
            import Mapgis7.WebService.BasLib.Dot_2D;
            import components.Mark;
            import mx.controls.Alert;
            import mx.core.UIComponent;
            import mx.managers.PopUpManager;
            import mx.rpc.events.ResultEvent;
            import zdims.control.IMSRoad;
            import zdims.control.Marker;
            import zdims.control.NetAnalyse;
            import zdims.event.IMSMapEvent;
            import zdims.map.IMSMap;
            import zdims.map.VectorMapDoc;
            import zdims.util.SpacialAnalyse;
            private var _map:IMSMap;
            private var netAnalyse:NetAnalyse;
            private var mapDoc:VectorMapDoc;
            [Inspectable(category="MapGisIMS")]
            public function set imsmap(m:IMSMap):void
            {
                this._map = m;
            }
            [Inspectable(category="MapGisIMS")]
```

```
        public function get imsmap():IMSMap
        {
        returnthis._map;
        }
        public function set imMapDoc(m:VectorMapDoc):void{
            this.mapDoc=m;
        }
        public function get imMapDoc():VectorMapDoc{
            returnthis.mapDoc;
        }
        public var GDBName:String="sample";
        [Inspectable("MapGisIMS")]
        public var GDBSvrName:String="MapGisLocal";
    [Inspectable("MapGisIMS")]
        public var Password:String="";
        [Inspectable("MapGisIMS")]
        public var User:String="";
        [Inspectable("MapGisIMS")]
        public var NetLayerName:String="ChinaRoad";
        private var road:IMSRoad;
        private var inputDot:Array=new Array();
        private var _spatial:SpacialAnalyse;
        private var pathDots:String="";
        private var pathDotsMiddle:String="";
        private var SuccessFlag:Boolean=true;
        public function init():void{
            Mapinit();
        }
        private function MapMoveTo(e:Event):void{
            this._map.panTo(114.3026612133789,30.547545303710937);
            this._map.removeEventListener(IMSMapEvent.MAP_RE-
SIZEOVER,MapMoveTo);
        }
        public function startPoint():void{
                this._map.addEventListener(MouseEvent.CLICK,
pointCoorA);
        }
        private function pointCoorA(e:MouseEvent):void{
```

```
            addMarker(this._map.mouseDownLogicPnt.x,
    this._map.mouseDownLogicPnt.y,"路径点","起始点",
 "image/bus/qidian.gif");
                aXId.text=this._map.mouseDownLogicPnt.x.toString();
                aYId.text=this._map.mouseDownLogicPnt.y.toString();
                this._map.removeEventListener(MouseEvent.CLICK,
pointCoorA);
            }
            public function endPoint():void{
                this._map.addEventListener(MouseEvent.CLICK,
pointCoorB);
            }
            private function pointCoorB(e:MouseEvent):void{
            addMarker(this._map.mouseDownLogicPnt.x,this._map.
mouseDownLogicPnt.y,
"路径点","目的点","image/bus/zhongdian.gif");
                bXId.text=this._map.mouseDownLogicPnt.x.toString();
                bYId.text=this._map.mouseDownLogicPnt.y.toString();
                this._map.removeEventListener(MouseEvent.CLICK,
pointCoorB);
            }
            public function middlePoint():void{
                this._map.addEventListener(MouseEvent.CLICK,
pointCoorC);
            }
   private function pointCoorC(e:MouseEvent):void{
            addMarker(this._map.mouseDownLogicPnt.x,
   this._map.mouseDownLogicPnt.y,"路径点","路径点","img/widget/
v1.gif");
                pathDotsMiddle+=this._map.mouseDownLogicPnt.x+","+
                        this._map.mouseDownLogicPnt.y+",";
                var x:int=new int(midCount.text);
                midCount.text=(++x).toString();
                this._map.removeEventListener(MouseEvent.CLICK,
pointCoorC);
            }
            public function addMarker
   (x:Number,y:Number,name:String,toolTip:String,iconSrc:String,
```

```
            showInfo:Boolean=false,markerConTitle:String="",
    markerContent:String="",markerConResponse:
    String="",markerConReTel:String=""):void{
            var pathDot1:Marker = new Marker();
            pathDot1.imsmap=this._map;
            pathDot1.logicX =x;
            pathDot1.logicY =y;
            pathDot1.name = name;
            pathDot1.toolTip=toolTip;
            pathDot1.setIconSrc(iconSrc);
            if(showInfo){
                pathDot1.setMarkerContent("",
        markerConTitle,markerContent,markerConResponse,marker-
ConReTel);
            }
            this._map.addChildAt(pathDot1,this._map.layerLenth);
        }
        private function  Mapinit():void{
            this.road=new IMSRoad();
                this.road.imsmap=this._map;
        }
        public function submit():void
        {
        if(aXId.text=="" ‖ aYId.text=="" ‖ bXId.text=="" ‖ bY-
Id.text=="")
{
                Alert.show("请设置起始点、终止点。","操作提示:");
            }else{
                pathDots=aXId.text+","+aYId.text+","+pathDots
Middle+
                bXId.text+","+bYId.text+",";
                if(SuccessFlag==true){
                    if (this.mapDoc == null ‖ ! this.mapDoc.
isLoadSucc)
                    {
        Alert.show("没有激活的图层。或者地图服务器未启动", "提示");
                        return;
                    }
```

```
                this.road.clear();
                this.road._roadCoorArr = "";
    this.road._stopIcon = new Array();
                var pathPnts:String=new String();
            pathPnts=this.pathDots.substring(0, this.
pathDots.length -2);
                var gdb:CGdbInfo = new CGdbInfo();
                gdb.GDBName = this.GDBName;
                gdb.GDBSvrName = this.GDBSvrName;
                gdb.Password = this.Password;
                gdb.User = this.User;
                var obj:CNetAnalyse = new CNetAnalyse();
                obj.GdbInfo = gdb;
                obj.NetLayerName = this.NetLayerName;
                obj.RequestDots = pathPnts;
                obj.BarrierDots ="";
                obj.FlgType = "line";
                this._spatial = new SpacialAnalyse(this.mapDoc);
                this._spatial.netAnalyse(obj,onSubmit);
                SuccessFlag=false;
            }
        }
    }
    public function onSubmit(e:Event):void{
        var obj:CPathAnalyzeResult=new CPathAnalyzeResult();
        obj= this._spatial.onNetAnalyse(e);
        Alert.show("路线分析成功。");
        if (obj == null || obj.Paths == null)
        {
            Alert.show("路线分析失败。");
            return;
        }
        var path:CNetPath = CNetPath(obj.Paths[0]);
        var edgeNum:int = path.Edges.length;
        this.road._roadCoorArr+=aXId.text+","+aYId.text+",";
        for (var i:int = 0; i < edgeNum; i++)
        {
            var edge:CNetEdge = CNetEdge(path.Edges[i]);
```

```
                    var dotNum:int = edge.Dots.length;
                    for (var j:int = 0; j < dotNum; j++)
                    {
                        var dot:Dot_2D = Dot_2D(edge.Dots[j]);
                        this.road._roadCoorArr += dot.x + "," + dot.y +
",";
                    }
                }
                        this.road._roadCoorArr += bXId.text + "," +
bYId.text+",";
this.road._roadCoorArr = this.road._roadCoorArr.substring(0,
                this.road._roadCoorArr.length -2);
                this.road.drawRoad();
                SuccessFlag=true;
            }
            private function onResult(evt:ResultEvent):void{
                if(evt.result as Boolean){
                    Alert.show("数据提交成功。");
                }
            }
        public function clearRoad():void{
                this.road.clear();
                this.road._roadCoorArr = "";
                this.road._stopIcon = new Array();
                aXId.text="";
                aYId.text="";
                bXId.text="";
                bYId.text="";
                midCount.text="0";
                clearPathDot();
            }
            public function clearPathDot():void
            {
                var mark:Array = this._map.getChildren();
                for (var i:int = 0; i < mark.length; i++)
                {
        if (mark[i] is Marker&&UIComponent(mark[i]).name == "路径
点")
```

```
                this._map.removeChild(mark[i]);
            }
            SuccessFlag=true;
        }
        protected function startClear(event:MouseEvent):void
        {
            aXId.text="";
            aYId.text="";
            this.clearPathDotByToolTip("起始点");
        }
        protected function endClear(event:MouseEvent):void
        {
            bXId.text="";
            bYId.text="";
            this.clearPathDotByToolTip("目的点");
        }
    protected function middleClear(event:MouseEvent):void
        {
            pathDotsMiddle="";
            midCount.text="0";
            this.clearPathDotByToolTip("路径点");
        }
          public function clearPathDotByToolTip(tip:String):
void
        {
            var mark:Array = this._map.getChildren();
            for (var i:int = 0; i < mark.length; i++)
            {
                if (mark[i] is Marker&&UIComponent(mark[i]).
name ==
    "路径点"&&UIComponent(mark[i]).toolTip==tip)
                this._map.removeChild(mark[i]);
            }
        }
    ]]>
</fx:Script>
```

交互界面操作如图 28.9 所示。

图 28.9　网络分析功能实现

第 29 章　MapGIS for Flex 高级开发

29.1　Flex 与 Web 服务器交互

Flex 可以通过与 Web 服务器交互实现与远程数据的同步调用与存储，可以快速方便地实现用户交互。Flex 提供了 3 个类以实现与服务器端的通信：HTTPService、RemoteObject 和 WebService。另外，还可以根据外部中间插件来实现 Flex 与服务器端的通信。

①HTTPService 类用于超文本传输协议(HTTP)实现与服务器的通信。Flex 应用程序用 GET 或 POST 请求将数据发送到服务端，并处理该请求所返回的 XML 或字符串。使用这个 HTTPService 类，可以与 PHP 页面、ColdFusion 页面、JSP 页面、Javaservlet、RubyonRails 以及 MicrosoftASP 页面进行通信。

②RemoteObject 类可以与服务器之间通过 ActionScriptMessageFormat(AMF)对象进行通信。通常情况下，也可以把 Blazeds 与 Lcds 归于这一类。RemoteObject 也可以与 java 或 coldFusion 远程网关进行通信，或者通过开源项目(例如 AMFPHP. SabreAMF 或 WebORB)与 . NET 和 PHP 进行通信。

③WebService 类与 Web 服务进行通信，使用基于 SOAP 的 XML，Web 服务通过 Web 服务描述语言(WSDL)定义其接口。

④Socket 类可以实现直接与应用程序通信，而不需要在 Web 的基础上进行。通信的方法与 Java 或 . NET 之间进行 Socket 的方法大致是一样的。

29.1.1　基于 FluorineFx 模板的服务器搭建

采用基于 FluorineFx 的通信方式的优点是，这种方式使得客户端与服务器端交互变得简单高效，且有效地实现了功能与数据的分离。

① 安装 VS2010 的项目模板：下载链接：http：// download. csdn. net/detail/naruto12345/4082301。安装模板，建立项目(项目建立后，点击【引用】文件夹，发现 Fluorine. dll 等几个动态库都出现黄色警示，此时需要手动添加，需手动重新添加的动态库即在 Fluorine 的安装目录下的 framework \ 3. 5 下找到)。

②在解决方案资源管理器中右击【Console. aspx】选择【在浏览器中查看】，效果如图 29. 1 所示。

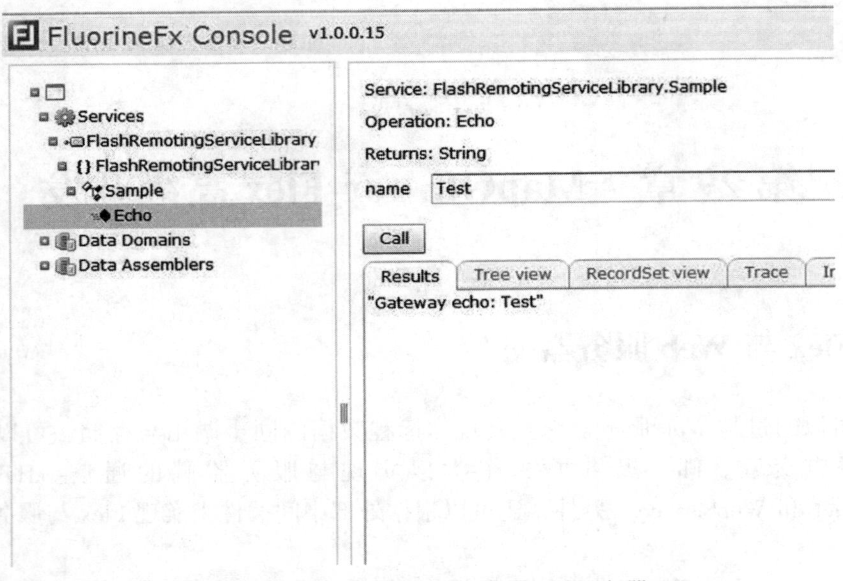

图 29.1 搭建基于 FluorineFx 模板的服务器

③打开 Adobe Flash builder4，首先创建(或修改项目属性)Flex 项目，并将项目路径指向前面建立的 FluorineFx 网站的根路径：FlashRemotingWebApplication 文件夹，应用程序类型为 Web，应用程序服务器类型为 ASP. NET，Web 应用程序 URL，其端口如图 29. 2 中框出部分，不同电脑的情况不同，如图 29. 3 所示。

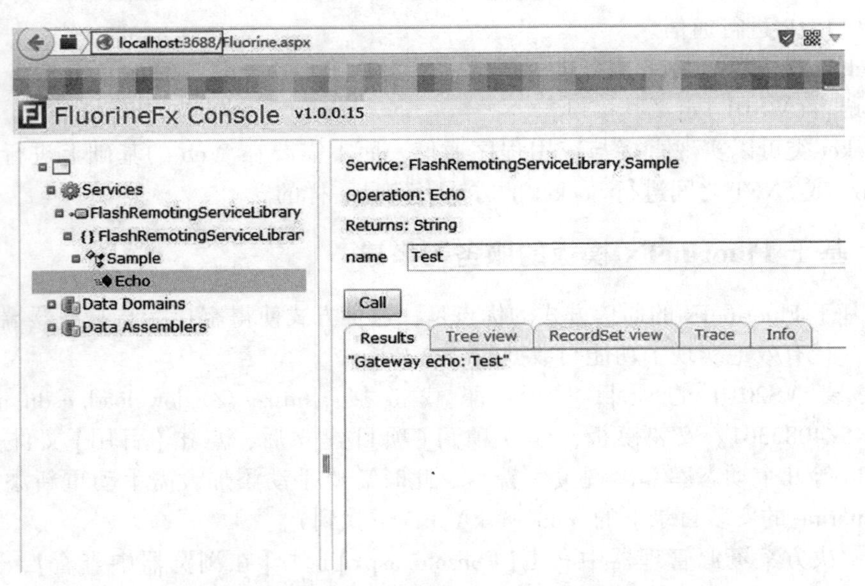

图 29.2 Web 应用程序 URL 端口

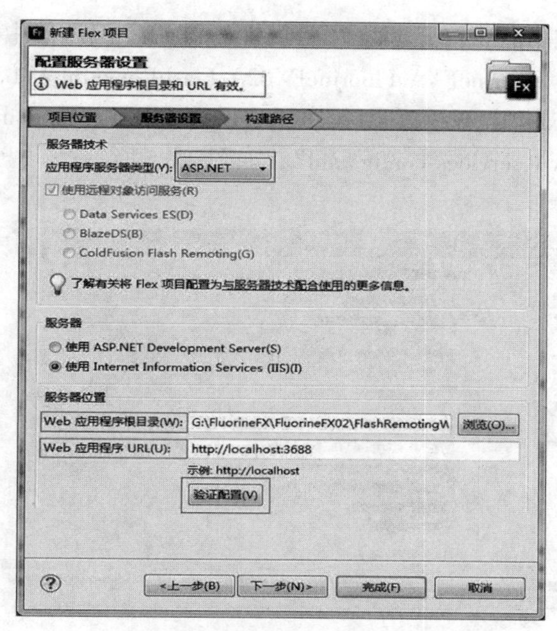

图 29.3　修改项目属性

若是工程修改，则应该修改工程属性中 Flex 服务器设置，如图 29.4 所示。

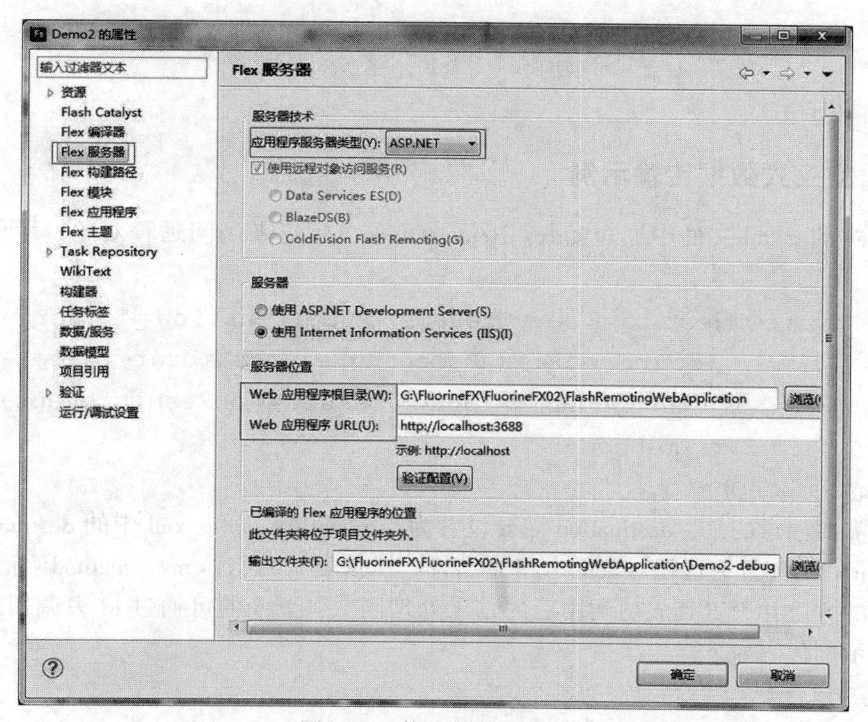

图 29.4　修改服务器设置

④Flex 程序建立完毕后，打开 Flex 项目属性，对 Flex 编译器中的附加编译参数进行修改，-services "G：\ FluorineFX \ FluorineFX02 \ FlashRemotingWebApplication \ WEB-INF \ flex \ services-config. xml"。Web 应用程序根目录还是 FlashRemotingWebApplication 文件夹下"\ WEB-INF \ flex \ services-config. xml"，如图 29. 5 所示。

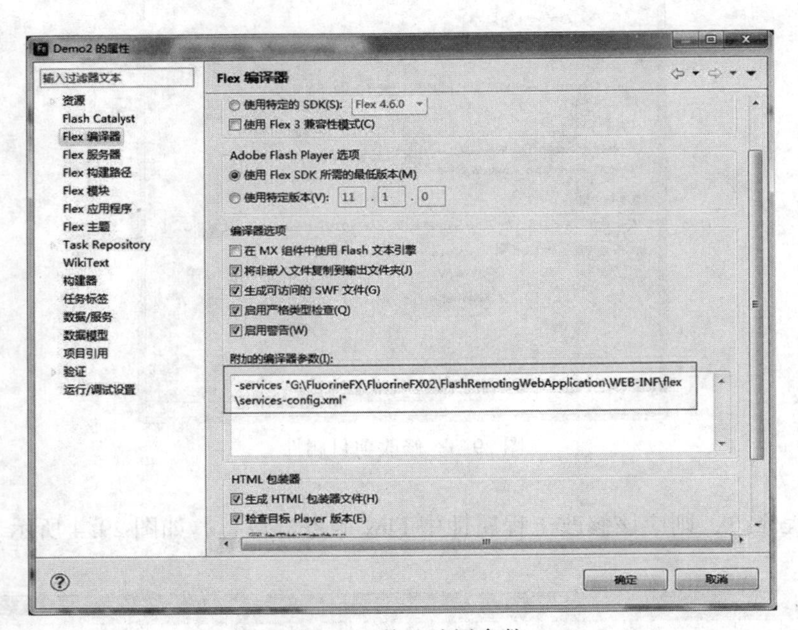

图 29. 5　修改编译参数

29. 1. 2　交互式数据传输示例

在 Flex 的 mxml 文件中通过<mx：RemoteObject>标签来访问远程对象，示例代码如下：

```
<mx:RemoteObject id="serviceRemote" destination="fluorine"
            source="FlashRemotingServiceLibrary.Sample">
        <mx:method name="Echo" result="onResult(event)">
        </mx:method>
</mx:RemoteObject>
```

这里需要注意的是，destination 需要设置为与 remoting-config. xml 中的 destination 的 id 一致，source 则配置为远程对象的全路径(名称空间+类)，通过<mx：method>标签配置远程对象下的方法并设置其成功调用后的结果处理函数，下面便可通过 id 去调用远程方法了。实现示例代码如下：

```
<fx:Script>
        <![CDATA[
            import mx.controls.Alert;
```

```
        import mx.rpc.events.ResultEvent;
        internal function onTest():void
        {
            serviceRemote.Echo("Test the comunicate of user and
service.");
        }
        internal function onResult(evt:ResultEvent):void
        {
          Alert.show(evt.result.toString());
        }
    ]]>
</fx:Script>
```

应用程序 Application 设置 creationComplete="{onTest()}"，运行程序，如图 29.6 所示。

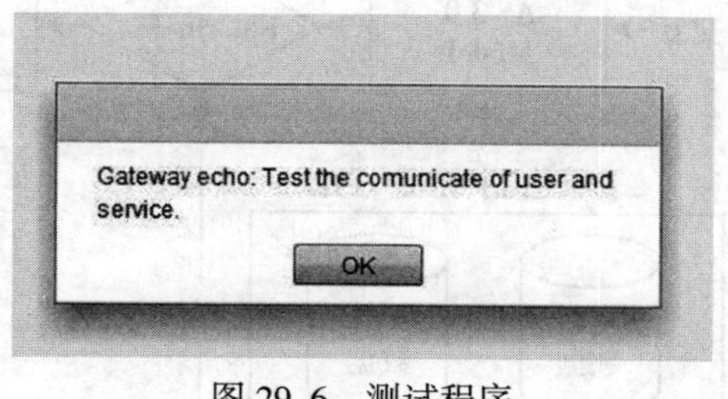

图 29.6　测试程序

29.2　“安全农产品服务系统”案例

“安全农产品服务系统”(研发者：董水峰、魏旭晨、柳诚、王恒，指导教师：柳林)为 2014 年度“MapGIS 杯”全国 GIS 技能大赛获奖作品，下面以此为案例来演示基于 MapGIS 平台进行 Web 应用开发的过程和方法。“安全农产品服务系统”案例更多内容请查阅随书所赠光盘。

随着国民经济的快速发展，地区之间的产品流通加速，跨地级市、省(或者国际)商品交易量迅速上升。同时，电子商务系统的成熟发展，提高了我国国民线上交易的信任度。农业代表广大农村老百姓的经济产业，所生产的农产品需要得到很好的销售利润。城市化使得城市的农产品需求量正逐年上升，并且需要获得本地以外的农产品输入。另外，消费者越来越注重农产品质量安全问题，希望消费到健康、绿色的产品。政府也致力于农产品的质量监管，农产品的资源配置。跨区域的信息发布、统计、处理，以及空间信息可视化表达的需求越来越强烈。当前的农产品发布信息系统，一是没有兼顾食品安全的概念，二是信息不具有可视化(基本是文本操作、图表展示)。地理信息系统现已成熟地应用于诸多行业，具有空间信息可视化，地理数据处理、集成、存储、空间分析、发布服务等优势。借助 GIS 构建安全农产品服务系统将使得此软件系统更加具有操作性、直观性、完善性。

29.2.1　开发环境的选择与配置

安全农产品服务系统是基于 Web 的面向全国多类型用户的地理信息系统，地图服务器采用 MapGIS IGServer for Flex，系统采用 MVC 开发模式，前台用户交互采用 Flex 4.5，模型层采用 Action Script 3.0，控制层采用 C# 类控制。关系型数据库采用 SQL Server，模型层与控制层通信采用 FluorineFx。辅助开发工具使用 Flash Builder 4.6、VS2010、SQL Server Management Studio 2008，如图 29.7 所示。

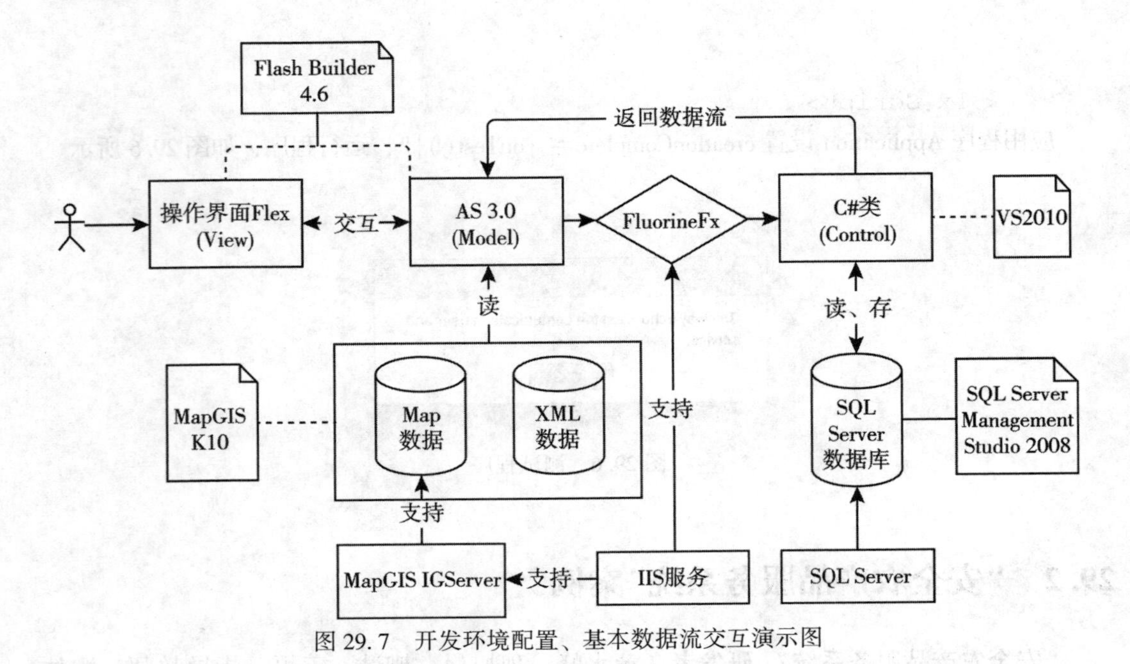

图 29.7　开发环境配置、基本数据流交互演示图

环境配置说明：MapGIS IGServer For Flex 配置参照官方文档，FluorineFx 请参照 FB+VS2010+FluorineFx 环境搭建文档。

29.2.2　农产品供求分布查询实现

1. 需求分析

提供两种粒度的查询(省、市)方式，几何选择提供点、多边形两种方式，根据选择的查询条件进行查询。首先进行地图文档中省或市的查询，然后根据查询得到的图形与条件查询的数据点进行拓扑分析，位置为内部的点即为符合要求的数据。最后将符合要求的数据地图标记显示，面板列表显示。用户点击地图上的基地图标可移动缩放地图至此位置，用户点击基地列表也可移动缩放地图到相应位置，且都能联动显示基地详细信息。

2. 数据组织

地图数据(省、市两级区数据)加上农产品供需信息(SQL Server)就构成了系统的数据组织，如图 29.8 所示。

产品发布信息表			productInfoRelease		
编码	<pi>	Integer	id	<pi>	Integer
用户名		Text	uName		Text
发布类别		Text	demandCategory		Text
联系人姓名		Text	relationPersonName		Text
联系人电话		Integer	relationTel		Integer
产品类别		Text	productClass		Text
产品名称		Text	productName		Text
量		Integer	count		
量单位		Text	countUnit		Text
价格A		Float	priceBottom		Float
价格B		Float	priceTop		Float
价格单位		Text	priceUnit		Text
地址		Text	provideDetailAddr		Text
附加信息		Text	addProductInfo		Text
发布时间		Date	publishTime		Date
有效时间A		Date	validTimeA		Date
有效时间B		Date	validTimeB		Date
地理X		Float	x		Float
地理Y		Float	y		Float
区		Text	area		Text
省		Text	province		Text
市		Text	city		Text
主标识符 <pi>			主标识符 <pi>		

图 29.8 数据组织图

3. 功能实现流程

功能实现流程如图 29.9 所示。

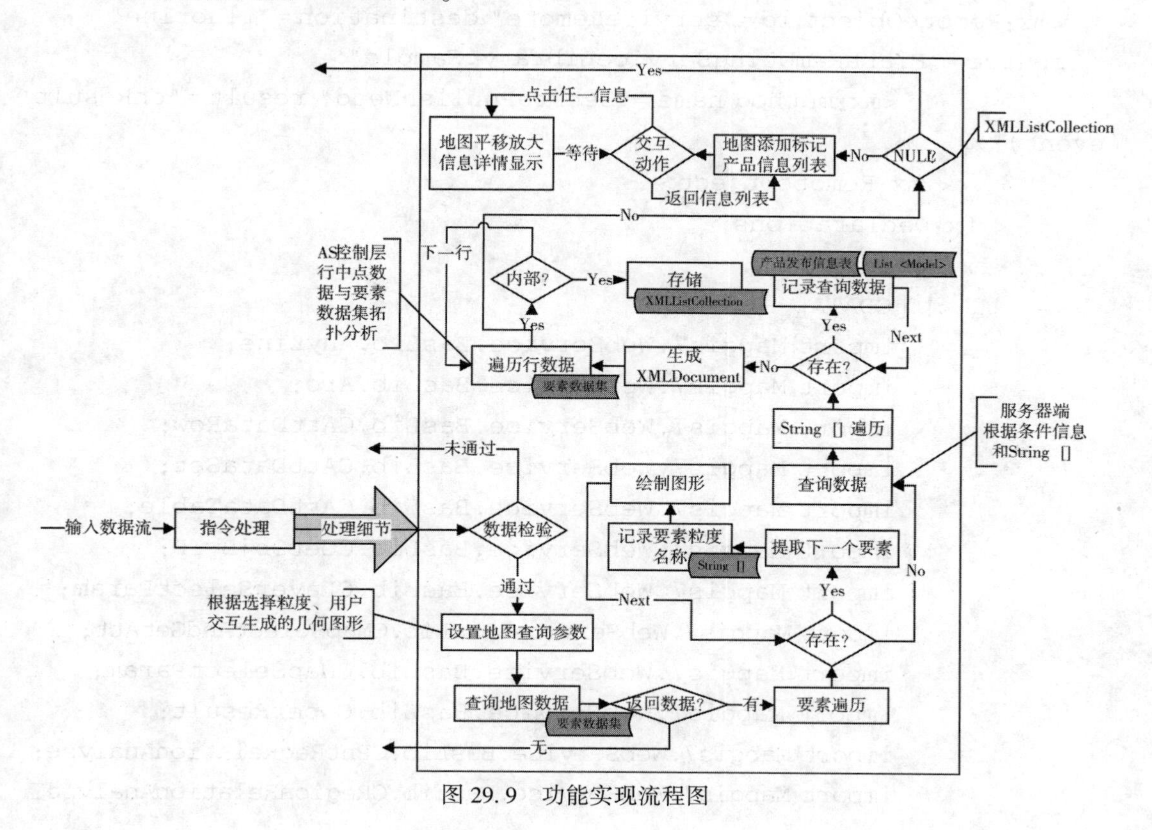

图 29.9 功能实现流程图

4. 案例代码

①功能交互界面：ProProvideNeedGeometry. mxml。具体代码如下：

```
<? xml version="1.0" encoding="utf-8"? >
<s:TitleWindow xmlns:fx="http://ns.adobe.com/mxml/2009"
               xmlns:s="library://ns.adobe.com/flex/spark"
               xmlns:mx="library://ns.adobe.com/flex/mx"
    width="382" height="180" fontSize="15" textAlign="center"
title="农产品供求分布查询"
               height.State2="312" title.State2="农产品供求查询结
果列表"
               height.State3="200" title.State3="农产品供求查询详情">
    <s:states>
        <s:State name="State1"/>
        <s:State name="State2"/>
        <s:State name="State3"/>
    </s:states>
    <fx:Declarations>
        <! -- 将非可视元素(例如服务、值对象)放在此处 -->
  <mx:RemoteObject id="serviceRemote" destination="fluorine"
  source="FlashRemotingServiceLibrary.Sample">
            <mx:method name="GetProPublishNeed" result="onResult
(event)"/>
        </mx:RemoteObject>
    </fx:Declarations>
    <fx:Script>
        <! [CDATA[
            import Mapgis7.WebService.BasLib.AnyLine;
            import Mapgis7.WebService.BasLib.Arc;
            import Mapgis7.WebService.BasLib.CAttDataRow;
            import Mapgis7.WebService.BasLib.CAttDataSet;
            import Mapgis7.WebService.BasLib.CAttDataTable;
            import Mapgis7.WebService.BasLib.CGetObjByID;
            import Mapgis7.WebService.BasLib.CLayerSelectParam;
            import Mapgis7.WebService.BasLib.CMapSelectAndGetAtt;
            import Mapgis7.WebService.BasLib.CMapSelectParam;
            import Mapgis7.WebService.BasLib.COperResult;
            import Mapgis7.WebService.BasLib.CPntRegRelationAnalyse;
            import Mapgis7.WebService.BasLib.CRegionRelationAnalyse;
```

```
import Mapgis7.WebService.BasLib.CWebSelectParam;
import Mapgis7.WebService.BasLib.Dot_2D;
import Mapgis7.WebService.BasLib.ESelectionType;
import Mapgis7.WebService.BasLib.EWebGeomType;
import Mapgis7.WebService.BasLib.EnumLayerStatus;
import Mapgis7.WebService.BasLib.GPoint;
import Mapgis7.WebService.BasLib.GRegion;
import Mapgis7.WebService.BasLib.IWebGeometry;
import Mapgis7.WebService.BasLib.Polygon;
import Mapgis7.WebService.BasLib.SFeature;
import Mapgis7.WebService.BasLib.SFeatureGeometry;
import mx.collections.ArrayCollection;
import mx.collections.XMLListCollection;
import mx.controls.Alert;
import mx.core.UIComponent;
import mx.rpc.events.ResultEvent;
import zdims.TextMarker;
import zdims.control.MapDocDataViewer;
import zdims.drawing.CoordinateType;
import zdims.drawing.DrawingType;
import zdims.drawing.GraphicsLayer;
import zdims.drawing.IMSPolygon;
import zdims.event.IMSMapMouseEvent;
import zdims.interfaces.IGraphics;
import zdims.interfaces.control.INavigationBar;
import zdims.map.IMSMap;
import zdims.map.VectorMapDoc;
import zdims.mark.IMSMark;
import zdims.mark.MarkLayer;
import zdims.util.IMSOperType;
import zdims.util.SpacialAnalyse;
private var _map:IMSMap;
private var mapDoc:VectorMapDoc;
private var graphicsLayer:GraphicsLayer
private var mapDocDataViewer:MapDocDataViewer;
private var _spatial:SpacialAnalyse;
private var layerIndex:int=2;
[Bindable]
```

```
        public var xmlList01:XMLListCollection;
        private var parameRemote1:String = "SELECT * FROM pro-
ductInfoRelease WHERE";
        private var parameRemote2:String = "";
        private var markLayer:MarkLayer;
        [Bindable]
        private var bd:XML;
        public function set imMap(m:IMSMap):void{
            this._map=m;
        }
        public function get imMap():IMSMap
        {
            if(_map==null)
                imMap=this.parent as IMSMap;
            returnthis._map;
        }
        public function set imVectorMapDoc(m:VectorMapDoc):void{
            this.mapDoc=m;
        }
        public function set imGraphicsLayer(m:GraphicsLayer):void{
            this.graphicsLayer=m;
        }
        public function set imMapDocDataViewer(m:MapDocData-
Viewer):void{
            this.mapDocDataViewer=m;
        }
        public function set imMarkLayer(m:MarkLayer):void{
            this.markLayer=m;
        }
        private function onResult(evt:ResultEvent):void{
            xmlList01=new XMLListCollection(evt.result.row);
            if(xmlList01.length<1){
                Alert.show("未查找到相关信息");
            }else{
                this.currentState="State2";
                drawMarks(xmlList01);
            }
        }
```

```
private function drawMarks(bl:XMLListCollection):void
{
    selectCount.text="共查询记录:"+bl.length.toString
()+"条";
    var i:int=0;
    for each(var xl:XML in bl){
        addMark(new Number(xl.x),new Number(xl.y),i);
        i++;
    }
    markLayer.enableMarkHiden=false;
}
[Bindable]
[Embed("img/widget/point.png")]
public static var icon:Class;
private function addMark(x:Number,y:Number,j:int):void{
    var mark:IMSMark;
    var mark1:IMSMark;
    var img:Image;
    img=new Image();
    img.source=icon;
    mark=new IMSMark(img,CoordinateType.Logic);
    mark.x=x;
    mark.y=y;
    mark.offsetX=16;
    mark.offsetY=32;
    mark.mouseClickCallback=changeBaseContent;
    markLayer.addMarkAt(mark,j);
}
private function changeBaseContent(mark:IMSMark,event:
MouseEvent):void{
    bd=new XML(xmlList01[markLayer.indexOf(mark)]);
    this._map.panTo(new Number(bd.x),new Number(bd.y));
    this._map.setLevel(3);
    this.currentState="State3";
}
public function onclick(event:MouseEvent):void
{
    clearPathDot();
```

```
                this.graphicsLayer.removeAllElements();
                parameRemote1="SELECT * FROM productInfoRelease WHERE";
                parameRemote2="";
                var proProivde:String="";
                var proNeed:String="";
                var selectScale:int=selectScaleId.selectedIndex;
                var timeA:String=timeAId.text;
                var timeB:String=timeBId.text;
                var className:String="";
                var selecteMethod:int=selecteMethodId.selectedIndex;
                var provideneed:String="";
                if(proPublishId.selected){
                    proProivde="1";
                    provideneed="provide";
                }
                if(proNeedId.selected){
                    proNeed="1";
                    provideneed+="need";
                }
                if(selectScale==0){
                    layerIndex=2;
                }else{
                    layerIndex=3;
                }
                if(timeAId.text! =""&&timeBId.text! =""){
    timeA=timeAId.text.split("/")[2]+timeAId.text.split("/")[0]+
timeAId.text.split("/")[1];
    timeB=timeBId.text.split("/")[2]+timeBId.text.split("/")[0]+
timeBId.text.split("/")[1];
                    var a:Number=new Number(timeA);
                    var b:Number=new Number(timeB);
                    if(a>b){
                        Alert.show("结束时间不应该小于起始时间。");
                        timeBId.text="";
                        return;
                    }else{
                        parameRemote1+="((validTimeA >= "+timeA+"
 AND validTimeA <= "+timeB+") OR (validTimeB >= "+timeA+" AND valid-
```

```
TimeB <= "+timeB+") OR (validTimeA<="+timeA+" AND validTimeB>="+
timeB+"))";
                }
            }
            if(provideneed.indexOf("provideneed")==-1){
                parameRemote1+=" AND (demandCategory='"+provi-
deneed+"')";
            }
            if(classId.selectedIndex! =-1){
                className=classId.selectedItem.className;
                parameRemote1+="AND(productClass='"+className+"')";
            }
            if((timeA=="" || timeB=="") ||
(proProivde==""&&proNeed=="")){
                Alert.show("时间起止(必选)、供需(至少选一)");
            }else{
                this.mapDoc.getMapLayerInfo(layerIndex).LayerStatus=
  EnumLayerStatus.Selectable;
                this.mapDoc.updateAllLayerInfo();
                this._map.activeMapDoc=this.mapDoc;
            this._map.activeMapDoc.activeLayerIndex=layerIn-
dex;  //激活图层索引号为?的图层
                //画点的操作
                if(selecteMethod==0){
                    this.graphicsLayer.drawingType=DrawingType.Point;
                }else{
                this.graphicsLayer.drawingType=DrawingType.Polygon;
                }
                this.graphicsLayer.drawingOverCallback=Drawing-
OverCallback;
            }
        }
        public function DrawingOverCallback
  (gLayer:GraphicsLayer,graphics:IGraphics,pntArr:Vector.<Point
>):void
        {
            var websel:CWebSelectParam=new CWebSelectParam();
 //定义查询参数对象
```

```
                websel.SelectionType = ESelectionType.SpatialRange;
  //设置查询参数对象

              if(pntArr.length>2){
                  var pol:Polygon=new Polygon();
                  //将多边形的信息用点的信息来表示
                  for(var i:int=0;i<pntArr.length;i++)
                  {
                      pol.Dots[i]=new Dot_2D(pntArr[i].x,pntArr
[i].y);
                  }
                  pol.Dots[pol.Dots.length-1]=pol.Dots[0];
                  websel.GeomType=EWebGeomType.Polygon;//设置查
询的类型为点击查询
                  websel.Geometry=pol;//获取鼠标点击的点
              }else{
                  var pnt:Dot_2D=new Dot_2D();//获取点的位置参数
                  pnt.x=pntArr[0].x;
                  pnt.y=pntArr[0].y;
                  websel.GeomType=EWebGeomType.Point;//设置控件
查询的类型为点击查询
                  websel.Geometry=pnt;//获取鼠标点击的点
              }
              websel.CompareRectOnly=this.mapDoc.compareRectOnly;
              websel.MustInside=this.mapDoc.mustInside;
              var mapsel:CMapSelectParam=new CMapSelectParam();//
初始化查询参数
              mapsel.SelectParam=websel;    //设置查询参数
              mapsel.PageCount = 0;
          this.mapDoc.select(mapsel,callBack);//使用查询参数对象进行
地图文档查询
            }
            public function callBack(e:Event):void{
                this._map.activeMapDoc=this.mapDoc;    //激活当前文档
             this._map.activeMapDoc.activeLayerIndex=layerIndex;
  //激活图层索引号为?的图层
        var result:CMapSelectAndGetAtt = this.mapDoc.onSelect(e);//
获取查询结果的属性集
```

```
            if(result==null||result.Count[0][layerIndex]==0)
            {
                Alert.show("未查到要素,请重新点击查询");
                return;
            }
            else
            {
        var adt:CAttDataTable=(result.AttDS[0] as CAttDataSet)
.attTables[layerIndex] as CAttDataTable;
            for(var j:int=0;j<adt.Rows.length;j++){
                var targetObj:CGetObjByID=new CGetObjByID();
    //targetObj.FeatureID=(((result.AttDS[0] as CAttDataSet).
attTables[layerIndex] as
                CAttDataTable).Rows[0] as CAttDataRow).FID;
                targetObj.FeatureID=(adt.Rows[j] as CAttDataRow).FID;
                targetObj.LayerIndex=layerIndex;
                if(layerIndex==2){
    parameRemote2+=(adt.Rows[j]as CAttDataRow).Values[3].toString
()+"|";
                }else{
    parameRemote2+=(adt.Rows[j]as CAttDataRow).Values[0].toString
()+"|";
                }
                //Alert.show(targetObj.FeatureID.toString());
                this.mapDoc.getFeatureByID(targetObj,onGetFeature);
            }
            //获取后台点数据
            //Alert.show(parameRemote1+"="+parameRemote2);
            serviceRemote.GetProPublishNeed(parameRemote1,pa-
rameRemote2);
            this.graphicsLayer.drawingType=DrawingType.None;
            this._map.setCurOper(IMSOperType.Drag);
        }
    }
    public function onGetFeature(e:Event):void
    {
        var sf:SFeature=this.mapDoc.onGetFeatureByID(e);
        var sfg:SFeatureGeometry=sf.fGeom;
```

```
            drawGetFeatrue(0x00FF00,0xFFFF66,sfg);
            this.graphicsLayer.flashGeo(sfg);
            var titleMarker:TextMarker=new TextMarker();
            titleMarker.name="名称";
            titleMarker.imsmap=this._map;
            titleMarker.logicX =(sf.bound.xmax+sf.bound.xmin)/2;
            titleMarker.logicY =(sf.bound.ymax+sf.bound.ymin)/2;
            titleMarker.enableShowName=true;
            if(layerIndex==2){
                //Alert.show(sf.AttValue[3].toString());
                titleMarker.IMarkName=sf.AttValue[3].toString();
                //parameRemote2+=sf.AttValue[3].toString()+"|";
            }else{
                //Alert.show(sf.AttValue[0].toString());
                titleMarker.IMarkName=sf.AttValue[0].toString();
                //parameRemote2+=sf.AttValue[0].toString()+"|";
            }
            this._map.addChildAt(titleMarker,this._map.layerLenth);
        }
        private function innerPoint(x:Number,y:Number,polygon:
GRegion):Boolean{
            var xArr:Array=new Array();
            var yArr:Array=new Array();
            var arcLine:AnyLine=polygon.GetRings(0);
            var arc:Arc=arcLine.GetArcs(0);
            //获取要素边界的点
            var ArcPntNum:int=arc.Dots.length;
            for(var i:int=0;i<ArcPntNum;i++){
                xArr.push(arc.Dots[i].x);
                yArr.push(arc.Dots[i].y);
            }
            if(pointInPolygon(x,y,xArr,yArr,ArcPntNum)){
                returntrue;
            }else{
                returnfalse;
            }
        }
        private function
```

```
    pointInPolygon(x:Number,y:Number,polyX:Array,polyY:Array,edge:
int):Boolean
    {
            var i:int=edge-1;
            var j:int=edge-1 ;
            var oddNodes:Boolean=false;
            for (i=0;i<edge; i++) {
        if((polyY[i]< y&&polyY[j]>=y || polyY[j]<y&&polyY[i]>=y)&&
(polyX[i]<=x || polyX[j]<=x))
                {
            if(polyX[i]+(y-polyY[i])/(polyY[j]-polyY[i]) * (polyX
[j]-polyX[i])<x)
                    {
                        oddNodes=! oddNodes;
                    }
                }
                j=i;
            }
            return oddNodes;
        }
        //绘制显示要素
        public function drawGetFeatrue
    (boderColor:uint,colorTemo:uint,sfeatureTemp:SFeatureGeometry):
IMSPolygon{
        var reg:GRegion=sfeatureTemp.GetRegGeom(0);
        var arcLine:AnyLine=reg.GetRings(0);
        var arc:Arc=arcLine.GetArcs(0);
        //获取要素边界的点
        var ArcPnt:Array=arc.Dots;
        var ArcPntNum:int=arc.Dots.length;
        //绘制多边形
        var poly:IMSPolygon=new IMSPolygon(CoordinateType.Logic);
        graphicsLayer.addGraphics(poly);
        for(var i:int=0;i<ArcPntNum;i++){
            poly.points.push(new Point(arc.GetDots(i).x, arc.
GetDots(i).y));
        }
        //Alert.show(ArcPntNum.toString());
```

```
        //设置边界线颜色及大小
        poly.lineStyle(2,boderColor);
        //设置填充区的颜色及透明度
        poly.beginFill(colorTemo,0.6);
        //绘制图形
        poly.draw();
        //poly.flicker();
        return poly;
        }
        public function clear():void{
            clearPathDot();
            this.markLayer.removeAllElements();
            this.graphicsLayer.removeAllElements();
        }
        public function clearPathDot():void
        {
            var mark:Array = this._map.getChildren();
            for (var i:int = 0; i < mark.length; i++)
            {
                if (mark[i] is TextMarker&&UIComponent(mark[i])
.name == "名称")
                    this._map.removeChild(mark[i]);
            }
        }
        [Bindable]
      private var classArr:ArrayCollection = new ArrayCollection
([{id:1,className:"粮油"},{id:2,className:"果蔬"},
   {id:3,className:"花卉"},{id:4,className:"林产品"},{id:5,class-
Name:"畜禽产品"},
                  {id:6,className:"水产品"},{id:7,className:"其他农副
产品"}]);
            [Bindable]
            private var ld:ArrayCollection=
   new ArrayCollection([{name:"省级单位"},{name:"市级单位"}]);
   [Bindable]
            private var sM:ArrayCollection
   =new ArrayCollection([{name:"点"},{name:"多边形"}]);
            public function showDetail():void{
```

```
                bd=new XML(list1.selectedItem);
                this._map.panTo(new Number(bd.x),new Number(bd.y));
                this._map.setLevel(4);
                this.currentState="State3";
            }
        ]]>
    </fx:Script>
    <s:Label x="15" y="14" text="农产品"
            visible.State1="true"
            visible.State2="false"
            visible.State3="false"/>
    <s:CheckBox id="proPublishId" x="90" y="10" label="发布" se-
lected="true"
              x.State1="73" y.State1="10"
              visible.State2="false"
              visible.State3="false"/>
    <s:CheckBox id="proNeedId" x="190" y="10" label="需求"
              x.State1="139" y.State1="10"
              visible.State2="false"
              visible.State3="false"/>
    <s:ComboBox id="selectScaleId" includeIn="State1" x="276"
y="10" width="92" dataProvider="{ld}"labelField="name" selectedIn-
dex="0"/>
    <s:Label x="15" y="46" text="时间"
            visible.State2="false"
            visible.State3="false"/>
    <mx:DateField id="timeAId" x="58" y="39" width="132"
                visible.State1="true" showToday.State1="true"
                visible.State2="false"
                visible.State3="false"/>
    <mx:DateField id="timeBId" x="239" y="39" width="132"
                showToday.State1="true"
                visible.State2="false"
                visible.State3="false"/>
    <s:Label x="207" y="46" text="至"
            visible.State2="false"
            visible.State3="false"/>
    <s:Label x="15" y="81" text="产品类别"
```

```
                visible.State2="false"
                visible.State3="false"/>
        <s:DropDownList id="classId" x="84" y="77" width="106" data-
Provider="{classArr}"
                        labelField="className"
                        visible.State2="false"
                        visible.State3="false"></s:DropDownList>
        <s:Label x="196" y="80" text="选择方式"
                visible.State1="true"
                visible.State2="false"
                visible.State3="false"/>
        <s:ComboBox id="selecteMethodId" x="261" y="75" width="109"
dataProvider="{sM}" labelField="name" visible.State1="true" se-
lectedIndex.State1="0"
        visible.State2="false"visible.State3="false"/>
        <mx:LinkButton x="146" y="111" label="清除标注" textDecora-
tion="underline"
        visible.State1="true" x.State1="207" y.State1="114"visible.
State2="false"
        visible.State3="false" click="clear()"/>
        <mx:LinkButton includeIn="State1" visible="true" x="80" y="
114" label="信息查询"
                        textDecoration="underline" click="onclick(event)"/>
          < mx: Canvas  y = " 10 "  width = " 364 "  height = " 260 "
backgroundAlpha="0.35" backgroundColor="#CCCCCC" borderColor="#
B7BABC"borderStyle="solid"
    cornerRadius="16" horizontalCenter="-2"visible.State1="false"
    visible.State2="true" visible.State3="false">
        <s:List id="list1" includeIn="State2,State3" x="10" y="20"
width = " 342 "  height = " 228 "  dataProvider = " { xmlList01 }"
itemRenderer="ItemRenderer.ProductPublishGeo" change="showDetail
()"></s:List>
        <mx:Image width="20" height="20" buttonMode="true"
                click="currentState = (currentState =='State2')?'
State1':'State2'"
                source="@ Embed('img/widget/m_minimize.png')" tool-
Tip="返回主菜单"
                x.State2="11" y.State2="0"
```

```
                x.State3="-197" y.State3="-82"/>
        <s:Label id="selectCount" includeIn="State2" x="84" y="
2" width="194" text="0"/>
    </mx:Canvas>
    <mx:Image width="20" height="20" buttonMode="true"
                click="currentState = (currentState =='State2')?'
State3':'State2'"
                source="@Embed('img/widget/m_minimize.png')" tool-
Tip="返回列表菜单"
                visible.State1="false"
                visible.State2="false" x.State2="-197" y.State2="-82"
                x.State3="3" y.State3="2"/>
    <s:Label includeIn="State3" x="37" y="12" text="类别"/>
  <s:Label includeIn="State3" x="75" y="12" width="65" color="#1351CF"
  text="{bd.demandCategory}"/>
    <s:Label includeIn="State3" x="158" y="10" text="名称"/>
    <s:Label includeIn="State3" x="196" y="10" width="174"
text="{bd.productName}"/>
    <s:Label includeIn="State3" x="37" y="35" text="量"/>
    <s:Label includeIn="State3" x="62" y="35" width="74" text="
{bd.count}"/>
    <s:Label includeIn="State3" x="144" y="34" text="价格"/>
  <s:Label includeIn="State3" x="196" y="33" width="162"
  text="{bd.priceBottom}-{bd.priceTop}"/>
    <s:Label includeIn="State3" x="37" y="58" text="信息有效时间"/>
    <s:Label includeIn="State3" x="146" y="57" width="193"
  text="{bd.validTimeA}-{bd.validTimeB}"/>
    <s:Label includeIn="State3" x="37" y="81" text="联系人姓名"/>
    <s:Label includeIn="State3" x="120" y="80" width="68"
  text="{bd.relationPersonName}"/>
    <s:Label includeIn="State3" x="203" y="80" text="联系电话"/>
    <s:Label includeIn="State3" x="274" y="80" width="96"
text="{bd.relationTel}"/>
    <s:Label includeIn="State3" x="37" y="104" text="具体地址"/>
    <s:Label includeIn="State3" x="106" y="103" width="252"
  text="{bd.provideDetailAddr}"/>
    <s:Label includeIn="State3" x="39" y="127" text="附加信息"/>
    <s:RichText includeIn="State3" x="103" y="126" width="255"
```

```
height="31"
    text="{bd.addProductInfo}"/>
        <s:Label includeIn="State1" x="206" y="14" text="选择粒度"/>
    </s:TitleWindow>
```

②FlashRemotingServiceLibrary. Sample 文件 GetProPublishNeed 方法如下：

```
public XmlDocument GetProPublishNeed(string sql, string parame)
{
GeoSearch gs = new GeoSearch();
return gs.getProProvideNeedPoints(sql,parame);
}
```

③FlashRemotingServiceLibrary. GeoSearch 文件 getProProvideNeedPoints 方法如下：

```
#region 获取发布农产品供求信息
public XmlDocument getProProvideNeedPoints(string sql, string parame)
{
XmlDocument doc = new XmlDocument();
doc.AppendChild ( doc.CreateXmlDeclaration ( " 1.0 ", " utf- 8 ",
null));//声明 XML 开头,根元素
XmlElement xmlnode = doc.CreateElement("Publish");
string[] slip=parame.Split('|');
string SQLS = sql;
for (int i = 0; i < slip.Length; i++)
    {
if (slip[i]! ="")
        {
            SQLS = sql;
            SQLS +="AND((province LIKE '% " + slip[i].Trim() + "%')OR
(city LIKE '% "+slip[i].Trim()+"%')) ORDER BY id DESC";
List<ProPublish> pp=getPublishProList(SQLS);
if (pp = = null)
            {
continue;
            }
else
            {
foreach (ProPublish r in pp)
                {
XmlElement xmlelem = doc.CreateElement("", "row", "");
XmlElement xmlelem1 = doc.CreateElement("x");
```

```
XmlText xmltext = doc.CreateTextNode(r.X);
                xmlelem1.AppendChild(xmltext);
                xmlelem.AppendChild(xmlelem1);
                xmlelem1 = doc.CreateElement("y");
                xmltext = doc.CreateTextNode(r.Y);
                xmlelem1.AppendChild(xmltext);
                xmlelem.AppendChild(xmlelem1);
                xmlelem1 = doc.CreateElement("demandCategory");
                xmltext = doc.CreateTextNode(r.DemandCategory);
                xmlelem1.AppendChild(xmltext);
                xmlelem.AppendChild(xmlelem1);
                xmlelem1 = doc.CreateElement("relationPersonName");
                xmltext = doc.CreateTextNode(r.RelationPersonName);
                xmlelem1.AppendChild(xmltext);
                xmlelem.AppendChild(xmlelem1);
                xmlelem1 = doc.CreateElement("relationTel");
                xmltext = doc.CreateTextNode(r.RelationTel);
                xmlelem1.AppendChild(xmltext);
                xmlelem.AppendChild(xmlelem1);
                xmlelem1 = doc.CreateElement("productName");
                xmltext = doc.CreateTextNode(r.ProductName);
                xmlelem1.AppendChild(xmltext);
                xmlelem.AppendChild(xmlelem1);
                xmlelem1 = doc.CreateElement("count");
                xmltext = doc.CreateTextNode(r.Count);
                xmlelem1.AppendChild(xmltext);
                xmlelem.AppendChild(xmlelem1);
                xmlelem1 = doc.CreateElement("priceBottom");
                xmltext = doc.CreateTextNode(r.PriceBottom);
                xmlelem1.AppendChild(xmltext);
                xmlelem.AppendChild(xmlelem1);
                xmlelem1 = doc.CreateElement("priceTop");
                xmltext = doc.CreateTextNode(r.PriceTop);
                xmlelem1.AppendChild(xmltext);
                xmlelem.AppendChild(xmlelem1);
                xmlelem1 = doc.CreateElement("provideDetailAddr");
                xmltext = doc.CreateTextNode(r.ProvideDetailAddr);
                xmlelem1.AppendChild(xmltext);
```

```
                        xmlelem.AppendChild(xmlelem1);
                        xmlelem1 = doc.CreateElement("addProductInfo");
                        xmltext = doc.CreateTextNode(r.AddProductInfo);
                        xmlelem1.AppendChild(xmltext);
                        xmlelem.AppendChild(xmlelem1);
                        xmlelem1 = doc.CreateElement("validTimeA");
                        xmltext = doc.CreateTextNode(r.ValidTimeA);
                        xmlelem1.AppendChild(xmltext);
                        xmlelem.AppendChild(xmlelem1);
                        xmlelem1 = doc.CreateElement("validTimeB");
                        xmltext = doc.CreateTextNode(r.ValidTimeB);
                        xmlelem1.AppendChild(xmltext);
                        xmlelem.AppendChild(xmlelem1);
                        xmlnode.AppendChild(xmlelem);
                    }
                }
            }
        }
    doc.AppendChild(xmlnode);
return doc;
}
#endregion
```

④FlashRemotingServiceLibrary. GeoSearch 文件 getPublishProList 方法如下:

```
#region 根据 sql 语句获取相应的供求信息
Public List<ProPublish> getPublishProList(string sql)
{
List<ProPublish> c = new List<ProPublish>();
string strConnection = "data source=localhost;uid="+SqlLogin.uid+";
pwd="+SqlLogin.pwd+";database="+SqlLogin.database;
SqlConnection objConnection = newSqlConnection(strConnection);
SqlCommand objCommand = newSqlCommand("", objConnection);
    objCommand.CommandText = sql;
try
    {
if (objConnection.State == System.Data.ConnectionState.Closed)
        {
            objConnection.Open();
        }
```

```
SqlDataAdapter mydata = new SqlDataAdapter(objCommand);
DataSet ds = new DataSet();
      mydata.Fill(ds,"code");
SqlCommandBuilder objcmdBuilder = new SqlCommandBuilder(mydata);
DataTable dt = ds.Tables["code"];
if (dt.Rows.Count != 0)
      {
for (int i = 0; i < dt.Rows.Count; i++) //遍历行
         {
DataRow dr = dt.Rows[i];
ProPublish rn = new ProPublish();
            rn.DemandCategory = dr[2].ToString();
            rn.RelationPersonName = dr[3].ToString();
            rn.RelationTel = dr[4].ToString();
            rn.ProductName = dr[6].ToString();
            rn.Count = dr[7].ToString()+dr[8].ToString();
            rn.PriceBottom = dr[9].ToString()+dr[11].ToString();
            rn.PriceTop = dr[10].ToString() + dr[11].ToString();
            rn.ProvideDetailAddr = dr[12].ToString();
            rn.AddProductInfo = dr[13].ToString();
            rn.ValidTimeA = dr[15].ToString();
            rn.ValidTimeB = dr[16].ToString();
            rn.X = dr[17].ToString();
            rn.Y = dr[18].ToString();
            c.Add(rn);
         }
      }
return c;
   }
catch (SqlException e)
   {
Console.Write(e.Message.ToString());
returnnull;
   }
finally
   {
if (objConnection.State == System.Data.ConnectionState.Open)
      {
```

```
            objConnection.Close();
        }
    }
}
#endregion
```

5. 功能截图

功能实现交互操作界面如图 29.10 至图 29.15 所示。

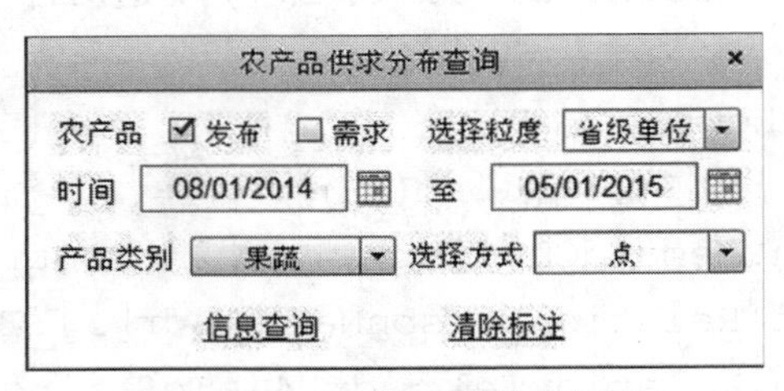

图 29.10　省级粒度查询界面

图 29.11　省级粒度查询结果

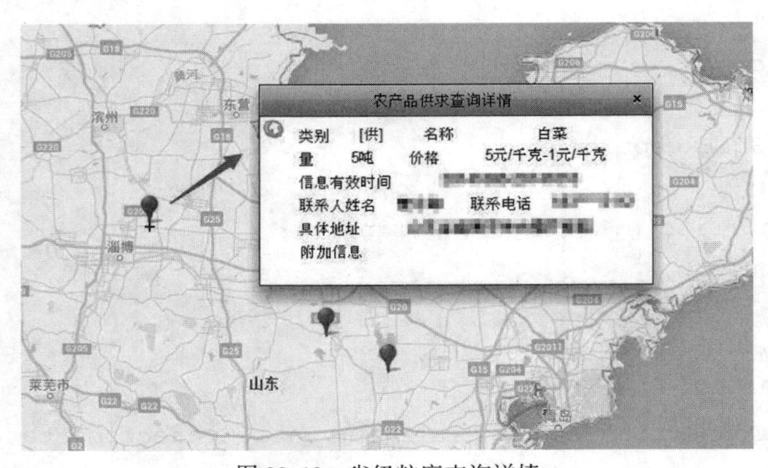

图 29.12　省级粒度查询详情

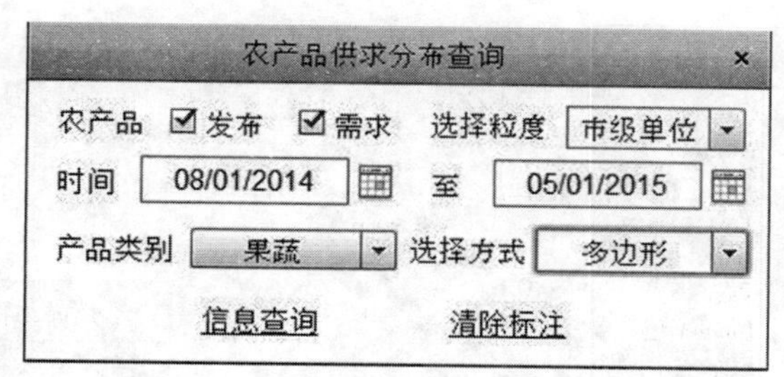

图 29.13　市级粒度查询界面

图 29.14　市级粒度查询结果

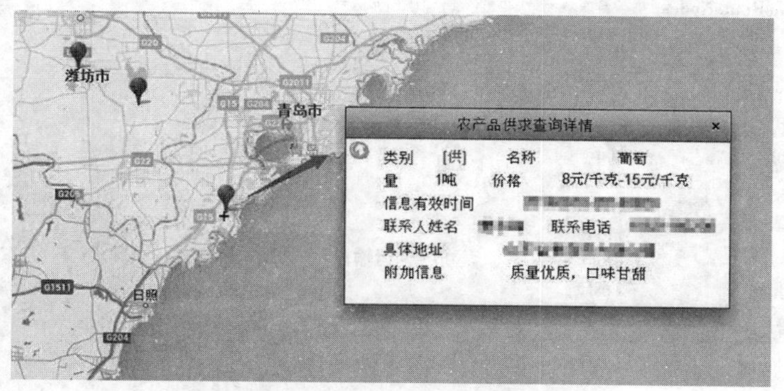

图 29.15　市级粒度查询详情

29.2.3　农产品轨迹查询功能实现

1. 需求分析

根据输入的产品编号，进行数据库(产品运输节点表、产品运输段信息表、产品运输节点表)联合查询，将获取的数据分类进行前台显示、交互。点击任何一个节点可查看其相应信息(生产地、加工地、运输商、销售地)。

2. 数据组织

产品运输轨迹信息(SQL Server)如图 29.16 所示。

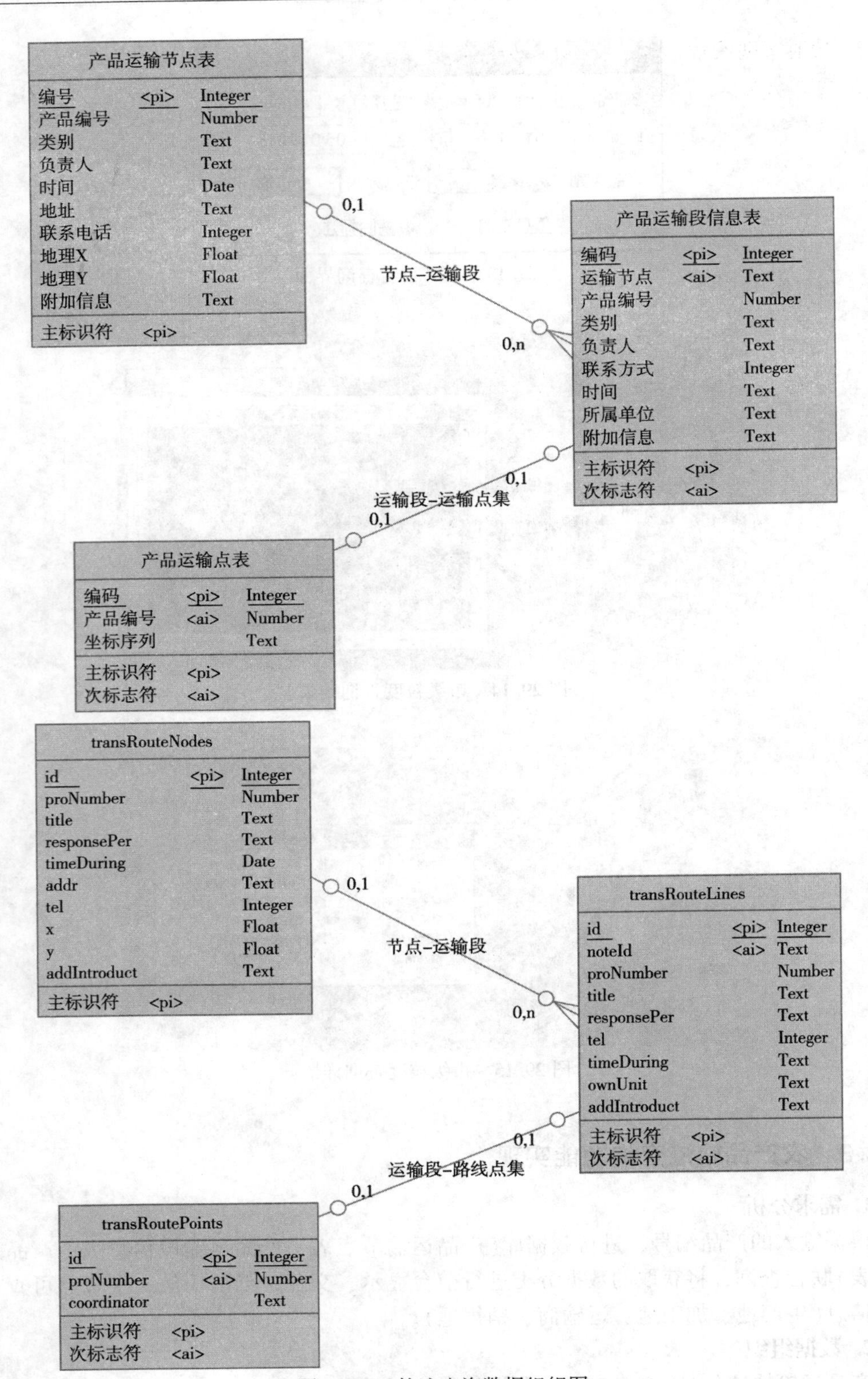

图 29.16　轨迹查询数据组织图

3. 功能实现流程

功能实现流程如图 29.17 所示。

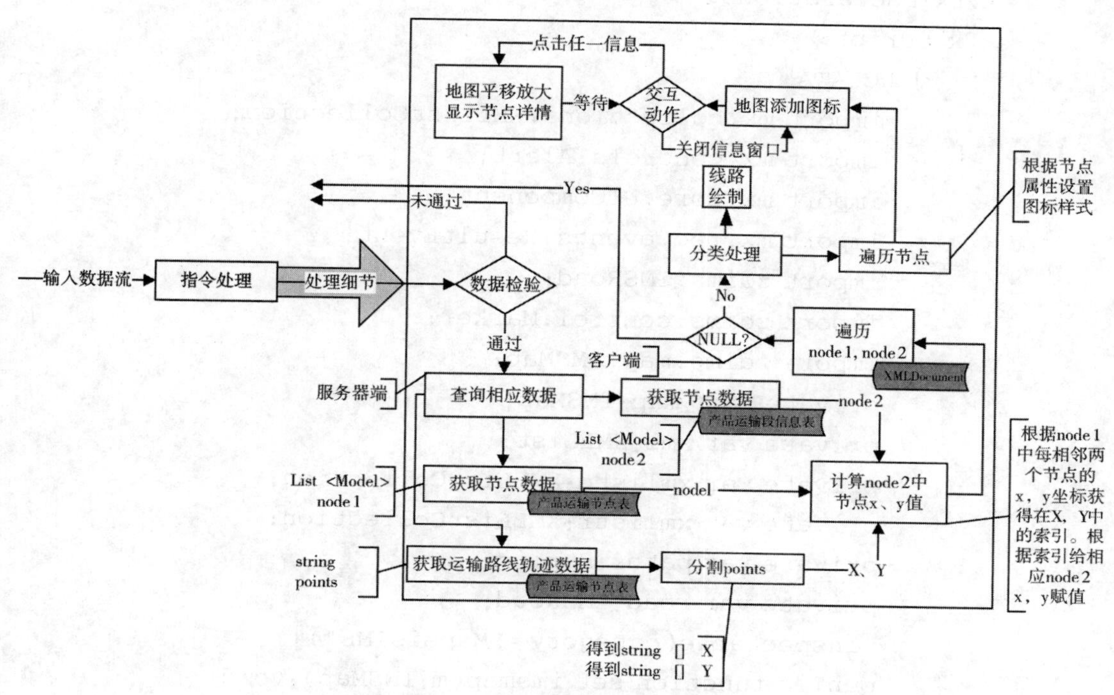

图 29.17 轨迹查询流程图

4. 案例代码

①功能交互界面：ProductRouteSearch. mxml。具体代码如下：

```
<? xml version="1.0" encoding="utf-8"? >
<s:TitleWindow xmlns:fx="http://ns.adobe.com/mxml/2009"
               xmlns:s="library://ns.adobe.com/flex/spark"
               xmlns:mx="library://ns.adobe.com/flex/mx"
        width="400" height="146" backgroundAlpha="0.67" corner-
Radius="13" enabled="true" fontSize="20" horizontalCenter="0" tex-
tAlign="center" title="农产品采购、运输、销售过程查询"
               verticalCenter="0" creationComplete="init()">
    <fx:Declarations>
        <! -- 将非可视元素(例如服务、值对象)放在此处 -->
        <! --点击查找运输路线  1. 弹出具体路线 2. 地图显示路线-->
    <mx:RemoteObject id="serviceRemote" destination="fluorine"
source="FlashRemotingServiceLibrary.Sample">
            <mx:method name="RequireRouteDeatil" result="onRe-
```

```
sult(event)"/>
            </mx:RemoteObject>
        </fx:Declarations>
        <fx:Script>
            <![CDATA[
                import mx.collections.XMLListCollection;
                import mx.controls.Alert;
                import mx.core.UIComponent;
                import mx.rpc.events.ResultEvent;
                import zdims.IMSRoad;
                import zdims.control.Marker;
                import zdims.map.IMSMap;
                private var _map:IMSMap;
                private var xml:XMLList;
                private var xmlistc:XMLListCollection;
                private var xmlistr:XMLListCollection;
                private var paths:String;
                private var road:IMSRoad;
                [Inspectable(category="MapGisIMS")]
                public function set imsmap(m:IMSMap):void
                {
                    this._map = m;
                }
                [Inspectable(category="MapGisIMS")]
                /**
                 * 给地图容器赋值
                 */
                public function get imsmap():IMSMap
                {
                    returnthis._map;
                }
                private function init():void{
                    this.road=new IMSRoad();
                    this.road.imsmap=this._map;
                }
                private function onResult(evt:ResultEvent):void{
                    xml=new XMLList(evt.result);
                    if(xml==null){
```

```
                    Alert.show("未查询到相关数据。");
                    return;
                }
                xmlistc=new XMLListCollection(xml[0].row);
                xmlistr=new XMLListCollection(xml[0].low);
                paths=xml[0].points[0].value;
                //绘制路线、添加节点
                var i:int=0;
                var j:int=0;
                var imgsrc:String="image/bus/car.png";
                 if(new Number(xmlistc[0].x)<new Number(xmlistc
[xmlistc.length-1].x)){
                    imgsrc="image/bus/carR.png";
                }
                for(i;i<xmlistc.length;i++)
                {
                    if(i==0){
    addMarker(xmlistc[i].x,xmlistc[i].y,xmlistc[i].title,"路径点",
xmlistc[i].timeDuring,"image/bus/farmer.png",
    true,xmlistc[i].responsePer,xmlistc[i].addr,
    xmlistc[i].tel,xmlistc[i].addIntroduct);
    this._map.panTo(new Number(xmlistc[i].x),newNumber(xmlistc[i].y));
                    }else if(i==(xmlistc.length-1)){
    addMarker(xmlistc[i].x,xmlistc[i].y,xmlistc[i].title,"路径点",
xmlistc[i].timeDuring,"image/bus/shopping.png",
    true,xmlistc[i].responsePer,xmlistc[i].addr,
    xmlistc[i].tel,xmlistc[i].addIntroduct);
                    }else{
    addMarker(xmlistc[i].x,xmlistc[i].y,xmlistc[i].title,"路径点",
xmlistc[i].timeDuring,"image/bus/_factory.png",
    true,xmlistc[i].responsePer,xmlistc[i].addr,
    xmlistc[i].tel,xmlistc[i].addIntroduct);
                    }
                }
                for(j;j<xmlistr.length;j++)
                {
    addMarker(xmlistr[j].x,xmlistr[j].y,xmlistr[j].title+"("+(j+
1)+")","路径",xmlistr[j].timeDuring,imgsrc,true,xmlistr[j].respon-
```

```
sePer,
   xmlistr[j].addr,xmlistr[j].tel,xmlistr[j].addIntroduct);
                    }
                    this.road._roadCoorArr=paths;
                    this.road.drawRoad();
                    this._map.setLevel(5);
                }
                public function addMarker
   (x:Number,y:Number,markTitle:String,name:String,toolTip:String,
iconSrc:String,
   showInfo:Boolean=false,markerConTitle:String="",markerContent:
String="",
   markerConResponse:String="",markerConReTel:String=""):void{
                    var pathDot1:Marker = newMarker();
                    pathDot1.imsmap=this._map;
                    pathDot1.logicX =x;
                    pathDot1.logicY =y;
                    pathDot1.name = name;
                    pathDot1.imTitle=markTitle;
                    pathDot1.toolTip=toolTip;
                    pathDot1.setIconSrc(iconSrc);
                    pathDot1.enableShowName=true;
                    if(showInfo){
                        pathDot1.setMarkerContent(toolTip,markerConTitle,
   markerContent,markerConResponse,markerConReTel);
                    }
                    this._map.addChildAt(pathDot1,this._map.layerLenth);
                }
                private function search():void{
                    if(proNumberId.text=="" ||proNumberId.text.length! =20){
                        Alert.show("请输入商品编号,并确认正确。","操作提示:");
                    }else{
                        this.clear();
                        serviceRemote.RequireRouteDeatil(proNumberId.text);
                    }
                }
                public function clear():void{
                    this.road.clear();
```

```
            this.road._roadCoorArr = "";
            clearPathDot();
        }
        public function clearPathDot():void
        {
            var mark:Array = this._map.getChildren();
            for (var i:int = 0; i < mark.length; i++)
            {
                if (mark[i] isMarker&&UIComponent(mark[i]).
name == "路径点")
                    this._map.removeChild(mark[i]);
            }
        }

    ]]>
    </fx:Script>
    <s:TextInput id = "proNumberId" x = "170" y = "24" width = "208"
fontSize="12"/>
    <s:Label x="36" y="25" fontSize="18" text="输入商品序列号"/>
    <s:Button x="39" y="70" label="查找产品信息" click="search()"/>
    <s:Button x="208" y="70" label="清除地图信息" click="clear()"/>
</s:TitleWindow>
```

②FlashRemotingServiceLibrary. Sample 文件 RequireRouteDeatil 方法如下：

```
public XmlDocument RequireRouteDeatil(string proNumber)
{
GeoSearch gs = newGeoSearch();
return gs.getProductRoute(proNumber);
}
```

③FlashRemotingServiceLibrary. GeoSearch 文件 getProductRoute 方法如下：

```
#region 产品路线追踪
public XmlDocument getProductRoute(string number1)
{
string points = "";
List<RouteNode> rows = newList<RouteNode>();
List<RouteNode> lowP = newList<RouteNode>();
List<RouteNode> lows = newList<RouteNode>();
//得到节点 row 相关信息
    rows = getRowsByProNum(number1);
```

```
//得到路段 low 相关信息预处理
    lowP = getLowsByProNum(number1);
//获取运输路线坐标序列
    points =this.productPoints(number1);
string[] xy = points.Split(',');
string[] xs = newstring[xy.Length /2];
string[] ys = newstring[xy.Length /2];
int j = 0;
int k = 0;
for (int i = 0; i < xy.Length; i++)
    {
if (i % 2 == 0 && i! =1)
        {
            xs[j] = xy[i].Trim();
            j++;
        }
else
        {
            ys[k] = xy[i].Trim();
            k++;
        }
    }
//得到带 x/y 坐标的 lows
    lows = getLowContainXY(rows, lowP, xs, ys);
if (points == "")
    {
returnnull;
    }
else
    {
XmlDocument doc = new XmlDocument();
doc.AppendChild ( doc.CreateXmlDeclaration ( " 1.0 ",  " utf-8 ",
null));//声明 XML 开头,根元素
XmlElement xmlnode = doc.CreateElement("Route");
//为 XML 文档加入元素/加入一个根元素
XmlElement xmlelem = doc.CreateElement("","points","");
//增加一个子元素
XmlElement xmlelem1 = doc.CreateElement("value");
```

```
XmlText xmltext = doc.CreateTextNode(points);
       xmlelem1.AppendChild(xmltext);
       xmlelem.AppendChild(xmlelem1);
       xmlnode.AppendChild(xmlelem);
foreach(RouteNode r in rows)
       {
          xmlelem = doc.CreateElement("", "row", "");
          xmlelem1 = doc.CreateElement("x");
          xmltext = doc.CreateTextNode(r.X);
          xmlelem1.AppendChild(xmltext);
          xmlelem.AppendChild(xmlelem1);
          xmlelem1 = doc.CreateElement("y");
          xmltext = doc.CreateTextNode(r.Y);
          xmlelem1.AppendChild(xmltext);
          xmlelem.AppendChild(xmlelem1);
          xmlelem1 = doc.CreateElement("title");
          xmltext = doc.CreateTextNode(r.Title);
          xmlelem1.AppendChild(xmltext);
          xmlelem.AppendChild(xmlelem1);
          xmlelem1 = doc.CreateElement("timeDuring");
          xmltext = doc.CreateTextNode(r.TimeDuring);
          xmlelem1.AppendChild(xmltext);
          xmlelem.AppendChild(xmlelem1);
          xmlelem1 = doc.CreateElement("responsePer");
          xmltext = doc.CreateTextNode(r.ResponsePer);
          xmlelem1.AppendChild(xmltext);
          xmlelem.AppendChild(xmlelem1);
          xmlelem1 = doc.CreateElement("addr");
          xmltext = doc.CreateTextNode(r.Addr);
          xmlelem1.AppendChild(xmltext);
          xmlelem.AppendChild(xmlelem1);
          xmlelem1 = doc.CreateElement("tel");
          xmltext = doc.CreateTextNode(r.Tel);
          xmlelem1.AppendChild(xmltext);
          xmlelem.AppendChild(xmlelem1);
          xmlelem1 = doc.CreateElement("addIntroduct");
          xmltext = doc.CreateTextNode(r.AddIntroduct);
          xmlelem1.AppendChild(xmltext);
```

```
            xmlelem.AppendChild(xmlelem1);
            xmlnode.AppendChild(xmlelem);
        }
foreach (RouteNode n in lows)
        {
            xmlelem = doc.CreateElement("", "low", "");
            xmlelem1 = doc.CreateElement("x");
            xmltext = doc.CreateTextNode(n.X);
            xmlelem1.AppendChild(xmltext);
            xmlelem.AppendChild(xmlelem1);
            xmlelem1 = doc.CreateElement("y");
            xmltext = doc.CreateTextNode(n.Y);
            xmlelem1.AppendChild(xmltext);
            xmlelem.AppendChild(xmlelem1);
            xmlelem1 = doc.CreateElement("title");
            xmltext = doc.CreateTextNode(n.Title);
            xmlelem1.AppendChild(xmltext);
            xmlelem.AppendChild(xmlelem1);
            xmlelem1 = doc.CreateElement("timeDuring");
            xmltext = doc.CreateTextNode(n.TimeDuring);
            xmlelem1.AppendChild(xmltext);
            xmlelem.AppendChild(xmlelem1);
            xmlelem1 = doc.CreateElement("responsePer");
            xmltext = doc.CreateTextNode(n.ResponsePer);
            xmlelem1.AppendChild(xmltext);
            xmlelem.AppendChild(xmlelem1);
            xmlelem1 = doc.CreateElement("addr");
            xmltext = doc.CreateTextNode(n.Addr);
            xmlelem1.AppendChild(xmltext);
            xmlelem.AppendChild(xmlelem1);
            xmlelem1 = doc.CreateElement("tel");
            xmltext = doc.CreateTextNode(n.Tel);
            xmlelem1.AppendChild(xmltext);
            xmlelem.AppendChild(xmlelem1);
            xmlelem1 = doc.CreateElement("addIntroduct");
            xmltext = doc.CreateTextNode(n.AddIntroduct);
            xmlelem1.AppendChild(xmltext);
            xmlelem.AppendChild(xmlelem1);
```

```
            xmlnode.AppendChild(xmlelem);
        }
        doc.AppendChild(xmlnode);
    return doc;
    }
}
#endregion
```

④FlashRemotingServiceLibrary. GeoSearch 文件 productPoints 方法如下：

```
#region 获取运输路线坐标序列
public string productPoints(string number2)
{
string c = "";
string strConnection = "data source=localhost;uid="+SqlLogin.uid+";
pwd="+SqlLogin.pwd+";database="+SqlLogin.database;
SqlConnection objConnection = newSqlConnection(strConnection);
SqlCommand objCommand = newSqlCommand("", objConnection);
string k = number2.Trim();
    objCommand.CommandText = " SELECT id, proNumber, coordinator
FROM transRoutePoints WHERE (proNumber = '"+k+"')";
try
  {
if (objConnection.State == System.Data.ConnectionState.Closed)
    {
        objConnection.Open();
    }
SqlDataAdapter mydata = newSqlDataAdapter(objCommand);
DataSet ds = newDataSet();
    mydata.Fill(ds,"asa");
SqlCommandBuilder objcmdBuilder = newSqlCommandBuilder(mydata);
DataTable dt = ds.Tables["asa"];
if (dt.Rows.Count ! = 0)
    {
DataRow dr = dt.Rows[0];
        c = dr[2].ToString();
    }
return c;
  }
catch (SqlException e)
```

```
    {
  Console.Write(e.Message.ToString());
  return c;
    }
  finally
    {
  if (objConnection.State == System.Data.ConnectionState.Open)
        {
            objConnection.Close();
        }
    }
  }
  #endregion
```

⑤FlashRemotingServiceLibrary. GeoSearch 文件 getRowsByProNum 方法如下：

```
#region 获取 row 节点信息
Private List<RouteNode> getRowsByProNum(string num)
{
List<RouteNode> c = new List<RouteNode>();
string strConnection = "data source=localhost;uid="+SqlLogin.uid+";
pwd="+SqlLogin.pwd+";database="+SqlLogin.database;
SqlConnection objConnection = new SqlConnection(strConnection);
SqlCommand objCommand = new SqlCommand("", objConnection);
objCommand.CommandText ="Select * from transRouteNodes where
(proNumber = '" + num.Trim() + "')";
try
  {
if (objConnection.State == System.Data.ConnectionState.Closed)
      {
          objConnection.Open();
      }
SqlDataAdapter mydata = new SqlDataAdapter(objCommand);
DataSet ds = newDataSet();
      mydata.Fill(ds,"code");
SqlCommandBuilder objcmdBuilder = new SqlCommandBuilder(mydata);
DataTable dt = ds.Tables["code"];
if (dt.Rows.Count ! = 0)
      {
for (int i = 0; i < dt.Rows.Count; i++) //遍历行
```

```
            {
DataRow dr = dt.Rows[i];
RouteNode rn = new RouteNode();
                rn.Title = dr[2].ToString();
                rn.ResponsePer = dr[3].ToString();
                rn.TimeDuring=dr[4].ToString();
                rn.Addr = dr[5].ToString();
                rn.Tel = dr[6].ToString();
                rn.X = dr[7].ToString();
                rn.Y = dr[8].ToString();
                rn.AddIntroduct = dr[9].ToString();
                c.Add(rn);
            }
        }
return c;
    }
catch (SqlException e)
    {
Console.Write(e.Message.ToString());
returnnull;
    }
finally
    {
if (objConnection.State == System.Data.ConnectionState.Open)
        {
            objConnection.Close();
        }
    }
}
#endregion
```

⑥FlashRemotingServiceLibrary. GeoSearch 文件 getLowsByProNum 方法如下：

```
#region 得到路段 low 相关信息预处理
privateList<RouteNode> getLowsByProNum(string num)
{
List<RouteNode> c = newList<RouteNode>();
string strConnection = "data source=localhost;uid="+SqlLogin.uid+";
pwd="+SqlLogin.pwd+";database="+SqlLogin.database;
SqlConnection objConnection = newSqlConnection(strConnection);
```

```
SqlCommand objCommand = newSqlCommand("", objConnection);
objCommand.CommandText ="Select * from transRouteLines where
(proNumber ='" + num.Trim() + "')";
try
    {
if (objConnection.State == System.Data.ConnectionState.Closed)
        {
            objConnection.Open();
        }
SqlDataAdapter mydata = newSqlDataAdapter(objCommand);
DataSet ds = newDataSet();
        mydata.Fill(ds,"code");
SqlCommandBuilder objcmdBuilder = newSqlCommandBuilder(mydata);
DataTable dt = ds.Tables["code"];
if (dt.Rows.Count ! = 0)
        {
for (int i = 0; i < dt.Rows.Count; i++)//遍历行
            {
DataRow dr = dt.Rows[i];
RouteNode rn = new RouteNode();
                rn.Title = dr[3].ToString();
                rn.ResponsePer = dr[4].ToString();
                rn.Tel = dr[5].ToString();
                rn.Addr = dr[6].ToString();
                rn.TimeDuring = dr[7].ToString();
                rn.X ="";
                rn.Y ="";
                rn.AddIntroduct = dr[8].ToString();
                c.Add(rn);
            }
        }
return c;
    }
catch (SqlException e)
    {
Console.Write(e.Message.ToString());
returnnull;
    }
```

```
finally
    {
    if (objConnection.State == System.Data.ConnectionState.Open)
            {
                objConnection.Close();
            }
    }
}
#endregion 预处理
```

⑦FlashRemotingServiceLibrary.GeoSearch 文件 getLowContainXY 方法如下：

```
#region 得到带 x/y 坐标的 lows
private List<RouteNode>  getLowContainXY(List<RouteNode> rows,
List<RouteNode> lowP,string[] xs,string[] ys)
{
if (lowP.Count == (rows.Count - 1))
    {
for (int i = 0; i < rows.Count - 1; i++)
        {
int a=getIndex(rows[i].X, rows[i].Y, xs, ys);
int b=getIndex(rows[i+1].X, rows[i+1].Y, xs, ys);
if (a == -1 || b == -1)
            {
continue;
            }
else
            {
                lowP[i].X = xs[(b + a) /2];
                lowP[i].Y = ys[(b + a) /2];
            }
        }
return lowP;
    }
else
    {
Return null;
    }
}
#endregion
```

FlashRemotingServiceLibrary. GeoSearch 文件 getIndex 方法如下：

```
#region 获取 x y index
privateint getIndex(string x, string y, string[] xs, string[] ys)
{
for (int i = 0; i < xs.Length; i++)
    {
if (xs[i].Trim() = = x.Trim()&&ys[i].Trim()= =y.Trim())
        {
return i;
        }
    }
return -1;
}
#endregion
```

5. 功能实现截图

功能实现交互操作界面如图 29. 18 所示。

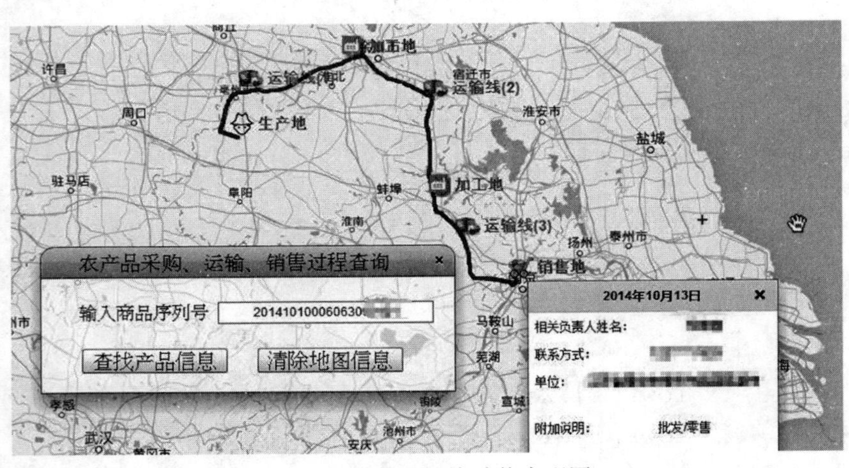

图 29. 18 轨迹查询功能实现图

29. 2. 4 大众评价等级专题图制作

1. 需求分析

提供用户两种粒度(省、市)的数据统计，若选市，则应选择相应省，统计等级可以选择 3~6，提交服务器端进行处理，统计分析的模型为：统计此地区近五年评价数据，最近两年权重各为 0. 3，中间一年为 0. 2，最远两年各为 0. 1。

评价数据统计模型为：

$S = m + p - r \times 10$　　$m =$ 中+好×2−差×2　　$p =$ 合适+便宜×2−贵×2

$W=(S_1+S_2)\times 0.3+S_3\times 0.2+S_4\times 0.1+S_5\times 0.1$

式中，m 为质量，p 为价格，r 为举报。S 为地区当年评价统计量，W 为地区五年评价统计量，S_1 代表当前年份，依次类推。

2. 数据组织

地图数据(省、市两级区数据)、农产品评价信息(SQL Server)和中国行政区存储数据库(XML)一起构成系统的数据组织。评价功能数据组织如图 29.19 所示，数据组织关系如图 29.20 所示。

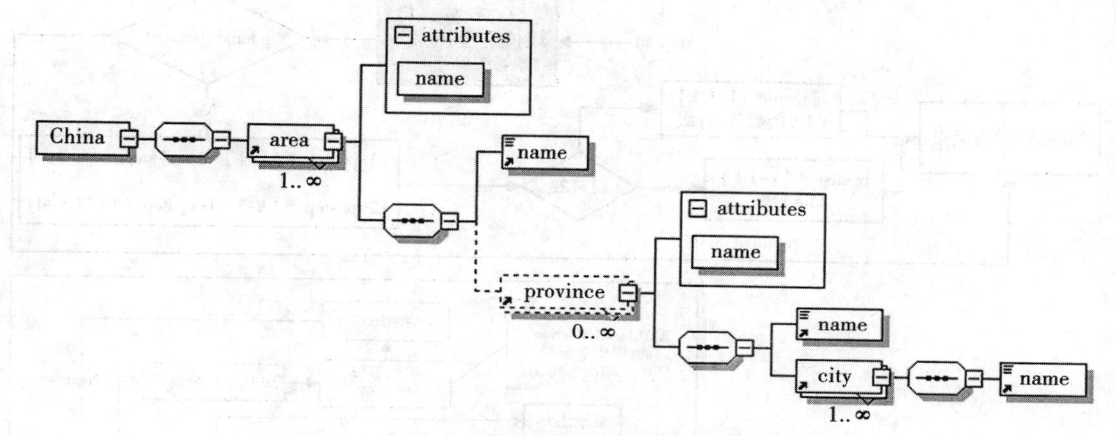

图 29.19　评价功能数据组织图

产品评价表		
编码	<pi>	Integer
产品编号		Number
名称		Text
质量评价		Text
价格评价		Text
描述		Text
发布时间		Date
举报		Integer
产品地址		Text
真名		Text
联系方式		Text
主标识符 <pi>		

productionEvaluation		
id	<pi>	Integer
productCode		Number
userName		Text
quality		Text
price		Text
description		Text
time		Date
isReport		Integer
productAddr		Text
trueName		Text
trueTel		Text
主标识符 <pi>		

图 29.20　评价功能数据组织关系

3. 功能实现流程

功能实现流程如下：

根据所有地区的 W 值进行分类划分：假设分为 k 类，首先对 W 排序 List<W>，然后筛选不重复数据 string[ ] p，若 p<k，则 k=p，若 p>=k，求得步长{p. length/k} min，若(p. length%k)= 0 ，则 W 分类将 List<W>按步长进行差异值划分即可，若(p. length%k)= 1，则 W 前(k-1)分类将 List<W>按步长进行差异值划分即可，后一数据需要比较 p 中前后数据，划分数据差的等级，最后数据为 k 等级。

算法过程如图 29. 21 所示。

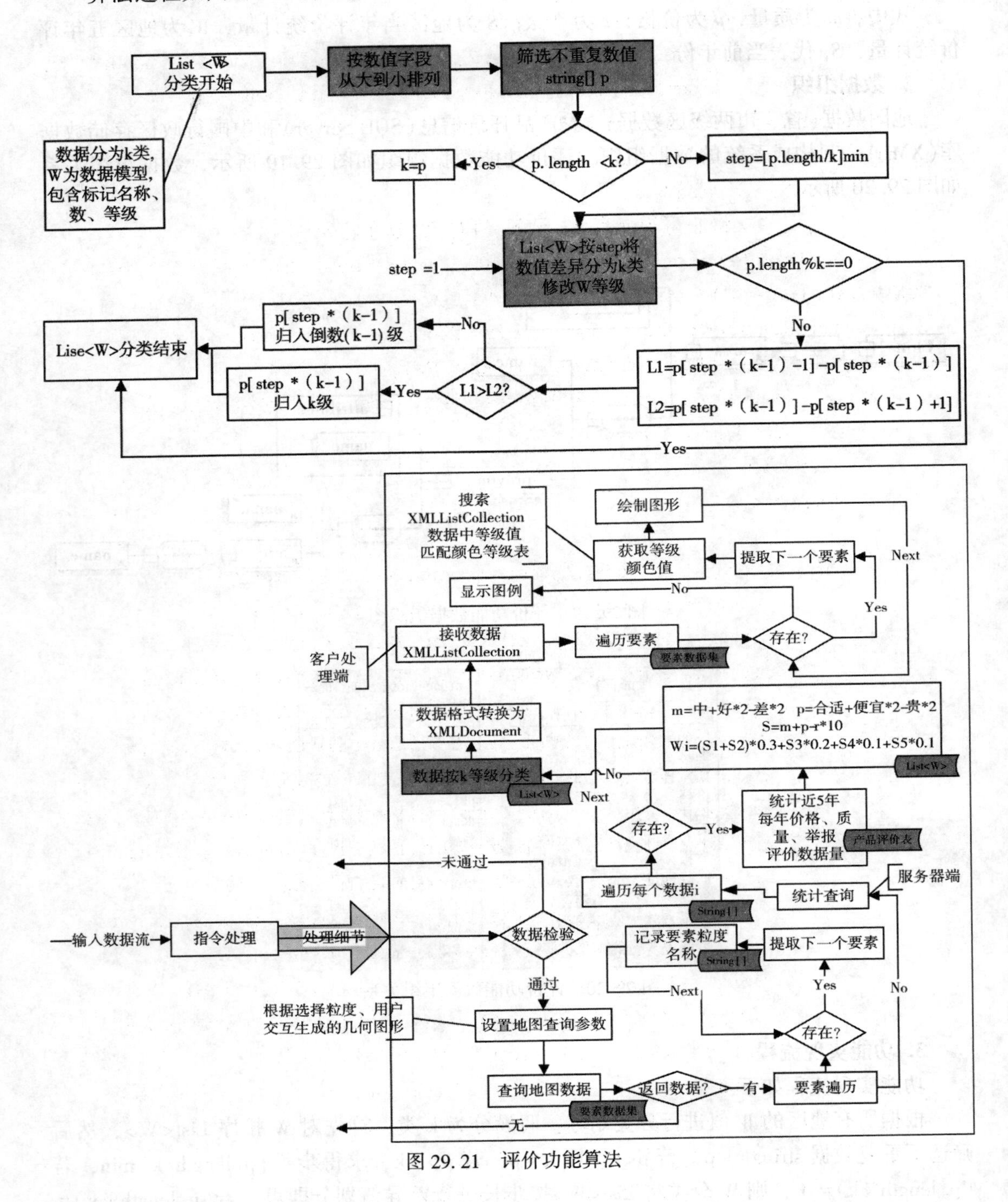

图 29. 21　评价功能算法

4. 案例代码

①功能交互界面：DegreeThemeControl. mxml，具体代码如下：

```
<?xml version="1.0" encoding="utf-8"?>
<s:TitleWindow xmlns:fx="http://ns.adobe.com/mxml/2009"
               xmlns:s="library://ns.adobe.com/flex/spark"
               xmlns:mx="library://ns.adobe.com/flex/mx"
    width="290" height="142" fontSize="16" textAlign="center"
  creationComplete="requireXMLData()">
    <fx:Declarations>
        <!-- 将非可视元素(如服务、值对象)放在此处 -->
    <s:HTTPService id="httpXml01" url="base/China.xml" result="
getXMLData(event)" resultFormat="e4x"/>
    <mx:RemoteObject id="serviceRemote" destination="fluorine"
source="FlashRemotingServiceLibrary.Sample">
        <mx:method name="GetDegreeTheme" result="onResult(event)"/>
        </mx:RemoteObject>
    </fx:Declarations>
    <fx:Script>
        <![CDATA[
            import Mapgis7.WebService.BasLib.AnyLine;
            import Mapgis7.WebService.BasLib.Arc;
            import Mapgis7.WebService.BasLib.CAttDataRow;
            import Mapgis7.WebService.BasLib.CAttDataSet;
            import Mapgis7.WebService.BasLib.CAttDataTable;
            import Mapgis7.WebService.BasLib.CGetObjByID;
            import Mapgis7.WebService.BasLib.CMapSelectAndGetAtt;
            import Mapgis7.WebService.BasLib.CMapSelectParam;
            import Mapgis7.WebService.BasLib.CWebSelectParam;
            import Mapgis7.WebService.BasLib.ESelectionType;
            import Mapgis7.WebService.BasLib.EnumLayerStatus;
            import Mapgis7.WebService.BasLib.GRegion;
            import Mapgis7.WebService.BasLib.SFeature;
            import Mapgis7.WebService.BasLib.SFeatureGeometry;
            import mx.collections.ArrayCollection;
            import mx.collections.ArrayList;
            import mx.collections.XMLListCollection;
            import mx.controls.Alert;
            import mx.core.UIComponent;
            import mx.messaging.AbstractConsumer;
            import mx.rpc.events.ResultEvent;
```

```
            import spark.events.IndexChangeEvent;
            import zdims.TextMarker;
            import zdims.drawing.CoordinateType;
            import zdims.drawing.GraphicsLayer;
            import zdims.drawing.IMSPolygon;
            import zdims.interfaces.control.INavigationBar;
            import zdims.map.IMSMap;
            import zdims.map.VectorMapDoc;
            import zdims.mark.MarkLayer;
            private var _map:IMSMap;
            private var mapDoc:VectorMapDoc;
            private var graphicsLayer:GraphicsLayer
            private var markLayer:MarkLayer;
            [Bindable]
            public var xmlList01:XMLListCollection;
            [Bindable]
            public var xmlPList01:XMLListCollection;
            public var xl:XMLList;
            private var degreeThemeList:XMLListCollection;
            private var layerIndex:int;
            private var classParam:int;
            private var pageNumber:int=0;
            private var whereString:String="";
            private var panScale:Boolean=true;
            private var panX:Number=-1;
            private var panY:Number=-1;
       private var colorArray:ArrayCollection = new
ArrayCollection([{id:1,name:"0xf1d61d"},{id:2,name:"0xf1b31d"},
{id: 3, name:"0xed8b0b"}, {id: 4, name:"0xf6bdba"}, {id: 5, name:"
0xfa5b4f"},{id:6,name:"0xf90a04"}]);
            public function set imMap(m:IMSMap):void{
              this._map=m;
            }
            public function get imMap():IMSMap
            {
              if(_map==null)
              imMap=this.parent as IMSMap;
              returnthis._map;
```

```
        }
        Public function set imVectorMapDoc(m:VectorMapDoc):void{
            this.mapDoc=m;
        }
        public function set imGraphicsLayer(m:GraphicsLayer):void
{
            this.graphicsLayer=m;
        }
        publicfunctionset imMarkLayer(m:MarkLayer):void{
            this.markLayer=m;
        }
        public function set imClassParam(m:int):void{
            this.classParam=m;
        }
        private function requireXMLData():void{
            httpXml01.send();
        }
        Private function getXMLData(event:ResultEvent):void{
            xmlList01=new XMLListCollection(event.result.area);
        }
        private function provinceListChange(event:IndexChan-
geEvent):void{
            xl=new XMLList(xmlList01[areaList.selectedIndex]);
            xmlPList01=new XMLListCollection(xl.province);
            provinceList.dataProvider=xmlPList01;
        }
        private function onResult(evt:ResultEvent):void{
            degreeThemeList=new XMLListCollection(evt.result.row);
        }
        private function enableStatu1():void{
            if(selectProvinceId.selected)
            {
                selectCityId.selected=false;
                areaList.enabled=false;
                provinceList.enabled=false;
            }else{
                selectCityId.selected=true;
                areaList.enabled=true;
```

```
            provinceList.enabled=true;
        }
    }
    private function enableStatu2():void{
        if(selectCityId.selected)
        {
            selectProvinceId.selected=false;
            areaList.enabled=true;
            provinceList.enabled=true;
        }else{
            selectProvinceId.selected=true;
            areaList.enabled=false;
            provinceList.enabled=false;
        }
    }
    private function analysis():void{
        this.clear();
        this.pageNumber=0;
        this.panScale=true;
        panX=-1;
        panY=-1;
        if(selectCityId.selected){
            if(provinceList.selectedIndex! =-1){
                this.layerIndex=3;
                this.whereString="ProvinceName LIKE '% "+
provinceList.selectedItem.name.substr(0,2).toString()+"%'";
                paramSelect();
            }else{
                Alert.show("请选择相应省份。");
            }
        }else{
            this.layerIndex=2;
            this.whereString="1=1";
            paramSelect();
        }
        if(degreeId.value>=3&&degreeId.value<=6){
            var stl:DegreeThemeLegend=new DegreeThemeLegend();
            stl.currentState="State"+(degreeId.value-2);
```

```
                stl.name="legend";
                stl.x=400;
                stl.y=410;
                if(classParam==1){
                stl.imIntro="数值越大,处罚次数越多";
                }else{
                    stl.imIntro="数值越大,综合欢迎度越高";
                }
                this._map.addElement(stl);
            }
        }
        public function paramSelect():void
        {
            this.mapDoc.getMapLayerInfo(layerIndex).LayerStatus
=EnumLayerStatus.Selectable;
            this.mapDoc.updateAllLayerInfo();
            this._map.activeMapDoc=this.mapDoc;
    this._map.activeMapDoc.activeLayerIndex=layerIndex;   //激活
图层索引号为? 的图层
            var mapsel:CMapSelectParam=new CMapSelectParam();
            var websel:CWebSelectParam=new CWebSelectParam();
            websel.CompareRectOnly=this.mapDoc.compareRectOnly;
            websel.Geometry=null;
            websel.MustInside=this.mapDoc.mustInside;
    websel.SelectionType=ESelectionType.Condition;//设置查询的类
型为条件查询
            websel.NearDistance=this.mapDoc.nearDistanse;
            websel.WhereClause=this.whereString;//设置查询的条件
            mapsel.SelectParam=websel;
            while(pageNumber<2){
                mapsel.PageCount=pageNumber;
                this.mapDoc.select(mapsel,callBack);
                pageNumber++;
            }
        }
        public function callBack(e:Event):void{
            this._map.activeMapDoc=this.mapDoc;    //激活当前文档
      this._map.activeMapDoc.activeLayerIndex=layerIndex;  //
```

激活图层索引号为？的图层

```
            var result: CMapSelectAndGetAtt = this.mapDoc.onSelect
(e);//获取查询结果的属性集
              if((result==null || result.Count[0][layerIndex]==0)
&&pageNumber!=2)
                    {
                        Alert.show("未查到要素,请重新点击查询");
                        return;
                    }
                    else
                    {
        var adt: CAttDataTable = (result.AttDS[0] as CAttDataSet)
.attTables[layerIndex] as CAttDataTable;
                          for(var j:int=0;j<adt.Rows.length;j++){
                          var targetObj:CGetObjByID=new CGetObjByID();
                              targetObj.FeatureID=(adt.Rows[j] as
CAttDataRow).FID;
                              targetObj.LayerIndex=layerIndex;
                                this.mapDoc.getFeatureByID(targetObj,
onGetFeature);
                          }
                    }
                }
                publicfunction onGetFeature(e:Event):void
                {
                    var sf:SFeature=this.mapDoc.onGetFeatureByID(e);
                    var sfg:SFeatureGeometry=sf.fGeom;
        drawGetFeatrue(0x00FF00,colorArray[new int(Math.random()*
this.degreeId.value)].name,sfg);
                    var titleMarker:TextMarker=new TextMarker();
                    titleMarker.name="名称";
                    titleMarker.imsmap=this._map;
                            titleMarker.logicX = (sf.bound.xmax +
sf.bound.xmin)/2;
                            titleMarker.logicY = (sf.bound.ymax +
sf.bound.ymin)/2;
                    titleMarker.enableShowName=true;
                    if(panScale&&this.layerIndex==3){
```

```
            panX=(sf.bound.xmax+sf.bound.xmin)/2;
            panY=(sf.bound.ymax+sf.bound.ymin)/2;
            this.panScale=false;
        }
        if(layerIndex==2){
            titleMarker.IMarkName=sf.AttValue[3].toString();
        }else{
            titleMarker.IMarkName=sf.AttValue[0].toString();
        }
        this._map.addChildAt(titleMarker,this._map.
layerLenth);
    }
    //绘制显示要素
    public function drawGetFeatrue
  (boderColor:uint,colorTemo:uint,sfeatureTemp:SFeatureGeometry):
IMSPolygon
  {
        var reg:GRegion=sfeatureTemp.GetRegGeom(0);
        var arcLine:AnyLine=reg.GetRings(0);
        var arc:Arc=arcLine.GetArcs(0);
        //获取要素边界的点
        var ArcPnt:Array=arc.Dots;
        var ArcPntNum:int=arc.Dots.length;
        //绘制多边形
        var poly:IMSPolygon=new IMSPolygon(Coordinate-
Type.Logic);
        graphicsLayer.addGraphics(poly);
        for(var i:int=0;i<ArcPntNum;i++){
  poly.points.push(new Point(arc.GetDots(i).x, arc.GetDots(i).y));
        }
        //设置边界线颜色及大小
        poly.lineStyle(2,boderColor);
        //设置填充区的颜色及透明度
        poly.beginFill(colorTemo,0.8);
        //绘制图形
        poly.draw();
        return poly;
    }
```

```
                public function clear():void{
                    clearPathDot();
                    this.markLayer.removeAllElements();
                    this.graphicsLayer.removeAllElements();
                }
                public function clearPathDot():void
                {
                    var mark:Array = this._map.getChildren();
                    for (var i:int = 0; i < mark.length; i++)
                    {
                        if (mark[i] is TextMarker&&UIComponent(mark
[i]).name == "名称")
                            this._map.removeChild(mark[i]);
                        if(mark[i] is DegreeThemeLegend&&UIComponent
(mark[i]).name
    =="legend")
                            this._map.removeChild(mark[i]);
                    }
                }
            ]]>
        </fx:Script>
        <s:CheckBox id="selectProvinceId" x="88" y="7" label="省(直
辖市)" selected="true" click="enableStatu1()"/>
        <s:CheckBox id="selectCityId" x="209" y="8" label="地级市"
click="enableStatu2()"/>
        <s:Label x="9" y="11" text="最小单位"/>
        <s:DropDownList id="areaList" x="69" y="34" width="85" ena-
bled = " false " dataProvider = " {xmlList01}" labelField = " name "
change="provinceListChange(event)"></s:DropDownList>
        <s:DropDownList id="provinceList" x="166" y="34" enabled="
false" labelField="name"></s:DropDownList>
        <s:Label x="10" y="36" enabled="false" text="选择省"/>
        <s:Label x="10" y="71" text="统计等级"/>
        <s:NumericStepper id="degreeId" x="91" y="61" width="53"
height="33" maximum="6" minimum="3"/>
        <mx:LinkButton x="153" y="59" label="执行统计" color="#
1081B6" fontSize="15" fontWeight="bold"
    textDecoration="underline" click="analysis()"/>
```

```
<mx:LinkButton x="192" y="81" label="清除操作" color="#
1081B6" fontSize="15" fontWeight="bold"
textDecoration="underline" click="clear()"/>
</s:TitleWindow>
```

②图例界面：DegreeThemeLegend.mxml，具体代码如下：

```
<? xml version="1.0" encoding="utf-8"? >
<mx:Canvas xmlns:fx="http://ns.adobe.com/mxml/2009"
           xmlns:s="library://ns.adobe.com/flex/spark"
           xmlns:mx="library://ns.adobe.com/flex/mx"
           width="148" height="114"  backgroundColor="#27E160"
           backgroundAlpha="0.7">
    <mx:states>
        <s:State name="State1"/>
        <s:State name="State2"/>
        <s:State name="State3"/>
        <s:State name="State4"/>
    </mx:states>
    <fx:Declarations>
        <! -- 将非可视元素(如服务、值对象)放在此处 -->
    </fx:Declarations>
    <fx:Script>
        <! [CDATA[
            [Bindable]
            private var info:String;
            public functionset imIntro(m:String):void{
                info=m;
            }
        ]]>
    </fx:Script>
    <s:Label x="48" y="27" width="28" height="15" backgroundCol-
or="#F1D61D" textAlign="left"verticalAlign="middle" x.State2="11"
y.State2="37"
    x.State3="18" y.State3="27" x.State4="18" y.State4="27"/>
    <s:Label x="48" y="50" width="28" height="15" backgroundCol-
or="#F1B31D" textAlign="left" verticalAlign="middle"x.State2="81"
y.State2="37"
    x.State3="18" y.State3="50" x.State4="18" y.State4="50"/>
    <s:Label x="48" y="73" width="28" height="15" backgroundCol-
```

```
or="#ED8B0B" textAlign="left" verticalAlign="middle" x.State2="
11" y.State2="59"
       x.State3="18" y.State3="73" x.State4="18" y.State4="73"/>
       <s:Label visible="false" x="81" y="48" width="28" height="
15" backgroundColor="#FA5B4F" visible.State3="true" x.State3="81"
y.State3="48"
       visible.State4="true" x.State4="81" y.State4="48"/>
       <s:Label visible="false" x="81" y="25" width="28" height="
15" backgroundColor="#F6BDBA" visible.State2="true" x.State2="81"
y.State2="59"
       visible.State3="true" visible.State4="true"/>
       <s:Label visible="false" x="81" y="75" width="28" height="
15" backgroundColor="#F90A04" visible.State4="true"/>
       <s:Label x="56" y="6" color="#0E587B" fontFamily="中易宋体"
fontSize="14" fontStyle="normal"fontWeight="bold" text="图例"/>
       <s:Label id="introduce" x="4" y="96" text="{info}"/>
       <s:Label x="86" y="28" text="1" x.State2="49" y.State2="38"
       x.State3="56" y.State3="28" x.State4="56" y.State4="28"/>
        <s:Label visible="false" x="119" y="28" text="4"
visible.State2="true"
        x.State2="119" y.State2="62" visible.State3="true"
visible.State4="true"/>
       <s:Label visible="false" x="119" y="51" text="5"
          visible.State2="false" visible.State3="true"
visible.State4="true"/>
       <s:Label visible="false" x="119" y="76" text="6"
          visible.State2="false" visible.State3="false"
visible.State4="true"/>
       <s:Label x="86" y="51" text="2"
       x.State2="119" y.State2="38" x.State3="56" y.State3="51"
       x.State4="56" y.State4="51"/>
       <s:Label x="86" y="76" text="3"
       x.State2="49" y.State2="62" x.State3="56" y.State3="76"
    x.State4="56" y.State4="76"/>
    </mx:Canvas>
```

5. 功能实现截图

功能实现交互操作界面如图 29.22、图 29.23 所示。

图 29.22　大粒度评价功能实现

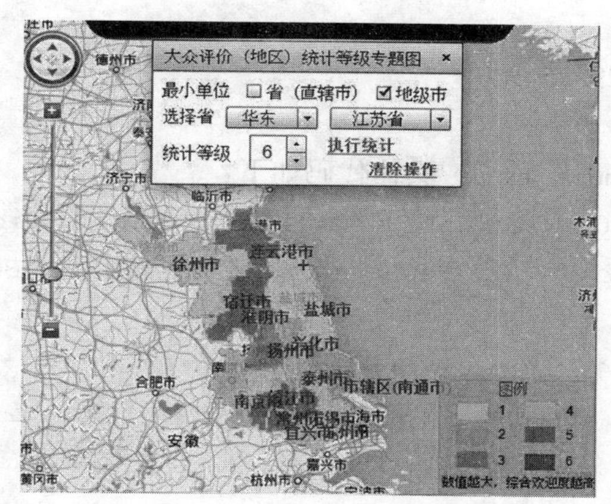

图 29.23　小粒度评价功能实现

29.2.5　用户地址搜索之灰色匹配

1. 需求分析

用户输入信息比较具有随意性，所以应该设计一个流程能够最大限度地获得用户满意结果。

若数据库中有以下几个地址：

山东省青岛经济技术开发区前湾港路 579 号 山东科技大学；

山西省太原市万柏林区窊流路 66 号太原科技大学；

山东省青岛市松岭路 99 号 青岛科技大学；

山东省青岛市宁夏路 308 号 青岛大学。

用户输入“山科大”或“山科大青岛”想查找的是第一条数据，输入“青科大”或“青岛科技大学”想查找第三条数据，输入“青岛大学”想查找第四条数据，当然用户还会输入很多其他的信息。

现在所面临的问题是如何通过用户输入的有限信息中提取到用户最终想要得到的数据。

2. 数据组织

中国地点地理信息表(SQL Server)如图 29.24 所示。

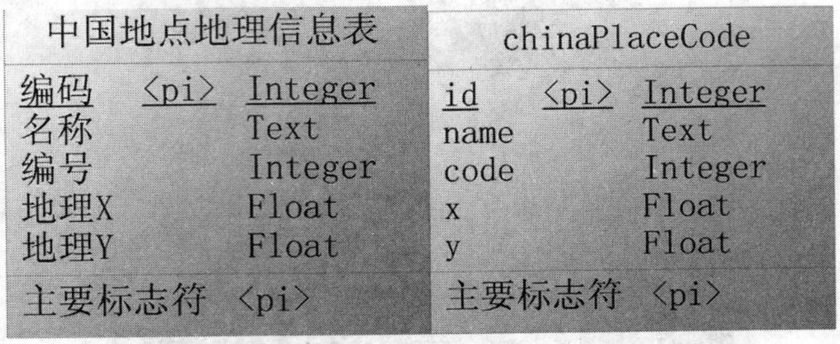

中国地点地理信息表			chinaPlaceCode		
编码	<pi>	Integer	id	<pi>	Integer
名称		Text	name		Text
编号		Integer	code		Integer
地理X		Float	x		Float
地理Y		Float	y		Float
主要标志符	<pi>		主要标志符	<pi>	

图 29.24　匹配功能数组组织

3. 功能实现流程

功能实现流程如图 29.25 所示。

4. 案例代码

①功能交互界面：index. mxml，具体代码如下：

```
<mx:Canvas y="80" width="296" height="30" backgroundAlpha="0.35"
backgroundColor="#CCCCCC" borderColor="#B7BABC" borderStyle=
"solid"
cornerRadius="16" horizontalCenter="0">
<s:ComboBox id = "positionSearch" x = "10" y = "0" width = "252"
height="27"
change="positionSearch_changeHandler(event)" dataProvider=
"{placeListTopFive}"
labelField = " addrName "  textInput = " positionSearch _ textIn-
putHandler(event)"
fontFamily="Arial" fontSize="16"/>
<s:Image x="262" y="2" width="25" height="25" buttonMode="true"
click="PositionFind()" source="@ Embed('img/widget/search.png
')" toolTip="信息查找"/>
</mx:Canvas>
```

Flex AS 控制代码如下：

```
public var SQLS:String;
public var characters:String;
public var extract:int;
```

```
private var drawMark:Boolean=false;
[Bindable]
public var placeList:XMLListCollection;
[Bindable]
public var placeListTopFive:XMLListCollection;
private var inputTextPre:String="";
[Bindable]
[Embed("img/widget/local.png")]
public static var testIcon:Class;
public function PositionFind():void{
    var inputText:String=positionSearch.textInput.text;
    preDeal(inputText);
    if(inputText=="" || (characters==""&&extract==0)){
        this.mapContainer.panTo(11574400.571054548,4363637.0707441345);
        this.mapContainer.setLevel(0);
        this.mk.removeAllChildren();
    }else{
        //提交服务器端获取数据
        serviceRemote.GetAddrPlaceList(SQLS,characters);
        drawMark=true;
    }
}
protected function positionSearch_textInputHandler(event:TextEvent):void
{
    //TODO Auto-generated method stub
    var inputText:String=positionSearch.textInput.text;
    preDeal(inputText);
    if(inputText=="" || (characters==""&&extract==0)){
        this.mapContainer.panTo(11574400.571054548,4363637.0707441345);
        this.mapContainer.setLevel(0);
        this.mk.removeAllChildren();
    }else{
        //提交服务器端获取数据
        serviceRemote.GetAddrPlaceList(SQLS,characters);
    }
}
private function onResultGetAddPlace(evt:ResultEvent):void{
```

```
        placeList=new XMLListCollection(evt.result.row);
        if(placeList! =null&&placeList.length<=5){
            placeListTopFive=placeList;
        }elseif(placeList! =null&&placeList.length>5){
             var placeListTopPreFive:XMLListCollection = new XMLList-
Collection();
            placeListTopPreFive.addItem(placeList[0]);
            placeListTopPreFive.addItem(placeList[1]);
            placeListTopPreFive.addItem(placeList[2]);
            placeListTopPreFive.addItem(placeList[3]);
            placeListTopPreFive.addItem(placeList[4]);
            placeListTopFive=placeListTopPreFive;
        }
        this.mapContainer.panTo(11574400.571054548,4363637.0707441345);
        this.mapContainer.setLevel(0);
        this.mk.removeAllChildren();
        if(drawMark&&placeList! =null){
            //———地图绘制标注点———
            for(var b:int=0;b<placeList.length;b++)
            {
                    addMark ( new Number ( placeList [ b ] .x ), new Number
(placeList[b].y));
            }
            drawMark=false;
        }
    }
    protected function positionSearch_changeHandler ( event: Index-
ChangeEvent):void
    {
        //TODO Auto-generated method stub
        if(inputTextPre! =positionSearch.selectedItem.addrName){
            inputTextPre=positionSearch.selectedItem.addrName;
            this.mk.removeAllChildren();
            addMark(new Number(placeList[positionSearch.selectedIndex].
x),new
    Number(placeList[positionSearch.selectedIndex].y));
            mk.enableMarkHiden=false;
            this.mapContainer.panTo(new
```

```
    Number(placeList[positionSearch.selectedIndex].x),new
Number(placeList[positionSearch.selectedIndex].y));
            this.mapContainer.setLevel(4);
        }
    }
    private function addMark(x:Number,y:Number):void{
        var mark:IMSMark;
        var img:mx.controls.Image;
        img=new mx.controls.Image();
        img.source=testIcon;
        mark=new IMSMark(img,CoordinateType.Logic);
        mark.x=x;
        mark.y=y;
        mark.offsetX=16;
        mark.offsetY=32;
        mark.enableAnimation=true;
        mk.addMark(mark);
    }
    public function preDeal(m:String):void{
        //(1)去掉空格/标点符号
        var j:String=" |,|.|。|\"|“|、|”|[|]|'|‘|:|:|\\|/|!|!|`|(|)|(|)|\
||;|;";
        var pureString:String=m;
        var charact:Array=new Array();
        var splitCharact:Array=j.split('|');
        var myPattern:RegExp = /\s/gi;
        var myPattern1:RegExp = /[^\u4e00-\u9fa5a-zA-Z0-9]/gi;
        pureString=pureString.replace(myPattern1,"");
        //(2)将数字/字母整体提取等待 d s 等待 4 | 5 等待 we 等待 2 1 等待
        myPattern1 = /\d+/gi;
        var mathArr:Array=pureString.match(myPattern1);
        //Alert.show(mathArr[0]+"--"+mathArr[1]);
        for(var k:int=0;k<mathArr.length;k++){
            charact.push(mathArr[k]);
            pureString=pureString.replace(mathArr[k], "");
        }
        myPattern1 = /[A-Za-z]+/gi;
        mathArr=pureString.match(myPattern1);
```

```
    for(var k:int=0;k<mathArr.length;k++){
        charact.push(mathArr[k].toString().toLocaleLowerCase
());
        pureString=pureString.replace(mathArr[k], "");
    }
    //(3)将剩余字符去掉重复值
    var filter:String="";
    for(var p:int=0;p<pureString.length;p++){
        if(filter.match(pureString.charAt(p))==null)
        {
            filter=filter.concat(pureString.charAt(p));
        }
    }
    //(4)生成 SQL 语句
    var SQL:String="SELECT * FROM chinaPlaceCode Where";
    if(charact.length>0 || filter.length>0){
        for(var c:int=0;c<charact.length;c++){
            if(filter.length==0&&c==charact.length-1)
  {
    SQL+="((name LIKE '% "+charact[c]+"%') OR (english LIKE '% "
  +charact[c]+"%'))";
            }else{
                SQL+="((name LIKE '% "+charact[c]+"%') OR (english
 LIKE '% "
   +charact[c]+"%')) AND ";
            }
        }
        for(var f:int=0;f<filter.length;f++){
            if(f==filter.length-1){
                SQL+="(name LIKE '% "+filter.charAt(f)+"%')";
            }else{
                SQL+="(name LIKE '% "+filter.charAt(f)+"%') AND ";
            }
        }
    }
    SQLS=SQL;
    extract=charact.length;
    characters=pureString;
```

```
}
```

②Flex<fx：Declarations> 远程交互代码如下：

```
<mx:RemoteObject id="serviceRemote" destination="fluorine"
source="FlashRemotingServiceLibrary.Sample">
<mx:method name="GetAddrPlaceList" result="onResultGetAddPlace
(event)"/>
</mx:RemoteObject>
```

③FlashRemotingServiceLibrary. Sample 文件 GetAddrPlaceList 方法如下：

```
Public XmlDocument GetAddrPlaceList(string SQLS, string characters)
{
GeoSearch gs = newGeoSearch();
return gs.getAddrList(SQLS, characters);
}
```

④FlashRemotingServiceLibrary. BlurAddr 文件如下：

```
using System;
using System.Collections.Generic;
using System.Linq;
using System.Text;
namespace FlashRemotingServiceLibrary
{
Public class BlurAddr
    {
Private int wordcount = 0;
Private int wordLong = 0;
Private string x;
Private string y;
Private string name;
public BlurAddr()
        {
        }
public BlurAddr(int wordc, int wordl, string x, string y, string name)
        {
this.wordcount = wordc;
this.wordLong = wordl;
this.x = x;
this.y = y;
this.name = name;
        }
```

```
public string X
        {
set{this.x = value;}
get { returnthis.x; }
        }
public string Y
        {
set { this.y = value; }
get { returnthis.y; }
        }
public int Wordcount
        {
set { this.wordcount += value; }
get { returnthis.wordcount; }
        }
public int WordLong
        {
set { this.wordLong = value; }
get { returnthis.wordLong; }
        }
public string Name
        {
set { this.name = value; }
get { returnthis.name; }
        }
    }
}
```

⑤FlashRemotingServiceLibrary. GeoSearch 文件相关方法如下：

```
#region 获取用户输入的匹配地址
Public XmlDocument getAddrList(string SQLS, string characters)
{
DataTable dt = requireSQLAddr(SQLS);
List<BlurAddr> addList = newList<BlurAddr>();
if (dt ! = null&& dt.Rows.Count > 0)
    {
for (int i = 0; i < dt.Rows.Count; i++)
            {
if (dt.Rows[i][1].ToString() = = "" || dt.Rows[i][3].ToString
```

```
() = = "" ‖ dt.Rows[i][4].ToString() = = "")
                {
 continue;
                }
 else
                {
 BlurAddr ba = coreDealFunction(dt.Rows[i], characters);
 if (ba ! = null)
                       {
                            insertSort(ref addList, ba);
                       }
 else
                       {
 continue;
                       }
                }
            }
 return createXMlD(addList);
     }
 else
     {
 returnnull;
     }
 }
 #endregion
 #region 根据 SQL 初步获取数据
 publicDataTable requireSQLAddr(string SQLS)
 {
 string strConnection = "data source=localhost;uid=" + SqlLogin.uid + ";
 pwd=" + SqlLogin.pwd + ";database=" + SqlLogin.database;
 SqlConnection objConnection = newSqlConnection(strConnection);
 SqlCommand objCommand = newSqlCommand("", objConnection);
     objCommand.CommandText =SQLS;
 try
     {
 if (objConnection.State == System.Data.ConnectionState.Closed)
         {
             objConnection.Open();
```

```
        }
    SqlDataAdapter mydata = new SqlDataAdapter(objCommand);
    DataSet ds = new DataSet();
          mydata.Fill(ds,"BluAddr");
    SqlCommandBuilder objcmdBuilder = new SqlCommandBuilder(mydata);
    DataTable dt = ds.Tables["BluAddr"];
    return dt;
       }
    catch (SqlException e)
       {
    Console.Write(e.Message.ToString());
    returnnull;
       }
    finally
       {
    if (objConnection.State == System.Data.ConnectionState.Open)
          {
              objConnection.Close();
          }
       }
    }
    #endregion
    #region 生成 XMLDocument 参数 DataTable dt
    Public XmlDocument xmlFunction(DataTable dt)
    {
    XmlDocument doc = new XmlDocument();
        doc.AppendChild(doc.CreateXmlDeclaration("1.0", "utf-8",
null));//声明 XML 开头,根元素

    XmlElement xmlnode = doc.CreateElement("BlurAddr");
    if (dt.Rows.Count != 0)
       {

    for (int i = 0; i < dt.Rows.Count; i++)//遍历行
          {
    DataRow dr = dt.Rows[i];
    //为 XML 文档加入元素/加入一个根元素
    XmlElement xmlelem = doc.CreateElement("", "row", "");
```

```
 //增加一个子元素
 XmlElement xmlelem1 = doc.CreateElement("addrName");
 XmlText xmltext = doc.CreateTextNode(dr[1].ToString());
           xmlelem1.AppendChild(xmltext);
           xmlelem.AppendChild(xmlelem1);

 XmlElement xmlelem2 = doc.CreateElement("x");
           xmltext = doc.CreateTextNode(dr[3].ToString());
           xmlelem2.AppendChild(xmltext);
           xmlelem.AppendChild(xmlelem2);

 XmlElement xmlelem3 = doc.CreateElement("y");
           xmltext = doc.CreateTextNode(dr[4].ToString());
           xmlelem3.AppendChild(xmltext);
           xmlelem.AppendChild(xmlelem3);

           xmlnode.AppendChild(xmlelem);
        }
        doc.AppendChild(xmlnode);
    }
return doc;
}
#endregion
#region 词匹配统计
Public BlurAddr coreDealFunction(DataRow dr,string target)
{
BlurAddr ba = new BlurAddr();
    ba.Name = dr[1].ToString();
    ba.X =dr[3].ToString();
    ba.Y =dr[4].ToString();
int wordLong = 0;
int wordcount = 0;
string standart = ba.Name;

//计算 wordcount wordLong
for (int k = 0; k < target.Length -1; k++)
    {
string key = target.Substring(k, 1);
```

```
List<int> lp = newList<int>();
        requireIndex(ref lp, key, standart);
if (lp.Count ! = 0)
        {
int innerWordLong = 0;
int innerWordcount = 0;
int s = standart.Length;
foreach (var item in lp)
            {
//standart 从 index=item 开始忽略下面字的多处
int v = 1;
int bit = 1;
string biaozhi = "无";
int jianluePre = -1;
for (int m = 1; m < s -item -1; m++)
                {
if (target.Substring(k + v, 1) = = standart.Substring(item + m, 1))
                    {
if (jianluePre = = m -1)
                        {//前后两个词无空隙则退出
                            innerWordcount += m;
break;
                        }
if (m = = 1)
                        {//连词
                            biaozhi ="连词";
                        }
else if (biaozhi ! = "连词")
                        {//简略词
                            biaozhi ="简略词";
                            jianluePre = m;
                        }
                        bit++;
                        v++;
if (v = = target.Length -k)
                        {
if (biaozhi = = "连词")
```

```
                                    {
 if (innerWordLong < bit)
                                        {
                                            innerWordLong = bit;
                                        }
                                        innerWordcount += bit;
                                    }
else if (biaozhi == "简略词")
                                    {
                                        innerWordcount += m;
                                    }
break;
                                }
                            }
if (biaozhi == "连词")
                            {
//标准顺延词数与输入顺延词数不同退出
if (m ! = bit -1)
                                {
if (innerWordLong < bit)
                                    {
                                        innerWordLong = bit;
                                    }
                                    innerWordcount += bit;
break;
                                }
                            }
if (biaozhi == "连词"&&m == (s -item -2))
                            {
//标准顺延词数与输入顺延词数不同退出
if (innerWordLong < bit)
                                {
                                    innerWordLong = bit;
                                }
                                innerWordcount += bit;
                            }
if (biaozhi == "简略词"&& m == (s -item -2))
```

```
                    {
                        innerWordcount += m;
                    }
                }
            }
    if (wordLong < innerWordLong)
            {
                wordLong = innerWordLong;
            }
            wordcount += innerWordcount;
        }
    else
        {
    continue;
        }
    }
        ba.Wordcount = wordcount;
        ba.WordLong = wordLong;
    return ba;
    }
    #endregion
    #region 数据插入排序
    Public void insertSort(refList<BlurAddr> addList, BlurAddr ba)
    {
    Boolean ok = false;
    if (addList.Count > 0)
        {
    for (int i = 0; i < addList.Count; i++)
        {
    if (ba.WordLong > addList[i].WordLong)
            {
                addList.Insert(i, ba);
                ok =true;
    break;
            }
    else if (ba.WordLong == addList[i].WordLong && ba.Wordcount >=
addList[i].Wordcount)
```

```
                    {
                        addList.Insert(i, ba);
                        ok =true;
    break;
                    }
                }
    if (! ok)
                {
                    addList.Add(ba);
                }
            }
    else
            {
                addList.Add(ba);
            }
    }
    #endregion
    #region 生成 XMLDocument 参数 List<BlurAddr>
    publicXmlDocument createXMlD(List<BlurAddr> bl)
    {
    XmlDocument doc = newXmlDocument();
        doc.AppendChild(doc.CreateXmlDeclaration("1.0", "utf-8",
null));//声明 XML 开头,根元素
    XmlElement xmlnode = doc.CreateElement("BlurAddr");
    if (bl.Count! = 0)
        {
    for (int i = 0; i < bl.Count; i++) //遍历行
            {
    BlurAddr ba = bl[i];
    //为 XML 文档加入元素/加入一个根元素
    XmlElement xmlelem = doc.CreateElement("", "row", "");
    //增加一个子元素
    XmlElement xmlelem1 = doc.CreateElement("addrName");
    XmlText xmltext = doc.CreateTextNode(ba.Name);
                xmlelem1.AppendChild(xmltext);
                xmlelem.AppendChild(xmlelem1);
    XmlElement xmlelem2 = doc.CreateElement("x");
```

```
            xmltext = doc.CreateTextNode(ba.X.ToString());
            xmlelem2.AppendChild(xmltext);
            xmlelem.AppendChild(xmlelem2);
    XmlElement xmlelem3 = doc.CreateElement("y");
            xmltext = doc.CreateTextNode(ba.Y.ToString());
            xmlelem3.AppendChild(xmltext);
            xmlelem.AppendChild(xmlelem3);
            xmlnode.AppendChild(xmlelem);
        }
        doc.AppendChild(xmlnode);
      }
    return doc;
    }
    #endregion
    #region 获取位置数组
    public void requireIndex (refList < int > lp, string key, string
standart)
    {
    int index = 0;
    foreach(var item in standart){
    if (item.ToString() = = key)
        {
            lp.Add(index);
        }
        index++;
      }
    }
    #endregion
```

5. 功能实现截图

功能实现交互操作界面如图 29.26～图 29.32 所示。模块可位于网站顶部，用户可以输入地址信息，灰色匹配进行查询，返回最优结果。

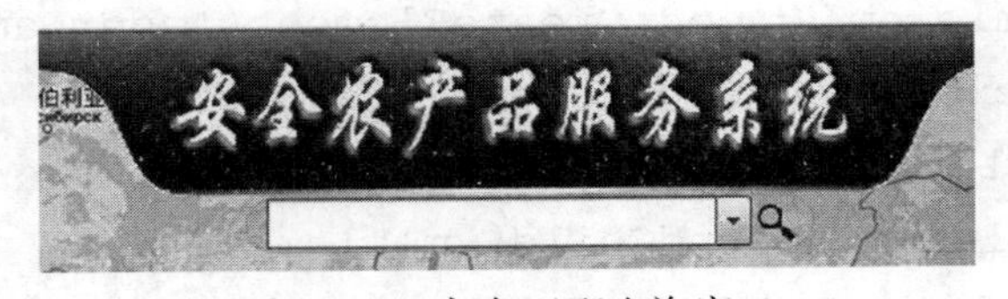

图 29.26 灰色匹配查询窗口

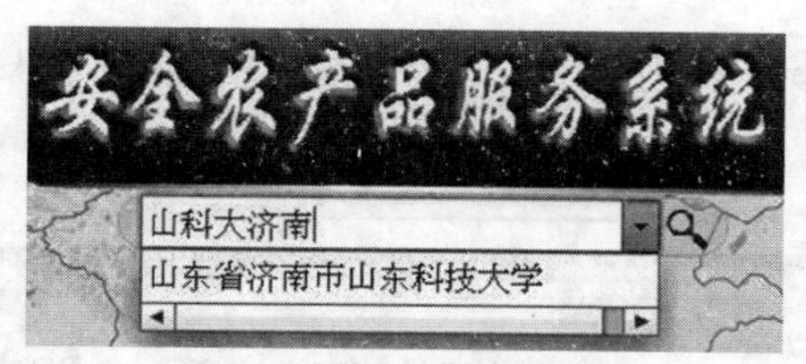

图 29.27　输入“山科大济南”

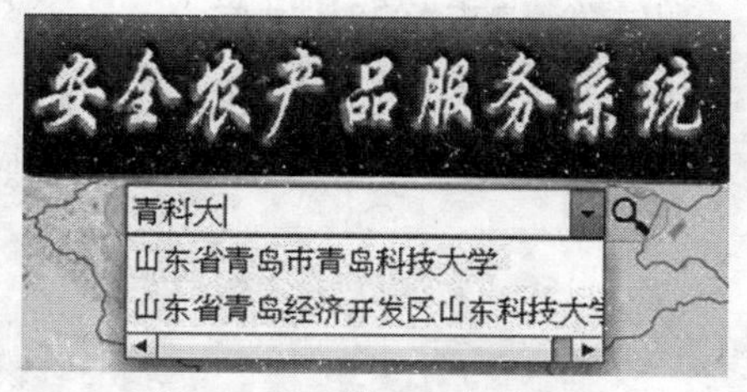

图 29.28　输入“青科大”

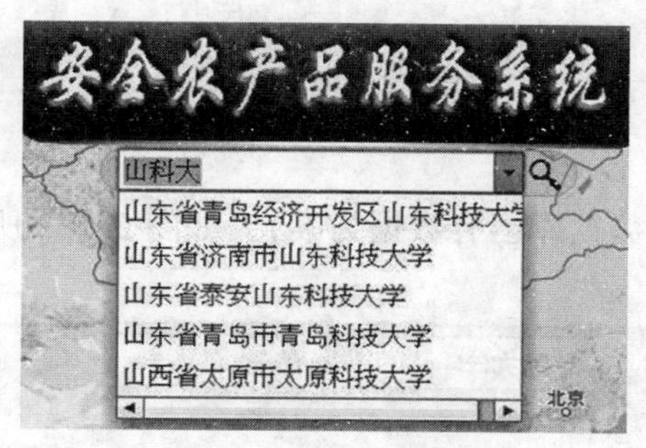

图 29.29　输入“山科大”

点击第一条信息(或其他信息)，则地图会放大移动到相应位置，且此位置有图标显示，如图 29.30 所示。

图 29.30　匹配到相应位置

输入“青岛大学”点击🔍，地图会缩放至全图，移动至中央(数据以数据库已有数据为准)如图 29. 31 所示。

图 29. 31 缩放居中显示

放大查看详情(数据以数据库已有数据为准)，如图 29. 32 所示。

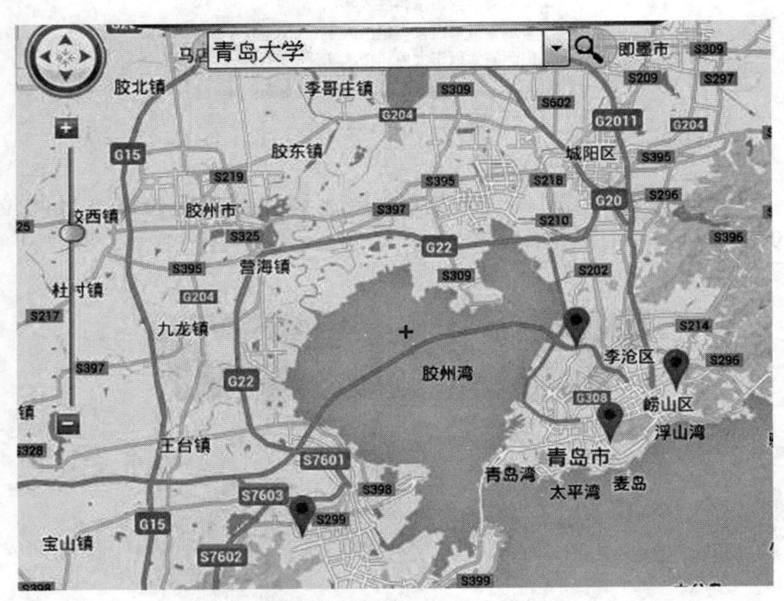

图 29. 32 匹配功能实现

输入框清空，点击🔍，即可清除地图标注数据。且地图缩放至全图，移动至中央。

第六编　SuperMap for JavaScript 开发

第 30 章　SuperMap for JavaScript 初级开发

30.1　平台简介

SuperMap iClient 7C for JavaScript 是一款在服务式 GIS 架构体系中面向 HTML 5 的应用开发，支持多终端、跨浏览器的客户端开发平台。SuperMap iClient 7C for JavaScript 采用 HTML + CSS + JavaScript 的开发组合，无需安装任何插件，便可在终端浏览器上实现美观的地图呈现、动态实时的要素标绘，以及与多源 GIS 服务的高效交互，快速构建内容丰富、响应迅速、体验流畅的地图应用，同时支持离线存储与访问地图功能，满足用户在离线状态下的地图应用。

SuperMap iClient 7C for JavaScript 目录结构如图 30.1 所示，包括以下几个部分：

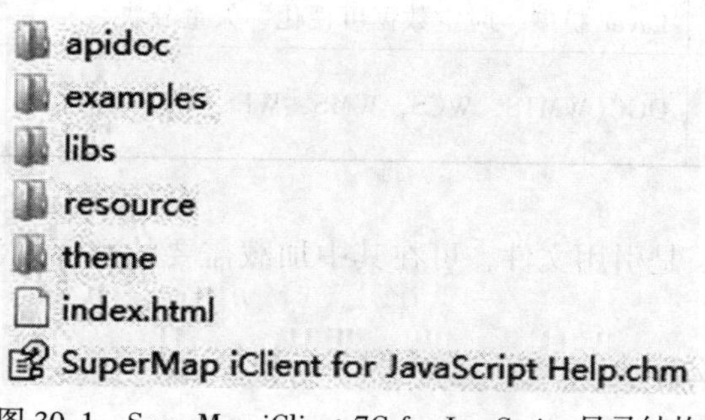

图 30.1　SuperMap iClient 7C for JavaScript 目录结构

1. apidoc

apidoc 存放产品的类参考，点击 index. html 文件可以查看产品的控件和所有接口的列表。

2. examples

examples 存放 SuperMap iClient 7C for JavaScript 产品页面及其相关资源，产品页面包括产品介绍、开发指南、示范程序、类参考、技术专题文档，其中：

①产品介绍。给用户介绍了产品是什么，其体系架构、功能和产品包结构等。该部分主要是从整体上帮助使用者了解产品。

②产品开发指南。介绍了产品包获取方法以及产品包的使用方法和基本开发流程，方便用户快速掌握产品的基本开发方法。

③示范程序。提供了产品所有示范代码的使用，给开发者提供了案例参考。

④类参考。提供所有控件、对象的接口列表，为开发者提供了接口使用参考。

⑤技术专题。介绍了产品的一些关键技术，主要涉及高性能矢量渲染、动态分段、离线缓存与 App 等。

3. libs

libs 文件夹存放产品的库文件、引用文件以及用于国际资源化的文件，见表 30. 1。其中包括 SuperMap. js、SuperMap_Basic. js、SuperMap_IServer. js、SuperMap_Visualization. js、SuperMap_OGC. js 和 SuperMap. Includ. js，以及 lang 文件夹。SuperMap. js 为总库文件，是 4 个分库的总和，支持所有功能。

表 30. 1　**库文件分类**

分库名称	实 现 功 能
基础库 SuperMap_Basic. js	基础地图显示、基础地图操作、添加矢量要素及标记物、添加控件、添加信息框、地图查询
iServer 服务库 SuperMap_IServer. js	数据集查询、服务器专题图、空间分析、网络分析、交通换乘分析
可视化库 SuperMap_Visualization. js	客户端专题图、热点图、聚散点图层、UTFGrid 图层、麻点图、Elements Layer 扩展、时空数据可视化、矢量分块
OGC 库 SuperMap_OGC. js	OGC(WMTS，WCS，WMS，WFS)，KML

SuperMap. Includ. js 是引用文件，可在其中加载需要的类库资源文件，例如，加载基础库和 iServer 服务库：

```
//加载类库资源文件
function loadSMLibs() {
        inputScript(baseurl+'SuperMap_Basic.js');
        inputScript(baseurl+'SuperMap_IServer.js');
        inputCSS('style.css');
        inputCSS('google.css');
}
//引入汉化资源文件
function loadLocalization() {
        inputScript(baseurl +'Lang/zh-CN.js');
}
```

4. lang

lang 文件夹存放汉化资源文件。

5. resource

resource 为产品使用的资源。目前包括 PhoneGap、WinRTApp、AppPackages 文件夹。

①PhoneGap 存放 PhoneGap 框架以及基于此框架实现的离线存储范例。通过 PhoneGap 开发框架实现离线存储与访问地图功能，满足用户在离线状态下的地图应用。

②WinRTApp 存放基于 Windows 8 JavaScript 版 Windows 8 应用商店程序的开发的源码文件，该源码程序基于 SuperMap 云服务图层，实现基本的地图浏览、缩放及量测功能。

③AppPackages 存放基于 Windows 8 JavaScript 版 Windows 8 应用商店程序的打包文件，下载后可以直接在 Windows 8 系统的 PC 直接安装。

6. theme

theme 文件夹存放产品使用的主题文件，包括 default、image 两个文件夹，其中：

①default 存放产品类库默认样式文件；②image 存放控件的图像资源。

7. index. html

产品首页，启动 index. html 文件可查看产品兼容性和产品变更信息，通过首页，可以链接到产品介绍，产品开发指南，示范程序，类参考以及技术专题等页面。

30. 2 创建第一个应用

30. 2. 1 获取开发包下载

从 SuperMap 官方网站上下载 SuperMap iClient 7C for JavaScript 软件包，然后将软件包解压到本地磁盘。

30. 2. 2 创建 HTML 页面

在磁盘上任意位置新建文件夹并自定义该文件夹，本例命名为“MyFirst”；在“MyFirst”文件夹下用文本编辑工具(如 Notepad++)新建一个“Demo. html”的 html 页面，注意将该 html 页面保存为 UTF-8 编码格式，并添加以下代码：

```
<! DOCTYPE HTML>
<html>
<body onload="init()">
<! --地图显示的 div-->
<div id = " map " style = " position: absolute; left: 0px; right: 0px; width:800px;height:500px;" >
</div>
</body>
</html>
```

30.2.3　引用资源文件

引用资源分两步进行：

①将第一步得到的 theme 文件夹拷贝到"MyFirst"文件夹下；

②拷贝 libs 文件夹到"MyFirst"文件夹下。

30.2.4　添加地图代码

在<html>和<body>之间添加如下代码，实现创建地图功能：

```
<head>
<title>SuperMap iClient for JavaScript:TiledDynamicRESTLayer</
title>
<script src="./libs/SuperMap.Include.js"></script>
<script type="text/javascript">
    //声明变量 map、layer、url
    var map, layer,
     url = " http:// localhost: 8090/iserver/services/map-world/
rest/maps/World";
    //创建地图控件
    function init() {
        map = new SuperMap.Map ("map");
        //其中"world"为图层名称,url 图层的服务地址
        layer = new SuperMap.Layer.TiledDynamicRESTLayer("World", url,
        null, {maxResolution:"auto"});
        layer.events.on({"layerInitialized": addLayer});
    }
    function addLayer() {
        //将 Layer 图层加载到 Map 对象上
        map.addLayer(layer);
        //出图,map.setCenter 函数显示地图
        map.setCenter(new SuperMap.LonLat(0, 0), 0);
    }
</script>
</head>
```

30.2.5　运行程序

使用浏览器运行程序，结果如图 30.2 所示。

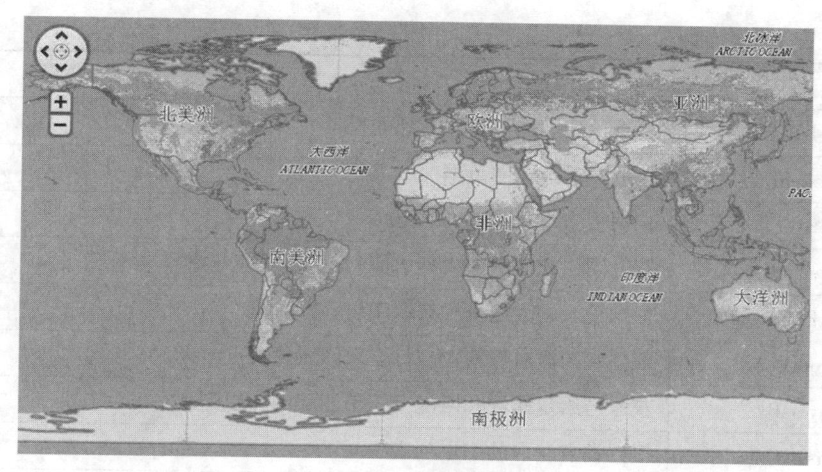

图 30.2　程序运行界面

30.3　地图基本操作

地图常用操作包括放大、缩小、平移、设置缩放级别、坐标转换、地图切换等。

30.3.1　地图方法

SuperMap. Map 地图类，用于实例化 map 类创建一个新地图，实现地图在客户端的交互操作，可通过给创建的 map 添加图层和控件来扩展应用，在创建地图时，如果没有添加指定的控件，则默认 Navigation、PanZoomBar 控件。地图类方法见表 30.2。

表 30.2　**SuperMap. Map 地图类方法列表**

方　法	说　明
addControl	为地图添加控件。可选的位置参数用来指定控件的像素位置
addControls	将控件添加到 map 上。options 的第二个数组通过像素对象控制控件的位置两个数组应该匹配，如果 pixel 设为 null，控件会显示在默认位置
addLayer	向地图中添加指定的单个图层
addLayers	向地图中添加指定的多个图层
addPopup	在地图中添加弹出窗口
destroy	销毁地图(注意，如果从 DOM 中移除 map 容器，需要在 destroy 前执行)
getBy	获取一个对象数组，并通过给定的属性匹配其中的项
getCenter	获取当前地图的中心点坐标
getControl	通过 id 值获取控件对象
getControlsBy	根据给定的属性和匹配字符串匹配到的控件列表

续表

方 法	说 明
getControlsByClass	根据给定类的类名匹配到的控件列表
getExtent	获取当前地图的范围
getLayer	根据传入参数 id 获取图层
getLayerIndex	获取图层在地图上的索引值(索引值从零开始)
getLayerPxFromLonLat	根据传入的大地坐标获取图层坐标对象
getLayerPxFromViewPortPx	根据视图窗口像素点坐标获取图层像素点坐标
getLayersBy	获取与给定属性和匹配字符串匹配的图层列表
getLayersByClass	根据类名获取的图层列表
getLayersByName	获取根据名称匹配得到的图层列表
getLonLatFromPixel	根据相对于地图窗口左上角的像素位置，返回其在地图上的地理位置。依据当前 baselayer 转换成 lon/lat (经度/纬度)形式
getMaxExtent	获取地图的最大范围
getMaxResolution	返回 baseLayer 最大的分辨率
getNumLayers	获取地图上的图层数量
getNumZoomLevels	获取当前底图的缩放级别总数，在底图存在的情况下与底图缩放级别(numZoomLevels)相同
getPixelFromLonLat	获取地图上的像素坐标。依照当前 baselayer，将指定的地理点位置坐标，转换成其相对于地图窗口左上角点的像素坐标
getProjection	该方法返回代表投影的字符串
getProjectionObject	返回 baseLayer 的投影
getResolution	获取当前地图的分辨率
getResolutionForZoom	根据缩放级别获取对应的分辨率
getScale	获取当前地图的缩放比例
getSize	获取当前地图容器大小
getTileSize	获取地图瓦片的大小
getUnits	获取地图的当前单位
getViewPortPxFromLayerPx	
getViewPortPxFromLonLat	根据指定地理位置，返回其相对于当前地图窗口左上角的像素位置
getZoom	获取当前地图的缩放比例级别
getZoomForExtent	通过给定的范围获取比例级别
getZoomForResolution	根据分辨率获取对应的缩放级别

续表

方　法	说　　明
pan	根据指定的屏幕像素值平移地图
panTo	平移地图到新的位置，如果新的位置在地图的当前范围内，地图将平滑地移动
raiseLayer	通过给定的增量值(delta 参数)来改变给定图层的索引值。如果增量值为正，图层就会在图层堆栈中向上移；如果增量值为负，图层就会向下移。这个方法同样不能将底图移动到叠加图层之上
removeControl	移除控件
removeLayer	通过删除可见元素(即 layer. div 属性)来移除地图上的图层。然后从地图的图层列表中移除该图层，同时设置图层的 map 属性为 null
removePopup	移除指定的弹出窗口
render	在指定的容器中渲染地图
setBaseLayer	允许用户指定当前加载的某一图层为地图新的底图
setCenter	设置地图中心点
setLayerIndex	移动图层到图层列表中的指定索引值(索引值从零开始)的位置。改变它在地图显示时的 z-index 值。使用 map. getLayerIndex( )方法查看图层当前的索引值。注意该方法不能将底图移动到叠加图层之上
setOptions	设置地图的 options
updateSize	通过动态调用 updateSize( )方式，来改变地图容器的大小
zoomOut	在当前缩放级别的基础上缩小一级
zoomIn	在当前缩放级别的基础上放大一级
zoomTo	缩放到指定的级别
zoomToExtent	缩放到指定范围，重新定位中心点
zoomToMaxExtent	缩放到最大范围，并重新定位中心点
zoomToScale	缩放到指定的比例

为一个特殊的事件注册一个监听对象使用下面的方法(详见 30.4 节)：

```
map.events.register(type, obj, listener);
```

30.3.2　地图功能示例

监听对象将会作为事件对象的参考，事件的属性将取决于所发生的事情。地图事件类型包含以下几种：

1. preaddlayer

图层被添加之前被触发。事件对象包含 layer 属性来指明添加的图层。当监听函数返回“false”表示无法添加图层。

2. addlayer

图层在被添加之后触发。事件对象包含 layer 属性来指明添加的图层。

3. preremovelayer

图层被移除之前触发。事件对象包含 layer 属性来指明要被移除的图层。当监听函数返回“false”，表示无法移除该图层。

4. removelayer

图层被移除后触发。事件对象包含 layer 属性来指明要被移除的图层。

5. changelayer

当一个图层的名称(name)、顺序(order)、透明度(opcity)、参数个数(params)、可见性(visibility)(由可见比例引起)或属性(attribution)发生改变时被触发。监听者会接收到一个事件对象，该事件对象包含 layer 和 property 属性，其中 layer 属性 指明发生变化的图层，property 属性是代表变化属性的关键字(name，order，opacity，params，visibility or attribution)。

6. movestart

drag，pan 或 zoom 操作开始时会触发该事件。

7. move

drag，pan 或 zoom 操作开始后被触发。

8. moveend

drag，pan 或 zoom 操作完成时被触发。

9. zoomend

zoom 操作完成后被触发。

10. mouseover

鼠标移至地图时被触发。

11. mouseout

鼠标移出地图时被触发。

12. mousemove

鼠标在地图上移动时被触发。

13. changebaselayer

改变基础图层时触发事件。

调用表 30.2 中的相应方法和上述事件可以轻松实现地图常用功能。示例实现代码如下：

```
<! DOCTYPE html>
<html>
<head>
<meta http-equiv="Content-Type" content="text/html; charset=utf-8" />
<title>地图基本操作</title>
<style type="text/css">
```

```
    body {
        margin: 0;
        overflow: hidden;
        background: #fff;
    }
    #map {
        position: relative;
        height: 520px;
        border: 1px solid #3473b7;
    }
    #toolbar {
        padding-left:500px;
        height: 33px;
        padding-top: 5px;
    }
    </style>
    <link href="CSS/index.css" rel="stylesheet" type="text/css" />
    <script src='libs/SuperMap.Include.js' charset="utf-8"></script>
    <script src="JS/jquery-1.7.2.min.js" type="text/javascript"></
script>
    <script type="text/javascript">
    var map, layer;
    var layerDay, layerNight, bt = false;
    var url1 = "http://localhost:8090/iserver/services/map-world/
rest/maps/世界地图_Day";
    var url2 = "http://localhost:8090/iserver/services/map-world/
rest/maps/世界地图_Night";
    function init() {
        map =new SuperMap.Map("map", {
            controls : [ new SuperMap.Control.Navigation() ]
        });
        map.minScale = 1.3732579019271662e-8;
        map.numZoomLevels = 6; //设置地图缩放级别的数量
        layerDay =new SuperMap.Layer.TiledDynamicRESTLayer("世界地图
_Day", url1, {
            transparent : true,
            cacheEnabled : true
        });
```

```
        layerDay.events.on({
            "layerInitialized" : addLayer1
        });
        map.events.on({
            "mousemove" : getMousePositionPx
        });
        setposition();
        addHandler(window,"resize", setposition);
    }
    function setposition() {
        var width = map.getSize().w;
        document.getElementById( "mousePositionDiv") .style.left =
width /2-160 + "px";
    }
    function addLayer1() {
        layerNight =new SuperMap.Layer.TiledDynamicRESTLayer( "世界地
图_Night",
                url2,{
                    transparent : true,
                    cacheEnabled : true
                });
        layerNight.events.on({
            "layerInitialized" : addLayer2
        });
    }
    function addLayer2() {
        layerDay.isBaseLayer =true;
        layerNight.isBaseLayer =true;
        map.addLayers([ layerDay, layerNight ]);
        map.setCenter(new SuperMap.LonLat(116, 40), 3);
    }
    function changeMap() {
        if (bt == false){
            layerDay.setVisibility(false);
            layerNight.setVisibility(true);
            map.setBaseLayer(layerNight);
            bt = true;
        }else if (bt == true) {
```

```
            layerDay.setVisibility(true);
            layerNight.setVisibility(false);
            map.setBaseLayer(layerDay);
            bt = false;
        }
    }
    function mapenlarge() {
        map.zoomIn();
    }
    function mapreduce() {
        map.zoomOut();
    }
    function mapPan() {
        map.pan(-20, -8);
    }
    //地图 div 左上角,鼠标像素坐标(x,y)应为(0,0)
    //所以应该减去 map div 的左上角的 offsetLeft、offsetTop 和 boder 宽度
    function getMousePositionPx(e) {
        var left_x = document.getElementById("map").offsetLeft + 1;
        var left_y = document.getElementById("map").offsetTop + 1;
         var lonlat = map.getLonLatFromPixel(new SuperMap.Pixel(e-
.clientX -left_x, e.clientY -left_y));
        var newHtml = "地图坐标转换:像素坐标与地理位置坐标转换<br>鼠标像素
坐标:" + "x="
                + Math.floor(e.clientX -left_x) + "," + "y=" + Math.
floor(e.clientY -left_y)
                + "<br>位置坐标:" + "lon=" + lonlat.lon.toFixed(5) +",
" + "lat="
                + lonlat.lat.toFixed(5);
        document.getElementById("mousePositionDiv").innerHTML = newHtml;
    }
    function addHandler(element, type, handler) {
        if (element.addEventListener) {
            element.addEventListener(type, handler, false);
        }else if (element.attachEvent) {
            element.attachEvent("on" + type, handler);
        }else {
            element["on" + type] = handler;
```

```
        }
    }
    </script>
    </head>
    <body onload="init()">
    <div id="toolbar">
    <input type="button" value="放大" onclick="mapenlarge()" />
    <input type="button" value="缩小" onclick=" mapreduce()" />
    <input type="button" value="平移" onclick="mapPan()" />
    <input id = " btn " type = " button " value = " 底图切换 " onclick = "
changeMap()" />
    </div>
    <div id="mousePositionDiv"></div>
    <div id="map"></div>
    </body>
    </html>
```

程序运行结果如图 30.3 所示。将鼠标放在“北京”上面，可以查看位置坐标是否准确。(参考：北京市界经纬度——北纬 39°26′至 41°03′，东经 115°25′至 117°30′)

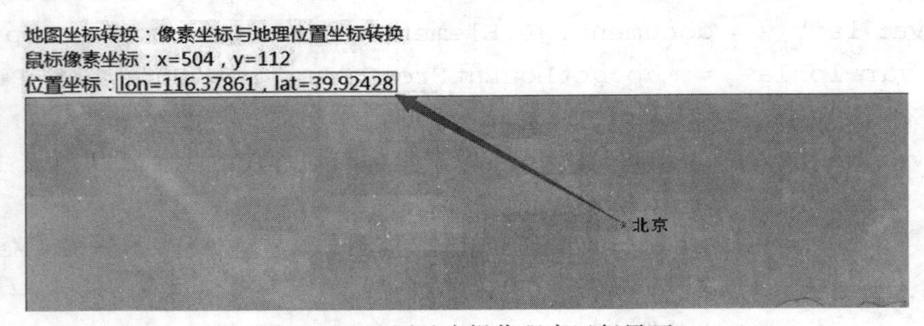

图 30.3 地图基本操作程序运行界面

30.4 地图事件

浏览器机制中的 JavaScript 事件驱动是完成用户交互的主要方法。通过监听事件，并在事件上注册事件监听的处理方法，可以在事件被激活的时候获取用户的操作信息，调用事件监听器上的代码进行响应处理。SuperMap iClient for JavaScript 拥有自定义事件模型，用来处理地理 API 中定义的事件(如 layer 的 layerInitialized 事件)。使用 Events 来处理自定义事件的过程和普通的浏览器事件处理流程基本相似。

SuperMap iClient for Javascript 地图 API 提供了公共的方法来帮助开发者解决注册监听浏览器事件时候的跨浏览器兼容问题。使用 SuperMap. Event. observe ( ) 和 SuperMap.

Event. stopObserving( ) 方法来处理对浏览器事件的监听，使用 SuperMap. Function. bind( ) 和 SuperMap. Function. bindAsEventListener( ) 方法来解决事件处理在跨浏览器时候的参数和 this 对象不一致的问题。

30. 4. 1　基础事件

下面示范代码是对一个 div 对象注册监听鼠标点击事件：

```
function init(){
    var div = document.createElement("div");
    document.body.appendChild(div);
    div.innerText = "点击测试";
    var observeFun = SuperMap.Function.bindAsEventListener
(clickHandler, div);
    //监听 div 的 click 事件,处理函数为 observeFun,即最终的 clickHandler
    SuperMap.Event.observe(div, "click",observeFun);
}
function clickHandler(event){
    alert(this.innerText);
}
```

30. 4. 2　自定义事件

自定义事件机制的核心类是 SuperMap. Events，该类提供了事件监听、触发的相关方法：

1. register( )

注册事件监听到事件处理队列的尾端。

2. registerPriority( )

注册优先事件监听，即添加到事件处理队列的最前面。

3. on( )

注册事件监听，参数为一个由事件/事件处理函数组成的键值对对象。

4. unregister( )

移除事件监听。

5. un( )

移除事件监听，参数同 on( )。

6. triggerEvent( )

触发(派发)自定义事件。

在 Map、Layer、SelectFeature 控件等以及其他可操作类中，都内置了 events 对象来实现对自定义事件的注册监听处理，layer 的异步加载就是一个很好的示例。下面分别演示使用 register( ) 和 on( ) 方法来实现对图层异步加载完成事件的监听处理，最终效果一样：

```
//使用 on()接口监听事件
layer.events.on({"layerInitialized": addLayer});
```

```
//使用 register()接口监听事件
layer.events.register("layerInitialized", undefined, addLayer)
```

使用unregister()和un()来移除事件的使用方法也很简单,示范代码如下:

```
//使用 un()接口监听事件
layer.events.un({"layerInitialized": addLayer});
//使用 unregister()接口监听事件
//layer.events.unregister("layerInitialized", undefined, addLayer)
```

使用地图 API 中的自定义事件机制，开发者也可以定义属于自己的事件类型和处理。使用的时候需要初始化一个 events 对象，然后调用该对象完成事件的相关处理。下面示例中，点击 div 对象触发一个自定义的事件“userselect”，然后监听 events 来完成：

```
var div;
function init(){
    div = document.createElement("div");
    div.innerText = "点击测试";
    document.body.appendChild(div);
    //初始化一个 events。定义事件的监听器和自定义事件名称
    var events = new SuperMap.Events(div, div, ["userselect"]);
    //将事件对象保存到 div 对象上
    div.events = events;
    div.events.register("userselect", div, userselectHandler)
    //将处理函数和调用对象绑定,函数最终调用里面的 this 对象
    //即指向第二个参数指定的 object 对象
    var observeFun = SuperMap.Function.bindAsEventListener(clickHan-
dler, div);
    SuperMap.Event.observe(div, "click",observeFun);
}
function clickHandler(event){
    //this 对象指向绑定的 object,即 DIV 本身
    this.events.triggerEvent("userselect", event);
}
//响应自定义事件处理
function userselectHandler(event){
    alert(this.innerText + event.type);
}
```

30.5　地图查询

地图查询是 GIS 系统中的核心功能之一，通过该功能可以获得相应的空间对象及属性信息，并在地图上进行定位显示、信息列表，包括缓冲区查询、几何查询和 SQL 查询，

而实际应用中多采用 SQL 条件与空间关系或距离范围相结合的综合查询。本节将讲解不同的查询功能的开发实现。

30.5.1　缓冲区查询

本节将实现对世界各国首都的缓冲区查询，首先通过 GetFeaturesByBufferParameters 设置查询条件，然后由 GetFeaturesByBufferService.processAsync（GetFeaturesByBufferParameters）向服务端提供请求，待服务器处理完成并返回结果后，在结果处理的回调函数中解析查询结果 QueryEventArgs.result，同时在地图上进行图标显示。SuperMap.REST.GetFeaturesByBufferParameters 构造函数参数见表 30.3。

表 30.3　**SuperMap.REST.GetFeaturesByBufferParameters 构造函数参数**

参　数	说　明
bufferDistance	{Number} buffer 距离，单位与所查询图层对应的数据集单位相同
attributeFilter	{String} 属性查询条件
fields	{Array(String)} 设置查询结果返回字段，默认返回所有字段
geometry	{SuperMap.Geometry} 空间查询条件
dataSetNames	{Array(String)} 数据集集合中的数据集名称列表
returnContent	{Boolean} 是否直接返回查询结果
fromIndex	{Integer} 查询结果的最小索引号
toIndex	{Integer} 查询结果的最大索引号

SuperMap.REST.GetFeaturesByBufferService 数据服务缓冲区查询服务类构造函数。

①示例代码如下：

```
var myGetFeaturesByBufferService = new SuperMap.REST.(url, {
  eventListeners: {
    "processCompleted": GetFeaturesCompleted,
    "processFailed": GetFeaturesError
    }
});
function GetFeaturesCompleted(QueryEventArgs){ //todo};
function GetFeaturesError(QueryEventArgs){ //todo};
```

②前台交互界面设计代码如下：

```
<body onload="init()">
<div id="toolbar">
<input type="button" value="点" onclick="drawPointGeometry()"/>
<input type="button" value="线" onclick="drawLineGeometry()"/>
```

```
<input type="button" value="面" onclick="drawPolygonGeometry()"/>
<input type="button" value="清除" onclick="clearFeatures()"/>
</div>
<div id="map"></div>
</body>
```

③交互实现代码如下：

```
<script src='libs/SuperMap.Include.js'></script>
<script type="text/javascript">
varhost = document.location.toString().match(/file:\/\//)?"ht-
tp://localhost:8090":
                                      'http://' + document.location.host;
varmap, local, layer, vectorLayer, drawPoint, drawLine, drawPoly-
gon, markerLayer,
            style = {
                strokeColor:"#304DBE",
                strokeWidth: 1,
                pointerEvents:"visiblePainted",
                fillColor:"#304DBE",
                fillOpacity: 0.5,
                pointRadius:2
            },
            url1=host+"/iserver/services/map-world/rest/maps/World",
            url2=host+"/iserver/services/data-world/rest/data";
function init(){
                layer =new SuperMap.Layer.TiledDynamicRESTLayer
("World", url1, {transparent: true, cacheEnabled: true}, {maxResolu-
tion:"auto"});
                layer.events.on({"layerInitialized":addLayer});
                vectorLayer =new SuperMap.Layer.Vector("Vector Layer");
                markerLayer =new SuperMap.Layer.Markers("Markers");
                drawPoint =new SuperMap.Control.DrawFeature
(vectorLayer, SuperMap.Handler.Point);
                drawPoint.events.on({"featureadded": drawCompleted});
                drawLine =new SuperMap.Control.DrawFeature
(vectorLayer, SuperMap.Handler.Path);
                drawLine.events.on({"featureadded": drawCompleted})
                drawPolygon =new SuperMap.Control.DrawFeature
(vectorLayer, SuperMap.Handler.Polygon);
```

```
            drawPolygon.events.on({"featureadded": drawCompleted});
            map =new SuperMap.Map("map",{controls: [
newSuperMap.Control.LayerSwitcher(),
newSuperMap.Control.ScaleLine(),
newSuperMap.Control.Zoom(),
newSuperMap.Control.Navigation({
                dragPanOptions: {
                    enableKinetic:true
}}),
                drawPoint, drawLine, drawPolygon]
            });
        }
function addLayer() {
            map.addLayers([layer, vectorLayer, markerLayer]);
            map.setCenter(new SuperMap.LonLat(0, 0), 1);
        }
function drawPointGeometry() {
            //先清除上次的显示结果
            vectorLayer.removeAllFeatures();
            markerLayer.clearMarkers();
            drawPoint.activate();
        }
function drawLineGeometry() {
            //先清除上次的显示结果
            vectorLayer.removeAllFeatures();
            markerLayer.clearMarkers();
            drawLine.activate();
        }
function drawPolygonGeometry() {
            //先清除上次的显示结果
            vectorLayer.removeAllFeatures();
            markerLayer.clearMarkers();
            drawPolygon.activate();
        }
function drawCompleted(drawGeometryArgs) {
    drawPoint.deactivate();
    drawLine.deactivate();
    drawPolygon.deactivate();
```

```
var feature = new SuperMap.Feature.Vector();
            feature.geometry = drawGeometryArgs.feature.geometry,
            feature.style = style;
            vectorLayer.addFeatures(feature);
    var getFeatureParameter, getFeatureService;
             getFeatureParameter =new SuperMap.REST.GetFeatures-
ByBufferParameters({
                bufferDistance: 30,
                datasetNames: ["World:Capitals"],
                returnContent:true,
                geometry: drawGeometryArgs.feature.geometry
            });
             getFeatureService =new SuperMap.REST.GetFeatures-By-
BufferService(url2, {
                eventListeners: {
                    "processCompleted": processCompleted,
                    "processFailed": processFailed
                }
            });
            getFeatureService.processAsync(getFeatureParameter);
        }
    function processCompleted(getFeaturesEventArgs) {
    vari, len, features, result = getFeaturesEventArgs.result;
    if(result && result.features) {
        features = result.features;
    for(i=0, len=features.length; i<len; i++) {
    var point = features[i].geometry,
        size = new SuperMap.Size(44, 33),
        offset =new SuperMap.Pixel(-(size.w/2), -size.h),
        icon =new SuperMap.Icon("theme/images/marker.png", size, offset);
        markerLayer.addMarker(new SuperMap.Marker(new SuperMap.Lon
Lat(point.x, point.y), icon));
                }
            }
        }
    function processFailed(e) {
            alert(e.error.errorMsg);
        }
```

```
function clearFeatures() {
            vectorLayer.removeAllFeatures();
            markerLayer.clearMarkers();
        }
</script>
```

运行程序，单击【点】，然后在地图上单击一点，运行效果如图 30.4 所示；单击【线】，然后在地图上画一条线，运行效果如图 30.5 所示；单击【面】，然后在地图上画一个多边形，运行效果如图 30.6 所示。

图 30.4　点缓冲区查询

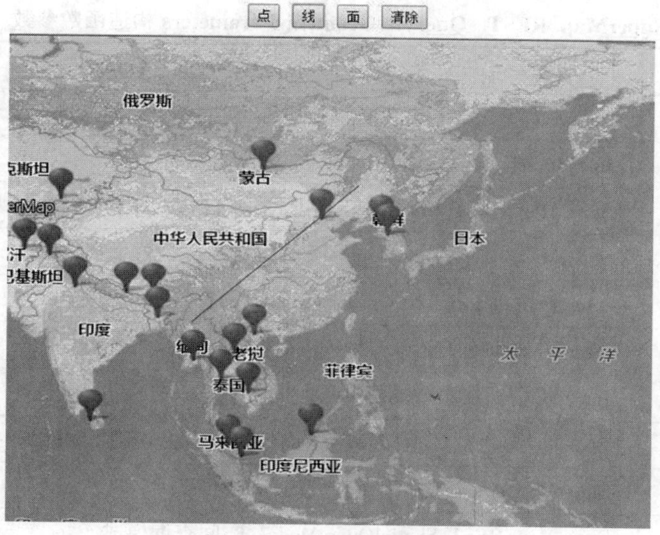

图 30.5　线缓冲区查询

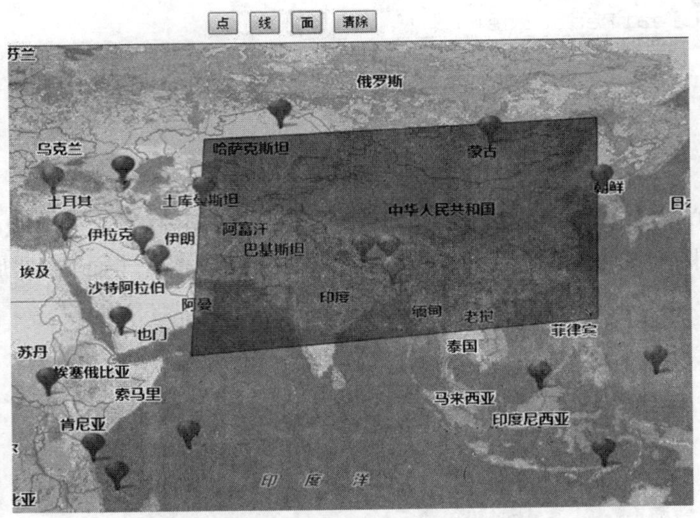

图 30.6　多边形缓冲区查询

30.5.2　几何查询

本节将实现对世界各国区域的几何查询，首先通过 QueryByGeometryParameters 设置查询条件，然后由 QueryByGeometryService. processAsync(QueryByGeometryParameters) 向服务端提供请求，待服务器处理完成并返回结果后在结果处理的回调函数中解析查询结果 QueryEventArgs. result ，同时在地图上进行图标显示。SuperMap. REST. QueryByGeometry Parameters 构造函数参数见表 30.4。SuperMap. REST. FilterParameter 构造函数参数见表 30.5。

表 30.4　　SuperMap. REST. QueryByGeometryParameters 构造函数参数

参　　数	说　　明
customParams	{String} 自定义参数，供扩展使用
expectCount	{Integer} 期望返回结果记录个数
networkType	{SuperMap. REST. GeometryType} 网络数据集对应的查询类型
queryOption	{SuperMap. REST. QueryOption} 查询结果类型枚举类
queryParams	{Array(SuperMap. REST. FilterParameter)} 查询过滤条件参数数组
startRecord	{Integer} 查询起始记录号
holdTime	{Integer} 资源在服务端保存的时间
returnContent	{Boolean} 是否立即返回新创建资源的表述还是返回新资源的 URI
geometry	{SuperMap. Geometry} 用于查询的几何对象
spatialQueryMode	{SuperMap. REST. SpatialQueryMode} 空间查询模式

表 30.5　**SuperMap. REST. FilterParameter 构造函数参数**

参　数	说　明
attributeFilter	{String}属性过滤条件
name	{String}查询数据集名称或者图层名称
joinItems	{Array(SuperMap. REST. JoinItem)} 与外部表的连接信息 JoinItem 数组
linkItems	{Array(SuperMap. REST. LinkItem)} 与外部表的关联信息 LinkItem 数组
ids	{Array(String)}查询 id 数组，即属性表中的 SmID 值
orderBy	{String}查询排序的字段，orderBy 的字段须为数值型的
groupBy	{String}查询分组条件的字段
fields	{Array(String)}查询字段数组

SuperMap. REST. QueryByGeometryService 查询服务类构造函数。

①示例代码如下：

```
var myQueryByGeometryService = new SuperMap.REST.QueryByGeometry-
Service(url, {
    eventListeners: {
      "processCompleted": queryCompleted,
      "processFailed": queryError
      }
});
function queryCompleted(QueryEventArgs){ //todo};
function queryError(QueryEventArgs){ //todo};
```

②前台交互界面示例设计代码如下：

```
<body onload="init()">
<div id="toolbar">
<input type="button" value="点" onclick="drawGeometry3()"/>
<input type="button" value="线" onclick="drawGeometry4()"/>
<input type="button" value="圆" onclick="drawGeometry1()"/>
<input type="button" value="多边形" onclick="drawGeometry2()"/>
<input type="button" value="清除" onclick="clearFeatures()"/>
</div>
<div id="map"></div>
</body>
```

③交互实现示例代码如下：

```
<script src='libs/SuperMap.Include.js' charset="utf-8"></script>
<script type="text/javascript">
```

```
    varmap, local, layer, vectorLayer, vectorLayer1, drawPolygon,
markerLayer,drawPoint, drawLine,
    style = {
        strokeColor:"#304DBE",
        strokeWidth: 1,
        pointerEvents:"visiblePainted",
        fillColor:"#304DBE",
        fillOpacity: 0.5
    },
    host = document.location.toString().match(/file:\/\//)?"http://
localhost:8090":
            'http://' + document.location.host,
    url=host+"/iserver/services/map-world/rest/maps/World";
    function init(){
        layer = new SuperMap.Layer.TiledDynamicRESTLayer ("World",
url, {transparent: true, cacheEnabled: true}, {maxResolution:" au-
to"});
        layer.events.on({"layerInitialized":addLayer});
        vectorLayer =new SuperMap.Layer.Vector("Vector Layer");
        vectorLayer1 =new SuperMap.Layer.Vector("Vector Layer1");
        markerLayer =new SuperMap.Layer.Markers("Markers");
        drawPolygon1 =new SuperMap.Control.DrawFeature(vectorLayer,
        SuperMap.Handler.RegularPolygon,{handlerOptions:{sides:50}});
        drawPolygon1.events.on({"featureadded": drawCompleted});
        drawPolygon2 =new SuperMap.Control.DrawFeature(vectorLayer,
SuperMap.Handler.Polygon);
        drawPolygon2.events.on({"featureadded": drawCompleted});
        drawPoint =new SuperMap.Control.DrawFeature(vectorLayer, Su-
perMap.Handler.Point);
        drawPoint.events.on({"featureadded": drawPointCompleted});
        drawLine =new SuperMap.Control.DrawFeature(vectorLayer, Su-
perMap.Handler.Path);
        drawLine.events.on({"featureadded": drawPointCompleted});
        map =new SuperMap.Map("map",{controls: [
            newSuperMap.Control.LayerSwitcher(),
            newSuperMap.Control.ScaleLine(),
            newSuperMap.Control.Zoom(),
            newSuperMap.Control.Navigation({
```

```
                    dragPanOptions: {
                        enableKinetic:true
            }}),
        drawPolygon1,drawPolygon2,drawPoint,drawLine]
    });
}
function ionaddLayer() {
    map.addLayers([layer, vectorLayer, vectorLayer1, markerLayer]);
    map.setCenter(new SuperMap.LonLat(0, 0), 0);
}
function clearStatus(){
    vectorLayer.removeAllFeatures();
    vectorLayer1.removeAllFeatures();
    markerLayer.clearMarkers();
}
function drawGeometry1(){
    clearStatus();
    drawPolygon1.activate();
}
function drawGeometry2(){
    clearStatus();
    drawPolygon2.activate();
}
function drawGeometry3(){
    clearStatus();
    drawPoint.activate();
}
function drawGeometry4(){
    clearStatus();
    drawLine.activate();
}
function drawCompleted(drawGeometryArgs){
var feature = new SuperMap.Feature.Vector();
    feature.geometry = drawGeometryArgs.feature.geometry,
    feature.style = style;
    vectorLayer.addFeatures(feature);
var queryParam, queryByGeometryParameters, queryService;
    queryParam  = new SuperMap.REST.FilterParameter({name: "Cap-
```

```
itals@ World.1"});
    queryByGeometryParameters =new SuperMap.REST.QueryByGeometry-
Parameters({
            queryParams: [queryParam],
            geometry: drawGeometryArgs.feature.geometry,
            spatialQueryMode: SuperMap.REST.SpatialQueryMode.INTERSECT
        });
        queryService =new SuperMap.REST.QueryByGeometryService(url, {
            eventListeners: {
                "processCompleted": processCompleted,
                "processFailed": processFailed
            }
        });
        queryService.processAsync(queryByGeometryParameters);
    }
    function drawPointCompleted(drawGeometryArgs) {
    var feature = new SuperMap.Feature.Vector();
        feature.geometry = drawGeometryArgs.feature.geometry,
        feature.style = style;
        vectorLayer.addFeatures(feature);
    var queryParam, queryByGeometryParameters, queryService;
        queryParam =new SuperMap.REST.FilterParameter({name: "Coun-
tries@ World"});
        queryByGeometryParameters =new SuperMap.REST.QueryByGeometry-
Parameters({
            queryParams: [queryParam],
            geometry: drawGeometryArgs.feature.geometry,
            spatialQueryMode: SuperMap.REST.SpatialQueryMode.INTERSECT
        });
        queryService =new SuperMap.REST.QueryByGeometryService(url,
{
            eventListeners: {
                "processCompleted": processCompleted,
                "processFailed": processFailed
            }
        });
        queryService.processAsync(queryByGeometryParameters);
    }
```

```
function processCompleted(queryEventArgs) {
drawPolygon1.deactivate();
drawPolygon2.deactivate();
drawPoint.deactivate();
drawLine.deactivate();
vari, j, result = queryEventArgs.result;
if(result && result.recordsets) {
for(i=0, recordsets=result.recordsets, len=recordsets.length;
i<len; i++) {
if(recordsets[i].features) {
for(j=0; j<recordsets[i].features.length; j++) {
varfeature = recordsets[i].features[j];
varpoint = feature.geometry;
if(point.CLASS_NAME == SuperMap.Geometry.Point.prototype.CLASS
_NAME){
varsize = new SuperMap.Size(44, 33),
offset = new SuperMap.Pixel(-(size.w/2), -size.h),
    icon =new SuperMap.Icon("theme/images/marker.png", size, offset);
        markerLayer.addMarker ( new  SuperMap.Marker ( new  Super-
Map.LonLat(point.x, point.y), icon));
                    }else{
                        feature.style = style;
                        vectorLayer1.addFeatures(feature);
                    }
                }
            }
        }
    }
}
function processFailed(e) {
    alert(e.error.errorMsg);
}
function clearFeatures() {
    vectorLayer.removeAllFeatures();
    vectorLayer1.removeAllFeatures();
    markerLayer.clearMarkers();
}
</script>
```

运行程序，单击【点】，然后在地图上单击一点，运行效果如图 30.7 所示；单击【线】，然后在地图上画一条线，运行效果如图 30.8 所示；单击【圆】，然后在地图上画一个圆，运行效果如图 30.9 所示；单击【多边形】，然后在地图上画一个多边形，运行效果如图 30.10 所示。

图 30.7　点几何查询

图 30.8　线几何查询

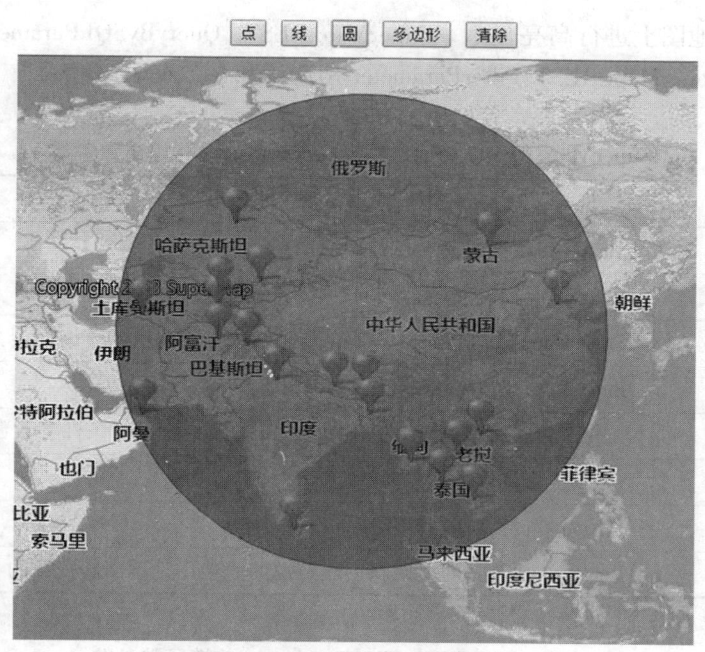

图 30.9　圆几何查询

图 30.10　多边形几何查询

30.5.3　SQL 查询

本节将实现对世界各国区域的 SQL 查询，首先通过 QueryBySQLParameters 设置查询条件，然后由 QueryBySQLService. processAsync (QueryBySQLParameters) 向服务端提供请求，待服务器处理完成并返回结果后在结果处理的回调函数中解析查询结果 QueryEventArgs.

result，同时在地图上进行高亮显示。SuperMap. REST. QueryBySQLParameters 构造函数参数见表 30. 6，SuperMap. REST. FilterParameter 构造函数参数见表 30. 7。

表 30. 6　　**SuperMap. REST. QueryBySQLParameters 构造函数参数**

参　数	说　明
customParams	{String} 自定义参数，供扩展使用
expectCount	{Integer} 期望返回结果记录个数
networkType	{SuperMap. REST. GeometryType} 网络数据集对应的查询类型
queryOption	{SuperMap. REST. QueryOption} 查询结果类型枚举类
queryParams	{Array(SuperMap. REST. FilterParameter)} 查询过滤条件参数数组
startRecord	{Integer} 查询起始记录号
holdTime	{Integer} 资源在服务端保存的时间
returnContent	{Boolean} 是否立即返回新创建资源的表述还是返回新资源的 URI

表 30. 7　　**SuperMap. REST. FilterParameter 构造函数参数**

参　数	说　明
attributeFilter	{String} 属性过滤条件
name	{String} 查询数据集名称或者图层名称
joinItems	{Array(SuperMap. REST. JoinItem)} 与外部表的连接信息 JoinItem 数组
linkItems	{Array(SuperMap. REST. LinkItem)} 与外部表的关联信息 LinkItem 数组
ids	{Array(String)} 查询 id 数组，即属性表中的 SmID 值
orderBy	{String} 查询排序的字段，orderBy 的字段须为数值型的
groupBy	{String} 查询分组条件的字段
fields	{Array(String)} 查询字段数组

①实例化查询服务类构造函数 SuperMap. REST. QueryBySQLService，其示例代码如下：

```
var queryParam = new SuperMap.REST.FilterParameter({
  name: "Countries@ World.1",
  atzributeFilter: "Pop_1994>1000000000 and SmArea>900"
});
var queryBySQLParams = new SuperMap.REST.QueryBySQLParameters({
  queryParams:[queryParam]
});
var myQueryBySQLService = new SuperMap.REST.QueryBySQLService
```

```
(url, {eventListeners: {
    "processCompleted": queryCompleted,
    "processFailed": queryError
    }
  });
  queryBySQLService.processAsync(queryBySQLParams);
  function queryCompleted(QueryEventArgs){//todo};
  function queryError(QueryEventArgs){//todo};
```

②前台交互界面设计示例代码如下：

```
<body onload="init()">
<div id="toolbar">
<input type="button" value="查询" onclick="queryBySQL()"/>
<input type="button" value="清除" onclick="clearFeatures()"/>
</div>
<div id="map"></div>
</body>
```

③交互示例代码如下：

```
<script src='libs/SuperMap.Include.js'></script>
<script type="text/javascript">
varmap, local, layer, vectorLayer,
style = {
            strokeColor: "#304DBE",
            strokeWidth: 1,
            fillColor: "#304DBE",
            fillOpacity: "0.8"
        },
    host = document.location.toString().match(/file:\/\//)?"http://
localhost:8090":
        'http://'+document.location.host,
    url=host+"/iserver/services/map-world/rest/maps/World";
    function init(){
    map = new SuperMap.Map("map",{controls: [
        newSuperMap.Control.LayerSwitcher(),
        newSuperMap.Control.ScaleLine(),
        newSuperMap.Control.Zoom(),
        newSuperMap.Control.Navigation({
                                    dragPanOptions: {
                                        enableKinetic:true
```

```
                                }
                        })]
            });
        layer = new SuperMap.Layer.TiledDynamicRESTLayer( "World",
url, {transparent: true, cacheEnabled: true},
    {maxResolution:"auto"});
        layer.events.on({"layerInitialized":addLayer});
        vectorLayer = new SuperMap.Layer.Vector("Vector Layer");
    }
    function addLayer() {
        map.addLayers([layer, vectorLayer]);
        map.setCenter(new SuperMap.LonLat(0, 0), 0);
    }
    function queryBySQL() {
    vectorLayer.removeAllFeatures();
    var queryParam, queryBySQLParams, queryBySQLService;
    queryParam =new SuperMap.REST.FilterParameter({
                    name:"Countries@ World.1",
                    attributeFilter:"Pop_1994>1000000000 and SmAr-
ea>900"
                });
    queryBySQLParams =new SuperMap.REST.QueryBySQLParameters({
                    queryParams: [queryParam]
                });
    queryBySQLService  = new  SuperMap.REST.QueryBySQLService ( url,
{eventListeners: { "processCompleted": processCompleted, "process-
Failed":processFailed}});
    queryBySQLService.processAsync(queryBySQLParams);
    }
    function processCompleted(queryEventArgs) {
    vari, j, feature,
    result = queryEventArgs.result;
    if(result && result.recordsets) {
    for(i=0; i<result.recordsets.length; i++) {
    if(result.recordsets[i].features) {
    for(j=0; j<result.recordsets[i].features.length; j++) {
                                feature = result.recordsets[i]
.features[j];
```

```
                        feature.style = style;
                        vectorLayer.addFeatures(feature);
                    }
                }
            }
        }
    }
function processFailed(e) {
        alert(e.error.errorMsg);
}
function clearFeatures() {
        //先清除上次的显示结果
        vectorLayer.removeAllFeatures();
        vectorLayer.refresh();
}
</script>
```

运行程序，点击【查询】，运行效果如图30.11所示；点击【清除】，则高亮数据清除。

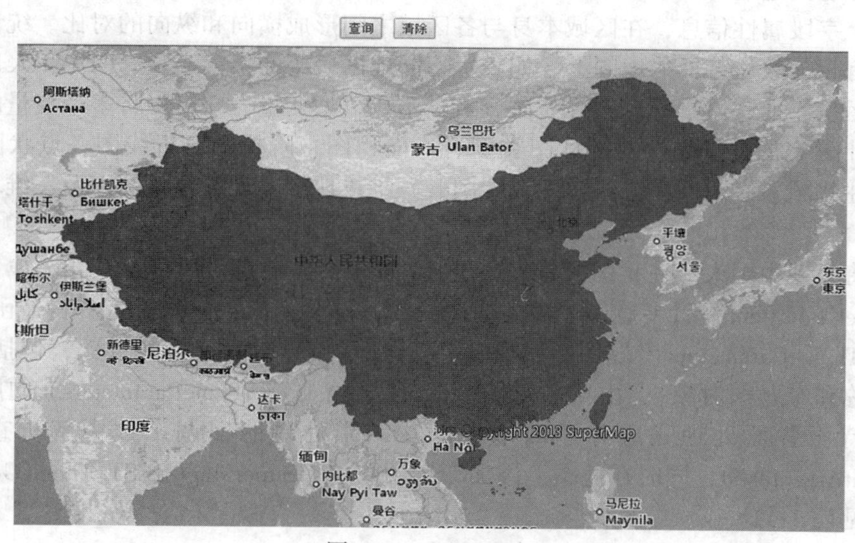

图30.11 SQL查询

第 31 章　SuperMap for JavaScript 中级开发

31.1　专题图

专题图是使用各种图形风格(如颜色、形状或填充模式)显示地图基础信息特征的地图，是空间数据的重要表达形式之一。制作专题图的本质是根据数据对现象的现状和分布规律及其联系进行渲染，从而充分挖掘利用数据资源，展现丰富数据内容，让用户直观地感受到数据的表达方式。专题图的形式有很多，本节将重点讲解统计专题图、标签专题图、栅格分段专题图的功能开发实现。

31.1.1　统计专题图

统计专题图通过为每个要素或记录绘制统计图来反映其对应的专题值的大小。它可同时表示多个字段属性信息，在区域本身与各区域之间形成横向和纵向的对比。统计专题图多用于具有相关数量特征的地图上，如表示不同地区多年的粮食产量、GDP、人口等，不同时段客运量、地铁流量等。允许一次分析多个数值型变量，即可以将多个变量的值绘制在一个统计图上。目前提供的统计图类型有：面积图、阶梯图、折线图、点状图、柱状图、三维柱状图、饼图、三维饼图、玫瑰图、三维玫瑰图、堆叠柱状图以及三维堆叠柱状图。

本节将实现中国人口柱状图，首先通过 ServerStyle 定义符号风格及其相关属性，然后设置 ThemeGraphItem 统计专题图子项类构造函数，再设置 ThemeGraphItem 构造函数参数，最后设置 ThemeParameters 专题图参数类(该类存储了制作专题所需的参数，包括数据源、数据集名称和专题图对象)，由 ThemeService. processAsync(ThemeParameters) 向服务端提供请求，待服务器处理完成并返回结果后在结果处理的回调函数中解析查询结果 ThemeEventArgs. result，同时在地图上添加专题图图层。SuperMap. REST. ThemeGraph 构造函数参数见表 31.1。

表 31.1　**SuperMap. REST. ThemeGraph 构造函数参数**

参　数	说　明
barWidth	{Number} 柱状专题图中每一个柱的宽度
flow	{SuperMap. REST. ThemeFlow} 统计专题图流动显示与牵引线设置

续表

参　数	说　明
graduatedMode	{SuperMap. REST. GraduatedMode} 统计图中地理要素的值与图表尺寸间的映射关系
graphAxes	{SuperMap. REST. ThemeGraphAxes} 统计图中坐标轴样式相关信息
graphSize	{SuperMap. REST. ThemeGraphSize} 统计符号的最大最小尺寸
graphSizeFixed	{Boolean}缩放地图时统计图符号是否固定大小
graphText	{SuperMap. REST. ThemeGraphText} 统计图上的文字是否可见以及文字标注风格
graphType	{SuperMap. REST. ThemeGraphType} 统计专题图类型
items	{Array(SuperMap. REST. ThemeGraphItem)} 统计专题图子项集合
memoryKeys	{Array(Integer)}以内存数组方式制作专题图时的键数组
negativeDisplayed	{Boolean}专题图中是否显示属性为负值的数据
offset	{SuperMap. REST. ThemeOffset} 统计图相对于要素内点的偏移量
overlapAvoided	{Boolean}统计图是否采用避让方式显示
roseAngle	{Number}统计图中玫瑰图或三维玫瑰图用于等分的角度
startAngle	{Number}饼状统计图扇形的起始角度

以中国 2010 年、2011 年、2012 年人口普查数据为基础，控制 Web 端制作人口柱状图。具体示例代码如下：

```
<body onload="init()">
<div id="toolbar">
<input type="button" value="创建专题图" onclick="addThemeGraph()"/>
<input type="button"value="移除专题图" onclick="removeTheme()"/>
</div>
<div id="map"></div>
</body>
<script src='libs/SuperMap.Include.js'></script>
<script type="text/javascript">
varmap, local, baseLayer, layersID, themeLayer,
host = document.location.toString().match(/file:\/\//)?"http://localhost:8090":
            'http://' + document.location.host,
url=host+"/iserver/services/map-China400/rest/maps/China400";
function init(){
```

```
            map =new SuperMap.Map("map",{controls:[
                new SuperMap.Control.LayerSwitcher(),
                new SuperMap.Control.ScaleLine(),
                new SuperMap.Control.Zoom(),
                new SuperMap.Control.Navigation({
                                        dragPanOptions:{
                                            enableKinetic:true
                                        }
                                    })]
            });
            baseLayer =new SuperMap.Layer.TiledDynamic-
            RESTLayer("全国人口分布图",url,{transparent: true,
            cacheEnabled: true},{maxResolution:"auto"});
            baseLayer.events.on({"layerInitialized":addLayer});
    }
    functionaddLayer(){
            map.addLayer(baseLayer);
            map.setCenter(new SuperMap.LonLat(117, 40), 0);
            map.allOverlays =true;
    }
    functionaddThemeGraph(){
            removeTheme();
          //创建统计专题图对象,ThemeGraph 必设 items
          //专题图参数 ThemeParameters 必设 theme(即以设置好的分段
专题图对象)
          //dataSourceName 和 datasetName
            varstyle1 = new SuperMap.REST.ServerStyle({
                        fillForeColor:new SuperMap.REST.ServerColor
(92,73,234)
                        lineWidth: 0.1
                    }),
                    style2 =new SuperMap.REST.ServerStyle({
                        fillForeColor:new SuperMap.REST.
ServerColor(211,111,240),
                        lineWidth: 0.1
                    }),
                    style3 =new SuperMap.REST.ServerStyle({
                        fillForeColor:new SuperMap.REST.
```

```
ServerColor(120,130,230),
                                lineWidth: 0.1
                            }),
                            item1 =new SuperMap.REST.ThemeGraphItem({
                                caption:"2010 年人口",
                                graphExpression:"Popu_2010",
                                uniformStyle: style1
                            }),
                            item2 =new SuperMap.REST.ThemeGraphItem({
                                caption:"2011 年人口",
                                graphExpression:"Popu_2011",
                                uniformStyle: style2
                            }),
                            item3 =new SuperMap.REST.ThemeGraphItem({
                                caption:"2012 年人口",
                                graphExpression:"Popu_2012",
                                uniformStyle: style3
                            }),
                    themeGraph  = new SuperMap.REST.ThemeGraph({
                                items:new Array(item1,item2,item3),
                                barWidth: 0.03,
                                graduatedMode: SuperMap.REST.
GraduatedMode.SQUAREROOT,
                                graphAxes:new SuperMap.REST.
ThemeGraphAxes({
                                    axesDisplayed:true
    }),
                                graphSize:new SuperMap.REST.ThemeGraphSize({
                                    maxGraphSize: 1,
                                    minGraphSize: 0.35
                                }),
                                graphText:new SuperMap.REST.ThemeGraphText({
                                    graphTextDisplayed:true,
                graphTextFormat:SuperMap.REST.ThemeGraphTextFormat.VALUE,
                graphTextStyle:new SuperMap.REST.ServerTextStyle({
                                    sizeFixed:true,
                                    fontHeight: 9,
                                    fontWidth: 5
```

```
                    })
                }),

            graphType: SuperMap.REST.ThemeGraphType.BAR3D
            }),
        //专题图参数对象

    themeParameters =new SuperMap.REST.ThemeParameters({
                themes: [themeGraph],
                dataSourceNames: ["China"],
                datasetNames: ["China_Province_pg"]
            }),
            //与服务端交互
            themeService=new SuperMap.REST.ThemeService
(url,{
                eventListeners: {
                    "processCompleted": ThemeCompleted,
                    "processFailed": themeFailed
                }
            });
        themeService.processAsync(themeParameters);
    }
//显示专题图。专题图在服务端为一个资源,每个资源都有一个 ID 号和一个 url
//要显示专题图即将资源结果的 ID 号赋值给图层的 layersID 属性即可
function ThemeCompleted(themeEventArgs){
if(themeEventArgs.result.resourceInfo.id){
themeLayer = new SuperMap.Layer.TiledDynamicRESTLayer("全国人口分
        布图_专题图", url,{cacheEnabled:false,transparent:
        true,layersID:
        themeEventArgs.result.resourceInfo.id},{maxResolu-
        tion:"auto"});
    themeLayer.events.on({"layerInitialized":addThemelayer});
    }
}
function addThemelayer(){
        map.addLayer(themeLayer);
}
function themeFailed(serviceFailedEventArgs){
```

```
            alert(serviceFailedEventArgs.error.errorMsg);
}
function removeTheme(){
if(map.layers.length > 1) {
                map.removeLayer(themeLayer,true);
        }
}
</script>
```

运行程序，专题图效果如图 31.1 所示。

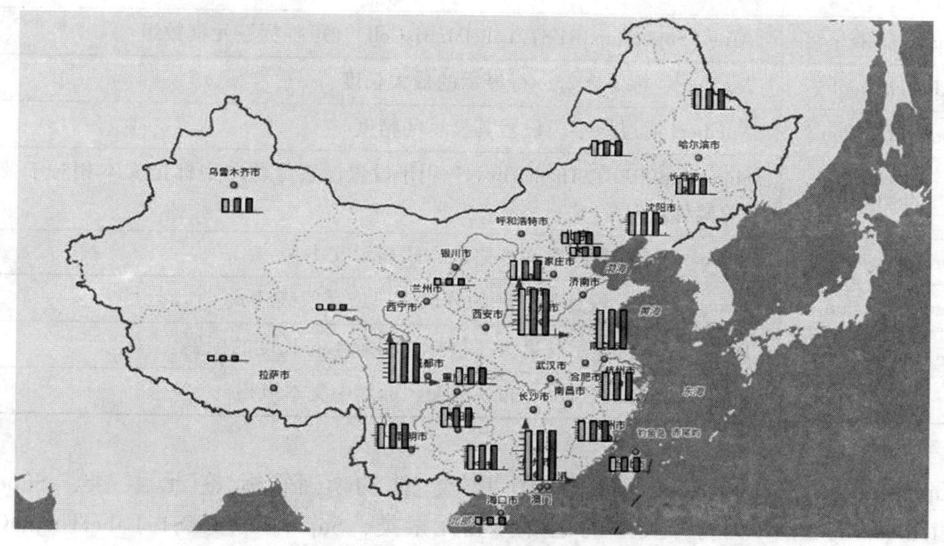

图 31.1　人口柱状图

31.1.2　标签专题图

标签专题图是用文本的形式在图层中显示点、线、面等对象的属性信息，一般将文本型或数值型字段标注于图层中，如地名、道路名称、河流等级、宽度等信息。这里需要注意的是，地图上一般还会出现图例说明、图名、比例尺，等等，这些都是制图元素，不属于标签专题图标注的范畴。标签专题图有两种表现形式：第一种统一标签专题图，即将指定图层的表达式的所有值使用统一的风格输出，labelExpression 用于设定标签专题图所使用的字段。第二种为分段标签专题图。它仍然使用 labelExpression 设定标签专题图显示的字段值，通过 rangeExpression 指定数字型的字段作为分段数据，items 中的每个子对象的[start，end]分段值必须来源于属性 rangeExpression 的字段值，并在 items 中为每个或部分分段子项自定义特殊的风格。SuperMap. REST. ThemeLabel 构造函数参数见表 31.2。

表 31.2　　　　　　**SuperMap. REST. ThemeLabel 构造函数参数**

参　数	说　明
alongLine	{SuperMap. REST. ThemeLabelAlongLine} 标签沿线标注方向样式类
background	{SuperMap. REST. ThemeLabelBackground} 标签专题图中标签的背景风格类
flow	{SuperMap. REST. ThemeFlow} 标签专题图标签流动显示与牵引线设置类
items	{Array(SuperMap. REST. ThemeLabelItem)} 分段标签专题图的子项数组
labelExpression	{String}标注字段表达式
labelOverLengthMode	{SuperMap. REST. LabelOverLengthMode} 标签专题图中超长标签的处理模式枚举类
matrixCells	{Array(SuperMap. REST. LabelMatrixCell)} 矩阵标签元素数组
maxLabelLength	{Number>}标签在每一行显示的最大长度
numericPrecision	{Number}通过该字段设置其显示的精度
offset	{SuperMap. REST. ThemeOffset} 用于设置标签专题图中标记文本相对于要素内点的偏移量对象
overlapAvoided	{Boolean}是否允许以文本避让方式显示文本
rangeExpression	{String}制作分段标签专题的分段字段或字段表达式
smallGeometryLabeled	{Boolean}是否显示长度大于被标注对象本身长度的标签
text	{SuperMap. REST. ThemeLabelText} 标签中文本风格

SuperMap. REST. LabelImageCell 为图片类型的矩阵标签元素类，SuperMap. REST. LabelSymbolCell 为符号类型的矩阵标签元素类，SuperMap. REST. LabelThemeCell 为专题图类型的矩阵标签元素类。以上都继承自 LabelMatrixCell 类，主要对矩阵标签中的专题图类型的矩阵标签元素进行设置。矩阵标签专题图是标签专题图(ThemeLabel)的一种，该类是这三种类型的矩阵标签元素其中的一种，用于定义符号类型的矩阵标签，如符号 ID 字段名称(符号 ID 与 SuperMap 桌面产品中点、线、面符号的 ID 对应)、大小等。用户在实现矩阵标签专题图时只需将定义好的矩阵标签元素赋值予 ThemeLabel. matrixCells 属性即可。matrixCells 是一个二维数组，每一维可以是任意类型的矩阵标签元素组成的数组(也可是单个标签元素组成的数组，即数组中只有一个元素)。SuperMap. REST. LabelImageCell 构造函数参数见表 31.3，SuperMap. REST. LabelSymbolCell 构造函数参数见表 31.4。

表 31.3　　　　　　**SuperMap. REST. LabelImageCell 构造函数参数**

参　数	说　明
height	{Number}设置图片的高度，单位为毫米
pathField	{String}设置矩阵标签元素所使用图片的路径

续表

参　　数	说　　明
rotation	{Number}图片的旋转角度。逆时针方向为正方向，单位为度，精确到 0.1 度。默认值为 0.0
width	{Number}设置图片的宽度，单位为毫米
sizeFixed	{Boolean}是否固定图片的大小。默认值为 false，即图片将随地图缩放

表 31.4　　**SuperMap. REST. LabelSymbolCell 构造函数参数**

参　　数	说　　明
style	{SuperMap. REST. ServerStyle} 获取或设置符号样式—— ServerStyle 对象
symbolIDField	{String}符号 ID 或符号 ID 所对应的字段名称

以 World 数据为基础，制作各国首都矩阵标签专题图。首先通过 ServerTextStyle 定义文本风格的相关属性，然后设置 ThemeLabelItem 分段标签专题图的子项类构造函数，LabelThemeCell 矩阵标签元素类构造函数，再设置 ThemeLabel 标签专题图构造函数，最后设置 ThemeParameters 专题图类参数(包括数据源、数据集名称和专题图对象)，由 ThemeService. processAsync(ThemeParameters) 向服务端提供请求，待服务器处理完成并返回结果后，在结果处理的回调函数中解析查询结果 ThemeEventArgs. result ，同时在地图上添加专题图图层。SuperMap. REST. LabelThemeCell 构造函数参数见表 31.5。具体示例代码如下：

表 31.5　　**SuperMap. REST. LabelThemeCell 构造函数参数**

参　　数	说　　明
themeLabel	{SuperMap. REST. ThemeLabel} 使用专题图对象作为矩阵标签的一个元素

```
<body onload="init()">
<div id="toolbar">
<input type="button" value="创建专题图" onclick="addThemeLabel
()" />
<input type="button" value="移除专题图" onclick="removeTheme()" />
</div>
<div id="map"></div>
</body>
<script src='libs/SuperMap.Include.js'></script>
<script type="text/javascript">
        var map, local, baseLayer, layersID, themeLayer,
```

```
    host = document.location.toString().match(/file:\/\//)?"http://
localhost:8090":
                                'http://' + document.location.host,
        url=host+"/iserver/services/map-world/rest/maps/World";
        function init(){
                map = new SuperMap.Map("map",{controls: [
                    new SuperMap.Control.LayerSwitcher(),
                    new SuperMap.Control.ScaleLine(),
                    new SuperMap.Control.Zoom(),
                    new SuperMap.Control.Navigation({
                        dragPanOptions: {
                            enableKinetic: true
                        }
                    })]
                });
                baseLayer = new SuperMap.Layer.TiledDynamicRESTLayer
("World", url,{transparent: true, cacheEnabled: true}, {maxResolu-
tion:"auto"});
                baseLayer.events.on({"layerInitialized":addLayer});
            }
        function addLayer() {
                map.addLayer(baseLayer);
                map.setCenter(new SuperMap.LonLat(0, 0), 0);
                map.allOverlays = true;
            }
        function addThemeLabel() {
                removeTheme();
                var themeService = new
SuperMap.REST.ThemeService(url,{eventListeners:{"processComplet-
ed": themeCompleted, "processFailed": themeFailed}}),
                    style1 = new SuperMap.REST.ServerTextStyle({
                            fontHeight: 4,
                            foreColor: new SuperMap.REST.ServerColor
(100,20,50),
                            sizeFixed: true,
                        }),
                        style2 = new SuperMap.REST.ServerTextStyle
({
```

```
                    fontHeight: 4,
                    foreColor: new SuperMap.REST.ServerColor
(250,0,0),
                    sizeFixed: true,
                    bold:true
                }),
                style3 = new SuperMap.REST.ServerTextStyle({
                    fontHeight: 4,
                    foreColor: new SuperMap.REST.ServerColor
(93,95,255),
                    sizeFixed: true,
                    bold:true
                }),
                themeLabelIteme1 = new SuperMap.REST.Theme-
LabelItem({
                    start: 0.0,
                    end: 7800000,
                    style: style1
                }),
                themeLabelIteme2 = new SuperMap.REST.Theme-
LabelItem({
                    start: 7800000,
                    end: 15000000,
                    style: style2
                }),
                themeLabelIteme3 = new SuperMap.REST.Theme-
LabelItem({
                    start: 15000000,
                    end:  30000000,
                    style: style3
                }),
                themeLabelIteme4 = new SuperMap.REST.Theme-
LabelItem({
                    start: 0.0,
                    end: 55,
                    style: style1
                }),
                themeLabelIteme5 = new SuperMap.REST.Theme-
```

```
LabelItem({
                    start: 55,
                    end: 109,
                    style: style2
                }),
                themeLabelIteme6 = new SuperMap.REST.Theme-
LabelItem({
                    start: 109,
                    end:  300,
                    style: style3
                }),
                themeLabelOne = new SuperMap.REST.
ThemeLabel({
                  labelExpression: "CAPITAL",
                    rangeExpression: "SmID",
                    numericPrecision: 0,
                    items: [themeLabelIteme4, themeLabelIteme5,
themeLabelIteme6]
                }),
                themeLabelTwo = new SuperMap.REST.ThemeLabel({
                    labelExpression: "CAP_POP",
                    rangeExpression: "CAP_POP",
                    numericPrecision: 0,
                    items: [themeLabelIteme1, themeLabelIteme2,
themeLabelIteme3]
                }),
                LabelThemeCellOne=new SuperMap.REST.
LabelThemeCell({
                    themeLabel: themeLabelOne
                }),
                LabelThemeCellTwo=new SuperMap.REST.
LabelThemeCell({
                    themeLabel: themeLabelTwo
                }),
                backStyle=new SuperMap.REST.ServerStyle({
                    fillForeColor: new SuperMap.REST.ServerColor
(255,255,0),
                    fillOpaqueRate: 60,
```

```
                                lineWidth: 0.1
                            }),
                            themeLabel = new SuperMap.REST.ThemeLabel({
                                matrixCells: [[LabelThemeCellOne],[La-
belThemeCellTwo]],
                                background:new SuperMap.REST.ThemeLabel
Background({
                                    backStyle: backStyle,
                                    labelBackShape:"RECT"
                                })
                            }),
                            themeParameters = new SuperMap.REST.
ThemeParameters({
                                themes: [themeLabel],
                                datasetNames: ["Capitals"],
                                dataSourceNames: ["World"]
                            });
                    themeService.processAsync(themeParameters);
                }
                function themeCompleted(themeEventArgs) {
                    if(themeEventArgs.result.resourceInfo.id) {
                        themeLayer = new SuperMap.Layer.TiledDynamicRESTLayer
("各国首都矩阵标签专题图", url, {cacheEnabled: false,transparent: true,
layersID: themeEventArgs.result.resourceInfo.id}, {"maxResolution":"au-
to"});
                        themeLayer.events.on({"layerInitialized":ad-
dThemelayer});
                    }
                }
                function addThemelayer() {
                    map.addLayer(themeLayer);
                }
                function themeFailed(serviceFailedEventArgs) {
                    //doMapAlert("",serviceFailedEventArgs.error.
errorMsg,true);
                    alert(serviceFailedEventArgs.error.errorMsg);
                }
                function removeTheme() {
```

```
            if (map.layers.length > 1) {
                map.removeLayer(themeLayer, true);
            }
        }
    </script>
```

运行程序，专题图效果如图 31.2 所示。

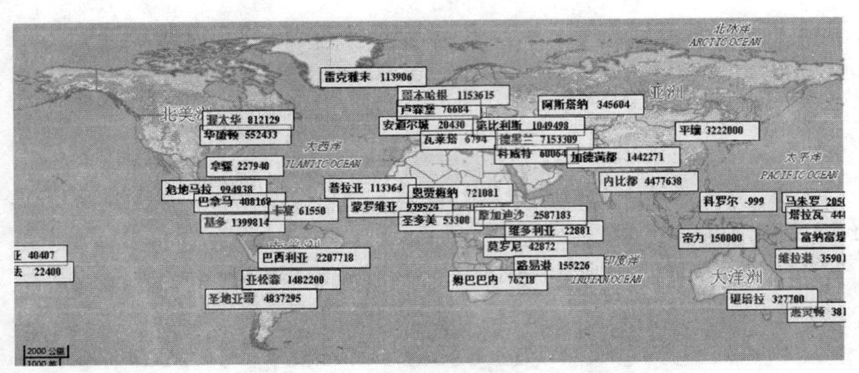

图 31.2　各国首都矩阵标签专题图

31.1.3　栅格分段专题图

栅格分段专题图，是将所有单元格的值按照某种分段方式分成多个范围段，值在同一个范围段中的单元格使用相同的颜色进行显示。栅格分段专题图一般用来反映连续分布现象的数量或程度特征。例如某年的全国降水量分布图，将各气象站点的观测值经过内插之后生成的栅格数据进行分段显示。该类类似于分段专题图类，不同点在于分段专题图的操作对象是矢量数据，而栅格分段专题图的操作对象是栅格数据。SuperMap. REST. ThemeGridRange 构造函数参数见表 31.6。

表 31.6　**SuperMap. REST. ThemeGridRange 构造函数参数**

参　　数	说　　明
items	{Array(SuperMap. REST. ThemeGridRangeItem)} 栅格分段专题图子项数组
reverseColor	{boolean} 是否对栅格分段专题图中分段的颜色风格进行反序显示
rangeMode	{SuperMap. REST. RangeMode} 分段专题图的分段模式
rangeParameter	{Number} 分段参数
colorGradientType	{SuperMap. REST. ColorGradientType} 渐变颜色枚举类

栅格分段专题图子项类 SuperMap. REST. ThemeGridRangeItem，在栅格分段专题图中，将栅格值按照某种分段模式被分成多个范围段。本类用来设置每个范围段的分段起始值、

终止值、名称和颜色等。每个分段所表示的范围为 [Start, End)。SuperMap. REST. ThemeGridRangeItem 构造函数参数见表 31.7。

表 31.7　　**SuperMap. REST. ThemeGridRangeItem 构造函数参数**

参　数	说　明
caption	{String} 栅格分段专题图子项的标题
color	{ServerColor} 栅格分段专题图中每一个分段专题图子项的对应的颜色
end	{Number} 栅格分段专题图子项的终止值
start	{Number} 栅格分段专题图子项的起始值

本节将以"Jingjin"数据为基础，制作京津地形高程分段专题图。首先通过 ServerColor 使用三原色(RGB)来表达颜色，然后设置 ThemeGridRangeItem 栅格分段专题图子项类构造函数，再设置 ThemeGridRange 构造函数参数，最后设置 ThemeParameters 专题图参数类(该类存储了制作专题所需的参数，包括数据源、数据集名称和专题图对象)，由 ThemeService. processAsync(ThemeParameters) 向服务端提供请求，待服务器处理完成并返回结果后在结果处理的回调函数中解析查询结果 ThemeEventArgs. result ，同时在地图上添加专题图图层。具体实现代码如下：

```
    <body onload="init()">
    <div id="toolbar">
    <input type=" button" value=" 创建专题图" onclick=" addTheme-
GridRange()" />
    <input type="button" value="移除专题图" onclick="removeTheme()" />
    </div>
    <divid="map"></div>
    </body>
    <script src='libs/SuperMap.Include.js'></script>
    <script type="text/javascript">
            var map, local, baseLayer, layersID, themeLayer,
            host = document.location.toString().match(/file:\/
\//)?"http://localhost:
                        8090":'http://' + document.location.host,
        url = host + "/iserver/services/map-jingjin/rest/maps/京津地
区人口分布图_专题图";
            function init(){
                map = new SuperMap.Map("map",{controls:[
                    new SuperMap.Control.LayerSwitcher(),
                    new SuperMap.Control.ScaleLine(),
```

```
            new SuperMap.Control.Zoom(),
            new SuperMap.Control.Navigation({
                dragPanOptions: {
                    enableKinetic: true
                }
            })]
        });
        baseLayer = new SuperMap.Layer.TiledDynamicRESTLayer
("Jingjin", url, {transparent: true, cacheEnabled: true}, {maxResolu-
tion:"auto"});
        baseLayer.events.on({"layerInitialized":addLayer});
    }
    function addLayer() {
        map.addLayer(baseLayer);
        map.setCenter(new SuperMap.LonLat(118, 40), 0);
        map.allOverlays = true;
    }
    function addThemeGridRange() {
        removeTheme();
        var themeService = new SuperMap.REST.ThemeService
(url,{eventListeners:{"processCompleted": themeCompleted,
                "processFailed": themeFailed}}
    ),
                color1 = new SuperMap.REST.ServerColor(198,
244,240),
                color2 = new SuperMap.REST.ServerColor(176,
244,188),
                color3 = new SuperMap.REST.ServerColor(218,
251,178),
                color4 = new SuperMap.REST.ServerColor(220,
236,145),
                color5 = new SuperMap.REST.ServerColor(96,
198,66),
                color6 = new SuperMap.REST.ServerColor(20,
142,53),
                color7 = new SuperMap.REST.ServerColor(85,
144,55),
                color8 = new SuperMap.REST.ServerColor(171,
```

```
168,38),
                        color9 = new SuperMap.REST.ServerColor(235,
165,9),
                        color10 = new SuperMap.REST.ServerColor(203,
89,2),
                        color11= new SuperMap.REST.ServerColor(157,
25,1),
                        color12= new SuperMap.REST.ServerColor(118,
15,3),
                        color13= new SuperMap.REST.ServerColor(112,
32,7),
                        color14= new SuperMap.REST.ServerColor(106,
45,12),
                        color15= new SuperMap.REST.ServerColor(129,
80,50),
                        color16= new SuperMap.REST.ServerColor(160,
154,146),
                        themeGridRangeIteme1 = new SuperMap.REST.
ThemeGridRangeItem({
                            start: -4,
                            end: 120,
                            color: color1
                        }),
                        themeGridRangeIteme2 = new SuperMap.REST.
ThemeGridRangeItem({
                            start: 120,
                            end: 240,
                            color: color2
                        }),
                        themeGridRangeIteme3 = new SuperMap.REST.
ThemeGridRangeItem({
                            start: 240,
                            end:  360,
                            color: color3
                        }),
                        themeGridRangeIteme4 = new SuperMap.REST.
ThemeGridRangeItem({
                            start: 360,
```

```
                    end: 480,
                    color: color4
                }),
                themeGridRangeIteme5 = new SuperMap.REST.
ThemeGridRangeItem({
                    start: 480,
                    end: 600,
                    color: color5
                }),
                themeGridRangeIteme6 = new SuperMap.REST.
ThemeGridRangeItem({
                    start: 600,
                    end: 720,
                    color: color6
                }),
                themeGridRangeIteme7 = new SuperMap.REST.
ThemeGridRangeItem({
                    start: 720,
                    end: 840,
                    color: color7
                }),
                themeGridRangeIteme8 = new SuperMap.REST.
ThemeGridRangeItem({
                    start: 840,
                    end: 960,
                    color: color8
                }),
                themeGridRangeIteme9 = new SuperMap.REST.
ThemeGridRangeItem({
                    start: 960,
                    end: 1100,
                    color: color9
                }),
                themeGridRangeIteme10 = new SuperMap.REST.
ThemeGridRangeItem({
                    start: 1100,
                    end: 1220,
                    color: color10
```

```
                        }),
                        themeGridRangeIteme11 = new SuperMap.REST.
ThemeGridRangeItem({
                            start: 1220,
                            end:  1340,
                            color: color11
                        }),
                        themeGridRangeIteme12 = new SuperMap.REST.
ThemeGridRangeItem({
                            start: 1340,
                            end:  1460,
                            color: color12
                        }),
                        themeGridRangeIteme13 = new SuperMap.REST.
ThemeGridRangeItem({
                            start: 1460,
                            end: 1600,
                            color: color13
                        }),
                        themeGridRangeIteme14 = new SuperMap.REST.
ThemeGridRangeItem({
                            start: 1600,
                            end: 1800,
                            color: color14
                        }),
                        themeGridRangeIteme15 = new SuperMap.REST.
ThemeGridRangeItem({
                            start: 1800,
                            end:  2000,
                            color: color15
                        }),
                        themeGridRangeIteme16 = new SuperMap.REST.
ThemeGridRangeItem({
                            start: 2000,
                            end:  2167,
                            color: color16
                        }),
                        themeGridRange = new SuperMap.REST.
```

```
ThemeGridRange({
                                    reverseColor:false,
                                    rangeMode: SuperMap.REST.RangeMode.
EQUALINTERVAL,
                                    items: [themeGridRangeIteme1,
                                        themeGridRangeIteme2,
                                        themeGridRangeIteme3,
                                        themeGridRangeIteme4,
                                        themeGridRangeIteme5,
                                        themeGridRangeIteme6,
                                        themeGridRangeIteme7,
                                        themeGridRangeIteme8,
                                        themeGridRangeIteme9,
                                        themeGridRangeIteme10,
                                        themeGridRangeIteme11,
                                        themeGridRangeIteme12,
                                        themeGridRangeIteme13,
                                        themeGridRangeIteme14,
                                        themeGridRangeIteme15,
                                        themeGridRangeIteme16
                                    ]
                                }),
                                themeParameters = new SuperMap.REST.
ThemeParameters({
                                    datasetNames: ["JingjinTerrain"],
                                    dataSourceNames: ["Jingjin"],
                                    joinItems: null,
                                themes: [themeGridRange]
                                });
                        themeService.processAsync(themeParameters);
                }
                function themeCompleted(themeEventArgs) {
                    if(themeEventArgs.result.resourceInfo.id) {
                            themeLayer = new SuperMap.Layer.TiledDynami-
cRESTLayer("京津地形高程分段专题图", url, {cacheEnabled:true,transpar-
ent: true, layersID: themeEventArgs.result.resourceInfo.id}, {"max-
Resolution":"auto"});
                            themeLayer.events.on({"layerInitialized":ad-
```

```
dThemelayer});
                    }
                }
                function addThemelayer() {
                    map.addLayer(themeLayer);
                }
                function themeFailed(serviceFailedEventArgs) {
                     //doMapAlert("",serviceFailedEventArgs.error.er-
rorMsg,true);
                    alert(serviceFailedEventArgs.error.errorMsg);
                }
                function removeTheme() {
                    if (map.layers.length > 1) {
                        map.removeLayer(themeLayer, true);
                    }
                }
        </script>
```

运行程序，京津地形高程分段专题图效果如图 31.3 所示。

图 31.3　京津地形高程分段专题图

31.2　空间分析

GIS 与一般电子地图最重要的区别之一，就是提供强大的查询统计、空间分析功能，而这些特性让其在各行业领域应用中发挥着重要作用，为生产生活提供了更多的便利与服务。

从 GIS 应用角度看，空间分析大致可以归纳为两大类：

①基于点、线、面基本地理要素的空间分析，通过空间信息查询与量测、缓冲区分析、叠加分析、网络分析、地理统计分析等空间分析方法挖掘出新的信息。

②地理问题模拟，解决应用领域对空间数据处理与输出的特殊要求，地理实体和空间关系通过专业模型得到简化和抽象，而系统则通过模型进行深入分析操作。

31.2.1 缓冲区分析

缓冲区分析是 GIS 中基本的空间分析，缓冲区分析实际上是在基本空间要素周围建立具有一定宽度的邻近区域。缓冲区分析多用在确定道路拓宽的范围、确定放射源影响的范围等方面。缓冲区分析可以应用在点、线和面状地物上，在对线状地物进行缓冲区分析时，可以设置地物的左、右侧缓冲距离，并且可以设置缓冲的端点类型为平头缓冲或者圆头缓冲，当对点和面状地物进行缓冲区分析时，则只需要设置地物的左侧缓冲距离即可，并且缓冲区端点类型只能为圆头缓冲。

SuperMap. REST. BufferAnalystService 缓冲区分析服务类，该类负责将客户设置的缓冲区分析参数传递给服务端，并接收服务端返回的缓冲区分析结果数据。缓冲区分析结果通过该类支持的事件的监听函数参数获取，参数类型为 {SuperMap. REST. BufferAnalystEventArgs} 获取的结果数据包括 originResult、result 两种，其中，originResult 为服务端返回的用 JSON 对象表示的缓冲区分析结果数据，result 为服务端返回的缓冲区分析结果数据，result 属性的类型与传入的参数类型相关，当缓冲区分析服务的参数为 {SuperMap. REST. DatasetBufferAnalystParameters} 时，返回的结果类型为 {SuperMap. REST. DatasetBuff erAnalystResult} ，当参数为 {SuperMap. REST. GeometryBufferAnalystParameters} 类型时，返回的结果类型为 {SuperMap. REST. GeometryBufferAnalystResult} 。

下面用一个实例画一条线，查询并显示出附近 300 米内的工厂。具体示例代码如下：

```
<body onload="init()">
<div id="toolbar">
<input type="button" value="绘线" onclick="addPath()"/>
<input type="button" value="缓冲区分析" onclick="bufferAnalyst-
Process()"/>
<input type="button" value="查询数据" onclick="queryByGeometry
()"/>
<input type="button" value="移除结果" onclick="clearElements()"/>
</div>
<div id="map"></div>
</body>
<script src='libs/SuperMap.Include.js'></script>
<script type="text/javascript">
```

```
    varhost = document.location.toString().match(/file: \/\//)?"ht-
tp://localhost:8090":'http://' + document.location.host;
    var spatialAnalystURL, local, map, layer, vectorLayer, resultLay-
er, markerLayer, pathLine, bufferResultGeometry,drawLine,
            styleLine = {
                strokeColor:"red",
                strokeWidth: 2,
                pointRadius: 3,
                pointerEvents:"visiblePainted",
                fill:false
    },
            styleRegion = {
                strokeColor:"#304DBE",
                strokeWidth: 2,
                pointerEvents:"visiblePainted",
                fillColor:"#304DBE",
                fillOpacity: 0.4
            },
    url=host+"/iserver/services/map-changchun/rest/maps/长春市区图",
    url2=host+"/iserver/services/spatialanalyst-changchun/restjsr/
spatialanalyst";
    functioninit(){
                vectorLayer =new SuperMap.Layer.Vector("Vector
Layer");
                resultLayer =new SuperMap.Layer.Vector("Result
Layer");
                markerLayer =new SuperMap.Layer.Markers("Markers
Layer");
               map =new SuperMap.Map("map",{controls:[
                      new SuperMap.Control.LayerSwitcher(),
                      new SuperMap.Control.ScaleLine(),
                      new SuperMap.Control.Zoom(),
                      new SuperMap.Control.Navigation({
                                    dragPanOptions:{
                                        enableKinet-
ic:true
                      }
                     })], units:"m"
```

```
                    });
                    vectorLayer.style=styleLine;
                      drawLine =new SuperMap.Control.DrawFeature(vec-
torLayer, SuperMap.Handler.Path, { multi: true});
                    drawLine.events.on({"featureadded": drawCompleted});
                    map.addControl(drawLine);
                      layer =new SuperMap.Layer.TiledDynamicRESTLayer
("Changchun", url, {transparent: true, cacheEnabled: true}, {maxReso-
lution: "auto"});
                    layer.events.on({"layerInitialized":addLayer});
                }
    function addLayer() {
                map.addLayers([layer, vectorLayer, resultLayer, mark-
erLayer]);
                map.setCenter(new SuperMap.LonLat(4503.6240321526, -
3861.911472192499), 0);
                }
    function addPath(){
                    drawLine.activate();
                }
    function drawCompleted(eventArgs) {
                var geometry = eventArgs.feature.geometry;
                    pathLine= geometry;
                    drawLine.deactivate();
                }
    function bufferAnalystProcess() {
    if(! pathLine){
      alert("请生成路径,用于缓冲区分析");
      return;
    }
    var bufferServiceByGeometry = new SuperMap.REST.BufferAnalystService
(url2),
                        bufferDistance =new SuperMap.REST.BufferDistance({
                            value: 300
                        }),
                        bufferSetting =new SuperMap.REST.BufferSetting({
                          endType: SuperMap.REST.BufferEndType.ROUND,
                            leftDistance: bufferDistance,
```

```
                    rightDistance: bufferDistance,
                    semicircleLineSegment: 10
                }),
                geoBufferAnalystParam =new SuperMap.REST.
GeometryBufferAnalystParameters({
                    sourceGeometry: pathLine,
                    bufferSetting: bufferSetting
                });
            bufferServiceByGeometry.events.on(
                {
                    "processCompleted": bufferAnalystCompleted
                });
            bufferServiceByGeometry.processAsync(geoBufferAn-
alystParam);
        }
    function bufferAnalystCompleted(BufferAnalystEventArgs) {
    var feature = new SuperMap.Feature.Vector();
            bufferResultGeometry = BufferAnalystEventArgs.
result.resultGeometry;
            feature.geometry = bufferResultGeometry;
            feature.style = styleRegion;
            resultLayer.addFeatures(feature);
        }
    //查询出信号影响范围内的工厂
    function queryByGeometry(){
    if(! bufferResultGeometry){
        alert("请先做缓冲区分析,以得到查询的范围");
        return;
    }
    var queryParam, queryByGeometryParameters, queryService;
                queryParam  = new SuperMap.REST.FilterParameter
({name: "Company@ Changchun.2"});
            queryByGeometryParameters =new SuperMap.REST.
QueryByGeometryParameters({
                queryParams: [queryParam],
                geometry: bufferResultGeometry,
                                        spatialQueryMode:
SuperMap.REST.SpatialQueryMode.INTERSECT
```

```
            });
            queryService =new SuperMap.REST.
QueryByGeometryService(url);
            queryService.events.on(
                {
                    "processCompleted": queryCompleted
                });
            queryService.processAsync(queryByGeometryParameters);
    }
    function queryCompleted(queryEventArgs) {
    vari, j, result = queryEventArgs.result;
    if(result && result.recordsets) {
    for(i = 0, recordsets = result.recordsets, len = recordsets.length;
i<len; i++) {
    if(recordsets[i].features) {
    for(j = 0; j<recordsets[i].features.length; j++) {
    var point = recordsets[i].features[j].geometry,
      size = new SuperMap.Size(44,40),
      offset = new SuperMap.Pixel(-(size.w/2), -size.h),
      icon = new SuperMap.Icon("theme/images/marker.png", size, offset);
                        markerLayer.addMarker(new Super-
                        Map.Marker ( new  SuperMap.LonLat
                        (point.x, point.y), icon));
                    }
                }
            }
        }
    }
    function clearElements() {
            pathLine=null;
            bufferResultGeometry=null;
            vectorLayer.removeAllFeatures();
            resultLayer.removeAllFeatures();
            markerLayer.clearMarkers();
        }
    </script>
```

运行程序，单击【绘线】在地图上画一条线，然后再单击【缓冲区分析】，生成缓冲多边形，最后点击【查询数据】，返回数据并以图标的方式添加到地图，实现效果如图 31.4 所示。

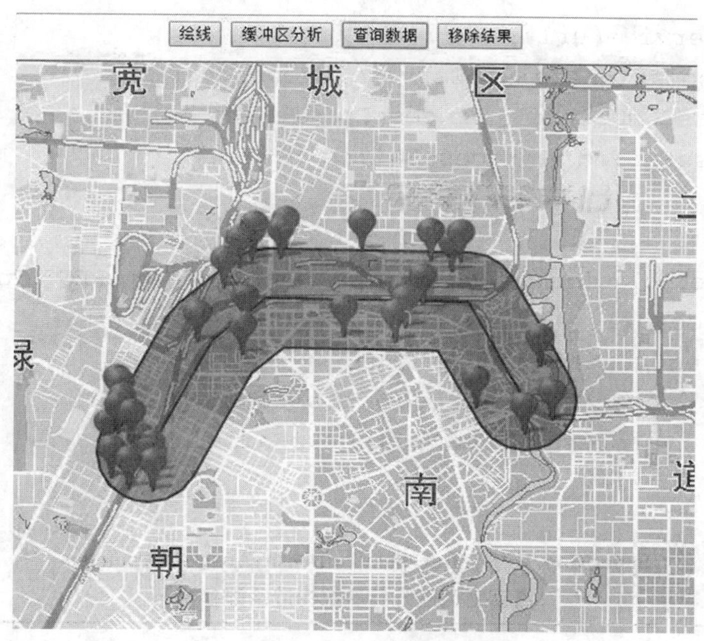

图 31.4　线缓冲区 300 米工厂查询

31.2.2　插值分析

一般情况下，采集的数据都是以离散点的形式存在，只有在这些采样点上有较为准确的数值，而其他未采集点上没有数值。然而，在实际应用中却很可能需要用到未采集数据点的值，这个时候就需要通过已采集的数据点值来推算未采集的点值。这样的一个运算过程就是插值运算。

插值分析是用于对离散的点数据进行插值得到栅格数据集。将某个区域的采样点数据插值生成栅格数据，实际上是将研究区域按照给定的格网尺寸(分辨率)进行栅格化，栅格数据中每一个栅格单元对应一块区域，栅格单元的值由其邻近的采样点的数值通过某种插值方法计算得到，因此，就可以预测采样点周围的数值情况，进而了解整个区域的数值分布情况。其中，插值方法主要有点密度插值法、距离反比权值插值法、克里金(Kriging)内插法、径向基函数 RBF(Radial Basis Function)插值法等，SuperMap. REST. InterpolationDensityAnalystParameters 点密度插值分析参数构造函数参数见表 31.8，SuperMap. REST. InterpolationIDWAnalystParameters IDW 分析参数类构造函数参数见表 31.9，SuperMap. REST. InterpolationKrigingAnalystParameters 克里金插值分析参数类构造函数参数见表 31.10。SuperMap. REST. Interpolation RBF AnalystParameters 样条插值分析参数类构造函数参数见表 31.11。

SuperMap. REST. InterpolationAnalystService 插值分析服务类构造函数，负责将客户端的查询参数传递到服务端。构造函数示例如下：

```
var myTInterpolationAnalystService =newSuperMap.REST.Interpola-
```

```
tionAnalystService(url);
    myTInterpolationAnalystService.events.on({
      "processCompleted": processCompleted,
      "processFailed": processFailed
      });
    或:
    varmyTInterpolationAnalystService =new SuperMap.REST.Interpolation-
AnalystService(url,{
          eventListeners: {
              "processCompleted": processCompleted,
              "processFailed": processFailed
      }});
```

表 31.8 **SuperMap. REST. InterpolationDensityAnalystParameters 点密度插值分析参数构造函数参数**

参 数	说 明
Bounds	{SuperMap. Bounds} 插值分析的范围，用于确定结果栅格数据集的范围
searchRadius	{Number}查找半径，即参与运算点的查找范围，与点数据集单位相同
zValueFieldName	{String}存储用于进行插值分析的字段名称，插值分析不支持文本类型的字段。必设参数
zValueScale	{Number}用于进行插值分析值的缩放比率，默认为 1
resolution	{Number}插值结果栅格数据集的分辨率，即一个像元所代表的实地距离，与点数据集单位相同
filterQueryParameter	{SuperMap. REST. FilterParameter} 属性过滤条件。必设参数
outputDatasetName	{String}插值分析结果数据集的名称。必设参数
pixelFormat	{String}指定结果栅格数据集存储的像素格式。必设参数
Dataset	{String}要用来做插值分析的数据源中数据集的名称。该名称用形如"数据集名称@数据源别名"形式来表示。必设参数

表 31.9 **SuperMap. REST. InterpolationIDWAnalystParameters IDW 分析参数类构造函数参数**

参 数	说 明
Power	{Number}距离权重计算的幂次
bounds	{SuperMap. Bounds} 插值分析的范围，用于确定结果栅格数据集的范围
searchMode	{String}插值运算时，查找参与运算点的方式，支持固定点数查找、定长查找。必设参数

续表

参　数	说　明
expectedCount	{Number}【固定点数查找】方式下，设置待查找的点数，即参与差值运算的点数
searchRadius	{Number}【定长查找】方式下，设置查找半径，即参与运算点的查找范围，与点数据集单位相同
zValueFieldName	{String} 存储用于进行插值分析的字段名称，插值分析不支持文本类型的字段。必设参数
zValueScale	{Number} 用于进行插值分析值的缩放比率，默认为 1
resolution	{Number} 插值结果栅格数据集的分辨率，即一个像元所代表的实地距离，与点数据集单位相同
filterQueryParameter	{SuperMap. REST. FilterParameter} 属性过滤条件。必设参数
outputDatasetName	{String} 插值分析结果数据集的名称。必设参数
outputDatasourceName	{String} 插值分析结果数据源的名称。必设参数
pixelFormat	{String} 指定结果栅格数据集存储的像素格式。必设参数
dataset	{String} 要用来做插值分析的数据源中数据集的名称。该名称用形如"数据集名称@数据源别名"形式来表示。必设参数

表 31.10　**SuperMap. REST. InterpolationKrigingAnalystParameters 克里金插值分析参数类构造函数参数**

参　数	说　明
type	{String} 克里金插值的类型。必设参数
mean	{Number}【简单克里金】类型下，插值字段的平均值
angle	{Number} 克里金算法中旋转角度值。默认值为 0
nugget	{Number} 克里金算法中块金效应值。默认值为 0
range	{Number} 克里金算法中自相关阈值，单位与原数据集单位相同。默认值为 0
sill	{Number} 克里金算法中基台值。默认值为 0
variogramMode	{String} 克里金插值时的半变函数类型，默认为球型(SPHERICAL)
exponent	{String}【泛克里金】类型下，用于插值的样点数据中趋势面方程的阶数，可选值为 exp1、exp2，默认为 exp1
bounds	{SuperMap. Bounds} 插值分析的范围，用于确定结果栅格数据集的范围
searchMode	{String} 插值运算时，查找参与运算点的方式，有固定点数查找、定长查找、块查找。必设参数

续表

参　　数	说　　明
expectedCount	{Number}【固定点数查找】方式下，设置待查找的点数，默认为 12；【定长查找】方式下，设置查找的最小点数，默认为 12
searchRadius	{Number}【定长查找】方式下，设置参与运算点的查找范围
maxPointCountForInterpolation	{Number}【块查找】方式下，设置最多参与插值的点数。默认为 200
maxPointCountInNode	{Number}【块查找】方式下，设置单个块内最多参与运算点数。默认为 50
zValueFieldName	{String}存储用于进行插值分析的字段名称，插值分析不支持文本类型的字段。必设参数
zValueScale	{Number}用于进行插值分析值的缩放比率，默认为 1
resolution	{Number}插值结果栅格数据集的分辨率，即一个像元所代表的实地距离，与点数据集单位相同
filterQueryParameter	{SuperMap. REST. FilterParameter} 属性过滤条件。必设参数
outputDatasetName	{String}插值分析结果数据集的名称。必设参数
pixelFormat	{String}指定结果栅格数据集存储的像素格式。必设参数
dataset	{String}要用来做插值分析的数据源中数据集的名称

表 31. 11　　　**SuperMap. REST. InterpolationRBFAnalystParameters 样条插值分析参数类构造函数参数**

参　　数	说　　明
smooth	{Number}光滑系数，该值表示插值函数曲线与点的逼近程度，值域为 0 到 1，默认值约为 0. 1
tension	{Number}张力系数，用于调整结果栅格数据表面的特性，默认为 40
bounds	{SuperMap. Bounds} 插值分析的范围，用于确定结果栅格数据集的范围
searchMode	{String}插值运算时，查找参与运算点的方式，有固定点数查找、定长查找、块查找。必设参数
expectedCount	{Number}【固定点数查找】方式下，设置参与差值运算的点数
searchRadius	{Number}【定长查找】方式下，设置参与运算点的查找范围
maxPointCountForInterpolation	{Number}【块查找】方式下，设置最多参与插值的点数。默认为 200
maxPointCountInNode	{Number}【块查找】方式下，设置单个块内最多参与运算点数。默认为 50
zValueFieldName	{String}存储用于进行插值分析的字段名称，插值分析不支持文本类型的字段。必设参数
zValueScale	{Number}用于进行插值分析值的缩放比率，默认为 1

续表

参　数	说　明
resolution	{Number} 插值结果栅格数据集的分辨率，即一个像元所代表的实地距离，与点数据集单位相同
filterQueryParameter	{SuperMap. REST. FilterParameter} 属性过滤条件。必设参数
outputDatasetName	{String} 插值分析结果数据集的名称。必设参数
pixelFormat	{String} 指定结果栅格数据集存储的像素格式。必设参数
dataset	{String} 要用来做插值分析的数据源中数据集的名称。必设参数

以全国温度信息收集站数据为基础，进行泛克里金插值，核心代码如下：

```
var url=host+"/iserver/services/map-temperature/rest/maps/全国温度变化图",
url2=host+"/iserver/services/spatialanalyst-sample/restjsr/spatialanalyst";
function interpolationUniversalKriging(){
    var interpolationParams = new SuperMap.REST.InterpolationKrigingAnalystParameters({
        dataset: "SamplesP@ Interpolation",
        outputDatasetName: "UniversalKriging_Result",
        outputDatasourceName: "Interpolation",
        pixelFormat: SuperMap.REST.PixelFormat.double,
        filterQueryParameter: {
            attributeFilter: ""
        },
        zValueFieldName: "AVG_TMP",
        searchRadius: "0",
        type: "UniversalKriging",
        angle: 0,
        nugget: 0,
        range: 0,
        sill: 0,
        variogramMode: "SPHERICAL",
        searchMode: "KDTREE_FIXED_COUNT",
        bounds:new SuperMap.Bounds(-2640403.6321084504, 1873792.1034850003,3247669.390292245, 5921501.395578556)
    });
    var interpolationService=new SuperMap.REST.Interpolation-
```

```
AnalystService(url2,{
        eventListeners: {
            "processCompleted": processCompleted,
            "processFailed": processFailed
        }});
    interpolationService.processAsync(interpolationParams);
}
```

//插值分析成功后,使用栅格分段专题图展示(部分代码省略)完全实现原理过程可参考31.1.3节栅格分段专题图

栅格分段专题图的示例代码如下:

```
function processCompleted(InterpolationAnalystEventArgs){
var color1 = new SuperMap.REST.ServerColor(170,240,231),
        color2 = new SuperMap.REST.ServerColor(176,244,188),
 ...
 ...
        color20= new SuperMap.REST.ServerColor(166,153,146),
        themeGridRangeIteme1 = new SuperMap.REST.ThemeGrid
RangeItem({
            start:-5,
            end:-3.4,
            color: color1
        }),
        themeGridRangeIteme2 = new SuperMap.REST.ThemeGrid
RangeItem({
            start: -3.4,
            end: -1.8,
            color: color2
        }),
 ...
 ...
        themeGridRangeIteme19 = new SuperMap.REST.ThemeGrid
RangeItem({
            start: 23.8,
            end:25.4,
            color: color19
        }),
        themeGridRangeIteme20 = new SuperMap.REST.ThemeGrid
RangeItem({
            start: 25.4,
```

```
                end:27,
                color: color20
            }),
            themeGridRange = new SuperMap.REST.ThemeGridRange({
                reverseColor:false,
                rangeMode: SuperMap.REST.RangeMode.EQUALINTERVAL,
                //栅格分段专题图子项数组
                items: [themeGridRangeIteme1,
                    themeGridRangeIteme2,
  ...
  ...

                    themeGridRangeIteme19,
                    themeGridRangeIteme20
                ]
            }),
            themeParameters = new SuperMap.REST.ThemeParameters({
                //制作专题图的数据集数组
    datasetNames:[InterpolationAnalystEventArgs.result.dataset.
split('@')[0]],
                //制作专题图的数据集所在的数据源数组
                dataSourceNames: ["Interpolation"],
                joinItems: null,
                //专题图对象列表
                themes: [themeGridRange]
            });
        var themeService = new SuperMap.REST.ThemeService(url, {even-
tListeners:{
                            "processCompleted": themeCompleted,
                            "processFailed": themeFailed}});
        themeService.processAsync(themeParameters);
    }
    //服务端成功返回专题图结果时调用
    function themeCompleted(themeEventArgs) {
        if(themeEventArgs.result.resourceInfo.id) {
            themeLayer = new SuperMap.Layer.TiledDynamicRESTLayer("插
值分析结果图", url, {cacheEnabled: true, transparent: true, layersID:
themeEventArgs.result.resourceInfo.id}, {"maxResolution":"auto"});
            themeLayer.events.on({"layerInitialized":addThemelayer});
        }
```

```
}
```

运行程序，图 31.5 为全国温度原始数据点图，图 31.6 为插值运算结束后栅格图。

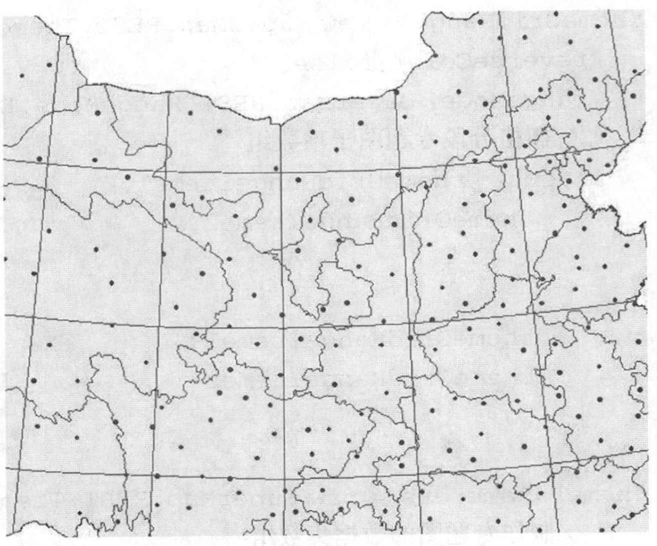

图 31.5　全国温度原始数据点图

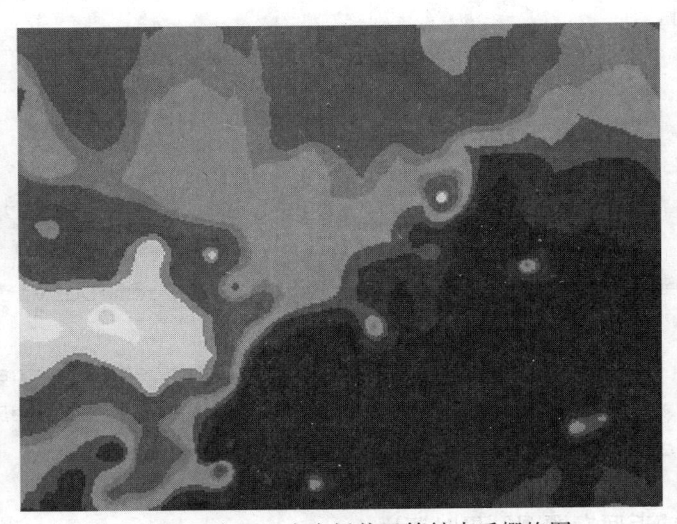

图 31.6　全国温度点插值运算结束后栅格图

31.2.3　表面分析

表面分析主要通过生成新数据集，如等值线、坡度、坡向、山体阴影等派生数据，获得更多的反映原始数据集中所暗含的空间特征、空间格局等信息。等值线是将相邻的具有相同值的点(如高程、温度、降水、污染或大气压力)连接起来的线。常见的有等温线、等压线、等高线、等势线等。等值线的分布反映了栅格表面上值的变化，等值线分布密集的地方，表示栅格表面值的变化比较剧烈。例如，如果为等高线，则越密集，坡度越陡；等值线分布较稀疏，表示栅格表面值的变化较小，若为等高线，则表示坡度很平缓。通过提取等值线，可以找到高程、温度、降水等的值相同的位置，同时等值线的分布状况也可以显示出变化的陡峭和平缓区。

SuperMap. REST. SurfaceAnalystService 表面分析服务类，该类负责将客户设置的表面分析服务参数传递给服务端，并接收服务端返回的表面分析服务分析结果数据。表面分析结果通过该类支持的事件的监听函数参数获取，参数类型为 {SuperMap. REST. SurfaceAnalystEventArgs}，获取的结果数据包括 originResult、result 两种，其中，originResult 为服务端返回的用 JSON 对象表示的表面分析结果数据，result 为服务端返回的表面分析结果数据，保存在 {SuperMap. REST. SurfaceAnalystResult} 对象中。

在 SurfaceAnalystService 类中处理所有事件的对象，支持 processCompleted、processFailed 两种事件，服务端成功返回表面分析结果时触发 processCompleted 事件，服务端返回表面分析结果失败时触发 processFailed 事件。示例代码如下：

```
    var mySurfaceAnalystService =newSuperMap.REST.Surface-
AnalystService(url);
    mySurfaceAnalystService.events.on({
       "processCompleted": surfaceAnalysCompleted,
       "processFailed": surfaceAnalysError
       }
    );
    function surfaceAnalysCompleted(surfaceAnalysEventArgs){//todo};
    function surfaceAnalysError(surfaceAnalysEventArgs){//todo};
```

表面分析参数设置类 SuperMap. REST. SurfaceAnalystParametersSetting，通过该类可以设置表面分析提取等值线、提取等值面的一些参数，包括基准值、等值距、光滑度、光滑方法等，主要参数见表 31. 12。SuperMap. REST. DatasetSurfaceAnalystParameters 数据集表面分析参数类构造函数参数见表 31. 13。

表 31. 12　　**表面分析参数设置类主要参数**

参　数	说　明
clipRegion	{SuperMap. Geometry} 获取或设置裁剪面对象，如果不需要对操作结果进行裁剪，可以使用 null 值取代该参数
datumValue	{Number} 获取或设置表面分析中提取等值线、提取等值面的基准值。基准值是作为一个生成等值线的初始起算值，并不一定是最小等值线的值。例如，高程范围为 220~1550 的 DEM 栅格数据，如果设基准值为 0，等值距为 50，则提取等值线时，以基准值 0 为起点，等值距 50 为间隔提取等值线，因为给定高程的最小值是 220，所以，在给定范围内提取等值线的最小高程是 250。提取等值线的结果是：最小等值线值为 250，最大等值线值为 1550

续表

参　数	说　明
expectedZValues	{Array(Number)}获取或设置期望分析结果的 Z 值集合。Z 值集合存储一系列数值，该数值为待提取等值线的值。即仅高程值在 Z 值集合中的等值线会被提取
interval	{Number}获取或设置等值距。等值距是两条等值线之间的间隔值
resampleTolerance	{Number}获取或设置重采样容限。容限值越大，采样结果数据越简化。当分析结果出现交叉时，可通过调整重采样容限为较小的值来处理
smoothMethod	{SuperMap. REST. SmoothMethod} 获取或设置光滑处理所使用的方法
smoothness	{Number}获取或设置表面分析中等值线或等值面的边界线的光滑度。以为 0~5 为例，光滑度为 0 表示不进行光滑操作，值越大表示光滑度越高。随着光滑度的增加，提取的等值线越光滑，当然光滑度越大，计算所需的时间和占用的内存也就越大。而且，当等值距较小时，光滑度太高会出现等值线相交的问题

表 31.13　**SuperMap. REST. DatasetSurfaceAnalystParameters 数据集表面分析参数类构造函数参数**

参　数	说　明
Dataset	{String}要用来做数据集表面分析的数据源中数据集的名称
filterQueryParameter	{SuperMap. REST. FilterParameter} 获取或设置查询过滤条件参数
zValueFieldName	{String}获取或设置用于提取操作的字段名称
extractParameter	{SuperMap. REST. SurfaceAnalystParametersSetting} 表面分析参数设置类。获取或设置表面分析参数
resolution	{Integer}获取或设置指定中间结果(栅格数据集)的分辨率
resultSetting	{SuperMap. REST. DataReturnOption} 结果返回设置类
surfaceAnalystMethod	{SuperMap. REST. SurfaceAnalystMethod} 获取或设置表面分析的提取方法，提取等值线和提取等值面

以全国温度信息收集站数据为基础，进行等值线绘制，核心代码如下：

```
function surfaceAnalystProcess() {
   resultLayer.removeAllFeatures();
   var surfaceAnalystService = new SuperMap.REST.
SurfaceAnalystService(url),
         surfaceAnalystParameters =new
            SuperMap.REST.SurfaceAnalystParametersSetting({
            datumValue: 0,
            interval: 2,
            resampleTolerance: 0,
```

```
                smoothMethod: SuperMap.REST.SmoothMethod.BSPLINE,
                smoothness: 3,
                clipRegion:null
            }),
            params =new SuperMap.REST.DatasetSurface-
AnalystParameters({
                extractParameter: surfaceAnalystParameters,
                dataset:"SamplesP@Interpolation",
                resolution: 3000,
                zValueFieldName:"AVG_TMP"
            });
    surfaceAnalystService.events.on ({"processCompleted": sur-
                                        faceAnalystCompleted,
                                        "processFailed": sur-
                                        faceAnalystFailed});
    surfaceAnalystService.processAsync(params);
}
function surfaceAnalystCompleted(args) {
    var features = args.result.recordset.features;
    for (var len = features.length, i = 0; i < len; i++) {
        style = {
            strokeColor:"#304DBE",
            fillOpacity: 0
        }
        features[i].style = style;
    }
    resultLayer.addFeatures(args.result.recordset.features);
}
```

运行程序，效果如图 31.7 所示。

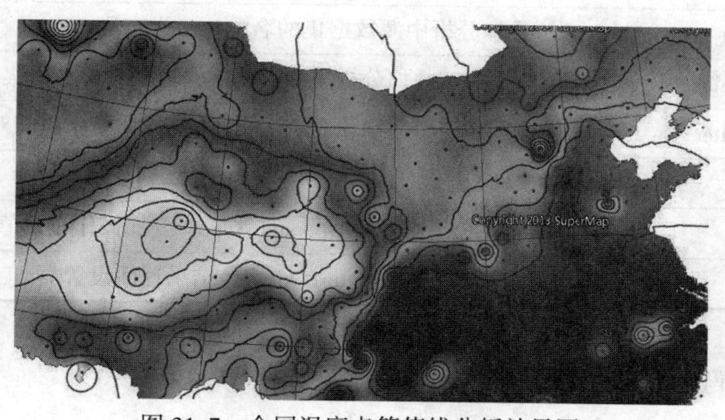

图 31.7　全国温度点等值线分析效果图

31.2.4 其他分析

1. 叠加分析

在统一空间参考系统下，通过对两个数据集进行一系列几何运算，提取用户需要的新的空间几何信息，产生新数据集的过程就是叠加分析。该空间处理支持剪裁、擦除、同一、求交、对称差、合并、更新等叠加分析操作。SuperMap 中的叠加分析涉及三个数据集，其中一个为源数据集，除合并运算和对称差运算必须是面数据集外，其他运算可以是点、线、面、CAD 数据集或者路由数据集；另一个数据集为叠加数据集，必须为面数据集；还有一个数据集就是叠加结果数据集，用于保存叠加分析得到的结果数据。

SuperMap. REST. OverlayAnalystService 类负责将客户设置的叠加分析参数传递给服务端，并接收服务端返回的叠加分析结果数据。叠加分析结果通过该类支持的事件的监听函数参数获取，参数类型为{SuperMap. REST. OverlayAnalystEventArgs}，获取的结果数据包括 originResult、result 两种，其中，originResult 为服务端返回的用 JSON 对象表示的量算结果数据，result 为服务端返回的量算结果数据，result 属性的类型与传入的参数类型相关，当缓冲区分析服务的参数为{SuperMap. REST. DatasetOverlayAnalystParameters}时，返回的结果类型为{SuperMap. REST. DatasetOverlayAnalystResult}；当参数为{SuperMap. REST. GeometryOverlayAnalystParameters}类型时，返回的结果类型为{SuperMap. REST. GeometryOverlayAnalystResult}。SuperMap. REST. DatasetOverlayAnalystParameters 参数见表 31.14。

表 31.14 **SuperMap. REST. DatasetOverlayAnalystParameters 参数**

参　　数	说　　明
operateDataset	{String}叠加分析中操作数据集的名称。必设字段
operateDatasetFields	{Array(String)} 叠加分析中操作数据集保留在结果数据集中的字段名列表
operateDatasetFilter	{SuperMap. REST. FilterParameter} 设置操作数据集中空间对象过滤条件
operateRegions	{Array(SuperMap. Geometry)} 操作区域。设置了操作区域后，仅对该区域内的对象进行分析
sourceDataset	{String}叠加分析中源数据集的名称。必设字段
sourceDatasetFields	{Array(String)} 叠加分析中源数据集保留在结果数据集中的字段名列表
sourceDatasetFilter	{SuperMap. REST. FilterParameter} 设置源数据集中空间对象过滤条件
tolerance	{Integer}容限
operation	{SuperMap. REST. OverlayOperationType} 叠加操作枚举值
resultSetting	{SuperMap. REST. DataReturnOption} 结果返回设置类

核心实现代码如下：

```
function overlayAnalystProcess() {
```

```
    var overlayServiceByDatasets = new SuperMap.REST.Overlay-
AnalystService(url),
        OverlayAnalystParameters =new SuperMap.REST.Dataset-
OverlayAnalystParameters({
            sourceDataset:"BaseMap_R@ Jingjin",
            operateDataset:"Neighbor_R@ Jingjin",
            tolerance: 0,
            operation: SuperMap.REST.OverlayOperationType.UNION
        });
    overlayServiceByDatasets.events.on ({ " processCompleted ":
overlayAnalystCompleted, "processFailed": overlayAnalystFailed });
    overlayServiceByDatasets.processAsync(OverlayAnalystParame-
ters);
  }
  function overlayAnalystCompleted(args) {
      var feature, features = [];
      for (var i = 0; i < args.result.recordset.features.length; i++) {
          feature = args.result.recordset.features[i];
          feature.style = style;
          features.push(feature);
      }
      resultLayer.addFeatures(features);
  }
```

2. 泰森多边形分析

泰森多边形可用于定性分析、统计分析、邻近分析等。例如，可以用离散点的性质来描述泰森多边形区域的性质；可用离散点的数据来计算泰森多边形区域的数据；判断一个离散点与其他哪些离散点相邻时，可根据泰森多边形直接得出，且若泰森多边形是 n 边形，则就与 n 个离散点相邻；当某一数据点落入某一泰森多边形中时，它与相应的离散点最邻近，无需计算距离。

泰森多边形的特性是：

①每个泰森多边形内仅含有一个离散点数据；

②泰森多边形内的点到相应离散点的距离最近；

③位于泰森多边形边上的点到其两边的离散点的距离相等。

SuperMap. REST. ThiessenAnalystService 类负责将客户设置的泰森多边形分析参数传递给服务端，并接收服务端返回的分析结果数据。泰森多边形分析结果通过该类支持的事件的监听函数参数获取，参数类型为 {SuperMap. REST. ThiessenAnalystEventArgs}；获取的结果数据包括 originResult、result 两种，其中，originResult 为服务端返回的用 JSON 对象表示的泰森多边形分析结果数据，result 为服务端返回的类型为 {SuperMap.

REST. ThiessenAnalystResult｝的泰森多边形分析结果数据对象。泰森多边形分析的参数支持两种，当参数为｛SuperMap. REST. DatasetThiessenAnalystParameters｝类型时，执行数据集泰森多边形分析，当参数为｛SuperMap. REST. GeometryThiessenAnalystParameters｝类型时，执行几何对象泰森多边形分析。

核心实现代码如下：

```
//数据集泰森多边形
function bufferAnalystProcess() {
    var myThiessenAnalystService = new SuperMap.REST.Thiessen-
AnalystService(url),
            dThiessenAnalystParameters =new
                SuperMap.REST.DatasetThiessenAnalystParameters({
                dataset: "Factory@ Changchun"
                  });
    myThiessenAnalystService.events.on({"processCompleted":
                                       thiessenAnalystCompleted,
                                       " processFailed ":  thies-
                                       senAnalystFailed});
    myThiessenAnalystService.processAsync ( dThiessenAnalystPa-
rameters);
}
//几何泰森多边形
function geometryAnalystProcess() {
    var points = [
    new SuperMap.Geometry.Point(5238.998556, -1724.229865),
    new SuperMap.Geometry.Point(4996.270055, -2118.538477),
    new SuperMap.Geometry.Point(5450.34263, -2070.794081),
    new SuperMap.Geometry.Point(5317.70775, -2521.162355),
    new SuperMap.Geometry.Point(5741.149405, -1970.130198),
    new SuperMap.Geometry.Point(4716.133098, -1575.858795),
    new SuperMap.Geometry.Point(5447.671615, -2255.928819)];
    var myThiessenAnalystService = new SuperMap.REST.Thiessen-
AnalystService(url);
      //初始化泰森多边形分析参数基类
  var gThiessenAnalystParameters = new
  SuperMap.REST.GeometryThiessenAnalystParameters({points: points});
      myThiessenAnalystService.events.on ({ " processCompleted":
thiessenAnalystCompleted, "processFailed": thiessenAnalystFailed});
      //向 iserver 发送请求
```

```
        myThiessenAnalystService.processAsync(gThiessenAnalystPa-
rameters);
    }
    function thiessenAnalystCompleted(serviceEventArgs) {
        resultLayer.removeAllFeatures();
        var feature, polygonFeature, features = [];
        for (var i = 0; i < serviceEventArgs.result.regions.length;
i++) {
            feature = serviceEventArgs.result.regions[i];
            polygonFeature =new SuperMap.Feature.Vector(feature);
            features.push(polygonFeature);
        }
        resultLayer.addFeatures(features);
    }
    function thiessenAnalystFailed(serviceEventArgs) {
        alert(serviceEventArgs.error.errorMsg);
    }
```

3. 核密度分析

核密度分析用于计算要素在其周围邻域中的密度，既可计算点要素的密度，也可计算线要素的密度；可用于测量建筑密度、获取犯罪情况报告，以及发现对城镇或野生动物栖息地造成影响的道路或公共设施管线。可根据要素的重要程度赋予某些要素比其他要素更大的权重，该字段还允许使用一个点表示多个观察对象。例如，一个地址可以表示一栋六单元的公寓，或者在确定总体犯罪率时可赋予某些罪行比其他罪行更大的权重。对于线要素，分车道高速公路可能比狭窄的土路产生更大的影响，高压线要比标准电线杆产生更大的影响。

核心实现代码如下：

```
    function densityKernelAnalyst() {
        //创建一个核密度分析服务实例
        var densityAnalystService = new SuperMap.REST.Density-
AnalystService(url, {
            eventListeners: {
                "processCompleted": KernelAnalystCompleted,
                "processFailed": KernelAnalystFailed
            }
        });
        //创建一个核密度分析参数示例
        var densityKernelAnalystParameters = new
            SuperMap.REST.DensityKernelAnalystParameters({
```

```
        //指定数据集
        dataset:"Railway@ Changchun",
        //指定范围
        bounds:new SuperMap.Bounds(3800, -3800, 8200, -2200),
        //指定数据集中用于核密度分析的字段
        fieldName:"SmLength",
        searchRadius: 50,
        //结果数据集名称
        resultGridName:"KernelDensity_Result",
        deleteExistResultDataset:true
    });
    densityAnalystService.processAsync(densityKernelAnalystPa-
rameters);
  }
  //用栅格专题图展示分析结果
  function KernelAnalystCompleted(densityKernelAnalysEventArgs) {
      var color1 = new SuperMap.REST.ServerColor(255, 212, 170),
          color2 =new SuperMap.REST.ServerColor(255, 127, 0),
          color3 =new SuperMap.REST.ServerColor(191, 95, 0),
          color4 =new SuperMap.REST.ServerColor(255, 0, 0),
          color5 =new SuperMap.REST.ServerColor(191, 0, 0),
          themeGridRangeIteme1 =new SuperMap.REST.Theme-
GridRangeItem({
              start: 0,
              end: 0.05,
              color: color1
          }),
          themeGridRangeIteme2 =new SuperMap.REST.ThemeGrid-
RangeItem({
              start: 0.05,
              end: 5,
              color: color2
          }),
            themeGridRangeIteme3 = new SuperMap.REST.ThemeGrid-
RangeItem({
              start: 5,
              end: 10,
              color: color3
```

```
            }),
              themeGridRangeIteme4 = new SuperMap.REST.ThemeGrid-
RangeItem({
                start: 10,
                end: 100,
                color: color4
            }),
              themeGridRangeIteme5 = new SuperMap.REST.ThemeGrid-
RangeItem({
                start: 100,
                end: 360,
                color: color5
            }),
            themeGridRange =new SuperMap.REST.ThemeGridRange({
                reverseColor:false,
                rangeMode: SuperMap.REST.RangeMode.EQUALINTERVAL,
                //栅格分段专题图子项数组
                items: [themeGridRangeIteme1,
                    themeGridRangeIteme2,
                    themeGridRangeIteme3,
                    themeGridRangeIteme4,
                    themeGridRangeIteme5
                ]
            }),
          themeParameters =new SuperMap.REST.ThemeParameters({
              //制作专题图的数据集(核密度分析的结果数据集)
              datasetNames: [densityKernelAnalysEventArgs.result.
dataset.split('@')[0]],
              dataSourceNames: ["Changchun"],
              joinItems:null,
              themes: [themeGridRange]
          });
      var themeService = new SuperMap.REST.ThemeService(url, {
          eventListeners: {
              "processCompleted": themeCompleted,
              "processFailed": themeFailed
          }
      });
```

```
    themeService.processAsync(themeParameters);
}
function themeCompleted(themeEventArgs) {
    if (themeEventArgs.result.resourceInfo.id) {
        resultLayer = new SuperMap.Layer.TiledDynamicRESTLayer
("核密度分析结果", url, { cacheEnabled: false, transparent: true, layer-
sID: themeEventArgs.result.resourceInfo.id }, { "maxResolution": "au-
to" });
        resultLayer.events.on({"layerInitialized": addThemelayer });
        resultLayer.setOpacity(0.8);
    }
}
```

第 32 章　SuperMap for JavaScript 高级开发

32.1　可视化图层

32.1.1　热点图

热点图是通过使用不同的标志将图或页面上的区域按照受关注程度的不同加以标注并呈现的一种分析手段。标注的手段一般采用颜色的深浅、点的疏密以及呈现比重的形式，不管使用哪种方式，最终得到的效果是一样的，那就是，眼前豁然开朗。

SuperMap. Layer. HeatMapLayer 表示热点图层，提供对热点信息的添加、删除等操作和渲染展示。由于使用 Canvas 绘制，所以不支持直接修改操作。其类结构见表 32.1。

表 32.1　**热点图层类结构**

属　性	介　绍
featureRadius	{String} 对应 feature. attributes 中的热点地理半径字段名称，feature. attributes 中热点地理半径参数的类型为 float
features	{Array(SuperMap. Feature. Vector)} 热点信息数组，记录存储图层上添加的所有热点信息
featureWeight	{String}对应 feature. attributes 中的热点权重字段名称，feature. attributes 中权重参数的类型为 float
maxWeight	{Number}当前权重最大值
radius	{Number}热点渲染的最大半径(热点像素半径)，默认为 50。热点显示的时候以精确点为中心点开始往四周辐射衰减，其衰减半径和权重值成比例
构造函数	**介　绍**
SuperMap. Layer. HeatMapLayer	创建一个热点图层
方　法	**介　绍**
addFeatures	添加热点信息
destroy	销毁图层，释放资源
refresh	强制刷新当前热点显示，在图层热点数组发生变化后调用，即使更新显示
removeAllFeatures	移除全部的热点信息
removeFeatures	移除指定的热点信息

具体实现代码如下：

```
var geometry = new SuperMap.Geometry.Point(120,60);
var attributes = {
    value":10,
    "radiusValue":10
};
var style = {
    strokeColor:"#339933",
    strokeOpacity:1,
    strokeWidth:3,
    pointRadius:6
};
var pointFeature = new SuperMap.Feature.Vector(geometry,attrib-
utes,style);
vectorLayer.addFeatures(pointFeature);
```

下面将实现一个简单的热点图，关键示例代码如下：

```
<body onload="init()">
<div id="toolbar">
<input type="button" value="渲染热点" onclick="createHeatPoints
()" />
<input type="button" value="清除" onclick="clearHeatPoints()" />
</div>
<div id="map"></div>
</body>
<script src='libs/SuperMap.Include.js'></script>
<script type="text/javascript">
var host = document.location.toString().match(/file:\/\//)?"ht-
tp://localhost:8090":'http://'+document.location.host;
var map, layer, heatMapLayer,
url=host+"/iserver/services/map-world/rest/maps/World";
function init(){
    map =new SuperMap.Map("map",{controls:[
        new SuperMap.Control.ScaleLine(),
        new SuperMap.Control.Zoom(),
        new SuperMap.Control.Navigation({
            dragPanOptions:{
                enableKinetic:true
            }
```

```
        })]
    });
    map.addControl(new SuperMap.Control.MousePosition());
    layer = new SuperMap.Layer.TiledDynamicRESTLayer ( "World",
url, {transparent: true, cacheEnabled: true}, {maxResolution:" au-
to"});
    heatMapLayer =new SuperMap.Layer.HeatMapLayer(
            "heatMap",
            {
                "radius":45,
                "featureWeight":"value",
                "featureRadius":"geoRadius"
            }
        );
    layer.events.on({"layerInitialized": addLayer});
}
function addLayer() {
    map.addLayers([layer,heatMapLayer]);
    map.setCenter(new SuperMap.LonLat(0, 0), 0);
}
function createHeatPoints(){
    clearHeatPoints();
    var heatPoints = [];
    heatPoints[0] =new SuperMap.Feature.Vector(
            new SuperMap.Geometry.Point(
                118,
                40
            ),
            {
                "value":60
            }
        );
    heatPoints[1] =new SuperMap.Feature.Vector(
            new SuperMap.Geometry.Point(
                110,
                40
            ),
            {
```

```
                "value":60
            }
    );
    heatPoints[2] =new SuperMap.Feature.Vector(
            new SuperMap.Geometry.Point(
                110,
                35
            ),
            {
                "value":50
            }
    );
    heatPoints[3] =new SuperMap.Feature.Vector(
        new SuperMap.Geometry.Point(
            114,
            36
        ),
        {
            "value":50
        }
    );
    heatPoints[4] =new SuperMap.Feature.Vector(
        new SuperMap.Geometry.Point(
            118,
            30
        ),
        {
            "value":35
        }
    );
    heatMapLayer.addFeatures(heatPoints);
}
function clearHeatPoints(){
    heatMapLayer.removeAllFeatures();
}
</script>
```

运行程序，点击【渲染热点】，结果如图 32.1 所示。

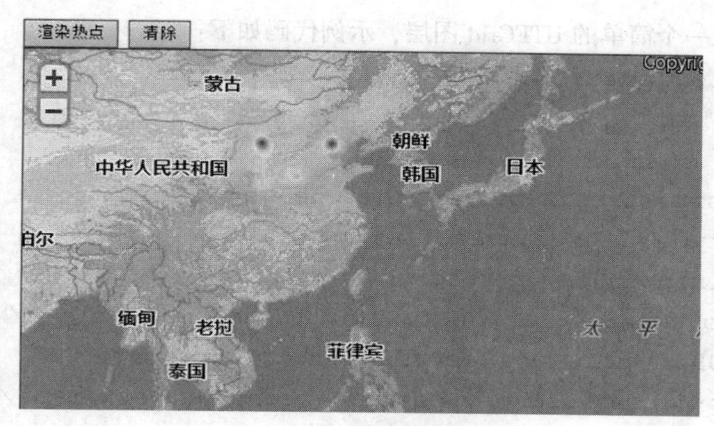

图 32.1　热点图

32.1.2　UTFGrid 图层

UTFGrid 是一个 JSON 格式的字符串，用来描述一张瓦片中每个像素所对应的地理要素信息。该 JSON 由三部分组成：grid、keys、data。这个图层从 UTFGrid 切片数据源读取数据。由于 UTFGrid 本质上是基于 JSON 的 ASCII 码"字符画"与属性数据的结合，所以它不能被可视化渲染。当需要实时查询地图上某些地物属性并且地物数量很大时，同时不希望实时地与服务器交互以获取属性信息，可以通过 UTFGrid 图层及时返回属性信息。例如：希望达到鼠标悬停或鼠标单击某一地物显示属性信息的快速交互。为了在地图中使用这个图层，用户必须同时添加 SuperMap. Control. UTFGrid 控件类，来控制触发事件类型。SuperMap. Control. UTFGrid 为 UTFGrid 控件类构造函数。这个控制类提供了与 UTFGrid 图层相关的行为。通过判断鼠标位置能够直接提取出 UTFGrid 图层对应要素的属性，不需要向服务端发送请求。这个控制类通过设置属性提供了 Mousemove(鼠标移动)，Hovering(鼠标悬停)以及 Click(鼠标点击)等事件来触发回调函数。最常见的例子可能就是当鼠标移动到 UTFGrid 图层上的要素的时候，可使用 DIV 显示该要素的属性。

SuperMap. Layer. UTFGrid 创建新的 UTFGrid 图层构造函数参数见表 32.2。

表 32.2　**UTFGrid 图层构造函数参数**

参　数	说　明
utfTileSize	{String} 瓦片的像素大小，默认 256 像素的正方形
pixcell	{Number} 瓦片中每个单元格的像素宽度，默认为 2，该属性与 options 属性中 utfgridResolution 属性对应
layerName	{String} 请求地图图层名称
isUseCache	{Boolean} 是否使用本地缓存策略。设置为 false 则不使用，默认使用
filter	{String} 过滤条件数组

下面将实现一个简单的 UTFGrid 图层，示例代码如下：

```
<body onload="init()">
<div id="map"></div>
</body>
<script src='libs/SuperMap.Include.js'></script>
<script type="text/javascript">
var map,infowin,layer,utfgrid,control,
host = document.location.toString().match(/file:\/\//)?"http://
localhost:8090":'http://'+document.location.host,
url=host+"/iserver/services/map-china400/rest/maps/China";
function init(){
    map =new SuperMap.Map("map",{controls:[
        new SuperMap.Control.ScaleLine(),
        new SuperMap.Control.Zoom(),
        new SuperMap.Control.LayerSwitcher(),
        new SuperMap.Control.Navigation({
            dragPanOptions:{ enableKinetic:true}
        })],
        projection:"EPSG:3857"
    });
    layer = new SuperMap.Layer.TiledDynamicRESTLayer("China",
url,{transparent:true},{useCanvas:true,maxResolution:"auto"});
    utfgrid =new SuperMap.Layer.UTFGrid("UTFGridLayer",url,
        {
            layerName:"China_Province_R@China400",
            utfTileSize:256,
            pixcell:8,
            isUseCache:true//设置是否使用缓存
        },
        {
            utfgridResolution:8//应该与pixcell的值相等
        });
    layer.events.on({"layerInitialized":addLayer});
    control =new SuperMap.Control.UTFGrid({
        layers:[utfgrid],
        callback:callback,
        handlerMode:"move"
    });
```

```
        map.addControl(control);
    }
    var callback = function (infoLookup, loc, pixel) {
        closeInfoWin();
        if (infoLookup) {
            var info;
            for (var idx in infoLookup) {
                info = infoLookup[idx];
                if (info && info.data) {
                    var dom = "<div style='font-size: 12px; color:
#000000;border: 0px solid #000000'>" + info.data.NAME + "</div>";
                    var xOff = (1 /map.getScale()) * 0.001;
                    var yOff = -(1 /map.getScale()) * 0.005;
                    var pos = new SuperMap.LonLat(loc.lon+xOff, loc.
lat+yOff);
                    infowin =new SuperMap.Popup.FramedCloud("Tip",
                        pos,
                        null,
                        dom,
                        null, true);
                    map.addPopup(infowin);
                }
            }
        }
    };
    function closeInfoWin() {
        if (infowin) {
            try {
                map.removePopup(infowin)
            }
            catch (e) {
            }
        }
    }
    function addLayer() {
        var center = new SuperMap.LonLat(11733502.481499, 4614406.
969325);
        map.addLayers([layer, utfgrid]);
```

```
    map.setCenter(center, 3);
}
</script>
```

运行程序，效果如图 32. 2 所示。

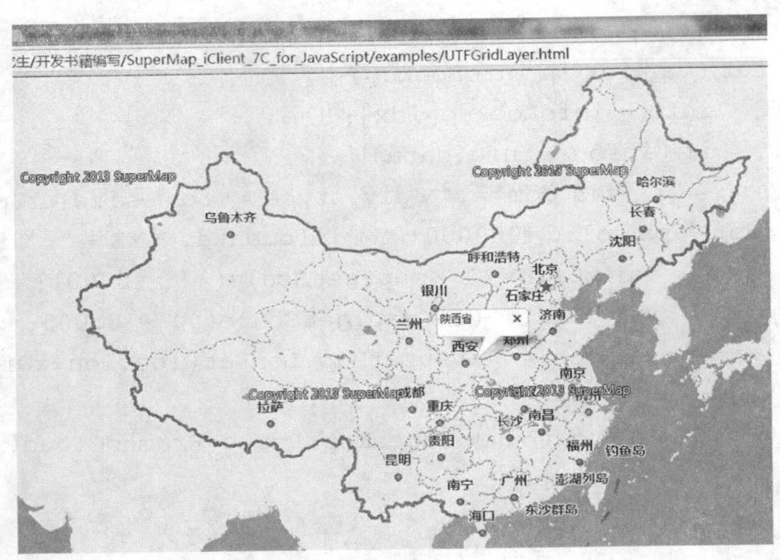

图 32. 2　UTFGrid 图

32. 1. 3　其他可视化图层

1. 聚类图层

聚类图层可以实现以聚合的方式显示大量点数据。这是一种渲染大数据量点的策略，当点非常多、非常密集的时候，将一定范围内的点聚合为一个大点(聚合点)，该聚合点反映了这些点的数量和大体位置信息。当放大地图的时候，聚合点会逐步散开为小点。SuperMap. Layer. ClusterLayer 类方法见表 32. 3。

表 32. 3　　**SuperMap. Layer. ClusterLayer 类**

方　法	说　明
addFeatures	聚合显示 features，将需要聚合显示 feature 传给该方法，便可以实现聚合显示
assembleFeature	组装散开后的要素，绘制之前会调用该方法，允许用户通过重写该方法自定义要素，通常在复杂要素的情况下使用，如三叶草对象
cancelDisplayFeatures	与 DisplayFeatures 方法相对应，还原该方法所显示的点
clearCluster	清除当前视图中已经绘制的要素，但是不清空存储，当平移缩放操作时仍然会重绘

续表

方　法	说　明
destroyCluster	彻底清除所有要素，平移缩放操作时不再重绘
displayFeatures	散开显示指定范围内的要素
getFeaturesByBounds	获取指定范围内的所有要素
refresh	刷新图层，清除当前已经绘制的 feature，重新进行聚合计算并绘制

SuperMap. Control. SelectCluster 类针对 SuperMap. Layer. ClusterLayer 要素选择控件，该控件实现在指定的图层上通过鼠标单击和悬浮选择矢量要素。

核心实现代码如下：

```
//创建聚散图层并添加 layers
clusterLayer =new SuperMap.Layer.ClusterLayer("Cluster");
layer = new SuperMap.Layer.TiledDynamicRESTLayer(" changchun",
url,{ transparent: true, cacheEnabled: true, redirect: true }, { max-
Resolution: "auto"});
map.addLayers([layer, clusterLayer]);
//创建聚散选择控件。该控件实现了聚散图层的鼠标事件
var select = new SuperMap.Control.SelectCluster(clusterLayer, {
    callbacks: {
        click:function (f) { //点击兴趣点,弹出信息窗口
            closeInfoWin();//云提示框注销
            if (! f.isCluster) { //当点击聚散点的时候不弹出信息窗口
                openInfoWin(f);//云提示框信息
            }
        },
        clickout:function () { //点击空白处关闭信息窗口
            closeInfoWin();//云提示框注销
        }
    }
});
//将控件添加到 map 上
map.addControl(select);
clusterLayer.events.on({
    "moveend": function (e) {
        //注册 moveend 事件,当缩放的时候关闭信息窗口
        if (e && e.zoomChanged) closeInfoWin();
    }
```

```
});
clusterLayer.events.on({
    "clusterend": function (e) {
    }
});
//激活控件
select.activate();
//往聚散图层中添加兴趣点
clusterLayer.addFeatures(fs1);
//fs1 为 SuperMap.Feature.Vector() SuperMap.Geometry.Point(x,y)
集合
```

2. 麻点图

这是一种很高效的 Web 端大数据量渲染解决方案。该功能支持大数据量、跨浏览器、事件响应，并且效率高。在大部分主流浏览器下都能快速渲染，且轻松漫游地图。创建 SuperMap. Layer. UTFGrid 对象，来让这些 POI 有鼠标事件响应。然后在最上层叠加一个 SuperMap. Layer. Markers，当鼠标移动到某个 POI 上时，就会在相应位置添加一个 Marker，高亮显示该点。SuperMap. GOIs 属性见表 32. 4，SuperMap. GOIs 事件见表 32. 5。

表 32. 4 **SuperMap. GOIs 属性**

属　性	介　绍
cacheEnabled	{Boolean} 是否使用服务端的缓存，默认为 true，即使用服务端的缓存
datasetName	{String} 所要显示的点数据集图层名称
datumAxis	{Number} 椭球体长半轴
dpi	{Number} 图像分辨率，表示每英寸内的像素个数
filter	{String} poi 的过滤条件
format	{String} 栅格图层图片格式
pixcell	{Number} UTFGrid 瓦片中每个单元格的像素宽度，默认为 8，详见 SuperMap. Layer. UTFGrid 的 pixcell 参数
projection	{SuperMap. Projection} or {String} 投影字符串，如“EPSG：900913”
resolutions	{Array} 分辨率数组，如果设置了 dpi，resolutions 和 scales 设置其一
scales	{Array} 比例尺数组，如果设置了 dpi，resolutions 和 scales 设置其一
style	{SuperMap. REST. ServerStyle} 图层中点的默认风格
units	{String} 地图坐标系统的单位
url	{String} 地图资源 url

表 32.5 **SuperMap. GOIs 事件**

事 件	说 明
SuperMap. Control. GOIs	麻点图控件，用于实现麻点图鼠标事件
onClick	{Function} 触发 poi 的 click 事件
onDblclick	{Function} 触发 poi 的 dbclick 事件
onMousedown	{Function} 触发 poi 的 mousedown 事件
onMousemove	{Function} 触发 poi 的 mousemove 事件
onMouseout	{Function} 触发 poi 的 mouseout 事件
onMouseover	{Function} 触发 poi 的 mouseover 事件
onMouseup	{Function} 触发 poi 的 mouseup 事件

核心实现代码如下：

```
function createLayer() {
    datasetName ="China_Town_P@ China400";
    url = host +"/iserver/services/map- china400/rest/maps/Chi-
na";
    //创建一个麻点图对象
    myGOIs =new SuperMap.GOIs({
        "url": url,
        "datasetName": datasetName,
        "style": new SuperMap.REST.ServerStyle({
            "markerSymbolID": 72,
            "markerSize": 4
        }),
        "pixcell": 16
    });
    myGOIs.events.on({
    "initialized": GOIsInitialized
    });
}
function GOIsInitialized() {
    var layers = myGOIs.getLayers();
```

```
    map.addLayers(layers);
    control =new SuperMap.Control.GOIs(layers, {
        onClick:function (evt) {
            var lonlat = evt.loc;
            var name = evt.data.NAME;
            openInfoWin(lonlat, name);//实例化 SuperMap.Popup.Framed-
Cloud 对象
        },
        highlightIcon:new SuperMap.Icon('images/circle.png', new
SuperMap.Size(16, 16), new SuperMap.Pixel(-8, -8)),
        isHighlight:true
    });
    map.addControl(control);
}
```

32.2　时空数据表达

同时具有时间维度和空间维度的数据，称为时空数据。传统的 GIS 表达静态的世界，人们通常通过修改或删除某一记录来反映当前变化，但是无法保存动态变化的信息，因此时态 GIS 应运而生。而对于如今的 GIS 行业来说，随着数据存储量越来越大，计算机的静态可视化已不能满足需求，迫切需要展现 GIS 数据在时间节点上的变化，从而诞生了时空数据。

时态 GIS 要求以各种形式在不同抽象层次上处理时空数据，支持矢量时空相关数据的高效管理，SuperMap iClient for JavaScript 提供了 SuperMap. Layer. AnimatorVector 来实现时空数据的管理和复杂效果。用户来自各行各业，数据来源各不相同，数据格式也五花八门，并且数据量越来越庞大，但不难发现，无论数据多么千奇百怪，总离不开 3 个共同的属性——ID、时间和位置。因此，只要保持原有的 feature 格式基础上添加了 ID 和时间属性来渲染数据。当数据通过查询在客户端展示时，只需在客户端设置 SuperMap. Layer. AnimatorVector 的属性 featureIdName 和 timeName（默认字段名为 FEATUREID 和 TIME），客户端就能识别、整理和渲染数据。时空数据可视化是基于浏览器 canvas2D 的一种渲染，可高效地渲染矢量数据，以下为时空数据可视化的应用模拟效果：

32.2.1　点闪烁

此范例模拟某一天从 0 时到 24 时之间全国一些重要城市的部分火车运行情况，黄色代表短途火车，紫色代表长途火车，可以按需求对尾巴和闪烁效果进行设置。效果展示如

图 32.3 所示。

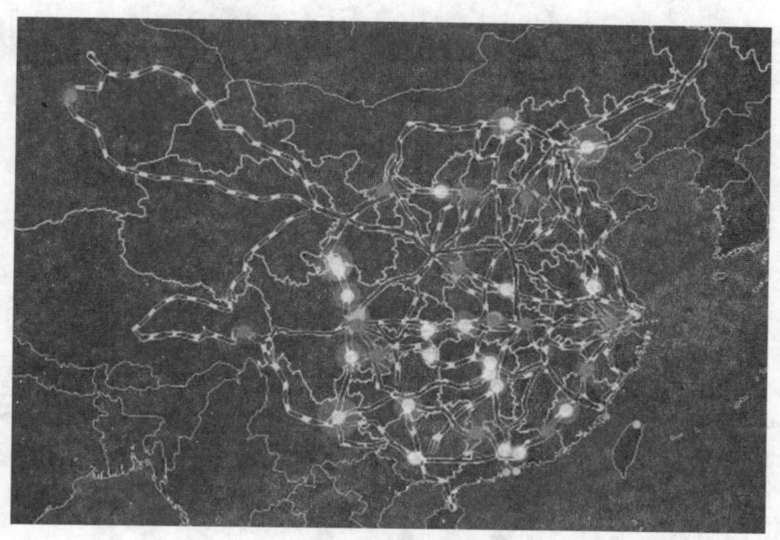

图 32.3　模拟火车监控

32.2.2　放射线

此范例展现了点数据以流动线的形式进行迁移的效果，模拟春运期间各主要城市的人口流动过程，可以看出北京、上海、广州、成都等城市在春运期间的人口流量变化情况(线越粗越长，表示迁移人口越多)，效果展示如图 32.4 所示。

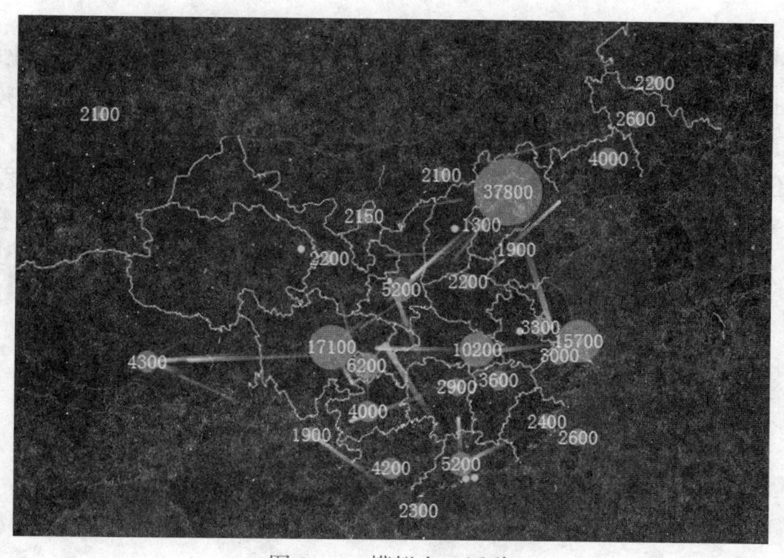

图 32.4　模拟人口迁移

32.2.3　伸缩线

此范例展示了线的延伸效果，模拟了北京地铁修建的过程。效果展示如图 32.5 所示。

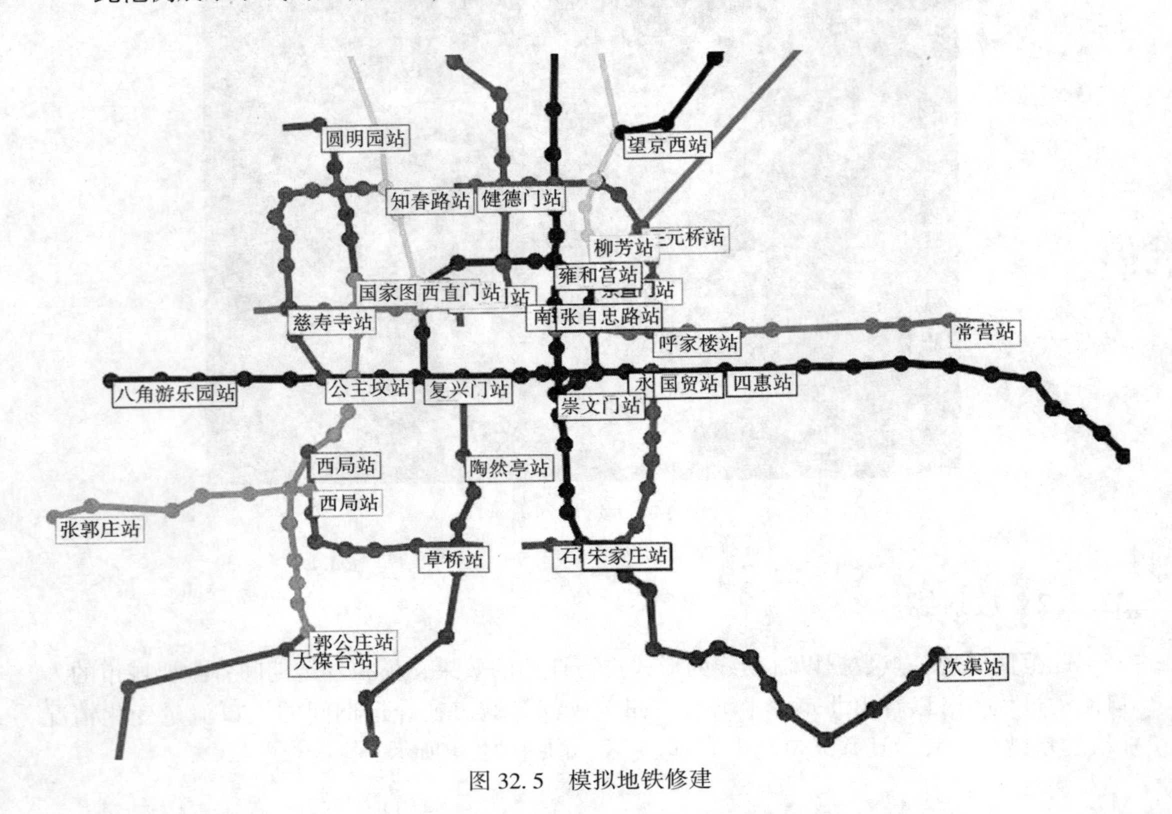

图 32.5　模拟地铁修建

以上是对时空数据动画的部分展示，许多效果也已经应用到了部分行业中，如气象、车辆运行实时监控等。时空数据可视化还可以应用到一些新的领域中。例如疫情严重区域和病毒扩散方向的实时监控。

此外，还可以轻松掌控地下管道流体运动路线和地铁、道路的修建进度，监控气流、等温线/面、等压线/面的变化，监控春运时人口大军的迁移规律，等等。

第 33 章 “旅印”系统开发案例

本章给出的“旅印”系统案例，是作者专门为本书研发的案例演示系统，案例更多内容参见随书所赠光盘。下面以“旅印”系统为案例讲解基于 SuperMap 平台进行 Web 开发的过程和方法。

33.1 需求分析

33.1.1 用户需求

近几年，外出旅游已成为大多数人闲暇时的首选。游客们每到一个景点会拍很多照片、写很多游记。“分享”是现代社会活动的显著特点，游客们喜欢分享自己的旅行，以此记录下自己的足迹。为了满足游客的需求，丰富分享的方式，因此，现结合 GIS(地理信息系统)开发一个 Web 系统——旅印，供用户使用。

借助 GIS 直观的可视化、强大的空间分析、友好的交互方式等特点，开发的系统应更具有普通 Web 网站所不能及的应用功能和竞争优势。系统可先面向国内的旅行用户应用，主要对象为游客。系统服务用户分为注册用户和未注册用户。注册用户需要提供的注册信息为昵称、密码、邮箱、设置问题答案(已备用户找回密码)，注册用户可以在网站内发布旅行印迹，一种是将可穿戴设备的旅行数据导入进行精确定位(系统需要转换目前市场主流的具备路线记录的可穿戴设备的数据为统一格式，并存入数据库)，另一种是用户给定几个地址节点进行大致描述。旅游类型分为跟团旅游、自驾游、徒步游、个人游、和驴友们旅游。游客可以在旅游印迹中添加旅游节点，每个旅游节点可以添加旅行记录(包括旅行照片、游记)。游记包括时间、天气状况、文字信息。已注册用户可以添加好友(默认好友为普通好友组内，此组不可被删除和修改组名)，并且可根据自己的需求进行好友分组，组别可以新建、删除、修改(包括组名、本组权限、组内成员)。已注册用户有权选择自己发布的旅行印迹、旅行日记是否对其他人可见或仅对某组内好友可见。已注册用户可以查看对自己可见的旅行印迹、旅行记录，并可对其旅行记录进行评论，评论的信息可以删除。未注册的用户只能查看那些对自己可见的旅行印迹、旅行记录，不能对旅行记录评论。所有用户都可对旅行印迹、旅行记录点赞。

所有用户都能够进行地图的基本操作(放大、缩小、平移、复位)。根据已注册用户的旅行数据生成旅游景点，用户可以查看景点的热度图。热度图是根据已注册用户的旅行数据为基础进行分析，可以根据时间维度的不同、区域的不同进行热度图的绘制，用户可根据颜色的深浅变化直观地感受景点的热度。所有用户都可以通过几何方式或缓冲区方式查询景点。

33.1.2　提取需求

根据用户需求描述，用户分为未注册用户和已注册用户。已注册用户具备未注册用户的所有功能，而未注册用户却不能拥有已注册用户的功能。

未注册用户能够拥有的功能有：查看旅行记录(扩展功能：点赞)、查看旅行印迹(扩展功能：点赞)、景点查询(包括：几何查询、缓冲区查询)、地图基本操作(包括：放大、缩小、平移、复位)、查看景点热度图、注册账号(扩展功能：找回密码)。

已注册用户能够拥有的功能有：登录(扩展功能：修改密码)、发布旅行印迹(包括：数据完整导入发布，设置节点发布；扩展功能：设置可见权限)、添加旅行记录(扩展功能：设置可见权限)、评论信息(扩展功能：删除评论)、好友管理(包括：添加好友、删除好友、好友分组)。

33.1.3　UML 用例图

根据用户的需求，制作系统用例图如图 33.1 所示。

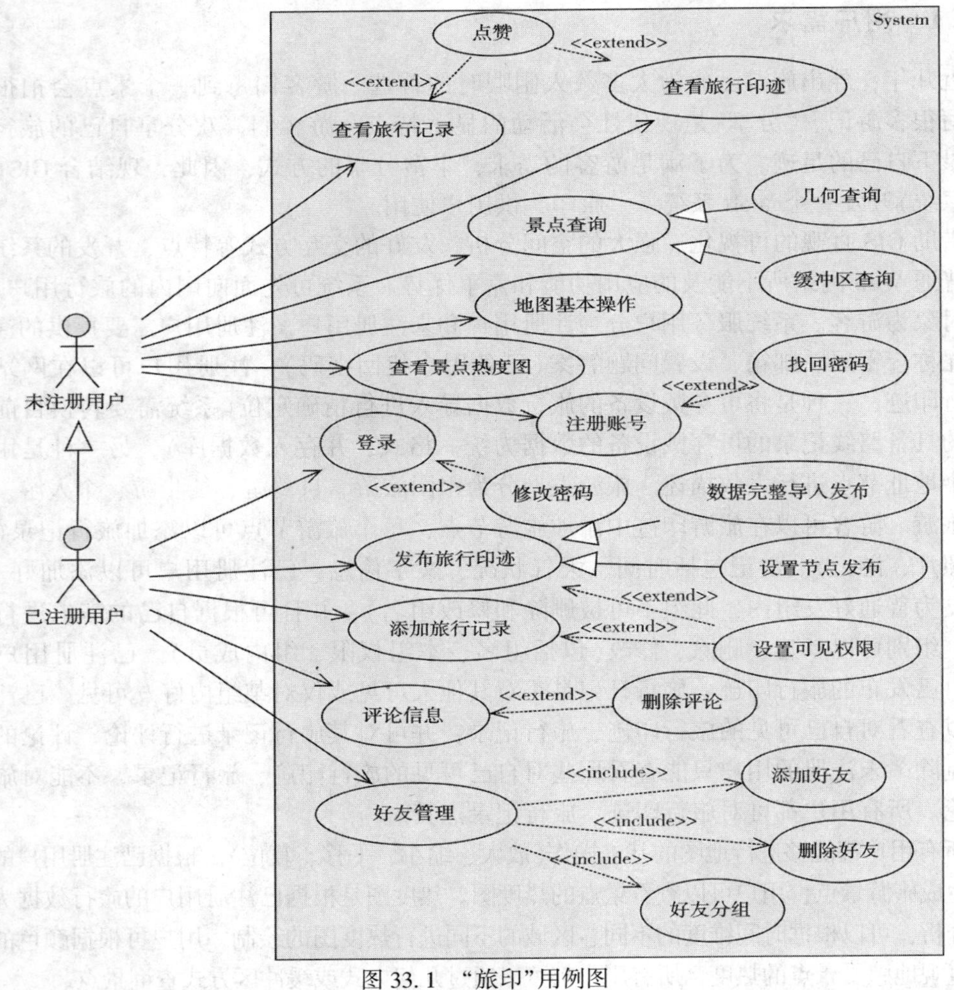

图 33.1　“旅印”用例图

33.2　系统设计

33.2.1　数据库设计

系统属性数据库采用名为 TouristTraceDB 的 Microsoft SQL Server 2008 数据库，它包括 11 个表：User 表、Question 表、ScenicSpot 表、TouristTrace 表、TouristStyle 表、TouristNote 表、TouristNode 表、TouristNoteComment 表、TouristImgArr 表、Friend 表、GroupClassify 表。

User 表保存用户的信息，它包含的字段及其说明见表 33.1。

表 33.1　**User 表**

字段名	数据类型	字段说明	键索引	备注
ID	int	标识符	PK	主键(自动增一)
NickName	varchar(50)	用户登录名		
PassWord	varchar(50)	用户登录密码		
Email	varchar(50)	注册邮箱		
Question_Index	int	问题索引	FK	引用 Question 表 ID 值
Answer	varchar(50)	问题答案		
Status	int	状态		

Question 表保存提问问题信息，它包含的字段及其说明见表 33.2。

表 33.2　**Question 表**

字段名	数据类型	字段说明	键索引	备注
ID	int	标识符	PK	主键(自动增一)
Content	varchar(50)	问题内容		

ScenicSpot 表保存景点信息，它包含的字段及其说明见表 33.3。

表 33.3　**ScenicSpot 表**

字段名	数据类型	字段说明	键索引	备注
ID	int	标识符	PK	主键(自动增一)
Name	varchar(50)	景点名称		
GeoX	float	地理 X		

续表

字段名	数据类型	字段说明	键索引	备注
GeoY	float	地理 Y		
Popular	int	受欢迎度		
District	varchar(50)	隶属行政区		

TouristTrace 表保存旅行的基本信息，它包含的字段及其说明见表 33.4。

表 33.4　**TouristTrace 表**

字段名	数据类型	字段说明	键索引	备注
ID	int	标识符	PK	主键(自动增一)
TraceTitle	varchar(100)	标题		
BriefIntro	varchar(100)	简介		
User_Index	int	用户索引	FK	引用 User 表 ID 值
Start_Time	varchar(50)	开始时间		
End_Time	varchar(50)	结束时间		
Good	int	点赞		

TouristStyle 表保存旅行类型，它包含的字段及其说明见表 33.5。

表 33.5　**TouristStyle 表**

字段名	数据类型	字段说明	键索引	备注
ID	int	标识符	PK	主键(自动增一)
Style	varchar(50)	旅游类型		

TouristNode 表保存旅行节点信息，它包含的字段及其说明见表 33.6。

表 33.6　**TouristNode 表**

字段名	数据类型	字段说明	键索引	备注
ID	int	标识符	PK	主键(自动增一)
TTrace_Index	int	旅印索引	FK	引用 TouristTrace 表 ID 值
TStyle_Index	int	旅行类型索引	FK	引用 TouristStyle 表 ID 值
Node_Name	varchar(50)	旅行节点名称		
Node_Intro	varchar(100)	旅行节点简介		

续表

字段名	数据类型	字段说明	键索引	备注
Start_Time	varchar(50)	开始时间		
End_Time	varchar(50)	结束时间		
GeoX	float	地理 X 值		
GeoY	float	地理 Y 值		

TouristNote 表保存旅行节点随记信息，它包含的字段及其说明见表 33.7(Group_Index 链接至 Friend 表中 Group_Index 查找表中好友索引是否有当前用户)。

表 33.7 **TouristNote 表**

字段名	数据类型	字段说明	键索引	备注
ID	int	标识符	PK	主键(自动增一)
TNode_Index	int	旅行节点索引	FK	引用 TouristNode 表 ID 值
Title	varchar(100)	随记标题		
PublishTime	varchar(50)	发布时间		
Weather	varchar(50)	天气状况		
Content	text	随记内容		
Visible	varchar(50)	可见对象		None | Regist | All | Group_Index
Good	int	点赞		

TouristImgArr 表保存旅行节点随记照片，它包含的字段及其说明见表 33.8。

表 33.8 **TouristImgArr 表**

字段名	数据类型	字段说明	键索引	备注
ID	int	标识符	PK	主键(自动增一)
TNote_Index	int	随记索引	FK	引用 TouristNote 表 ID 值
ImgUrl	varchar(100)	图片地址		
Illustrator	varchar(100)	描述		

TouristNoteComment 表保存旅行节点随记信息评论，它包含的字段及其说明见表 33.9。

表 33.9 **TouristNoteComment 表**

字段名	数据类型	字段说明	键索引	备注
ID	int	标识符	PK	主键(自动增一)
User_Index	int	用户索引	FK	引用 User 表 ID 值
TNote_Index	int	随记索引	FK	引用 TouristNote 表 ID 值
Comment	varchar(100)	内容		
Time	varchar(50)	时间		
Good	int	点赞		
Visible	varchar(50)	可见对象		None | Regist | All | Group_Index

Friend 表保存用户添加好友的信息，它包含的字段及其说明见表 33.10。

表 33.10 **Friend 表**

字段名	数据类型	字段说明	键索引	备注
ID	int	标识符	PK	主键(自动增一)
Friend_User_Index	int	好友用户标识	FK	引用 User 表 ID 值
Group_Index	int	组别标识	FK	引用 GroupClassify 表 ID 值

GroupClassify 表保存用户分组信息，它包含的字段及其说明见表 33.11。

表 33.11 **GroupClassify 表**

字段名	数据类型	字段说明	键索引	备注
ID	int	标识符	PK	主键(自动增一)
User_Index	int	用户标识	FK	引用 User 表 ID 值
GroupName	varchar(50)	组别名		

属性数据库实体关系如图 33.2 所示。

系统地图数据采用 Web 墨卡托的 2013 年中国地图数据、中国道路网数据、河流网、省级行政区、地级市行政区(关键属性字段：市名)、省名注记、地级市注记、省界线、地级市界线、旅游景点等。

33.2.2 功能设计

各功能基本操作流程框架如图 33.3 所示，具体的“指令处理”需要根据各功能具体信息进行处理设计。

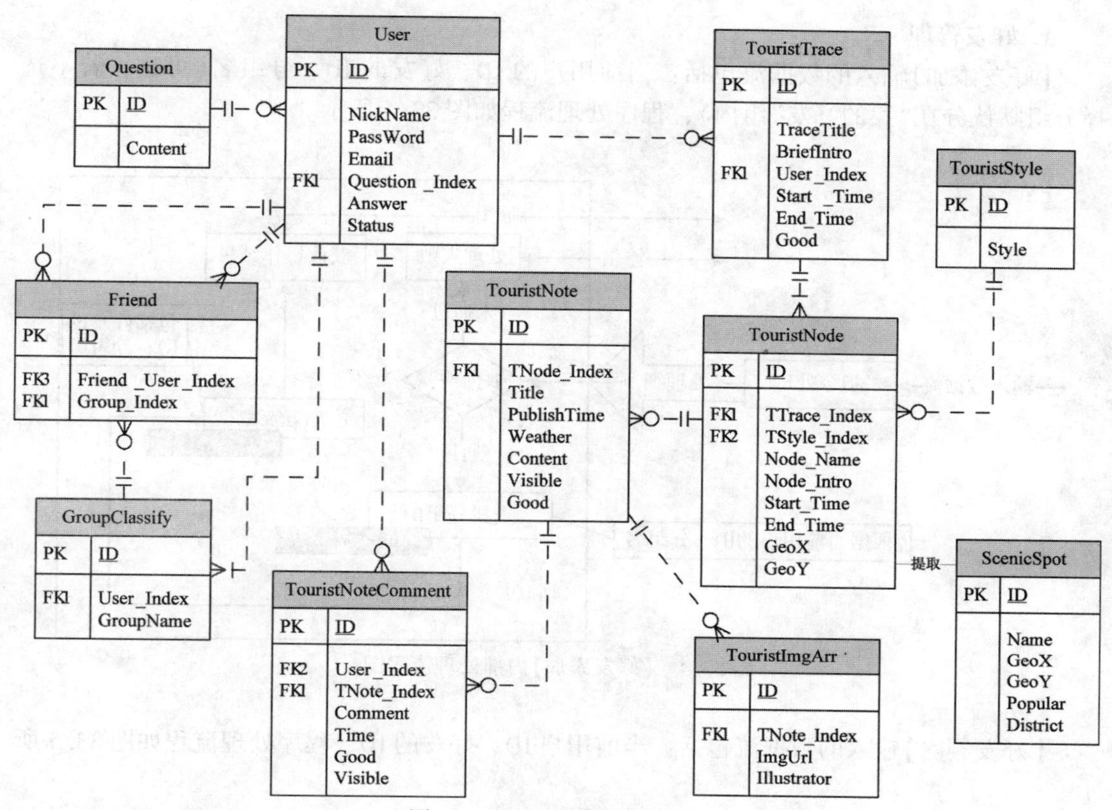

图 33.2 属性数据库关系图

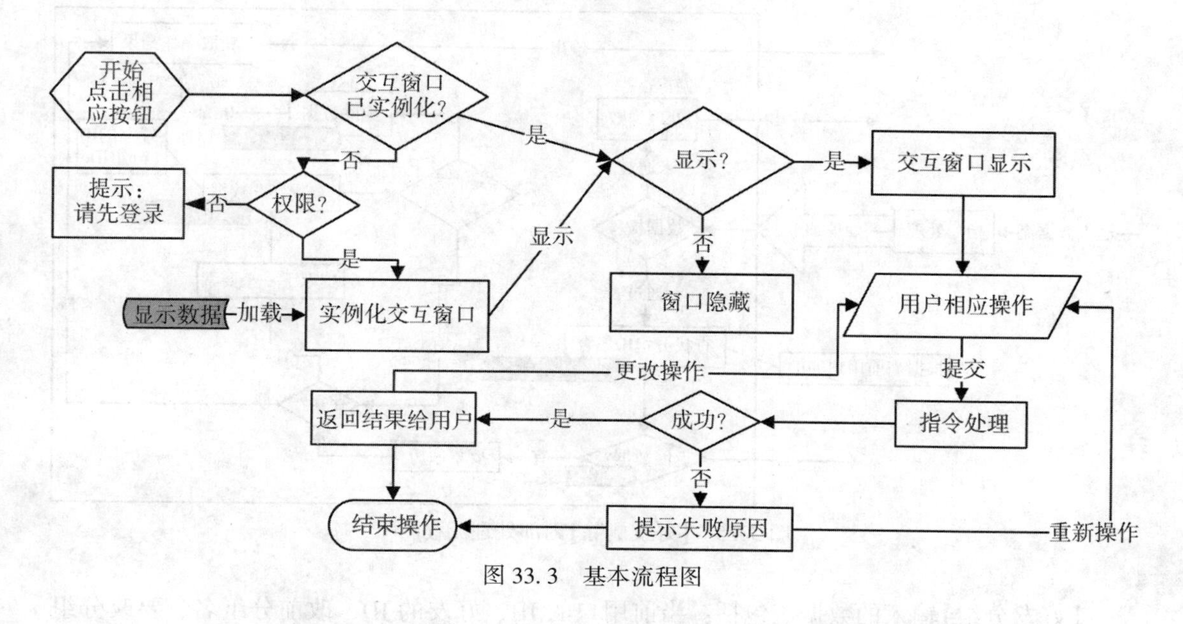

图 33.3 基本流程图

本节受篇幅限制在此仅对好友管理、查询景点、景点热度图、发布旅行印迹、查看旅行印迹功能进行“指令处理”实现设计说明。

1. 好友管理

【好友添加】输入的数据流包括：当前用户的 ID、好友的 ID、分组名(可选，若不选择分组默认分在“我的好友”组内)，程序处理流程如图 33.4 所示。

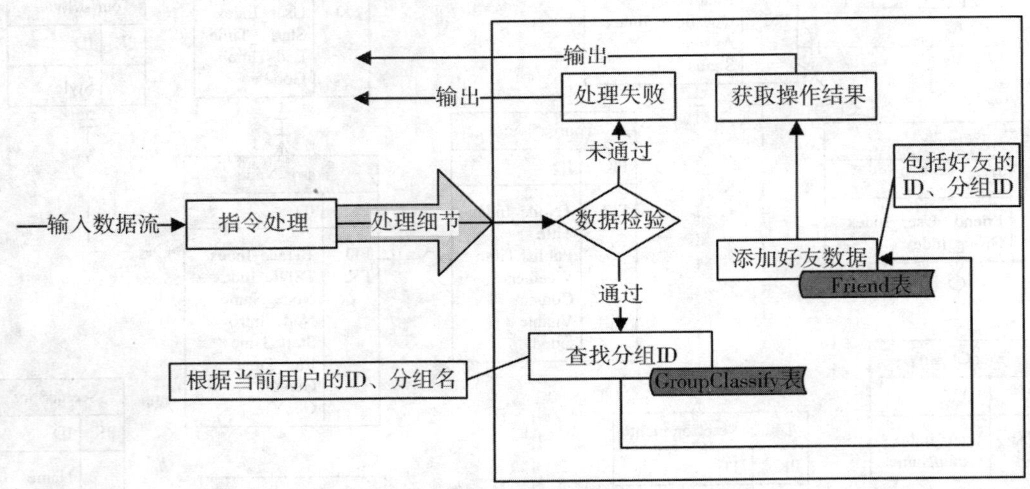

图 33.4　【好友添加】内部处理流程图

【好友删除】输入的数据流包括：当前用户 ID、好友的 ID，程序处理流程如图 33.5 所示。

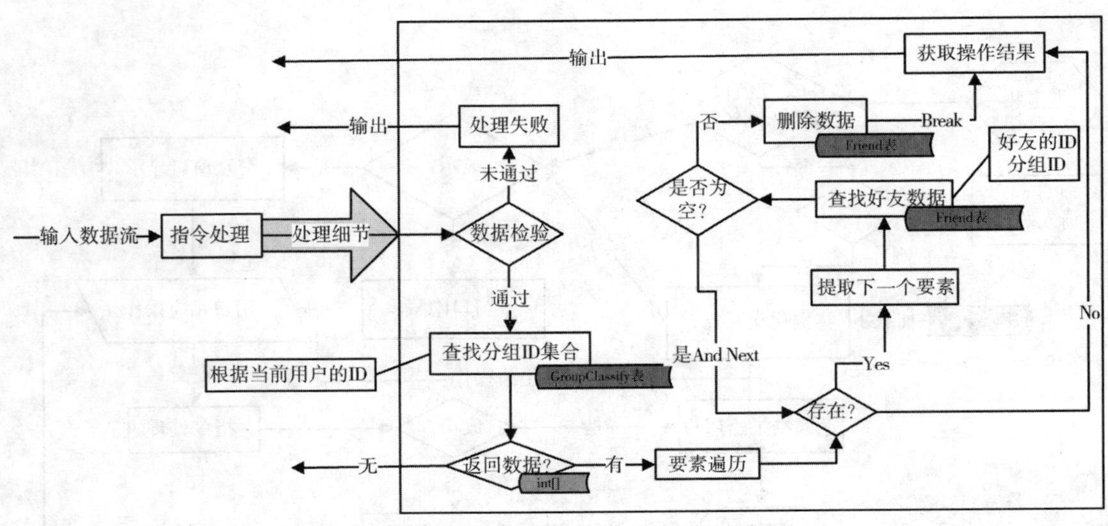

图 33.5　【好友删除】内部处理流程图

【好友分组】输入的数据流包括：当前用户的 ID、好友的 ID、改前分组名、要改分组名，程序处理流程如图 33.6 所示。

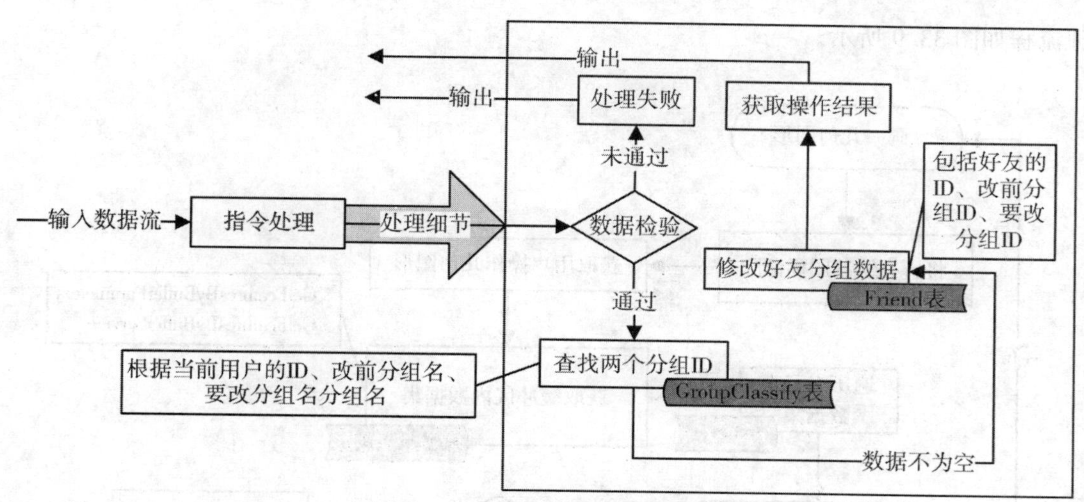

图 33.6 【好友分组】内部流程图

2. 查询景点

【几何数据集查询】根据输入的几何图形查询与指定几何对象符合一定空间关系的矢量要素(景点)，程序处理流程如图 33.7 所示。

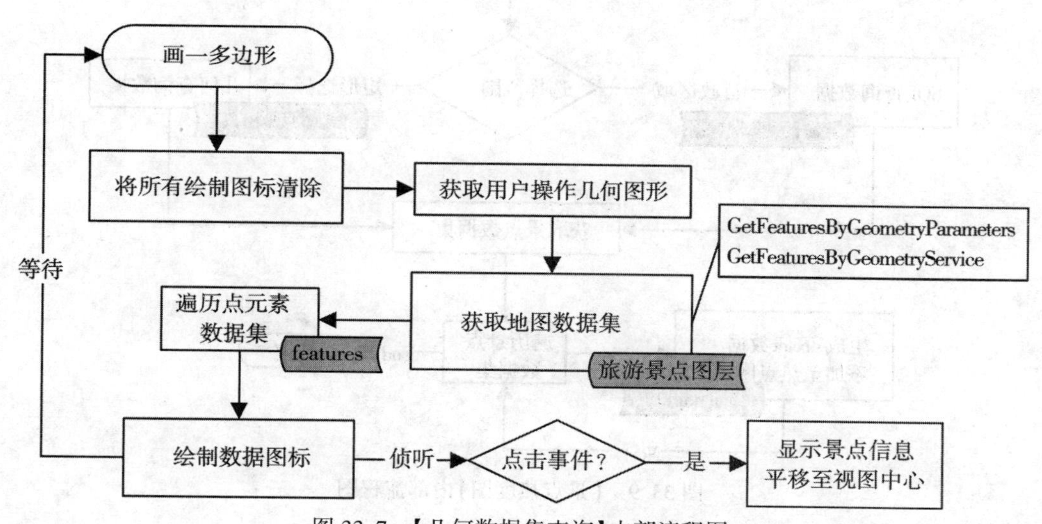

图 33.7 【几何数据集查询】内部流程图

【缓冲区数据集查询】对指定的几何对象进行一定距离的缓冲，从指定数据集集合中查询出与缓冲区区域相交的矢量数据(景点)，程序处理流程如图 33.8 所示。

3. 景点热度图

两种方式：一种是选择行政区，另一种是矩形选择区域。从旅游景点地图数据(或 ScenicSpot 表)中查出数据集，生成 SuperMap. Feature. Vector 点数据，并对 featureWeight、featherRadius 进行赋值，添加至提前生成的 SuperMap. Layer. HeatMapLayer 图层中，程序处理流程如图 33.9 所示。

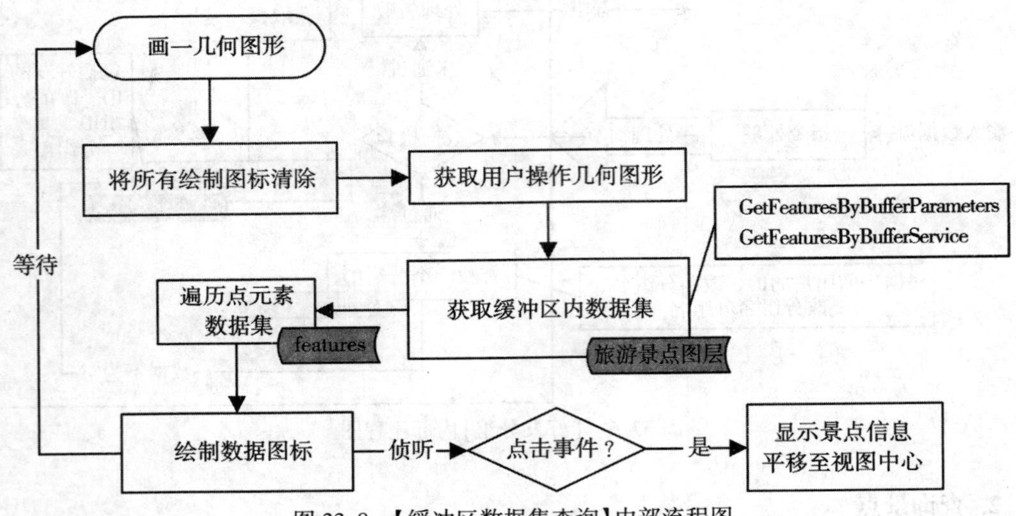

图 33.8 【缓冲区数据集查询】内部流程图

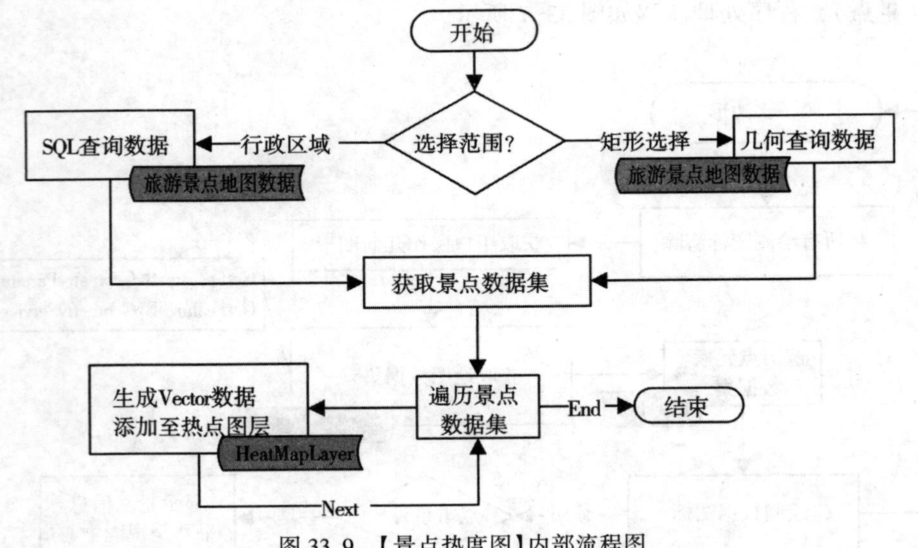

图 33.9 【景点热度图】内部流程图

4. 发布旅行印迹

用户要对本次的旅行进行总体描述以及提供一些必要信息，包括各旅游节点信息等，存储在 TouristTrace 表中。根据用户点击旅游节点的数量生成输入节点信息的相应数目，每个节点信息按用户的旅行顺序依次输入所需信息，提交存储在 TouristNode 表中。用户节点坐标通过选择地图标注节点获取，程序处理流程如图 33.10 所示。

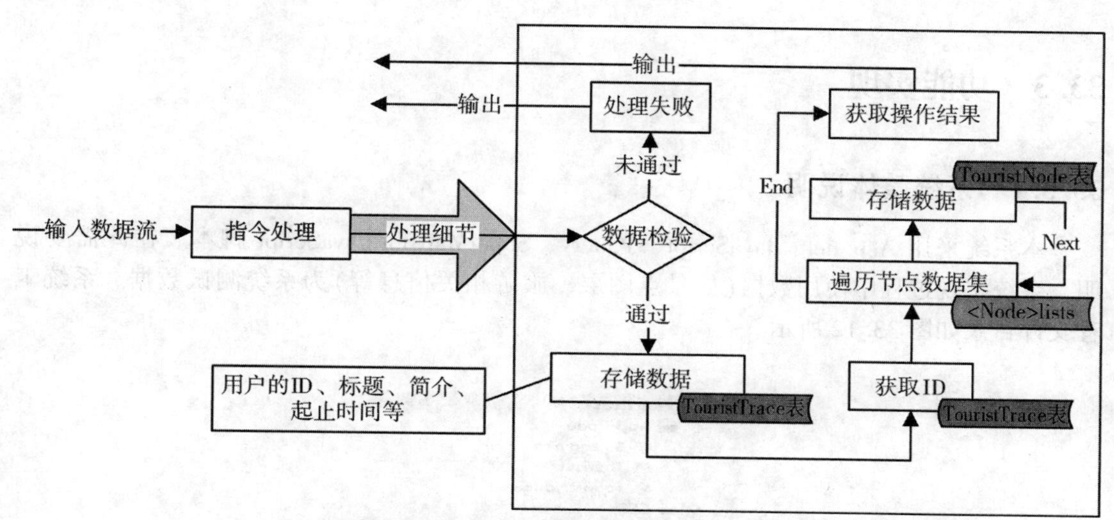

图 33.10 【发布旅行印迹】内部流程图

5. 查看旅行印迹

根据用户点击的旅行印迹 ID 可以查找到 TouristTrace 表和 TouristNode 表中的数据，根据这些数据进行前台的交互设计。根据 TouristNode 表返回的数据集生成 SuperMap. Feature. Vector 数据(补充相应属性)添加至 SuperMap. Layer. AnimatorVector 图层(展示运动效果)和 SuperMap. Layer. Vector 图层(显示旅行路线数据)，程序处理流程如图 33. 11 所示。

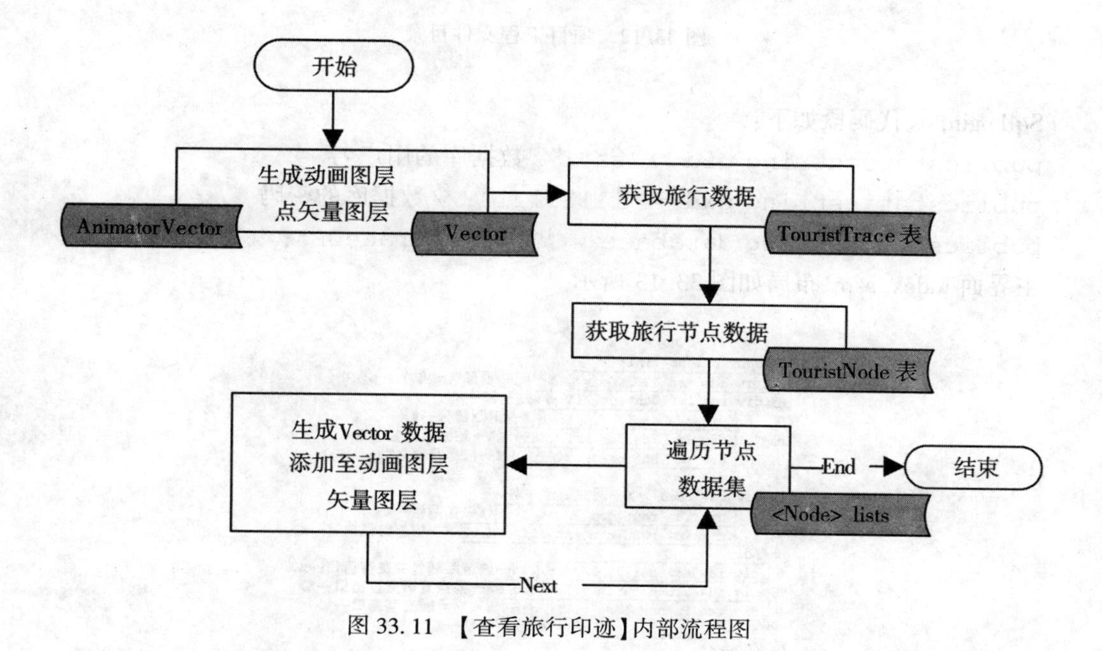

图 33. 11 【查看旅行印迹】内部流程图

33.3　功能实现

33.3.1　系统整体说明

本系统采用 Asp. net、JavaScript、JQuery、SuperMap for JavaScript 脚本。在此需要说明，系统实现过程中使用数据(如景点图层、旅游相关信息等)为系统测试数据。系统工程文件目录如图 33.12 所示。

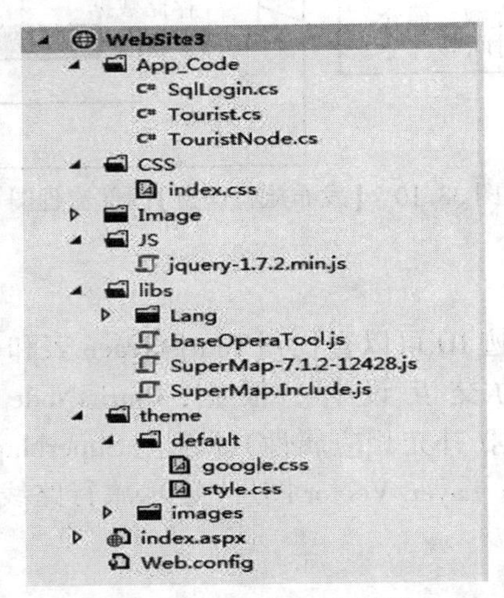

图 33.12　项目工程文件目录

SqlLogin. cs 代码段如下：

```
publicstaticstring uid = "sa";//数据库的用户名
publicstaticstring pwd = "11111111";//数据库的密码
publicstaticstring database = "TouristTraceDB";//数据库的名称
```

主界面 index. aspx 布局如图 33.13 所示。

```
<body onload="init()">
<div id="top">...</div> <%--网站顶部（菜单）--%>
<div id="leftContent"><%--网站左部--%>
    <div id="map">...</div><%--地图部分--%>
    <div id="ToolFun">...</div><%--地图工具栏--%>
    <div id="bottom">...</div><%--鼠标数据--%>
 </div>
<div id="road"><%--网站右部--%>
    <div id="road_top">...</div><%--标题栏--%>
    <div id="road_content">...</div><%--旅游路线内容--%>
</div>
<div id="moveBar">...</div> <%--旅游路线提交交互窗口--%>
<div id="SceneSearch">...</div><%--景点查询交互窗口--%>
<div id="SceneHotMap">...</div><%--热点图交互窗口--%>
<script>...</script> <%--实现DIV拖动--%>
</body>
</html>
```

图 33.13　主界面 index. asp 页面布局

界面整体设计、运行效果如图 33.14 所示。

图 33.14 “旅印”程序运行主界面

33.3.2 景点查询

功能实现代码如下：

①index.css 文件：

```
a {
    color:#0088DB;
    text-decoration:none;
    cursor:pointer
}
a:hover {
    color:#5e0daa
}
.addMark{
    color:white;
    text-decoration:none;
    cursor:pointer
}
.close {
    float:right;
    color:#999;
    padding-right:5px;
    font:bold14px/14pxsimsun;
    text-shadow:01px0#ddd
}
```

```
#SceneSearch{
    width:170px;
    height:130px;
    position: absolute;
    left:300px;
    top:120px;
    background: #4667d2;
    color:white;
    visibility:visible;
    text-align:center;
}
#SceneBanner{
    background: #4b9df1;
    cursor: move;
    height:30px;
    text-align:center;
    padding-top:5px;
}
```

②index. aspx 文件:

```
<div id="SceneSearch">
<div id="SceneBanner">景点查询
<a href="javascript:;" id="A1" onclick="controlStatueB()"
title="关闭" class="close">×</a></div>
<div id="GeoSearch" style="float:left;padding:3px;">几何查询
<a href="javascript:;" class="addMark" onclick="CircleSearch
()" title="圆形查询"><u>圆形</u></a>
<a href="javascript:;" class="addMark" onclick="PolygonSearch
()" title="多边形查询"><u>多边形</u></a>
</div>
<div style="clear:both;"></div>
<div id="Buffer" style="float:left;padding:3px;">
缓冲半径
<input type="text" id="BufferDistance" style="margin-left:5px;
margin-right:5px;" value="3000" size="5" />米
</div>
<div style="clear:both;"></div>
<div id="BufferSearch" style="float:left;padding:5px;">缓冲区查
询
```

```
<a href="javascript:;" class="addMark" onclick="drawPointGeome-
try()" title="点缓冲区查询"><u>点</u></a>
<a href="javascript:;" class="addMark" onclick="drawLineGeome-
try()" title="线缓冲区查询"><u>线</u></a>
</div>
</div>
<script>
    jQuery(document).ready(
function () {
                $('#SceneBanner').mousedown(
function (event) {
var isMove = true;
var abs_x = event.pageX - $('#SceneSearch').offset().left;
var abs_y = event.pageY - $('#SceneSearch').offset().top;
                        $(document).mousemove(function (event)
{
if (isMove) {
var obj = $('#SceneSearch');
                                        obj.css({'left': event.pageX -abs_
x,'top': event.pageY -abs_y });
                                    }
                                }
                            ).mouseup(
function () {
                                            isMove =false;
                                        }
                            );
                        }
                );
    );
</script>
```

③baseOperaTool.js 文件：

```
var map, layer, markLayer, vectorLayer, drawPolygon,drawCircle ,
marker, icon, framedCloud, host = "http://localhost:8090";
var drawPoint, drawLine, drawRectangle, queryBounds, result-
Globle = null;
var bt = false;
var url = host + "/iserver/services/map- China4002/rest/maps/
```

```
China400";
    var url2 = host + "/iserver/services/data-China4002/rest/data";
    var style = {
            strokeColor:"#304DBE",
            strokeWidth: 3,
            pointerEvents:"visiblePainted",
            fillColor:"#304DBE",
            fillOpacity: 0.3
        };
    function scenicSpotsSearch() {
            document.getElementById("SceneSearch").style.visibility
= "visible";
        }
    function controlStatueB() {
    if (document.getElementById("SceneSearch").style.visibility ==
"visible") {
                document.getElementById("SceneSearch").style.
visibility = "hidden";
                closeInfoWin();
                clearStatus();
                clearActive();
            }else {
                document.getElementById("SceneSearch").style.
visibility = "visible";
            }
        }
    function init() {
        layer =new SuperMap.Layer.TiledDynamicRESTLayer("中国地图",
url, {
            transparent : true,
            cacheEnabled : true
        }, { maxResolution:"auto" });
        layer.events.on({"layerInitialized": addLayer });

        vectorLayer =new SuperMap.Layer.Vector("Vector Layer");
        markLayer =new SuperMap.Layer.Markers("Markers");
        drawPolygon =new SuperMap.Control.DrawFeature(vectorLayer,
SuperMap.Handler.Polygon);
```

```
        drawPolygon.events.on({"featureadded": drawCompleted });
        drawCircle =new SuperMap.Control.DrawFeature(vectorLayer,
SuperMap.Handler.RegularPolygon, { handlerOptions: { sides: 50 } });
        drawCircle.events.on({"featureadded": drawCompleted });
        drawPoint =new SuperMap.Control.DrawFeature(vectorLayer, Su-
perMap.Handler.Point);
        drawPoint.events.on({"featureadded": drawCompleted2 });
        drawLine =new SuperMap.Control.DrawFeature(vectorLayer, Su-
perMap.Handler.Path);
        drawLine.events.on({"featureadded": drawCompleted2 });

        map =new SuperMap.Map("map", {
            controls: [new SuperMap.Control.ScaleLine(),
        new SuperMap.Control.Zoom(),
        new SuperMap.Control.Navigation({
                        dragPanOptions: {
                            enableKinetic:true
                        }
                    }), drawPolygon, drawCircle, drawPoint, draw-
Line], units:"m"
        });
        map.minScale = 1.3732579019271662e-8;
        map.numZoomLevels = 7;//设置地图缩放级别的数量
        map.events.on({
            "mousemove": getMousePositionPx
        });
    }

    function addLayer() {
        map.addLayers([layer, markLayer, vectorLayer]);
        map.setCenter(new SuperMap.LonLat(12161566.94829, 4601160.
54256), 4);
    }
    function clearStatus() {
        vectorLayer.removeAllFeatures();
        markLayer.clearMarkers();
        heatMapLayer.removeAllFeatures();
    }
```

```
function clearActive() {
    drawPolygon.deactivate();
    drawCircle.deactivate();
    drawPoint.deactivate();
    drawLine.deactivate();
}
function CircleSearch() {
    clearStatus();
    drawCircle.activate();
}
function PolygonSearch() {
    clearStatus();
    drawPolygon.activate();
}
function drawPointGeometry() {
    clearStatus();
    drawPoint.activate();
}
function drawLineGeometry() {
    clearStatus();
    drawLine.activate();
}
function drawCompleted(drawGeometryArgs) {
var feature = new SuperMap.Feature.Vector();
    feature.geometry = drawGeometryArgs.feature.geometry;
           feature.style = style;
    vectorLayer.addFeatures(feature);

var queryParam, queryByGeometryParameters, queryService;
    queryParam =new SuperMap.REST.FilterParameter({ name: "Scen-
icSpot@ China"});
    queryByGeometryParameters =new SuperMap.REST.
QueryByGeometryParameters({
        queryParams: [queryParam],
        geometry: drawGeometryArgs.feature.geometry,
        spatialQueryMode: SuperMap.REST.SpatialQueryMode.INTERSECT
    });
    queryService =new SuperMap.REST.QueryByGeometryService(url,
```

```
{
        eventListeners: {
            "processCompleted": processCompleted,
            "processFailed": processFailed
        }
    });
    queryService.processAsync(queryByGeometryParameters);
    //向服务端传递参数,然后服务端返回对象
  }
  function drawCompleted2(drawGeometryArgs) {
  var feature = new SuperMap.Feature.Vector();
      feature.geometry = drawGeometryArgs.feature.geometry,
          feature.style = style;
      vectorLayer.addFeatures(feature);

  var getFeatureParameter, getFeatureService;
      getFeatureParameter =new SuperMap.REST.GetFeaturesBy-
BufferParameters({
          bufferDistance: document.getElementById (" BufferDis-
tance").value,
          datasetNames: ["China:ScenicSpot"],
          returnContent:true,
          geometry: drawGeometryArgs.feature.geometry
      });
      getFeatureService =new SuperMap.REST.GetFeaturesBy-
BufferService(url2, {
          eventListeners: {
              "processCompleted": processCompleted2,
              "processFailed": processFailed
          }
      });
      getFeatureService.processAsync(getFeatureParameter);
  }
  function processCompleted(queryEventArgs) {
      clearActive();
  var i, j, result = queryEventArgs.result;
  if (result && result.recordsets) {
  for (i = 0, recordsets = result.recordsets, len = recordsets.
```

```
length; i < len; i++) {
    if (recordsets[i].features) {
    for (j = 0; j < recordsets[i].features.length; j++) {
    var feature = recordsets[i].features[j];
    var point = feature.geometry;
    if (point.CLASS_NAME == SuperMap.Geometry.Point.prototype.CLASS
_NAME) {
    var size = new SuperMap.Size(44,33),
                                  offset =new SuperMap.Pixel(-(size.w
/2), -size.h),
                                   icon =new SuperMap.Icon("../theme/
images/marker.png", size, offset);
                            marker =new SuperMap.Marker(new Super-
Map.LonLat(point.x, point.y), icon);
                            marker.Name = feature.attributes.Name;
                            marker.Popu = feature.attributes.Popular;
                            markLayer.addMarker(marker);
                            marker.events.on({"click": mouseMarker-
ClickHandler, "scope": marker});
    //map.events.un({ "click": addMarkerFun });
                        }else {
                            feature.style = style;
                            vectorLayer.addFeatures(feature);
                        }
                    }
                }
            }
        }
    }
    function processCompleted2(getFeaturesEventArgs) {
        clearActive();
    var i, len, features, result = getFeaturesEventArgs.result;
    if (result && result.features) {
            features = result.features;
    for (i = 0, len = features.length; i < len; i++) {
    var point = features[i].geometry,
            size =new SuperMap.Size(44, 33),
                                  offset =new SuperMap.Pixel(-(size.w
```

```
/2), -size.h),
                                        icon =new SuperMap.Icon("../theme/
images/marker.png", size, offset);
                        marker  = new SuperMap.Marker(new SuperMap.LonLat
(point.x, point.y), icon);
                marker.Name = features[i].attributes.NAME;
                marker.Popu = features[i].attributes.POPULAR;
                markLayer.addMarker(marker);
    //注册 click 事件,触发 mouseClickHandler()方法
                marker.events.on({"click": mouseMarkerClickHandler,
"scope": marker});
    //map.events.un({ "click": addMarkerFun });
            }
        }
    }
    function processFailed(e) {
        alert(e.error.errorMsg);
    }
    function mouseMarkerClickHandler(event) {
        closeInfoWin();
    var marker = this;
    var contentHTML = "<div style='font-size:12px;font-weight:bold ;
opacity: 0.8'>";
        contentHTML +="旅游点:" + marker.Name+"</br>";
        contentHTML +="受欢迎度:" + marker.Popu;
        contentHTML +="<hr/><div  style='text-align:center;'><a on-
click='showScene(\"" + marker.ID
                        +"\")'>查看景点</a></div>";
        contentHTML +="</div>";
    //初始化 FramedCloud 类
        framedCloud =new SuperMap.Popup.FramedCloud(
            "MarkerNode",
            marker.getLonLat(),
            null,
            contentHTML,
            icon,
            true,
```

```
        null,
        true
    );
    infowin = framedCloud;
    map.addPopup(framedCloud);
}
function closeInfoWin() {
if (infowin) {
try {
            infowin.hide();
            infowin.destroy();
        }
catch (e) { }
    }
}
```

运行程序，交互过程如下：

点击【景点查询】，弹出交互框，如图 33.15 所示。

图 33.15 【景点查询】交互框

点击【圆形】，在地图上画一个圆，可查询圆内的景点。点击任意景点图标都可以查看详细信息，如图 33.16 所示。

点击【多边形】，在地图上画一个多边形，可查询多边形内的景点。点击任意景点图标可以查看详细信息，如图 33.17 所示。

输入缓冲半径，点击【点】，在地图上画一个点，可查询点附近缓冲区内的景点。点击任意景点图标可以查看详细信息，如图 33.18 所示。

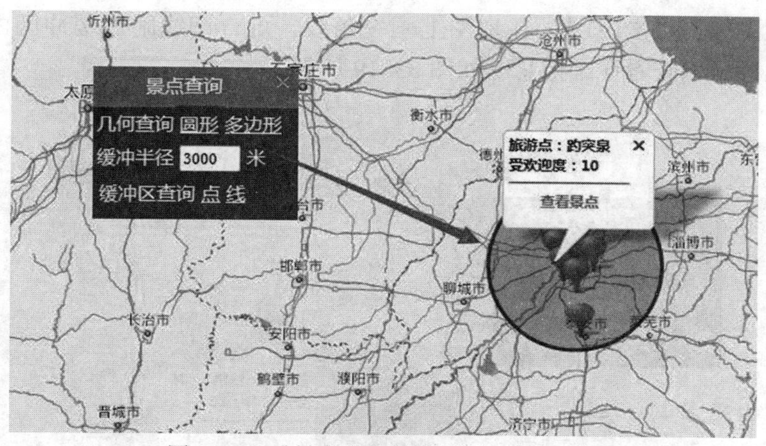

图 33.16 【景点查询】≫【圆形】几何查询

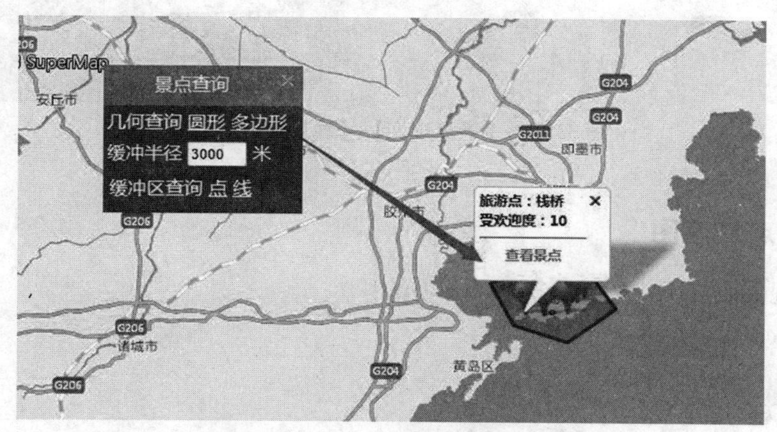

图 33.17 【景点查询】≫【多边形】几何查询

图 33.18 【景点查询】≫【点】缓冲区查询

输入缓冲半径，点击【线】，在地图上画一条线，可查询线附近缓冲区内的景点。点击任意景点图标可以查看详细信息，如图 33.19 所示。

图 33.19 【景点查询】≫【线】缓冲区查询

33.3.3 景点热度图

功能实现代码如下：

①index.css 文件：

```
a {
    color:#0088DB;
    text-decoration:none;
    cursor:pointer
}
a:hover {
    color:#5e0daa
}
.addMark{
    color:white;
    text-decoration:none;
    cursor:pointer
}
.close {
    float:right;
    color:#999;
    padding-right:5px;
```

```
    font:bold14px/14pxsimsun;
    text-shadow:01px0#ddd
}
#SceneHotMap{
    width:170px;
    height:130px;
    position: absolute;
    left:100px;
    top:120px;
    background: #4667d2;
    color:white;
    visibility:visible;
    text-align:center;
}
#HotMapBanner{
    background: #4b9df1;
    cursor: move;
    height:30px;
    text-align:center;
    padding-top:5px;
}
```

②index. aspx 文件:

```
<div id="SceneHotMap">
<div id="HotMapBanner">景点热度图<a href="javascript:;" id="A2" onclick="controlStatueC()" title="关闭" class="close">×</a></div>
<div id="Div3" style="float:left;padding-left:10px;padding-top:5px;">选择区域
<select id="heatMapArea" name="heatMapArea" style="margin-left:10px;width:60px;">
<option value="0">中国</option>
<option value="1">北京</option>
<option value="2">上海</option>
<option value="3">广东省</option>
<option value="4">江苏省</option>
<option value="5">浙江省</option>
<option value="6">贵州省</option>
<option value="7">云南省</option>
<option value="8">甘肃省</option>
```

```
<option value="9">陕西省</option>
<option value="10">山东省</option>
<% --<option>.. 在此省略 ..</option>--% >
</select>
</div>
<div style="float:left;padding-left:30px;padding-top:5px;">
或者<a href="javascript:;" class="addMark" style="margin-left:
20px;" onclick="drawRectangleCompleted()" title="矩形选择"><u>矩形区
域</u></a>
</div>
<div style="float:left;padding-left:30px;padding-top:5px;">
<a href="javascript:;" class="addMark" style="margin-top:20px;"
onclick="createHeatPoints()" title="生成景点热度图"><u>生成景点热度图
</u></a>
</div>
<div style="clear:both;"></div>
</div>
<script>
    jQuery(document).ready(
function () {
                $('#HotMapBanner').mousedown(
function (event) {
var isMove = true;
var abs_x = event.pageX -$('#SceneHotMap').offset().left;
var abs_y = event.pageY -$('#SceneHotMap').offset().top;
                        $(document).mousemove(function (event) {
if (isMove) {
var obj = $('#SceneHotMap');
                                obj.css({'left': event.pageX -abs_
x,'top': event.pageY -abs_y });
                            }
                        }
                        ).mouseup(
function () {
                                isMove =false;
                            }
                        );
                    }
```

```
                );
            }
    );
    </script>
```

③baseOperaTool.js 文件:

```
    function sceneHeataMap() {
            document.getElementById("SceneHotMap").style.
visibility = "visible";
        }
    function controlStatueC() {
    if (document.getElementById("SceneHotMap").style.visibility ==
"visible") {
                document.getElementById("SceneHotMap").style.
visibility = "hidden";
                closeInfoWin();
                clearStatus();
                clearActive();
            }else {
                document.getElementById("SceneHotMap").style.
visibility = "visible";
            }
        }
    var map, layer, markLayer, vectorLayer, marker, icon, framedCloud,
host = "http://localhost:8090";
    var heatMapLayer, queryBounds, resultGloble = null;
    var bt = false;
    var url = host + "/iserver/services/map-China4002/rest/maps/
China400";
    var url2 = host + "/iserver/services/data-China4002/rest/data";
    var style = {
            strokeColor:"#304DBE",
            strokeWidth: 3,
            pointerEvents:"visiblePainted",
            fillColor:"#304DBE",
            fillOpacity: 0.3
        };
    function init() {
        layer =new SuperMap.Layer.TiledDynamicRESTLayer("中国地图",
```

```
url, {
            transparent : true,
            cacheEnabled : true
        }, { maxResolution:"auto" });
        layer.events.on({"layerInitialized": addLayer});

        vectorLayer =new SuperMap.Layer.Vector("Vector Layer");
        markLayer =new SuperMap.Layer.Markers("Markers");
        heatMapLayer =new SuperMap.Layer.HeatMapLayer(
    "heatMap",
                                    {
    "radius": 45,
    "featureWeight": "value",
    "featureRadius": "geoRadius"
                                    }
                );
        map =new SuperMap.Map("map", {
            controls: [new SuperMap.Control.ScaleLine(),
                        new SuperMap.Control.Zoom(),
                        new SuperMap.Control.Navigation({
                                                dragPanOptions: {
                                                    enableKinetic:true
                                                }
                                            })], units:"m"
        });
        map.minScale = 1.3732579019271662e-8;
        map.numZoomLevels = 7;//设置地图缩放级别的数量
        map.events.on({
        //"click":getPosition,
            "mousemove" : getMousePositionPx
        });
    }

    function addLayer() {
        map.addLayers([layer, markLayer, heatMapLayer, vectorLayer]);
        map.setCenter(new SuperMap.LonLat(12161566.94829, 4601160.
54256), 4);
    }
```

```
function clearStatus() {
    vectorLayer.removeAllFeatures();
    markLayer.clearMarkers();
    heatMapLayer.removeAllFeatures();
}
//矩形
function drawRectangleCompleted() {
    clearStatus();
var control = new SuperMap.Control();
    SuperMap.Util.extend(control, { //Util 工具类  extend 指的是将
复制所有的属性的源对象到目标对象
        draw:function () {
this.box = new SuperMap.Handler.Box ( control, { " done ":
this.notice });//此句是创建一个句柄,Box 是一个处理地图拖放一个矩形的事件,这
个矩形显示开始于按下鼠标,然后移动鼠标,最后完成在松开鼠标
this.box.boxDivClassName = "qByBoundsBoxDiv";// boxDivClassName
用于绘制这个矩形状的图形
this.box.activate();//激活句柄
        },
//将拖动的矩形显示在地图上
        notice:function (bounds) {
this.box.deactivate();//处理关闭激活句柄

var ll = map.getLonLatFromPixel(new SuperMap.Pixel(bounds.left,
bounds.bottom)),//getLonLatFromPixel 从视口坐标获得地理坐标
            ur = map.getLonLatFromPixel(new SuperMap.Pixel
(bounds.right, bounds.top));
        queryBounds = new SuperMap.Bounds(ll.lon, ll.lat,
ur.lon, ur.lat);

var feature = new SuperMap.Feature.Vector();
        feature.geometry = queryBounds.toGeometry(),
            feature.style = style;
        vectorLayer.addFeatures(feature);

var queryParam, queryByBoundsParams, queryService;
        queryParam =new SuperMap.REST.FilterParameter({
            name:"ScenicSpot@ China"
```

```
            });//FilterParameter 设置查询条件,name 是必设的参数(图层
名称格式:数据集名称@ 数据源别名)
            queryByBoundsParams =new
SuperMap.REST.QueryByBoundsParameters({ queryParams:[queryParam],
bounds: queryBounds });//queryParams 查询过滤条件参数数组。bounds 查询范围
            queryService =new SuperMap.REST.QueryByBoundsService
(url, {
                eventListeners:{
                    "processCompleted": processCompleted0,
                    "processFailed": processFailed
                }
            });
            queryService.processAsync(queryByBoundsParams);//向
服务端传递参数,然后服务端返回对象
        }
    });
    map.addControl(control);
  }
  function processCompleted0(queryEventArgs) {
      clearActive();
      resultGloble = queryEventArgs.result;
  }
  function createHeatPoints() {
  if (! resultGloble){
  var sellect = document.getElementById ( " heatMapArea") [ docu-
ment.getElementById("heatMapArea").selectedIndex].innerHTML;
  var queryParam, queryBySQLParams, queryBySQLService;
        queryParam =new SuperMap.REST.FilterParameter({
            name:"ScenicSpot@ China",
            attributeFilter:"District ='" + sellect+"'"
        });
        queryBySQLParams =new SuperMap.REST.QueryBySQLParameters({
            queryParams:[queryParam]
        });
        queryBySQLService = new SuperMap.REST.QueryBySQLService
(url, {
            eventListeners: {"processCompleted": processComplet-
ed0, "processFailed": processFailed}
```

```
            });
        queryBySQLService.processAsync(queryBySQLParams);
      }
      HeatMapFun();
    }
    function HeatMapFun() {
    var heatPoints = [];
    var i, j, k = 0;
    if (resultGloble && resultGloble.recordsets) {
    for (i=0, recordsets=resultGloble.recordsets, len = recordsets.
length; i < len; i++) {
    if (recordsets[i].features) {
    for (j = 0; j < recordsets[i].features.length; j++) {
    var feature = recordsets[i].features[j];
    var point = feature.geometry;
    if (point.CLASS_NAME == SuperMap.Geometry.Point.prototype.CLASS
_NAME) {
                        heatPoints[k] =new SuperMap.Feature.Vector
(
    new SuperMap.Geometry.Point(
                                point.x,
                                point.y
                            ),
                            {
    "value": feature.attributes.Popular
                            }
                        );
                        k++;
                    }
                }
            }
        }
    }
        heatMapLayer.addFeatures(heatPoints);
        resultGloble =null;
    }
```

运行程序，交互过程如下：

点击【景点热度】，弹出交互框，如图33.20所示。

图 33.20 【景点热度】交互框

点击【矩形区域】，在地图上画一个矩形。点击【生成景点热度图】，生成矩形内热度图，如图 33.21 所示。

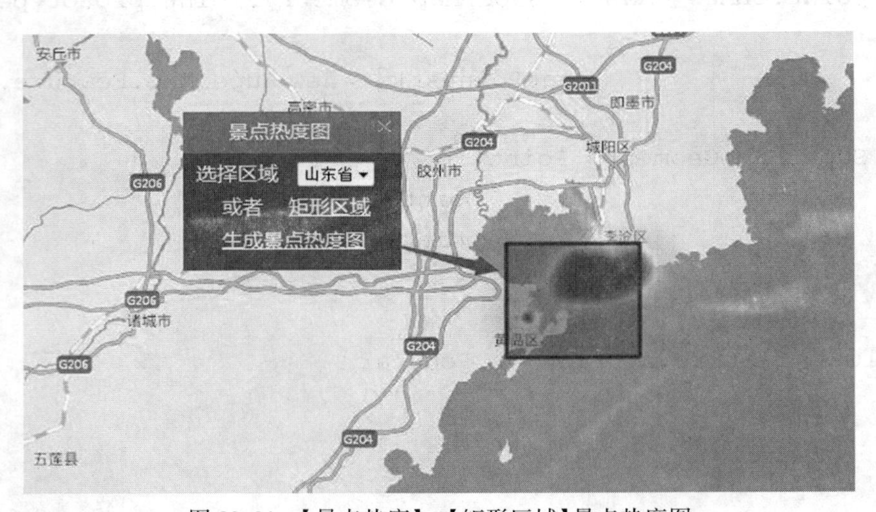

图 33.21 【景点热度】≫【矩形区域】景点热度图

点击下拉列表，选择行政区域，点击【生成景点热度图】，如图 33.22 所示。

33.3.4 发布旅行印迹

功能实现代码如下：

①index.css 文件：

```
#moveBar {
    position: absolute;
    right:520px;
```

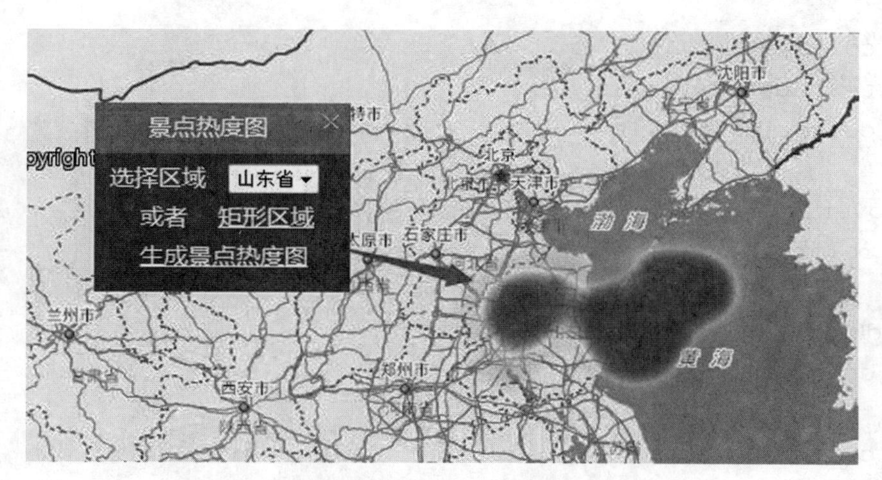

图 33.22 【景点热度】≫【选择区域】景点热度图

```
    top:120px;
    width: 250px;
    height: 360px;
    background: #4667d2;
    color:white;
    visibility:visible;
}
#banner {
    background: #4b9df1;
    cursor: move;
    height:30px;
    text-align:center;
    padding-top:5px;
}
#linkA {
    height:32px;
    text-align:center;
}
#content{
    height:170px;
    background-color:#5577e4;
    overflow:auto;
    width:100% ;
}
```

```
a {
    color:#0088DB;
    text-decoration:none;
    cursor:pointer
}
a:hover {
    color:#5e0daa
}
.addMark{
    color:white;
    text-decoration:none;
    cursor:pointer
}
.close {
    float:right;
    color:#999;
    padding-right:5px;
    font:bold14px/14pxsimsun;
    text-shadow:01px0#ddd
}
```

②index. aspx 文件：

```
<div id="moveBar" runat="server">
<div id="banner">上传旅游印迹信息<a href="javascript:;" id=
"close" onclick="controlStatue()" title="关闭" class="close">×</a>
</div>
<div id="TouristContent" style="padding:5px;text-align:center;">
<span style="padding-right:15px;">标题</span><input type="text"
id="toutistTitile" size="25"/><br/>
<span style="padding-right:15px;">开始时间</span><input type=
"text" id="startTime" size="20"/><br/>
<span style="padding-right:15px;">结束时间</span><input type=
"text" id="endTime" size="20"/><br/>
<span style="padding-right:15px;">简介</span><br/>
<textarea id="touristIntro" rows="2" cols="25"></textarea>
</div>
<div id="linkA" runat="server"><div style="width:70px;float:
left;padding:5px;">节点内容</div>
<div style="width:80px;float:left;">
```

```
<a href="javascript:;" class="addMark" onclick="NodeMarkAddInfo()" title="添加节点(请按顺序添加)"><img src="Image/button_plus_red.png" width="32" height="32"/></a>
<a href="javascript:;" class="addMark" onclick="NodeMarkRemoveInfoAll()" title="删除所有节点"><img src="Image/button_minus_red.png" width="32" height="32"/></a>
</div>
<form runat="server">
<asp:ScriptManager ID="ScriptManager1" runat="server" EnablePageMethods="true">
</asp:ScriptManager>
<div style="float:left;padding:5px;"><a href="javascript:;" class="addMark" onclick="SubmitData()" title="点击提交数据"><u>提交数据</u></a></div>
</form>
<div style="clear:both;"></div>
</div>
<div id="content">
<%--以下注释代码段为测试模板--%>
<%--<div id="Div1" style="padding:5px;text-align:center;">
<hr style="width:220px;"/>
<div style="width:80px;float:left;padding-top:5px;">节点标识:</div>
<div style="width:100px;float:left;">
<a href="javascript:;" class="addMark" onclick="NodeMarkRemoveInfoAll()" title="删除所有节点"><img src="Image/button_minus_red.png" width="32" height="32"/></a>
</div>
<div style="clear:both;"></div>
<span style="padding-right:15px;">标题</span><input type="text" id="Text1"  size="22"/><br/>
<span style="padding-right:15px;">开始时间</span><input type="text" id="Text2"  size="18"/><br/>
<span style="padding-right:15px;">结束时间</span><input type="text" id="Text3"  size="18"/><br/>
<span style="padding-right:10px;">类型</span>
<select id="mysel" name="mysel" style="width:170px;">
<option value="0">跟团旅游</option>
```

```
<option value="1">自驾游</option>
<option value="2">徒步游</option>
<option value="3">个人游</option>
<option value="4">和驴友们旅游</option>
</select><br/>
<span style="">简介</span><br />
<textarea id="Textarea1" rows="2" cols="19"></textarea>
</div> --%>
</div>
</div><%--旅游路线提交交互窗口--%>
<script>
  jQuery(document).ready(
function (){
              $('#banner').mousedown(
function (event) {
var isMove = true;
var abs_x = event.pageX -$('#moveBar').offset().left;
var abs_y = event.pageY -$('#moveBar').offset().top;
                    $(document).mousemove(function (event)
{
if (isMove) {
var obj = $('#moveBar');
                          obj.css({'left': event.pageX -abs_
x,'top': event.pageY -abs_y});
                   }
                  }
                  ).mouseup(
function () {
                          isMove =false;
                   }
                  );
               }
            );
     }
  );
</script><%--实现 DIV 拖动--%>
```

③baseOperaTool. js 文件：

```
function TouristTrace() {
```

```
        document.getElementById("moveBar").style.visibility =
"visible";
    }
    function controlStatue() {
    if (document.getElementById("moveBar").style.visibility ==
"visible") {
            document.getElementById("moveBar").style.
visibility = "hidden";
            closeInfoWin();
            clearStatus();
            clearActive();
        }else {
            document.getElementById("moveBar").style.
visibility = "visible";
        }
    }
    var map, layer, markLayer, vectorLayer ,marker, icon, framedCloud,
host = "http://localhost:8090";
    var url = host + "/iserver/services/map-China4002/rest/maps/
China400";
    var url2 = host + "/iserver/services/data-China4002/rest/data";
    var style = {
        strokeColor:"#304DBE",
        strokeWidth: 3,
        pointerEvents:"visiblePainted",
        fillColor:"#304DBE",
        fillOpacity: 0.3
    };
    function init() {
        layer =new SuperMap.Layer.TiledDynamicRESTLayer("中国地图",
url, {
            transparent : true,
            cacheEnabled : true
        }, { maxResolution:"auto" });
        layer.events.on({"layerInitialized": addLayer });

        vectorLayer =new SuperMap.Layer.Vector("Vector Layer");
        markLayer =new SuperMap.Layer.Markers("Markers");
```

```
        map =new SuperMap.Map("map", {
            controls: [new SuperMap.Control.ScaleLine(),
    new SuperMap.Control.Zoom(),
    new SuperMap.Control.Navigation({
                        dragPanOptions: {
                            enableKinetic:true
                        }
                    })], units:"m"
        });
        map.minScale = 1.3732579019271662e-8;
        map.numZoomLevels = 7;//设置地图缩放级别的数量
        map.events.on({
        //"click":getPosition,
            "mousemove" : getMousePositionPx
        });
    }

    function addLayer() {
        map.addLayers([layer, markLayer, vectorLayer]);
            map.setCenter  ( new  SuperMap.LonLat  ( 12161566.94829,
4601160.54256), 4);
    }
    var markIndex = 0;
    function NodeMarkAddInfo() {
        map.events.on({"click": addMarkerFun });
    }
    function addMarkerFun(e) {
    var lonlat = map
                .getLonLatFromPixel(new SuperMap.Pixel(e.clientX-10,
e.clientY-70));
        size =new SuperMap.Size(44, 33),
        offset =new SuperMap.Pixel(-(size.w /2), -size.h),
        icon =new SuperMap.Icon("../theme/images/marker.png", size,
offset);
        //初始化标记覆盖物类
        marker =new SuperMap.Marker(new SuperMap.LonLat(lonlat.lon,
lonlat.lat), icon);
        marker.Name ="MarkerNode" + markIndex;
```

```
        markIndex++;
        //添加覆盖物到标记图层
        markLayer.addMarker(marker);
        //注册 click 事件,触发 mouseClickHandler()方法
        marker.events.on({"click": mouseClickHandler, "scope": mark-
er});
        map.events.un({"click": addMarkerFun});

    var contentHTML =
    "<hr style='width:220px;'/>"+
    "<div style='width:190px;float:left;padding-top:5px;'>节点标识:"+
marker.Name+"</div>"+
    "<div style='width:32px;float:left;'>"+
    "<a href='javascript:;'class='addMark'onclick='NodeMarkRemoveIn-
fo(\""+marker.Name+"\")'title='删除此节点'><img src='Image/button_
minus_red.png'width='32'height='32'/></a>"+
    "</div>"+
    "<div style='clear:both;'></div>"+
    "<span style='padding-right:15px;'>标题</span><input type='text'
name='TitleNode'  size='22'/><br/>"+
    "<span style='padding-right:15px;'>开始时间</span><input type=
'text'name='StarTimeNode'  size='18'/><br/>"+
    "<span style='padding-right:15px;'>结束时间</span><input type=
'text'name='EndTimeNode'  size='18'/><br/>"+
    "<span style='padding-right:10px;'>类型</span>"+
    "<select id='TouristStyleNode'name='TouristStyleNode'style=
'width:170px;'>"+
    "<option value='0'>跟团旅游</option>"+
    "<option value='1'>自驾游</option>"+
    "<option value='2'>徒步游</option>"+
    "<option value='3'>个人游</option>"+
    "<option value='4'>和驴友们旅游</option>"+
    "</select><br/>"+
    "<span style=''>简介</span><br />"+
    "<textarea name='IntroduceNode'rows='2'cols='22'></textarea>";
    var newElement = document.createElement('div');
        newElement.setAttribute('id',marker.Name);
        newElement.setAttribute('style',"padding:5px;text-align:cen-
```

```
ter;");
        newElement.innerHTML = contentHTML;
        document.getElementById("content").appendChild(newElement);
    }
    function SubmitData(){
    var touristContent = document.getElementById ( " toutistTitile ")
.value+" |"+
                        document.getElementById("startTime").value + " |" +
                        document.getElementById("endTime").value + " |" +
                        document.getElementById("touristIntro").value;
    var nodeContent=[];
    for ( var i = 0; i < document.getElementsByName ( " TitleNode ")
.length; i++) {
    var node = document.getElementsByName("TitleNode")[i].value + " |" +
                    document.getElementsByName("StarTimeNode")[i]
.value + " |" +
                    document.getElementsByName("EndTimeNode")[i]
.value + " |" +
    document.getElementsByName ( " TouristStyleNode ") [ i ] [ document.
getElementsByName("TouristStyleNode")[i].selectedIndex].value +
" |" +
                    document.getElementsByName("IntroduceNode")[i]
.value + " |" +
                    markLayer.markers[i].lonlat.lon+" |" +
                    markLayer.markers[i].lonlat.lat;
        nodeContent[i] = node;
        }
    //JavaScript 异步调用定义在 ASP.Net 页面中的方法
    //1.将该方法声明为公有(public);
    //2.将该方法声明为类方法(C#中的 static,VB.NET 中的 Shared),而不是实例
方法;
    //3.将该方法添加【WebMethod】属性
    //4.将页面中 ScriptManager 控件的 EnablePageMethods 属性设置为 true;
    //5.在客户端使用如下 JavaScript 语法调用该页面方法
    // PageMethods.[MethodName](param1, param2, ..., callbackFunc-
tion);
    //6.为客户端异步调用指定回调函数,在回调函数中接受返回值并进一步处理;
    //7.添加 using System.Web.Services;
```

```
        PageMethods.SubmitDateShell(touristContent, nodeContent, on-
SubmitSucceeded);
    }
    function onSubmitSucceeded(result)//绑定的回调函数
    {
        alert(result);
    }
    var infowin = null;
    //定义mouseClickHandler函数,触发click事件会调用此函数
    function mouseClickHandler(event) {
        closeInfoWin();
    var marker = this;
    var contentHTML = "<div style='width:80px; font-size:12px;font-
weight:bold ; opacity: 0.8'>";
        contentHTML +="节点标识"+marker.Name;
        contentHTML +="</div>";
    //初始化FramedCloud类
        framedCloud =new SuperMap.Popup.FramedCloud(
    "touristNode",
            marker.getLonLat(),
    null,
            contentHTML,
            icon,
    true,
    null,
    true
        );
        infowin = framedCloud;
        map.addPopup(framedCloud);
    }

    function closeInfoWin() {
    if (infowin) {
    try {
                infowin.hide();
                infowin.destroy();
            }
    catch (e) {}
```

```
        }
    }
    function NodeMarkRemoveInfoAll() {
        closeInfoWin();
        clearStatus();
        document.getElementById("content").innerHTML = "";
        markIndex = 0;
    }
    function NodeMarkRemoveInfo(n) {
        closeInfoWin();
    if (markLayer.markers) {
    for (j = 0; j < markLayer.markers.length; j++) {
            marker = markLayer.markers[j];
    if (marker.Name == n) {
                markLayer.removeMarker(marker);
    break;
            }
        }
    }
    var thisNode = document.getElementById(n);
        thisNode.parentNode.removeChild(thisNode);
    }
    function clearStatus() {
        vectorLayer.removeAllFeatures();
        markLayer.clearMarkers();
    }
```

④index. aspx. cs 文件:

```
[WebMethod]//标示为Web服务方法属性
publicstaticstring SubmitDateShell(string touristContent,string
[] nodes)//注意函数的修饰符,只能是静态的
    {
string[] tourist=touristContent.Split('|');
        BasicTouristTrace(tourist[0], tourist[1], tourist[2],
tourist[3], 1);
int traceIndex=BasicTouristTrace(tourist[0], 1);
Boolean status= BasicTouristNode(traceIndex, nodes);
if (status)
        {
```

```
return"数据提交成功!";
        }
else
        {
return"数据提交失败!";
        }
    }
publicstaticBoolean BasicTouristTrace(string traceTitle, string
briefIntro, string start_Time, string end_Time,int userIndex)
    {
string strConnection = " data source = localhost; uid = " +
SqlLogin.uid + "; pwd = " + SqlLogin.pwd + "; database = " +
SqlLogin.database;
SqlConnection objConnection = newSqlConnection(strConnection);
SqlCommand objCommand = newSqlCommand("", objConnection);

        objCommand.CommandText ="SELECT * FROM TouristTrace";
try
        {
if (objConnection.State == System.Data.ConnectionState.Closed)
            {
                objConnection.Open();
            }
SqlDataAdapter mydata = newSqlDataAdapter(objCommand);
DataSet ds = newDataSet();
            mydata.Fill(ds,"TouristTrace");
DataTable dt = ds.Tables["TouristTrace"];
SqlCommandBuilder objcmdBuilder = newSqlCommandBuilder(mydata);
            mydata.UpdateCommand = objcmdBuilder.GetUpdateCommand();
            mydata.DeleteCommand = objcmdBuilder.GetDeleteCommand();
            mydata.InsertCommand = objcmdBuilder.GetInsertCommand();
            dt = ds.Tables["TouristTrace"];
DataRow dr = dt.NewRow();
            dr["TraceTitle"] = traceTitle;
            dr["BriefIntro"] = briefIntro;
            dr["Start_Time"] = start_Time;
            dr["End_Time"] = end_Time;
            dr["User_Index"] = userIndex;
```

```
            dr["Good"] = 0;
            dt.Rows.Add(dr);
    //将更新提交数据库
            mydata.Update(ds,"TouristTrace");
    returntrue;
        }
    catch (SqlException e)
        {
    Console.Write(e.Message.ToString());
    returnfalse;
        }
    finally
        {
    if (objConnection.State == System.Data.ConnectionState.Open)
            {
                objConnection.Close();
            }

        }
      }
    publicstaticint BasicTouristTrace(string traceTitle,int userIndex)
      {
    string strConnection = " data source = localhost; uid = " +
SqlLogin.uid + "; pwd = " + SqlLogin.pwd + "; database = " +
SqlLogin.database;
    SqlConnection objConnection = newSqlConnection(strConnection);
    SqlCommand objCommand = newSqlCommand("", objConnection);

        objCommand.CommandText ="SELECT * FROM TouristTrace WHERE
(TraceTitle='" + traceTitle + "') AND (User_Index=" + userIndex + ") OR-
DER BY ID DESC";
    try
        {
    if (objConnection.State == System.Data.ConnectionState.Closed)
            {
                objConnection.Open();
            }
    SqlDataAdapter mydata = newSqlDataAdapter(objCommand);
```

```
DataSet ds = newDataSet();
            mydata.Fill(ds,"TouristTrace");
SqlCommandBuilder objcmdBuilder = newSqlCommandBuilder(mydata);
DataTable dt = ds.Tables["TouristTrace"];
if (dt.Rows.Count ! = 0)
            {
DataRow dr = dt.Rows[0];
returnInt32.Parse(dr[0].ToString());
            }
else
            {
return 0;
            }
        }
catch (SqlException e)
        {
Console.Write(e.Message.ToString());
return 0;
        }
finally
        {
if (objConnection.State == System.Data.ConnectionState.Open)
            {
                objConnection.Close();
            }
        }
    }
publicstaticBoolean BasicTouristNode(int traceIndex, string[]
nodes)
    {
string strConnection = " data source = localhost; uid = " +
SqlLogin.uid + "; pwd = " + SqlLogin.pwd + "; database = " +
SqlLogin.database;
SqlConnection objConnection = newSqlConnection(strConnection);
SqlCommand objCommand = newSqlCommand("", objConnection);

        objCommand.CommandText = "SELECT * FROM TouristNode";
```

```
try
        {
if (objConnection.State == System.Data.ConnectionState.Closed)
            {
                objConnection.Open();
            }
SqlDataAdapter mydata = newSqlDataAdapter(objCommand);
DataSet ds = newDataSet();
            mydata.Fill(ds,"TouristNode");

DataTable dt = ds.Tables["TouristNode"];

SqlCommandBuilder objcmdBuilder = newSqlCommandBuilder(mydata);
            mydata.UpdateCommand = objcmdBuilder.GetUpdateCommand();
            mydata.DeleteCommand = objcmdBuilder.GetDeleteCommand();
            mydata.InsertCommand = objcmdBuilder.GetInsertCommand();
            dt = ds.Tables["TouristNode"];
foreach (string node in nodes)
            {
string[] nodeContent = node.Split('|');
DataRow dr = dt.NewRow();
                dr["TTrace_Index"] = traceIndex;
                dr["Node_Name"] = nodeContent[0];
                dr["Start_Time"] = nodeContent[1];
                dr["End_Time"] = nodeContent[2];
                dr["TStyle_Index"] = nodeContent[3];
                dr["Node_Intro"] = nodeContent[4];
                dr["GeoX"] = nodeContent[5];
                dr["GeoY"] = nodeContent[6];
                dt.Rows.Add(dr);
            }
//将更新提交数据库
            mydata.Update(ds,"TouristNode");
returntrue;
        }
catch (SqlException e)
        {
```

```
Console.Write(e.Message.ToString());
returnfalse;
        }
finally
        {
if (objConnection.State == System.Data.ConnectionState.Open)
            {
                objConnection.Close();
            }
        }
}
```

运行程序，交互过程介绍如下：

登录系统，选择【上传旅游印迹】，弹出交互框，如图 33.23 所示。

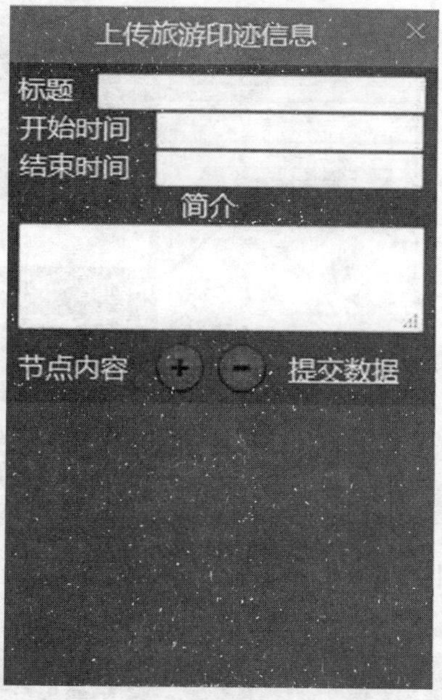

上传旅游印迹信息
标题
开始时间
结束时间
简介
节点内容
提交数据

图 33.23 【上传旅游印迹信息】交互框

每点击一次⊕，在地图上找到相应位置，点击一下即生成一个标注。点击⊕旁边的⊖，则标注、节点内容全部清空。点击每个节点后面的⊖，则会清除相应的标注和节点内容。可以通过点击标注查看节点标识来对应要输入信息的节点标识，如图 33.24 所示。

填写好相应内容后，点击【提交数据】，会得到交互结果，如图 33.25、图 33.26 所示。

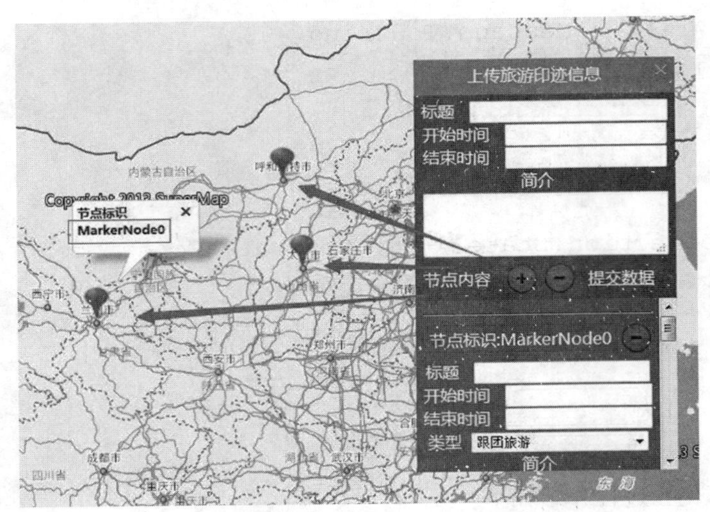

图 33.24 【添加节点】/【删除节点】

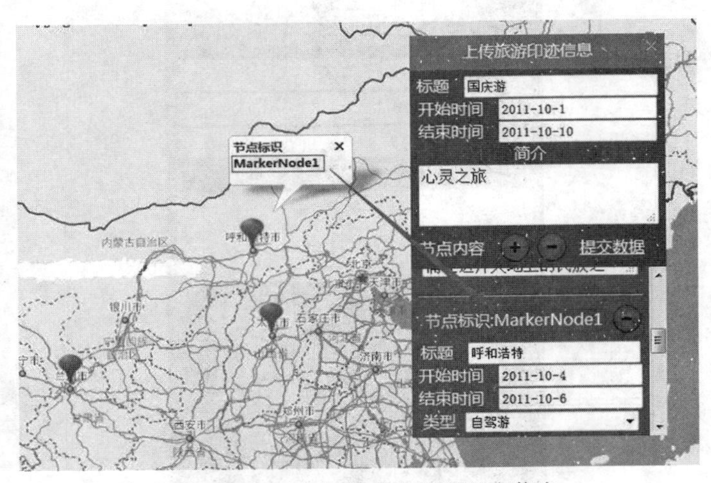

图 33.25 对应旅行节点填写相应信息

图 33.26 【提交数据】交互提示

33.3.5 查看旅行印迹

功能实现代码如下:

①index.css 文件:

```
#road {
    position: absolute;
    right:7px;
    height:570px;
    width:462px;
    background-image:url("../Image/rightContent.png");
}
#road_top
{
    height:10px;
    text-align:center;
    margin-top: -12px;
}
#road_content
{
    height: 515px;
    margin-top: 30px;
    margin-left: 30px;
    overflow:auto;
}
.RouteLine
{
    background-color:#e3d7a8;
    padding:10px;
    margin-bottom:10px;
}
.RouteLine_hover{
    background: #d4c37e;
    padding:10px;
    margin-bottom:10px;
}
```

②index.aspx 文件:

```
<div id="road"><% --网站右部--% >
<div id="road_top">
```

```
<h3 id="title_text" style="color:white;">热点路线</h3>
<%--标题栏--%>
</div><%--标题栏--%>
<div id="road_content">
<%--<div id="RouteLine" class="RouteLine"
onmouseover="this.className='RouteLine_hover'" onmouseout=
"this.className='RouteLine'" onclick="ShowRoute('aaa','aaaa')">
<div>江南7日游:安昌古镇——南浔古镇——乌镇——西塘——周庄</div>
<div>简介:江南风情</div>
<div>起止时间:2014-10-1至2014-10-7</div>
<div>点赞:1000</div>
</div>--%>
<%=HotRoadsDIV()%>
</div><%--旅游路线内容--%>
</div>
```

③baseOperaTool.js 文件:

```
var map, layer, markLayer, vectorLayer, marker, icon, framedCloud,
host = "http://localhost:8090";
var animatorVector;
var bt = false;
var url = host + "/iserver/services/map-China4002/rest/maps/China400";
var style = {
        strokeColor:"#304DBE",
        strokeWidth: 3,
        pointerEvents:"visiblePainted",
        fillColor:"#304DBE",
        fillOpacity: 0.3
    };
var vectorLineLayer,
    styleLine = {
        fillColor:"#cc0000",
        pointRadius: 3,
        strokeColor:"#cc0000",
        strokeWidth: 3
    },
    styleTrekking =
    {
        externalGraphic:"Image/trekking.png",
```

```
        graphicWidth: 32,
        graphicHeight: 32
    };
    styleBus =
    {
        externalGraphic:"Image/bus.png",
        graphicWidth: 32,
        graphicHeight: 32
    };
    styleCar =
    {
        externalGraphic:"Image/car.png",
        graphicWidth: 26,
        graphicHeight: 26
    };
  var touristType = [];
    touristType.push("跟团旅游"); touristType.push("自驾游");
touristType.push("徒步游");
    touristType.push("个人游"); touristType.push("和驴友们旅游");
  function init() {
  var broz = SuperMap.Util.getBrowser();
  if (! document.createElement('canvas').getContext) {
        alert('您的浏览器不支持 canvas,请升级');
  return;
    }elseif (broz.device === 'android') {
        alert('您的设备不支持高性能渲染,请使用 pc 或其他设备');
  return;
    }
    layer =new SuperMap.Layer.TiledDynamicRESTLayer("中国地图", url, {
        transparent : true,
        cacheEnabled : true
    }, { maxResolution:"auto" });
    layer.events.on({"layerInitialized": addLayer});

    vectorLayer =new SuperMap.Layer.Vector("Vector Layer");
    markLayer =new SuperMap.Layer.Markers("Markers");
    map =new SuperMap.Map("map", {
        controls: [new SuperMap.Control.ScaleLine(),
```

```
                    new SuperMap.Control.Zoom(),
                    new SuperMap.Control.Navigation({
                                dragPanOptions: {
                                    enableKinetic:true
                                }
                            })], units:"m"
    });
    map.minScale = 1.3732579019271662e-8;
    map.numZoomLevels = 7;//设置地图缩放级别的数量
    map.events.on({
    //"click":getPosition,
        "mousemove" : getMousePositionPx
    });
    //初始化路线图层
    vectorLineLayer =new SuperMap.Layer.Vector("Vector Line Layer", {
        styleMap:new SuperMap.StyleMap({
    "default": styleLine
        })
    });
    //初始化动画图层
    animatorVector = new SuperMap.Layer.AnimatorVector( " Anima-
tor", {}, {
        //设置速度为每帧播放 0.05 小时的数据
        speed: 0.001,
        //开始时间为 0 时
        startTime: 0,
        //结束时间设置为最后运行结束时间
        endTime: 10
    });
}

function addLayer() {
    map.addLayers([layer, markLayer, vectorLayer, vectorLineLay-
er, animatorVector]);
        map.setCenter ( new  SuperMap.LonLat  ( 12161566.94829,
4601160.54256), 4);
}
function ShowRoute(a,b) {
```

```
    var roads = b.split('|');
        animatorVector.animator.stop();
        clearStatus();
    var orientation = 1;
    var points = [];
    var features = [];
    var cars = [];
    var firstPoint;
    var icon;
    for (i = 0, len = roads.length; i < len-1; i+=9) {
          size =new SuperMap.Size(26, 26),
          offset =new SuperMap.Pixel(-(size.w /2), -size.h);
    if (roads[i + 7] == 0) {
              icon =new SuperMap.Icon("../Image/bus0.png", size, offset);
          }elseif (roads[i + 7] == 1) {
              icon =new SuperMap.Icon("../Image/car0.png", size, offset);
          }else {
                icon =new SuperMap.Icon("../Image/trekking0.png",
size, offset);
          }
          marker =new SuperMap.Marker(new SuperMap.LonLat(roads[i+
5], roads[i+6]), icon);
          marker.ID = roads[i + 0];
          marker.Name = roads[i + 1];
          marker.NodeInfo = roads[i + 2];
          marker.StartTime = roads[i + 3];
          marker.EndTime = roads[i + 4];
          marker.TouristType=roads[i + 7];
          markLayer.addMarker(marker);
          //注册 click 事件,触发 mouseClickHandler()方法
          marker.events.on({"click": mouseRouteMarkerClickHandler,
"scope": marker});
    var point = new SuperMap.Geometry.Point(parseFloat(roads[i +
5]), parseFloat(roads[i + 6]));
    if (i == 0) {
              firstPoint = point;
          }
          points.push(point);
```

```
    }
    points.push(firstPoint);
    roadLine =new SuperMap.Geometry.LineString(points);
    roadLine.style ="";
  var feature = new SuperMap.Feature.Vector();
    feature.geometry = roadLine;
    features.push(feature);
    vectorLineLayer.addFeatures(features);
  var car;
  for (var i = 0; i < points.length; i++) {
  if (roads[(i-1) * 9 + 7] == 0) {
          car =new SuperMap.Feature.Vector(points[i].clone(),
              {
                  FEATUREID: feature.id,
                  TIME: i
              }, styleBus
          );
      }elseif (roads[(i -1) * 9 + 7] == 1) {
          car =new SuperMap.Feature.Vector(points[i].clone(),
              {
                  FEATUREID: feature.id,
                  TIME: i
              }, styleCar
          );
      }else {
          car =new SuperMap.Feature.Vector(points[i].clone(),
              {
                  FEATUREID: feature.id,
                  TIME: i
              }, styleTrekking
          );
          }
      cars.push(car);
    }
    animatorVector.addFeatures(cars);
    animatorVector.animator.start();
  }
  //开始播放动画
```

```
    function startAnimator() {
        animatorVector.animator.start();
    }
    //暂停播放动画
    function pauseAnimator() {
        animatorVector.animator.pause();
    }
    //停止播放动画
    function stopAnimator() {
        animatorVector.animator.stop();
    }
    function mouseRouteMarkerClickHandler(event) {
    var marker = this;
    var contentHTML = "<div style='font-size:12px;font-weight:bold ;
opacity: 0.8'>";
        contentHTML += "标题:" + marker.Name + " |" + touristType
[marker.TouristType] + "</br>";
        contentHTML +="<hr/>简介:" + marker.NodeInfo + "</br>";
        contentHTML +="开始时间:" + marker.StartTime + "</br>";
        contentHTML +="结束时间:" + marker.EndTime + "</br>";
        contentHTML +="<hr/><div  style='text-align:center;'><a on-
click='showNotes(\"" + marker.ID
                    +"\")'>查看照片</a>|<a onclick='showPics("\" +
marker.ID + "\")'>查看日记</a></div>";
        contentHTML +="</div>";
    //初始化FramedCloud类
        framedCloud =new SuperMap.Popup.FramedCloud(
    "MarkerRouteNode",
            marker.getLonLat(),
    null,
            contentHTML,
            icon,
    true,
    null,
    true
        );
        infowin = framedCloud;
        map.addPopup(framedCloud);
```

```
}
function clearStatus() {
    vectorLayer.removeAllFeatures();
    markLayer.clearMarkers();
    vectorLineLayer.removeAllFeatures();
    animatorVector.removeAllFeatures();
}
```

④Tourist. cs 文件：

```
privatestring traceTitle;
privatestring briefIntro;
privateint user_Index;
privatestring start_Time;
privatestring end_Time;
privateint goods;
privateList<TouristNode> touristNode;

public Tourist(string title, string intro, int userIndex, string
startT, string endT, int goods, List<TouristNode> tourist_Node)
    {
this.TraceTitle=title;
this.BriefIntro = intro;
this.User_Index = userIndex;
this.Start_Time = startT;
this.End_Time = endT;
this.Goods = goods;
this.TouristNode = tourist_Node;
    }
publicstring TraceTitle
    {
get { return traceTitle; }
set { traceTitle = value; }
    }
publicstring BriefIntro
    {
get { return briefIntro; }
set { briefIntro = value; }
    }
publicint User_Index
```

```
    {
get { return user_Index; }
set { user_Index = value; }
    }
publicstring Start_Time
    {
get { return start_Time; }
set { start_Time = value; }
    }
publicstring End_Time
    {
get { return end_Time; }
set { end_Time = value; }
    }
publicint Goods
    {
get { return goods; }
set { goods = value; }
    }
publicList<TouristNode> TouristNode
    {
get { return touristNode; }
set { touristNode = value; }
    }
```

⑤TouristNode. cs 文件:

```
privateint id;
privateint tTrace_Index;
privateint tStyle_Index;
privatestring node_Name;
privatestring node_Intro;
privatestring start_Time;
privatestring end_Time;
privatestring geoX;
privatestring geoY;

public TouristNode ( int _id, int traceIndex, int tStyleIndex,
string nodeName, string nodeIntro,
string startTime, string endTime, string geoX, string geoY)
```

```
    {
this.Id = _id;
this.TTrace_Index = traceIndex;
this.TStyle_Index = tStyleIndex;
this.Node_Name = nodeName;
this.Node_Intro = nodeIntro;
this.Start_Time = startTime;
this.End_Time = endTime;
this.GeoX = geoX;
this.GeoY = geoY;
    }
publicint Id
    {
get { return id; }
set { id = value; }
    }
publicint TStyle_Index
    {
get { return tStyle_Index; }
set { tStyle_Index = value; }
    }
publicstring GeoY
    {
get { return geoY; }
set { geoY = value; }
    }
publicstring GeoX
    {
get { return geoX; }
set { geoX = value; }
    }
publicstring End_Time
    {
get { return end_Time; }
set { end_Time = value; }
    }
publicstring Start_Time
    {
```

```
get { return start_Time; }
set { start_Time = value; }
    }
publicstring Node_Intro
    {
get { return node_Intro; }
set { node_Intro = value; }
    }
publicstring Node_Name
    {
get { return node_Name; }
set { node_ Name = value; }
    }
publicint TTrace_Index
    {
get { return tTrace_Index; }
set { tTrace_Index = value; }
    }
```

⑥index. aspx. cs 文件:

```
publicstaticstring HotRoadsDIV()
    {
List<Tourist> touristList = getTouristTrace(100);
string tourist="";
string node="";
string intro="";
string firstDiv = "";
string secondDiv = "";
string thirdDiv = "";
string contentDiv = "";
string contentNodeDiv = "";

foreach (Tourist t in touristList)
        {
            tourist = t.TraceTitle +" |" + t.BriefIntro + " |"+t.
Start_Time + " |" + t.End_Time + " |" + t.Goods + " |" + t.User_Index;
            firstDiv = t.TraceTitle +":";
            intro ="简介:"+t.BriefIntro;
            secondDiv ="起止时间:" + t.Start_Time + "至" + t.End_Time;
```

```
            thirdDiv ="点赞:" + t.Goods;
            node ="";
    foreach(TouristNode n in t.TouristNode)
            {
                node += n.Id+" |"+n.Node_Name + " |" + n.Node_Intro + "
|" + n.Start_Time + " |" + n.End_Time + " |" + n.GeoX + " |" + n.GeoY + " |" +
n.TStyle_Index + " |" + n.TTrace_Index+" |";
                firstDiv += n.Node_Name+"----";
            }
            firstDiv=firstDiv.Substring(0, firstDiv.Length -4);
            contentNodeDiv ="<div id='RouteLine' class='RouteLine'
onmouseover='this.className= \"RouteLine_hover\"'" +
"onmouseout='this.className= \"RouteLine\"' onclick='ShowRoute( \"" +
tourist + "\",\"" + node + "\")'>" +
    "<div>" + firstDiv + "</div>" +
    " <div>"+intro+"</div>" +
    "<div>" + secondDiv + "</div>" +
    "<div>" + thirdDiv + "</div>" +
    "</div>";
            contentDiv += contentNodeDiv;
        }
    return contentDiv;
      }
    //读取路线数据
    publicstaticList<Tourist> getTouristTrace(int count)
      {
    List<Tourist> touristList =newList<Tourist>();
    string strConnection = "data source=localhost;uid=" + SqlLogin.
uid + ";pwd=" + SqlLogin.pwd + ";database=" + SqlLogin.database;
    SqlConnection objConnection = newSqlConnection(strConnection);
    SqlCommand objCommand = newSqlCommand("", objConnection);
          objCommand.CommandText ="SELECT TOP(" + count + ") *  FROM
TouristTrace ORDER BY Good DESC";
    try
          {
    if (objConnection.State == System.Data.ConnectionState.Closed)
            {
                objConnection.Open();
```

```
        }
    SqlDataAdapter mydata = newSqlDataAdapter(objCommand);
    DataSet ds = newDataSet();
            mydata.Fill(ds,"TouristTrace");
    SqlCommandBuilder objcmdBuilder = newSqlCommandBuilder(mydata);
    DataTable dt = ds.Tables["TouristTrace"];

    if (dt.Rows.Count ! = 0)
        {
    for (int i = 0; i < dt.Rows.Count; i++) //遍历行
            {
    DataRow dr = dt.Rows[i];
    List<TouristNode> touristNode=getTouristNode(Int32.Parse(dr[0]
.ToString()));
                    touristList.Add(newTourist(dr[1].ToString(),
dr[2].ToString(),
    Int32.Parse(dr[3].ToString()), dr[4].ToString(),
                                dr[5].ToString(),Int32.Parse(dr
[6].ToString()), touristNode));
            }
        }
    return touristList;
        }
    catch (SqlException e)
        {
    Console.Write(e.Message.ToString());
    returnnull;
        }
    finally
        {
    if (objConnection.State = = System.Data.ConnectionState.Open)
            {
                objConnection.Close();
            }
        }
    }
    publicstaticList<TouristNode> getTouristNode(int tTrace_Index)
    {
```

```
    List<TouristNode> touristNodeList = newList<TouristNode>();
    string strConnection = "data source=localhost;uid=" + SqlLogin.
uid + ";pwd=" + SqlLogin.pwd + ";database=" + SqlLogin.database;
    SqlConnection objConnection = newSqlConnection(strConnection);
    SqlCommand objCommand = newSqlCommand("", objConnection);
            objCommand.CommandText = " SELECT   * FROM TouristNode
WHERE TTrace_Index=" + tTrace_Index;
    try
            {
    if (objConnection.State == System.Data.ConnectionState.Closed)
                {
                    objConnection.Open();
                }
    SqlDataAdapter mydata = newSqlDataAdapter(objCommand);
    DataSet ds = newDataSet();
            mydata.Fill(ds,"TouristNode");
    SqlCommandBuilder objcmdBuilder = newSqlCommandBuilder(mydata);
    DataTable dt = ds.Tables["TouristNode"];

    if (dt.Rows.Count != 0)
                {
    for (int i = 0; i < dt.Rows.Count; i++) //遍历行
                    {
    DataRow dr = dt.Rows[i];
touristNodeList.Add(newTouristNode(Int32.Parse(dr[0].ToString()),
Int32.Parse(dr[1].ToString()),Int32.Parse(dr[2].ToString()),
dr[3].ToString(),dr[4].ToString(),dr[5].ToString(),dr[6].ToString
(),
                                    dr[7].ToString(),dr[8]
.ToString()));
                    }
                }
    return touristNodeList;
            }
    catch (SqlException e)
            {
    Console.Write(e.Message.ToString());
    returnnull;
```

```
        }
finally
        {
if (objConnection.State == System.Data.ConnectionState.Open)
            {
                objConnection.Close();
            }
        }
    }
```

运行程序，交互过程如下：点击“热点路线”栏下的信息(显示器显示红色框内)，绘制出路线图(动画展示)，根据旅游类型生成图标，如图 33.27 所示。

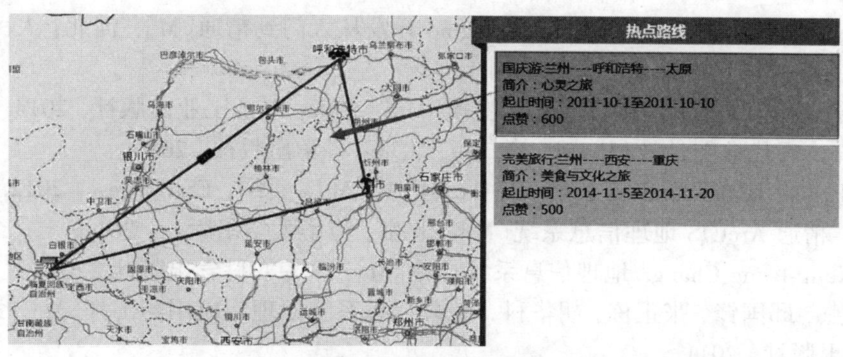

图 33.27　路线动画展示

点击“热点路线”栏下的信息(显示器高亮显示框内)，绘制出路线图(动画展示)，根据旅游类型生成图标，点击图标可以查看路线节点的信息，如图 33.28 所示。

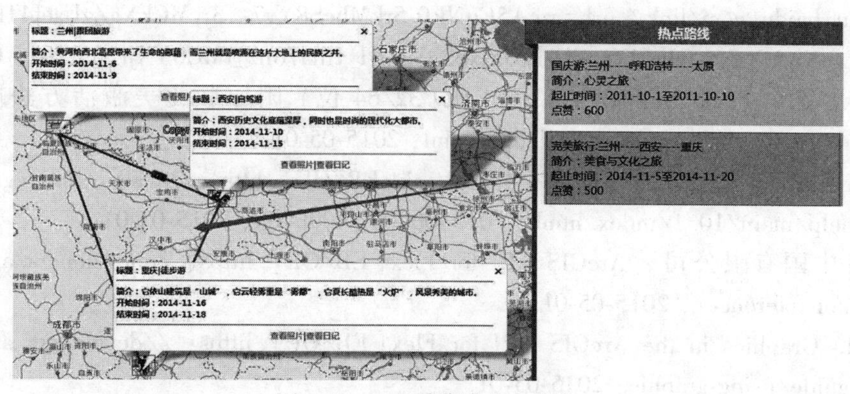

图 33.28　查看路线节点信息

参 考 文 献

[1][美]斯坦利·B. 李普曼. 深度探索 C++对象模型[M]. 侯捷，译. 北京：电子工业出版社，2012.

[2][美]Stephen Ritchie. .NET 最佳实践[M]. 黄灯桥，黄浩宇，李永，译. 北京：机械工业出版社，2014.

[3]百度百科. .NET[EB/OL]. http://baike.baidu.com/view/4294.htm，2015-05-01.

[4]邱洪钢，张青莲，陆绍强. ArcGIS Engine 开发从入门到精通[M]. 河北：人民邮电出版社，2010.

[5]王美玲，付梦印. 地图投影与坐标变换[M]. 北京：电子工业出版社，2014.

[6]袁勘省. 现代地图学教程[M]. 第二版. 北京：科学出版社，2014.

[7]ESRI Press. Understanding ArcSDE：ArcGIS 10[M]. 美国：ESRI Press，2012.

[8]池建. 精通 ArcGIS 地理信息系统[M]. 北京：清华大学出版社，2011.

[9][美]Kang-tsung Chang. 地理信息系统导论[M]. 北京：科学出版社，2010.

[10]郑春燕，邱国锋，张正栋，胡华科. 地理信息系统原理、应用与工程[M]. 武汉：武汉大学出版社，2011.

[11]蔡苑彬，刘露，陈荦，等. 基于地图制图脚本的交互式图例动态生成方法[J]. 地理空间信息，2014，05：154-157.

[12]王家耀. 地图学原理与方法[M]. 第二版. 北京：科学出版社，2014.

[13]ESRI中国信息技术有限公司. ArcGIS 10.1 for Server 体系架构[EB/OL]. http://wenku.baidu.com/link? url=rmA5CqNE0-5rkMhcS8Yy7-_3jyMCkXgZgUsef4ThtF8Wt6lt0GuSLgeXhAjX4x6-OIwMYi_ElX1k0uJxS2y3upiH_Rha78nz_rMC0YXjr1u，2015-05-01.

[14]脚本之家. Flash Builder 4.7 正式版(32/64 位)附原版完美激活方法[EB/OL]. http://www.jb51.net/softs/103858.html，2015-05-01.

[15]ESRI中国有限公司. ArcGIS 帮助 10.1[EB/OL]. http://resources.arcgis.com/zh-CN/help/main/10.1/index.html#//0154000003vt000000，2015-05-01.

[16]ESRI中国有限公司. ArcGIS API for Flex[EB/OL]. https://developers.arcgis.com/flex/api-reference/，2015-05-01.

[17]ESRI. Graphics in the ArcGIS API for Flex[EB/OL]. https://developers.arcgis.com/flex/guide/using-graphic，2015-05-01.

[18]CSDN 博客. findTask，queryTask，indentifyTask 之间的区别. [EB/OL]. http://blog.csdn.net/yiyuhanmeng/article/details/7816897，2015-05-01.

[19]Roy Thomas Fielding. Architectural Styles and the Design of Network-based Software

Architectures. 第五章. 2000.
[20] 百度百科. ArcGIS API for JavaScript [EB/OL]. http://baike.baidu.com/view/7770077.htm, 2015-05-01.
[21] ESRI 中国. ArcGIS API for JavaScript 帮助[EB/OL]. https://developers.arcgis.com/javascript/, 2015-05-01.
[22] ArcGIS 产品与技术专栏. 什么是 ArcGIS 影像服务[EB/OL]. http://m.blog.csdn.net/blog/arcgis_all/8239434, 2015-05-01.
[23] ESRI 中国. [EB/OL]. http://www.bamboosilk.org/Wssf/2003/chenjian03.htm, 2015.
[24] ESRI 中国信息技术有限公司. ArcGIS Runtime SDK for iOS[EB/OL]. http://www.esrichina-bj.cn/2015/0108/2854.html, 2015-05-01.
[25] ESRI 中国信息技术有限公司. ArcGIS Runtime SDK for Android [EB/OL]. http://www.esrichina-bj.cn/2015/0108/2848.html, 2015-05-01.
[26] 百度文库. ArcGIS Runtime SDK for Windows Phone 入门教程[EB/OL]. http://wenku.baidu.com/view/11990f40cf84b9d528ea7a88.html, 2015-05-01.
[27] 百度文库. Android 基本概念[EB/OL]. http://wenku.baidu.com/link? url=hK79u-37RTJM9TNyJDz97ipjyhekMDL6Lx7iurPM9et-r8GZsloGnxZrdj6-efFvtjhuHPOq1hUuixfz2rb47mxvZoz5ZHhSHsD0bLHu7tG, 2011-01-22.
[28] sunny2038. eclipse IDE 学习笔记[EB/OL]. https://developers.arcgis.com/android/api-http://blog.csdn.net/sunny2038/article/details/7762807, 2015-05-01.
[29] ESRI 中国. MapView[EB/OL]. https://developers.arcgis.com/android/api-reference/reference/com/esri/android/map/MapView.html, 2015-05-01.
[30] 中地数码集团. MapGIS IGServer Flex 开发手册. 中地数码集团, 2013-08-01.
[31] 中地数码集团. MapGIS IGServer Flex 入门手册. 中地数码集团, 2013-08-01.
[32] 北京超图软件股份有限公司. SuperMap iClient 7C for JavaScript 帮助文档. 北京超图软件股份有限公司, 2015-01-01.